3rd International Conference on Power and Renewable Energy (ICPRE 2018)

E3S Web of Conferences Volume 64 (2018)

Berlin, Germany
21-24 September 2018

Editors:

Huang Qi **Mohan Kolhe**

ISBN: 978-1-5108-7554-8

Printed from e-media with permission by:

Curran Associates, Inc.
57 Morehouse Lane
Red Hook, NY 12571

Some format issues inherent in the e-media version may also appear in this print version.

No changes have been made to the content of these proceedings. There may be changes to pagination and minor adjustments for aesthetics.

Printed by Curran Associates, Inc. (2019)

For additional information, please contact EDP Sciences – Web of Conferences at the address below.

EDP Sciences – Web of Conferences
17, Avenue du Hoggar
Parc d'Activité de Courtabœuf
BP 112
F-91944 Les Ulis Cedex A
France

Phone: +33 (0) 1 69 18 75 75
Fax: +33 (0) 1 69 28 84 91

contact-edps@webofconferences.org

Additional copies of this publication are available from:

Curran Associates, Inc.
57 Morehouse Lane
Red Hook, NY 12571 USA
Phone: 845-758-0400
Fax: 845-758-2633
Email: curran@proceedings.com
Web: www.proceedings.com

TABLE OF CONTENTS

Preface

2018 The 3rd International Conference on Power and Renewable Energy

ICPRE, September 21-24, 2018, at Berlin, Germany

The third International Conference on Power and Renewable Energy (ICPRE) has been organized and held on September 21-24 at Berlin (Germany). The ICPRE 2018 has been aimed for providing unique platform for researchers, scientists, engineers and professionals from all over the world to present their latest research result and innovative ideas in the fields of power and renewable energy. The objective of ICPRE is to bring together a diverse group of researchers, scientists, academics, and engineers in clean power and renewable energy from all over the world, who share much more than technical interests, among other things, like culture and history, for making sustainable development of the global community.

With the successful experience of the previous two years in ICPRE 2016 at Shanghai (China), and ICPRE 2017 at Chengdu (China), we have organized ICPRE 2018 at Berlin (Germany) with greater success and provided very good platform for the participants having fruitful discussion on role of clean power and renewable energy technologies for sustainable development. Proceedings of ICPRE 2016 and 2017 have been included in the IEEE Explore Digital Library (ICPRE 2016 – IEEE Catalog Number: CFP16H25-ART, ISBN: 978-1-5090-3068-2; and ICPRE 2017 - IEEE Catalog Number: CFP17H25-ART, ISBN: 978-1-5386-2156-1, IEEE Press).

The ICPRE 2018 has been organized in the program chapters as: Renewable Energy and Clean Energy, Solar Energy Development and Utilization, Photovoltaic Systems and Power Generation Technologies, Energy Storage Technology and Energy Engineering, Electrical Engineering and Mechatronics, Power System and Energy.

The ICPRE 2018 organizing committee is pleased to have 116 submissions from around 30 countries and regions, including Canada, Mexico, Egypt, Turkey, China, Taiwan, Japan, South Korea, India, Iran, Saudi Arabia, Germany, Australia, United Kingdom, South Africa, France, Brazil, Portugal, Lithuania, Ecuador, Romania, Poland, Algeria, Denmark, Morocco, Colombia, Italy, Indonesia, Latvia, Indonesia, Norway. Due to constrain of number of papers in the proceeding and suitability of the topics of the submissions, 74 submissions have been accepted for presentations and out of that 57 submissions are accepted as full papers to be published in the ICPRE2018 proceedings. It has really been a difficult task to select the most representative papers.

The ICPRE 2018 has been sponsored by the University of Electronic Science and Technology of China and Southeast University, China; Tongji University, China and Clean Energy published by Oxford University Press. It has been technically assisted by the University of Agder, Norway; University of Malaya, Malaysia; Near East University, Turkey.

The ICPRE 2018 peer-reviewed and accepted papers have been edited as conference proceedings to be published with E3S Web of Conferences (Open Access proceedings in Environment, Energy and Earth Sciences), which is indexed by Google Scholar, CAS, DOAJ, CPCI (Web of Science), EBSCO, ProQuest, Scopus, and Ei Compendex.

As editor of this special issue of the ICPRE 2018 proceedings, I wish to express my sincere gratitude to the plenary speakers, authors, participants, reviewers, program and technical committee members, and organizing members; who have contributed in organization of the conference and publication of this special issue. With high standard and high-quality submissions and presentations in the ICPRE, I am very much sure, one day ICPRE will become a leading conference in this specific academic area of 'Power and Renewable Energy'.

Professor Mohan Lal Kolhe
Editor of the ICPRE 2018 Proceedings
Professor in Smart Grid and Renewable Energy
Faculty of Engineering and Science,
University of Agder,
PO Box 422, NO 4604, Kristiansand, Norway
www.uia.no

sciences

Statement of Peer review

In submitting conference proceedings to E3S Web of Conferences, I certify to the Publisher that I adhere to the **Policy on Publishing Integrity** of the journal in order to safeguard good scientific practice in publishing.

1. All articles have been subjected to peer review administered by the proceedings editors.
2. Reviews have been conducted by expert referees, who have been requested to provide unbiased and constructive comments aimed, whenever possible, at improving the work.
3. Proceedings editors have taken all reasonable steps to ensure the quality of the materials they publish and their decision to accept or reject a paper for publication has been based only on the merits of the work and the relevance to the journal.

Title, date and place of the conference

2018 the 3rd International Conference on Power and Renewable Energy

Berlin, Germany

September 21-24, 2018

Proceedings editor(s):

Prof. Huang Qi, University of Electronic Science and Technology of China, China

Prof. Mohan Kolhe, University of Agder, Norway

Date and editor's signature

Turkey's Energy Strategy for 2023 Targets after 2000 MW Giant Renewable Energy Contract

Haci Sogukpinar[1], *Ismail* Bozkurt[2], and *Serkan* Cag[3]

[1]Department of Electric and Energy, Vocational School, University of Adiyaman, Adiyaman 02040, Turkey
[2]Department of Mechanical Engineering, Faculty of Engineering, University of Adiyaman Adiyaman 02040, Turkey
[3]Department of Machinery and Metal Technology, Vocational School, University of Adiyaman, Adiyaman 02040, Turkey

Abstract. Turkey wants to become the world's 10th largest economy in the 100th anniversary of the foundation of the republic of Turkey. In order to achieve this goal, there are many breakthroughs in the political, economic and in energy fields. Turkey's installed power capacity was 85000 MW in 2017 but installed power of 125.000MW is targeted to achieve the objective of 2023 targets. The government is aiming to increase the total production of renewable energy share by 30% in 2023, while foreseeing the increase in capacity due to nuclear and fossil fuel consumption. Targets for different technologies are 34000 MW hydroelectric, 20000 MW wind energy, 5000 MW solar energy (photovoltaic and condensed solar energy), 1000 MW geothermal energy and 1000 MW biomass. Capacity utilization in hydroelectricity is 62%, wind power is 14%, and geothermal power is 33%. The total installed capacity of Biogas, Biomass, Waste Heat and Pyrolytic Oil Power Plants is 530 MW. Theoretical total power capacity of the solar energy for Turkey as 300 TWh/year and reached 45% of the 2023 target in 2017 in the last three years. However, it is estimated that the targets of 2023 in solar energy can be exceeded. Government offers attractive incentive packages for renewable and other energy sector to achieve 2023 goals. In order to encourage domestic production, a total of 2000 MW wind and solar energy installation bid was carried out in 2017.This contract is expected to make Turkey as energy hub both in terms of installation and technology. In this study, Turkey's renewable energy potential, and energy strategies and breakthroughs for this were investigated and discussed.

1 Introduction

Turkey has world's 17th and Europe's 6th largest economy and wants to have the world's 10th largest economy in 2023. Turkey Statistical Institute (TSI), announced its 2017 economic growth data for Turkey and Turkey's economy recorded 7.4 percent growth in 2017 [1]. Therefore Turkey is experiencing a sharp increase in demand for energy. 55% of the total electricity production was provided from imported natural gas and coal in 2017. When considering heating in the houses and transportation costs, Turkey's dependence on foreign energy is about 70 percent [2]. It is inevitable to encourage the use of domestic and renewable energy sources under the assumption that this rate may increase in the future. On the one hand Turkey wants to meet at least 30 percent of the total electricity demand from renewable energy 2023. Considering electric vehicles will be a certain rate in the next decade, the transportation sector aims to meet 10% of energy demand from renewable energy [3]. Turkey, with regard to minimizing the negative environmental impact of industrial activity, became a member to the United Nations Framework Convention on Climate Change and has ratified the Kyoto Protocol on 28 May 2009[4]. Therefore Turkey has been implementing a series of measures against adverse environmental impacts since 2010. With the new incentive legislation, investments in the wind energy sector accelerated in 2006. Total installation was 20MW until 2005, but only in 2006 30MW was installed and installation capacity has increased very rapidly each fallowing year [5]. Total installation passed 3000 MW at the end of 2013 and approached 7000 MW in 2018. Turkey installed more than 2000 MW of new wind power capacity during 2016 and 2017[6]. Considering the 2023 targets, 35% of it has already been achieved and there is still much work to be done. However, the government made a 1000 MW installment bid in 2017 to speed up investments. Renewable energy (including hydropower) now accounts for 28.73% of Turkey's total installed power generation capacity. It is estimated that the consumption of electricity in 2023 will increase about twice compared to today and the share of renewable energy in electricity generation is aimed to be at least 30 percent [7]. European countries installed 15,680 MW of new wind power capacity during 2017 and Germany installed 42%

of the total EU new installation [8]. The share of renewable energy in the EU's total installed power capacity has increased from 26% in 2005 to 47% in 2017. Turkey is a candidate country to the European Union even if membership did not occur but the clean energy strategy of Europe has set the objective of turkey and is also working on the development of local technology. Turkey is trying to increase the appeal by introducing new incentive laws in order to increase the share of solar energy in electricity generation. Turkey's average annual sunshine duration is 2640 hours, which equals a total of 7.2 hours per day. The average annual total solar radiation is 1311 kWh/m² this equals to the daily total of 3.6 kWh/m².

The total installation was around 50 MW at the beginning of 2015, whereas in 2016 the installation was reached 442 MW [9] and passed 2250 MW at the end of 2017. Considering Turkey's solar potential, installation is not yet at the desired level but the government tries to accelerate the total installed power by making a bid for installation of 1000 MW in 2017. There are indigenous design and production technologies in the specifications of the tender for both solar and wind power installation. Of course, we should not only think about electricity generation, because energy is consumed quite high in order to meet our daily need for hot water. Roof type collectors have used widely in the last 40 years and studies on solar ponds have been continuing [10,11]. On the other hand Turkey is using 62% of the total capacity of hydroelectric power and the total installed capacity of 597 Plants in Turkey is 26,694.92 MW. Some of the remaining part is in the phase of installation and some of it is in the phase of licensing and project and when they are all completed total power will be 43,682 MW.

The Turkish government has issued two bid announcements with the aim of increasing installation and developing domestic technology. The winning companies will be able to set up a 1000 MW solar and 1000 MW wind power plant if they meet the requirements. In this study Turkey's renewable energy potential was investigated and 2023 goals were discussed and the effects of these two bid on 2023 targets were examined.

2 Wind power

Turkey is quite in a good condition in terms of wind energy potential. Total commercial wind power potential of Turkey is around 48,000 MW. Figure 1 shows that the offshore wind energy potential is concentrated in the Aegean and Marmara sea and onshore wind energy potential is extended along the coast of the Aegean sea to Marmara region. Other wind energy potential is concentrated around Hatay in Mediterranean Sea and around Sinop in Black sea region. In inner regions where highly elevated areas are usually complex terrains. The wind energy potential in the inner part are local winds originating from the mountainous structure especially in eastern Anatolia region. Turkey's wind potential is given in Table 1. The wind speed value more than 4 m/s is of particular interest since it is typically only at wind speeds above this threshold that turbines can operate effectively [12]. Turkey's wind energy potential between 7 m/s and 7,5 m/s is about 30,000 MW, and for 7.5 m/s and higher speed is around 18.000 MW. Electricity has been generated from a total of 6752 MW (14% of total potential in 2017 and 33% of 2023 target) and wind power plants under construction is 552 MW and wind power plants under the licensed and under project phase is 1500 MW. Considering that all of them including the project phase is completed, a 9000 MW wind power plant will have been established, which means that nearly half of the target will be realized. The government made a serious attempt to achieve these goals in 2017. Ministry of Energy and Natural Resources announced 1000 MW Wind Energy bid announcement. The main purpose here is to speed up the installation process while also ensuring the development of domestic production technology. On August 3, 2017, the German-Turkish consortium of Siemens-Türkerler-Kalyon offered the lowest bid at 0.0348 US$ per kWh for the delivery of electricity to the national grid. The factory to produce domestic wind turbines will have a production capacity of at least 150 units/year or 400 MW/year wind turbine. With the power plant going into operation, a minimum of 3 billion kWh of electricity will be generated each year and about 1.1 million homes will be covered by the annual electricity demand. The consortium will carry out R&D work, for ten years, on wind turbine blades, generator design, materials technology and production techniques, software and innovative gearbox areas. R&D activities will be carried out with 50 technical personnel consisting of 80 % domestic engineers, while budget of $ 5 million is allocated for R & D work every year. The installation period of the industry will be 21 months from the date of contract signing, and the license will last 30 years. With this and similar incentives, It is aimed development of the Turkish industry and the goals of taking part in the top 10 economies in 2023.

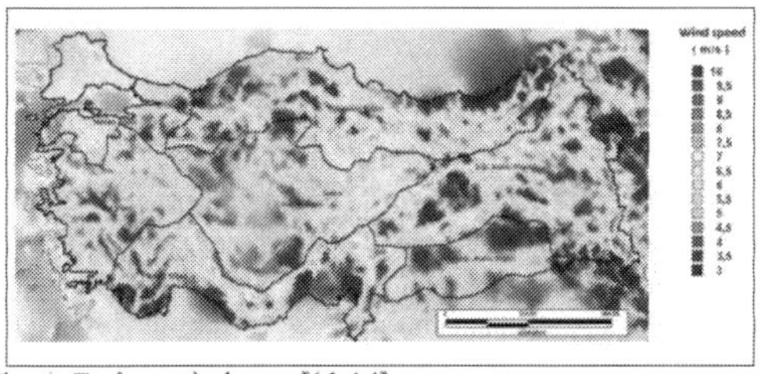

Fig. 1. Turkey wind map [13,14].

Table 1. Wind power potential vs. wind speed distribution [15]

The average annual wind speed (m/s)	Power density (W/m²)	Capacity (MW)
7.0-7.5	400-500	29,259.36
7.5-8.0	500-600	12,994.32
8.0-9.0	600-800	5,399.92
>9.0	>800	195.84
	Total	47,849.00
	Onshore (MW)	37,836.00
	Offshore(MW)	10,013.00

 https://doi.org/10.1051/e3sconf/20186401001

3 Solar power

Turkey has high solar energy potential due to its geographical location compared to many other European countries. Turkey's average annual sunshine duration were measured 2640 hours (a total of 7.2 hours per day), and the average total solar radiation was calculated as 1311 kWh/m-²year (daily total of 3.6 kWh /m² [9]) Turkey's solar map is given in the Figure 2 and average solar radiation and total sunshine durations is given in Table 2. East and south-east Anatolia region are two of the most solar potential of Turkey but the other five regions are in a quite good condition when compared to European countries [16]. The state provides a purchase guarantee for a 1-kWh electric energy generated from unlicensed solar power plants for a 10-year period of $ 13.3 cents and agreements are renewed every 10 years. The solar and wind energy systems to be built for houses in turkey have cost recovery times ranging from about 6 to 10 years [17,18].

Fig. 2. Turkey's solar map[19].

Table 2. Average solar radiation and total sunshine durations [19].

Region	Total Solar radiation Energy(kWh/m2-year)	Sunshine duration(h/year)
South East Anatolia	1460	2993
Mediterranean	1390	2956
Eastern Anatolia	1365	2664
Central Anatolia	1314	2628
Aegean	1304	2738
Marmara	1168	2409
Black sea	1120	1971

There is no storage system, so the amortization time for solar power plant is around 6 years. Investments related to solar energy have gained momentum by 2016 and the total installed volume has risen to 2250 MW in 2-3 years. The state has implemented the 1000 MW Karapinar Renewable Energy Resource Area power plant Karapinar bid to accelerate the installation and to improve the technological infrastructure in the Ministry of Energy and Natural Resources with the participation of 4 consorsium. Turkish Kalyon with South Korean partner Hanwha Group offered the lowest bid at 0.0699 US$ per kWh for the delivery of electricity to the national grid. The under the scope of the agreement, a factory to be constructed with a production capacity of 500 megawatts of photovoltaic modules. The consortium will carry out R&D work, for ten years, on module design and increasing cell efficiency. The goal is to have 60 percent of the domestic technology rate in the first 500 MW and 70 percent in the second 500 megawatt. The factory will be established in 21 months following the signing of the contract and the photovoltaic modules to be produced at the plant will be installed in the field. The project aims to increase the production of electricity from renewable source and also equipment to make technology transfers and to reduce the dependency on foreign companies in the coming years. Karapinar will be Turkey's largest solar power plant, which will be the first application based on the condition of indigenization in the energy sector and on the basis of price determination based on renewable energy sources. While 1.3 billion dollars investment is expected with the project, 1.7 billion kilowatt electricity will be generated from the plant and 600,000 house needs will be met. Solar cell production in Turkey are not yet in the commercial sense, but is made for R & D. Besides this, 18 companies have been producing solar module with good standards but after the 1000 MW bid, other domestic and foreign companies have begun licensing work for domestic cell production. It is estimated that the inner Anatolia region will be the solar technology center in the next 10 years.

4 Geothermal and biomass

Turkey's theoretical geothermal potential is given in the Figure 3 and total potential is 31,500 MW, and 78% of potential is in western Anatolia, 9% in Central Anatolia, 7% Marmara Region, 5% in Eastern Anatolia 1% are located in other regions. 90% of geothermal resources are low and medium temperature and suitable for direct applications (heating, thermal tourism, mineral waters etc.) and 10% for indirect applications (electric energy production) [20].

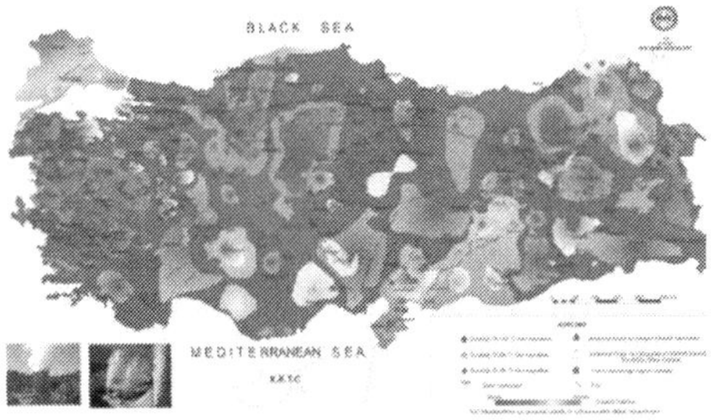

Fig. 3. Turkey's geothermal map [21].

There are 39 Geothermal Power Plants in Turkey with total installed capacity of 1,028 MW and 1,98 % of the electricity is produced from geothermal power plants in 2017 [21]. The first five countries for geothermal energy usage (for non-electrical) for direct geothermal applications are China, the USA, Sweden, Turkey and Iceland. Theoretical total power capacity of the solar energy for Turkey was calculated 300 TWh/year [22]. The total installed capacity of 82 Biogas, Biomass, Waste Heat and Pyrolytic Oil Power Plants is 530 MW and

1.27 % of the electricity is produced from these power plants in 2017.

5 Conclusion

Turkey generated 29,09 % of the electricity from renewable energy sources in 2017 and want to increase from 85,000 MW (2017) to 125.000 MW in 2023 with 30 % of renewable energy. In order to increase the use of clean energy resources, 20000 MW for wind power and 5000 MW for solar power have been targeted for the100th anniversary of republic of Turkey. In addition to this, Turkey is planning to rise among the world's top 10 economies in 2023. In order to achieve these goals, serious work is being done in areas such as the development of advanced technology, the design and production of high value added products and the reduction of external dependency on energy. Energy is a common problem in all countries of the world and it is important to meet the needs with local resources. However, global warming, which has been on the agenda for the past 30 years, has now ceased to be effective in everyday life and greenhouse gas reduction measures need to be taken. On the other hand, it is necessary to meet the increasing energy demand every year. Although hydropower has reached a certain saturation in the world but there is enough potential for wind and solar energy. Turkey to meet the growing need for energy on one hand making a serious investment on nuclear power plants, the fossil-fueled power plants and renewable energy sources, on the other hand offers attractive incentives to the private sector. Only 14% of the electricity was generated by state-owned power plants, while the remaining 86% was produced directly or indirectly by the private sector in the last year.

References

1. Turkish Statistical Institute, "Periodic Gross Domestic Product", Ankara, Turkey, available at:http://www.tuik.gov.tr.
2. E.T. Karagöl, I. Kavaz, "Dünyada Ve Türkiye'de Yenilenebilir Enerji", Volume: 19, April 2017
3. Ministry of Energy and Natural Resources General Directorate of Renewable Energy, "Turkish National Renewable Energy Action Plan", Ankara, 2014
4. Republic of Turkey Ministry of Foreign Affairs, "Kyoto Protocol", Ankara,Turkey, 2017
5. H. Sogukpinar, I. Bozkurt,"Wind energy utilization, potential and targets after 2013 in Turkey, Europe and the world", Adıyaman Üniversitesi Mühendislik Bilimleri Dergisi, vol. 1 (3), (2014), pp. 23-33.
6. TWEA (Turkish Wind Energy Association), (2017). "Türkiye Rüzgâr Enerjisi İstatistik Raporu". Available at:http://www.tureb.com.tr TUREB, 2017.
7. H. Sogukpinar, I. Bozkurt., S. Cag, "Economic Wind Potential and Effect of Feed in Tariff on Installation in Turkey", 2016 IEEE International Conference on Power and Renewable Energy,2016, pp 649-652.
8. Wind in power 2017, "Annual combined onshore and offshore wind energy statistics", February 2018.
9. H. Sogukpinar, I. Bozkurt., S. Cag, "Experimental and Numerical Analysis of Solar Potential and Feed in Tariff in Turkey", 2016 IEEE International Conference on Power and Renewable Energy, 2016, pp 594-597.
10. H. Sogukpinar, I. Bozkurt., S. Cag, "Numerical Evaluation of the Performance Increase for a Solar Pond With Glazed and Unglazed", 2016 IEEE International Conference on Power and Renewable Energy,2016, pp. 594-597.
11. H. Sogukpinar, I. Bozkurt, M. Karakilcik, "Performance Comparison of Aboveground and Underground Solar ponds", Thermal Science, Volume 22,2018, pp. 1-9
12. Europe's onshore and offshore wind energy potential, "An assessment of environmental and economic constraints", EEA Technical report No 6/2009.
13. EIE-Yenilenebilir Enerji genel Müdürlüğü, "Türkiye Rüzgâr Enerjisi Potansiyeli; 2015". Available at:<http:// www.eie.gov.tr//>
14. H. Sogukpinar, I. Bozkurt., "Adiyaman wind potential and statistical analysis, in Turkey",AIP Conference Proceedings 1815, 040006 (2017); doi: 10.1063/1.4976375
15. H. Sogukpinar, I. Bozkurt, M. F. Baran, H. Turkmenler, M. Pala, K.E. Engin, A.I. Kaya, "Micro-Turbine Design, Production and Testing", International Journal of Engineering Technologies Vol.1, No.4, 2015.pp. 141-145
16. H. Sogukpinar, I. Bozkurt., N. Calis, "A Comparison of Experimental and Estimated Data Analyses of Solar Radiation, in Adiyaman, Turkey", International Journal Of Engineering Technologies Vol.1, No.3, 2015,pp.113-117
17. H. Sogukpinar, I. Bozkurt, "Analysis on Meeting the Electric Energy Demand with Small Wind Turbine in Turkey", International Journal Of Engineering Technologies ,Vol.1, No.2, 2015,pp.51-55
18. H. Sogukpinar, I. Bozkurt, An Economic Analysis Of Residential Pv System For Adiyaman, Turkey, Uludağ University Journal of The Faculty of Engineering, Volume 20, Number 2, 2015,pp.111-118
19. Meteorology Regional Offices Turkish State Meteorological Service. "Turkey's solar map", Ankara, Turkey. https://www.mgm.gov.tr/ .[accessed 10 March 2018].
20. Ministry of Energy and Natural Resources, ankara,Turkey. Available at http://www.enerji.gov.tr/tr-TR/Sayfalar/Biyokutle.
21. The Foundation of the People Caring For The Future http://www.carfu.org/
22. H. Olgun, "Status Of Bioenergy In Turkey", Workshop on International cooperation in the field of bioenergy, 22-24 October, Moscow, Russia.

Supercritical Water Gasification of Eucalyptus Wood Chips Using $NiFe_2O_4$ as a Catalyst

Ane Caroline Pereira Borges[1], *Jude* A. Onwuidili[2], *Carine* Tondo Alves[3], *Andrew* Ingram[4], *Heloysa* Martins Carvalho Andrade[5], *Ednildo* Andrade Torres[1], and *Silvio* Alexandre Beisl Vieira de Melo[6]

[1]Interdisciplinary Center of Energy and Environmental (CIENAM), Federal University of Bahia (UFBA), Salvador, Brazil
[2]European Bioenergy Research Institute, Chemical Engineering & Applied Chemistry, Aston University, Birmingham, England
[3]Energy Engineering, Federal University of Reconcavo of Bahia (UFRB), Feira de Santana, Brazil
[4]Chemical Engineering, University of Birmingham, Birmingham, England
[5]Chemical Institute, Federal University of Bahia (UFBA), Salvador, Brazil
[6]Chemical Engineering, Federal University of Bahia (UFBA), Salvador, Brazil

Abstract. In this work, the supercritical water gasification of eucalyptus wood chips has been investigated in relation to reaction temperature and presence of catalyst. Experiments were performed in a batch reactor at 450 °C and 500 °C with two different feed concentrations. This work wanted to evaluate mainly the gases were formed during the reactions. The gas products were analysed by gas chromatography. According to the results, it was found that eucalyptus wood chips reacted to form mainly H_2, CH_4 gases with little yield of CO_2. Increasing the reaction temperature beyond 500 °C led to the increasing production of both CH_4 (around 31.1 mol%) and H_2 (up to 38 mol%) gases and the liquid sample and solid residue have decreased. Generally, this work suggests that the SCWG has improved significantly the production of H_2 but more experiments still necessary to verify the effects of other experimental parameters and to characterise the liquid sample and solid residue.

1 Introduction

Searching for renewable energy sources to ensure economic development with social inclusion without environmental aggression is a challenge all over the world. An alternative is biomass, a primary and renewable energy source. Brazil has 5,673,783 hectares of eucalyptus planted, 70.74 million m^3 of all this eucalyptus was consumed by the pulp and paper industry in Brazil in 2016 [1]. In 2016, Brazil generated 47.8 million tons of solid waste, with 14.1 million tons (29.5 %) from industrial activities (pulp and paper sector). In the paper industry, this waste is burned to generate energy, avoiding the use of fossil fuels but more waste is generated than can be used for the process energy requirements. Nevertheless, it is still necessary to discover new technologies to generate clean energy using this waste.

Hydrogen from waste biomass is considered to be a clean gaseous fuel and efficient for heat and power generation due to its high energy content [2]. A good choice to produce hydrogen-rich-gas is biomass gasification, involving a thermochemical conversion process of a solid into fuel gas, which maintains 70-80% of the original energetic fuel. The process occurs at high temperatures (700-1500 °C) and in the presence of a gasification agent (air or oxygen, steam, carbon dioxide). However, during the gasification process, some of the fuel is not converted into gas remaining as a condensate product (tar) [3]. Tar formation is one of the major problems to deal with during the biomass gasification process because it is a complex mixture of condensable hydrocarbons with low condensation temperatures [3-5]. A new and promising technology to ensure lower tar formation and higher hydrogen yield is the supercritical water gasification (SCWG). The main difference between SCWG and other gasification techniques is related to the gasification medium i.e., supercritical water compared to inert gas and/or steam. Supercritical water gasification is a process that can also convert wet biomass, which reduces pre-treatment costs. Further, the high solubility of organic compounds in supercritical water enhances their reactivity, which leads to an improvement of the SCWG performance. These advantages make SCWG an even more attractive technology, because one of the major disadvantages of the traditional gasification is the large production of tar. The SCWG also produces higher hydrogen content at high temperatures (> 700 °C) and higher methane content at lower temperatures (< 600 °C) [6,7]. Even though the SCWG increases the selectivity of hydrogen production, it requires high pressure and temperature for the reaction without catalysts. Catalytic supercritical water gasification (CSCWG) is a promising solution to improve the efficiency of the gasification process, producing a mixture of gases and values-added liquids products depending on the process conditions. The choice of catalyst to gasify lignocellulosic biomass in supercritical water depends on its capability to break C-C bonds, promote water gas shift (WGS) reaction, accelerate the cleavage of the C-O bond, and decrease the activation energy in SCWG reaction [8]. Heterogeneous catalyst has higher selectivity and recyclability compared with the homogeneous catalyst. The transition metals are typical catalysts for CSCWG because they offer high selectivity, high catalytic activity, and high level of carbon conversion to gas at lower temperature [9].

The present work aimed to perform the supercritical water gasification process of eucalyptus wood chips in the presence of a novel $NiFe_2O_4$ for increased hydrogen production using a batch reactor. The investigation involved study of the influences of biomass concentration in addition to reaction temperature on the gasification process.

2 Experimental

2.1 Raw Materials

The raw material used was *Eucalyptus grandis* chips with approximately 30 % moisture content and size of 710 μm obtained from Federal University of Bahia. The biomass was characterised before use. All analysis was performed according to the Standard Test Method (ASTM). The characteristics of the wood chips were determined by physic-chemical analyses as ultimate analysis; determination of immediate analysis and Higher Heating Value (HHV). The methodology of these analyses has been discussed in an earlier paper [10]. The results of these analyses are summarized in Table 1.

Table 1. Characteristics of the biomass used.

Sample	Moisture Content (%)	Proximate Analysis (%)			Ultimate Analysis (%)			
		Volatile Matter	Fixed Carbon	Ash	C	H	N	O
Eucalyptus wood chips	30	83.23	16.42	0.5	46.81	6.11	0.12	46.76

The catalyst used, $NiFe_2O_4$, was synthesized by combustion reaction method using ferric nitrate $Fe(NO_3)_3.9H_2O$, nickel nitrate $Ni(NO_3)_2.6H_2O$ and urea $CO(NH_2)_2$ as fuel. The methodology of that synthesis and characteristics of the catalyst has been discussed in an earlier paper [11].

2.2 Reactor system and experimental procedure

The supercritical water gasification experiments were performed in a non-stirred Hastelloy-C reactor obtained from Parr Instruments Co., Inc. USA. The reactor has 75 ml of volume capacity and the maximum operating conditions of temperature and pressure are 600 °C and 45 MPa, respectively.

In each experiment, first the feed material was loaded into the reactor containing deionized water, then the catalyst was loaded. The amount of water loading was 20 mL at 450 °C and 15 mL at 500 °C, giving corresponding operating pressures of 33 MPa and 28 MPa, respectively.

Experiments were performed using two different amounts of biomass, 2.22 g (10 wt %) and 5.0 g (20 wt %) and amounts of catalyst (1.0 g and 2.0 g). These were designed to evaluate some of the main parameters of the process (temperature, catalyst, feed concentration).

Once the reactor was loaded, it was sealed. It is placed into an electrical furnace fitted with temperature controller. The reactor was heated fast at a rate of 20 °C min^{-1} up to 450 and 500 °C, such that the reaction temperature was reached after just 15 minutes. The reaction time was mostly 30 min. At the end of each reaction, the reactor was rapidly withdrawn from the heater and quickly cooled to ambient temperature with compressed air. After cooling, the gas outlet valve was opened to collect the gas samples. After that, the reactor was opened to collect the liquid sample and solid residues.

The results have been obtained are only preliminary. It is necessary to carry out experiments in duplicates/triplicates to guarantee a good reproducibility in terms of product balance and physical experiments parameters. In this work only, the gases formed during the reactions were analyzed properly.

2.3 Gas products analysis

The gases were collected from the reactor using a gas sampling bag then 1 μL were sampled into a gas tight syringe and analysed using gas chromatograph (Shimadzu GC-2014 gas chromatograph) fitted with two columns and detectors were used. Usually, two samples were taken per experiment and the average reported. The hydrocarbon gases (methane, ethene, ethane, propene, propane, butane and butene) were separated on a 2 m long by 2 mm diameter and packed with 80 - 100 mesh Hayesep and analysed using a flame ionization detector (FID). The permanent gases (hydrogen, oxygen, nitrogen, and carbon monoxide) were separated on a 2 m packed molecular sieve (60 – 80 mesh) column and analyzed using a thermal conductivity detector (TCD) while a packed Hayesep column of similar dimensions was used for analysing carbon dioxide. The GC was first calibrated with standard gas mixtures (1 % H_2, 1 % CO, 1 % CH_4, 1 % CO_2, 1 % ethene, 1 % ethane, 1 % propane, 1 % propene, 1 % butane and 91 % N_2 as the swing gas) at levels similar to the composition of the produced gases. Argon was used as the carrier gas. The methods of analysis applied are in agreement with previous work [12]. The results from the GC were obtained in volume percent and were converted to moles of each gas using the general gas equation.

3 Results and discussion

The parameters investigated included the influence of feed concentration, feed catalyst and the temperature on the supercritical water gasification of eucalyptus wood chips. Mostly of the experiments were performed at 450 °C and 30 min of residence time, only the test 10 was carried out for 60 min. Only three experiments were performed at 500 °C. The experiments performed, and the gas analysis results obtained from the experiments can be observed in Table 2 and Table 3, correspondingly.

Table 2. List of experiments performed.

	Temperature								
	450 °C						500 °C		
Tests	1	2	3	4	5	6	7	8	9
Catalyst (g)	1.003	1.010	2.007	2.004	-	-	-	1.019	1.004
Biomass (g)	2.246	4.997	5.006	2.221	5.005	2.219	2.223	2.228	2.224

Table 3. Gas analysis results.

Gases	% mols								
	Test 1	**Test 2**	**Test 3**	**Test 4**	**Test 5**	**Test 6**	**Test 7**	**Test 8**	**Test 9**
C_2H_6	10.408	12.395	4.851	3.910	4.238	2.615	3.615	6.853	4.656
C_2H_2	1.159	0.798	0.319	0.405	0.302	0.372	0.121	0.144	0.188
C_3H_8	4.285	5.105	2.027	1.699	1.602	0.950	1.416	2.804	1.849
C_3H_6	2.429	1.788	0.715	0.985	0.622	0.727	0.257	0.304	0.392
C_4H_{10}	4.015	3.790	0.734	1.111	0.682	0.666	0.658	1.221	0.878
CH_4	**17.753**	**18.951**	**32.042**	**22.468**	**29.947**	**20.164**	**34.961**	**31.766**	**31.102**
H_2	**23.476**	**13.147**	**24.348**	**36.898**	**20.586**	**18.693**	**17.677**	**28.091**	**38.059**
CO_2	**17.507**	**32.991**	**7.522**	**3.890**	**10.534**	**1.931**	**7.893**	**5.540**	**2.383**
N_2	16.813	8.607	23.396	25.631	19.235	39.137	22.440	22.311	19.114
CO	2.156	2.428	4.046	3.004	12.252	14.746	10.962	0.966	1.378

3.1 Effect of feed concentration and catalyst concentration

The effect of feed concentration on gasification of eucalyptus wood chips, in supercritical water gasification was carried out in the presence of $NiFe_2O_4$ at 450 °C for 30 min reaction time. The wood sample concentrations used were 10 wt% and 20 wt%. The concentration of eucalyptus wood chips and the presence of catalyst especially have a significant effect on the gasification process. It can be observed that increasing feed concentration the CH_4 (33 %) gas product yield rises significantly (Table 3). In general, the concentration of the other gaseous products (CO, CO_2, and hydrocarbons) usually increase at higher concentrations of feed samples, as some researchers have reported [13,14,15]. But it was not noticed during the experiments. It could be explained by the presence of the catalyst, which probably helped to keep the predominance of water gas shift and methanation reactions instead of the alternative reaction pathways during the reaction at higher biomass.

Nickel catalyst is a good option not only because of the cost-effective but is also known promote water-gas shift reaction and to largely suppress tar formation by cracking it. Many researchers have been introducing it into a supercritical water gasification reaction to understand better the activity and stability of its. Most searches found that using nickel catalyst the conversion of biomass can be accelerated. Researchers [8, 16, 17] have done experiments to evaluate the hydrothermal gasification performance using Ni as a catalyst. According to these authors, chemical and physical structure of Ni catalyst importantly changes at hydrothermal conditions, these changes can improve the gasification reactions of biomass under supercritical even subcritical water conditions. According to the "Fig. 1 (A)" the presence of catalyst during the experiments at 450 °C increased potentially the H2 and CH4 gas product yields by 49.3 % and 10.2 %, respectively. The results were consistent with earlier literature reports [8, 12, 16, 17].

About the liquid obtained from each experiment, at lower feed concentration the liquid samples were clear. As the feed concentration was increased, a dark brownish oil layer was present in the liquid sample. This indicated that the range of reaction temperature and reaction time were not enough to complete degraded the biomass used during the experiments.

3.2 Effect of temperature

"Fig. 1 (B)" illustrates the effect of reaction temperature increasing from 450 to 500 °C on the composition of gas products from hydrothermal gasification of eucalyptus wood in the presence of $NiFe_2O_4$ catalyst. The feed of biomass was kept at 2.2 g (10 wt %). According to the "Fig 1 (B)", the temperature has a significant effect on

https://doi.org/10.1051/e3sconf/20186401002

the gas yields, the higher temperature favour gaseous production, especially H_2 and CH_4 yields due to the promotion of free-radical reactions which stimulate gas formation. The highest hydrogen yield (around 38 %) was obtained by conversion of eucalyptus wood chips at 500 °C. even during the experiments carried out without catalyst.

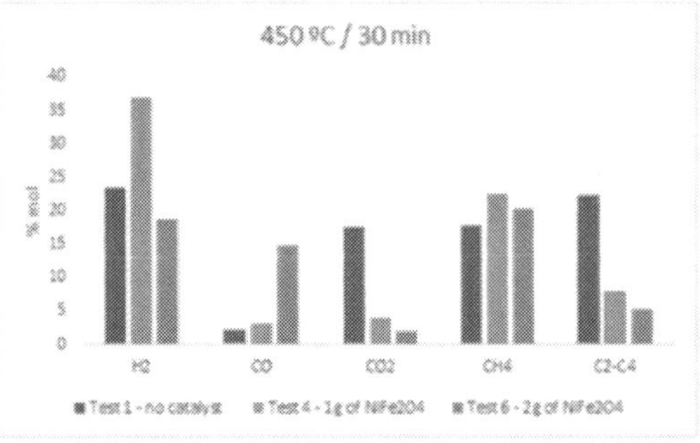

(A): Influence of NiFe₂O₄ at 450 °C for 30 min of residence time

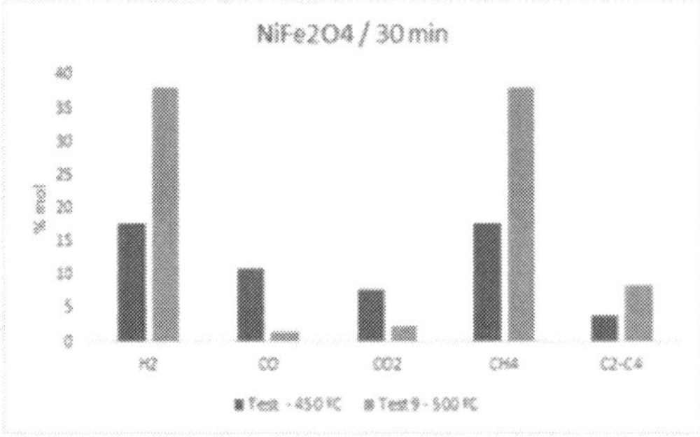

(B): Effect of reaction temperature

Fig. 1. Effect of the catalyst (A) and reaction temperature (B).

3.3 Possible uses of gas products

The gasification process led to the production of hydrogen and methane fuel gases, which have various existing applications. First, the gas product can be fed into a gas engine for electricity production and combusted directly for clean energy in the form of heat for the cellulose industry. Second, the gas products are obtained at high pressure, which makes separation using e.g. pressure-swing absorption or membranes, highly applicable. Hydrogen may be obtained pure and used in hydrogen fuel cells, while methane can be used in gas engines to produce electricity. Pure methane can also be fed into domestic gas networks for homes or converted to syngas for further processing. All these added value activities from the SCWG process can potentially increase the economic and environmental viability of the cellulose and other biomass process industries

4 Conclusions

According to the preliminary results, it was observed that the supercritical water gasification is a viable process for convert eucalyptus biomass in H_2 and CH_4. It was found that eucalyptus wood chips decompose to H_2, CH_4 gases with little yield of CO_2. Increasing the reaction temperature beyond 500 °C led to the increasing production of both CH_4 (around 31.1 %) and H_2 (around 38 %) gases and the liquid sample and solid residue have decreased. It was also observed the influence of the catalyst. The NiF_2O_4 increased the C-C bonds decomposition of eucalyptus. Using the nickel ferrite led to the increasing production of H_2 (49 %). This work proposes that the supercritical water gasification has improved expressively the production of H_2 but more experiments still necessary to verify the effects of other parameters and to characterise the liquid sample and solid residue.

Detailed feasibility studies will be carried in collaboration with the cellulose industry in Brazil, to evaluate the potential yields of these gases and associated costs of the process in a typical cellulose plant in consideration for technology take-up.

Acknowledgment

The authors would like to thank CAPES - Coordenação de Aperfeiçoamento de Pessoal de Nível Superior and FAPESB - Fundação de Amparo A Pesquisa do Estado da Bahia, of the Ministry for Education of Brazil, for their support of this research.

References

1. Indústria Brasileira de Árvores - IBÁ. Relatório Anual IBÁ 2017. Brasília, 2017.
2. Reddy, S.N.; Nanda, S.; Dalai, A.K.; Kozin, J.A. Supercritical water gasification of biomass for hydrogen production. *International Journal of Hydrogen Energy*, v.39, p. 6912-6926, 2014.
3. Hernández, J.J.; Ballesteros, R.; Aranda, G. Characterisation of tars from biomass gasification: Effect of the operating conditions. Energy, v. 50, p. 333-342, 2013.
4. Dudynski, M.; Van Dyk, J.; Kwiatkowski, K.; Sosnawska, M.. Biomass gasification: Influence of torrecation on syngas production and tar formation. Fuel Processing Technology, v. 131, p. 203-212, 2015.
5. Devi L, Ptasinski KJ, Janssen FJJG. A review of the primary measures for tar elimination in biomass gasification processes. Biomass Bioenergy, v. 24, 2003.
6. Kruse, A. Hydrothermal biomass gasification. *The journal of supercritical fluids*, n. 47, p. 391-399, 2009.
7. Behenia, I. Treatment of aqueous biomass and waste via supercritical water gasification for the production of CH_4 and H2. 2013. 108f.. University of Western Ontario London. Ontário-Canadá, 2013.
8. Elliot, D.C. Catalytic hydrothermal gasification of biomass. Biofuels Bioproducts and Biorefining 2(3), p. 254-65, 2008.

9. Guo,Y., Wang, S. Z., Xu, D. H., Gong, Y. M., Ma, H. H., Tang, X. Y. Review of catalytic supercritical water gasification for hydrogen production from biomass. Renewable and Sustainable Energy Reviews, v.14, p.334-343, 2010.
10. Borges, A. C. P.; Alves, C. T.; Torres, E. A. . Torrefied Eucalyptus Grandis Characterization as a Biomass to Using in Industrial Scale. Chemical Engineering Transactions, v. 49, p. 283-288, 2016.
11. Borges, A. C. P.; Alves, C. T. ; Fiuza, R. ; Andrade, H. M. C. ; Ingram, A. ; Vieira de Melo, S.A.B. ; Torres, E. A. . Preparation of heterogeneous catalysts by combustion reaction for water gas shift reaction. In: IMRC 2017 - XXVI International Materials Research Congress, 2017, Cancun. XXVI International Materials Research Congress, 2017.
12. Muangrat, R.; Onwudili, J.A.; Williams, P.T. Influence of alkali catalysts on the production of hydrogen-rich gas from the hydrothermal gasification of food processing waste. Applied Catalysis B: Environmental v. 100, p. 440–449, 2010.
13. Hao XH, Guo LJ, Mao X, Zhang XM, Chen XJ. Hydrogen production from glucose used as model compound of biomass gasified in supercritical water. Int J Hydrogen Energy v. 28, p. 55-64, 2003.
14. Onwudili JA, Williams PT. Role of sodium hydroxide in the production of hydrogen gas from the hydrothermal gasification of biomass. International journal of hydrogen energy, v. 34, p. 5645-5656, 2009.
15. Kruse A, Gawlik A. Biomass conversion in water at 330-410 °C and 30-50 MPa identification of key compounds for indicating different chemicals reaction pathways. Ind Eng Chem Res, v. 42, p. 267-279, 2003.
16. Elliot, D.C., Sealock, Jr.Lj, Baker, E.G.. Chemical processing in high-pressure aqueous environments. Development of catalyst for gasification. Industrial & Engineering Chemistry Reasearch, v. 32 (8), p. 1542-1548, 1993.
17. Waldner, M.H; Vogel, F. Renewable production of methane from woody biomass by catalytic hydrothermal gasification. Industrial & Engineering Chemistry Research, v. 44 (13), p. 4543-4551 2005.

Performance Analysis of a Hybrid Micro-Energy System for SA Data Centers

Sempe Thom Leholo[1], *Pius* Adewale Owolawi[1], *Kayode* Timothy Akindeji[2]

[1]Tshwane University of Technology, Department of Computer Systems Engineering, Pretoria, South Africa
[2]Durban University of Technology, Department of Electrical Power Engineering, Durban, South Africa

Abstract. The integration of hybridized renewable energy sources (RES) with AC/DC converters has become the focus of the 21st century for green Information Communication Technology (ICT) applications such as the data center. As the data traffic grows exponentially, the corresponding demand for energy to drive the growth becomes a great challenge and considering the environmental impact, a hybrid renewable energy system is favored for eco-sustainability and economic reasons. This is especially true for data centers which represent a dominant share of the total power in cellular networks. This paper evaluates the actual performance of a fuel cell in a renewable energy hybrid system considering the hybridization of photovoltaic (PV), Wind, Fuel Cell, and battery storage system with a choice of a half-grid mode. The reduction and the absence of available PV power by shading and rainy conditions will be easily reduced by the compensation of the other renewable sources. The modeling and simulations are performed using HOMER software. The results show the effectiveness of the proposed system as the energy supply is less intermittent and more stable.

1 Introduction

Most hybrid power systems today utilize wind, hydro and solar energy sources. Solar energy makes the majority of renewable energy utilized on earth today in multiple forms because photovoltaic panels are easily accessible and affordable in addition to the abundance of solar irradiance across the globe [1]. A Hybrid system combining all these different types of power sources has an advantage over PV in terms of balance, stability and satisfied energy outputs [1]. The resources that make-up a data center such as high-level servers, high capacity storage devices, network devices such as routers and switches must be secured with the uninterrupted power supply as this is critical to the business of data centers. It must be noted that most data centers consumed large amounts of energy as the data transaction traffic grows in billions of online users worldwide [2]. In fact, Industrial Development Corporation reports that server's energy consumption is growing by 9% per year globally attributed to the rise of performance which pushes demand for energy [3]. Data centers energy efficiency must, therefore, be improved upon to realize significant tangible energy savings improvements [4]. In South Africa, the rise in big data and Internet-of-Things (IoT) industries in recent years have spurred the continued construction of more data centers [5]. It has been confirmed that the carbon emissions emanating from ICT infrastructure are estimated to be within the range of two to four percent of the global accumulative carbon emissions [6].

Considering the operation of the data center it is observed that the aforementioned percent (2-4) would increase by 40-60 percent [7]. This is further predicted that by 2020 this exponential increase percentage of emission will be double. A huge part of this emission's, one-sixth of that will be attributed to telecommunication networks [7]. Fossil fuel resources have a finite future because of their increasing depletion rate combined with the environmental damage done atmospherically, like global warming, this has started to change the world's perspective on a global warming and it has driven the world economies to dwell more on technologies of renewable energy sources that are safe for humans and do not damage the environment. One of the concerns with this paper is to investigate the costs in line with the inclusion of hydrogen in the proposed hybrid system. The investigation revealed what configuration of PV and wind systems couple with the electrolyzer is feasible in order to produce power that's adequate to sustain constant power supplies to a data center.

The debate on the addition of fuel cell may not be seen as ideal for the short run of the project. Hence, the long run of the combination confirmed a reduction of fuel cell prices by at least 45% of their current capital cost, which shows the competitiveness for the remote data center system deployment and implementation. In most of data centers, the supply of electricity is erratic due to multiple factors such as environmental constraints, inaccessibility by developers and financial constraint as a result of high cost of installation over long-distance from the grid [8].

The data, air and power flow, as well as the interconnection of all components in a typical data center, are shown in Figure. 1. The cooling system is assumed to be based on chilled water. No matter what architecture a data center might have, it only has two types of connections with the outer world – electrical power and internet connections. The data load determines server utilization and the total power demand changes based on the data load. How do the interconnections of the components in a data center affect the total power consumption? To answer this question, there is a need to develop detailed models for all components and their interconnections.

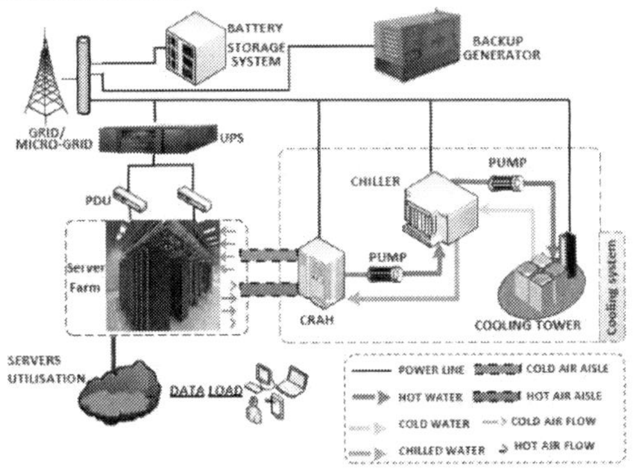

Fig. 1. The interconnection of components in a typical data center.

Comparative economic analyses on a power generating hybrid systems is one of HOMER software capabilities [9]. The chosen parameters serve as inputs to HOMER software to simulate every possible combination of chosen RES and rank in order of best performance. The analysis tests such as capital costs or cost of energy will be parameters of interest in making choice of the suitability of RES. The impact on the Levelized cost of energy (COE) by the cost of fuel like diesel or hydrogen can be done by "HOMER sensitivity analyses" which is often achieved when the input parameters of the renewable resources are optimized. This simulation test is further used to determine any economic feasibility study of energy production and comparability test of electricity generation systems and technologies adopted to achieve the chosen configuration [10]. In addition to the required input parameters to estimate COE, fuel cost, capital cost, operation, and maintenance cost coupled with other basic financing costs play a key role in the estimation [11].

In this work, the datasheet of the chosen systems and the available natural data from solar irradiance, wind speed, and other measurable parameters are used as the input parameters for the simulation to estimate the present, future performance and costs, and efficiencies of the chosen hybrid system.

2 Data collection

2.1 Topographical data

Modelling a system requires input data that is based on the site where the data center is located. These data is inserted into the HOMER software for modelling and analysis. The renewable energy resources amount available to support the system load are determined and estimated by using input data. Data like solar irradiance and wind speed are sampled and processed, before input into the HOMER Software.

Solar resources like solar irradiance data are required for our photovoltaic system. In the certain locations, particularly remote areas in South Africa where the data center will be located, it is important to determine the amount of sunlight available at a certain time, be hourly or daily. In the same vein, wind speed and other relevant data of the location where the wind turbine is to be installed must be obtained. Table 1 below displays the monthly average data for solar irradiances, clearness index and wind speed for the selected Cape Town data center location.

Table 1. Topographical Data

Month	solar irradiances kWh/m^2	Clearness index data	Wind speed m/s
Jan	3.96	0.394	4
Feb	5.1	0.491	4.5
Mar	6.1	0.581	4.9
April	7.02	0.688	5
May	7.68	0.795	5
Jun	8.33	0.895	4.9
July	8.19	0.868	4.7
Aug	7.59	0.765	4.5
Sep	6.78	0.656	4.4
Oct	5.7	0.55	4.3
Nov	4.32	0.428	3.9
Dec	3.59	0.362	4
Avg	6.197	0.623	4.51

2.2 Energy consumption: data center

Having a good understanding of data center energy consumption is critical to implementing a hybrid RES. In this analysis, a data center in a defined and chosen location of interest (typical area of 465 m2) is to be modelled. The modelled data center has a total energy consumption of 215.86 kWh/day and the energy use is classified as either demand aspect or supply aspect. The demand-side system includes is the server power supplies, processors, communication equipment, storage and other server components and accounts for 53% of the sum of consumption. Included in the supply-side systems is the UPS, cooling, power distribution, building switchgear and lighting all account for 47% of total consumption. In addition, it must be noted that a substantial energy is needed to remove the generated heat from the data center.

The total hourly data center power requirement is 8.994 kW, therefore, the total power requirement for the data center annually is 78.79 MW.

2.3 Load profile data

The load profile of the modelled data center is shown in Figure 2. In order to get a clear power demand of the data centers, the energy demand management concept and corrector factor will be used. Calculations in kWh for Energy produced is in relation to the hybrid power system in a daily allowance situation. Demand factor is used to scale down real daily energy demand to help with actual sizing of the RES in order to cover the data center energy needs. With the utilization of HOMER software and hourly load data, in turn, the data center load profile is obtained as displayed in Figure 2. From the profile, the scaled average for the day is 215kWh/d and the peak is 9.96kW with a load factor of 0.438. The electrical load factor explained as the ratio of the average load, or efficiency of electrical energy usage, in other words, it is the ratio of total energy (kWh) consumed in a certain billing period of hours to the maximum demand which happened at a certain period.

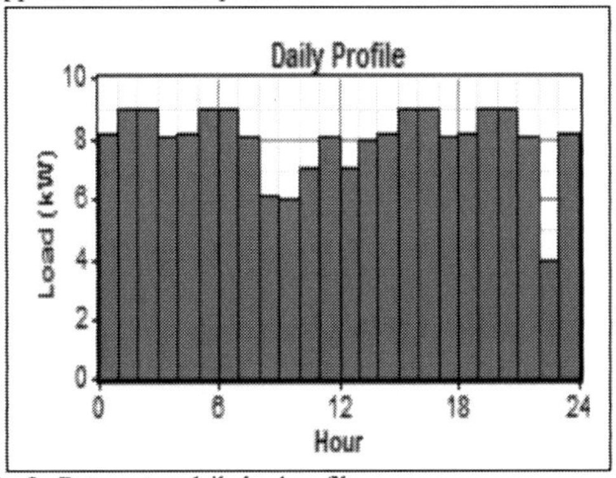

Fig. 2. Data centers daily load profile

2.4 Architectural design of hybrid power system

An RES consisting of solar PV, wind turbines, and a fuel cell integrated with AC–DC converter, electrolyzer, hydrogen tank and a storage battery bank is simulated using the HOMER software. Simulation of the different multivalent system to maximize the available RES and minimize total operation cost. The preferred mode for this work is non-grid-connected with an energy storage device (battery) that automatically and seamlessly connect to the load when RES is not enough to power the load. The schematic diagram of the RES used for the simulation is shown in Figure 3. In order to determine the optimal value of each RES, in order to reach the goal of the optimization process, component variables are considered. These variables can be chosen to determine the sizes of equipment needed to model the system and then HOMER can show multiple possible designs in its optimization process.

There are two goals for the simulation process. Determining the feasibility of the RES is the first step. If the hybrid system can adequately accommodate the data centers load and mitigate any constraint that arises during the selection of systems configurations. Secondly, the cost of the life-cycle of the RES is estimated, in which the installation costs and the lifetime operation costs of the RES are combined [12]. Comparing the economics of various system configurations is a convenient metric which is called the life-cycle cost [13].

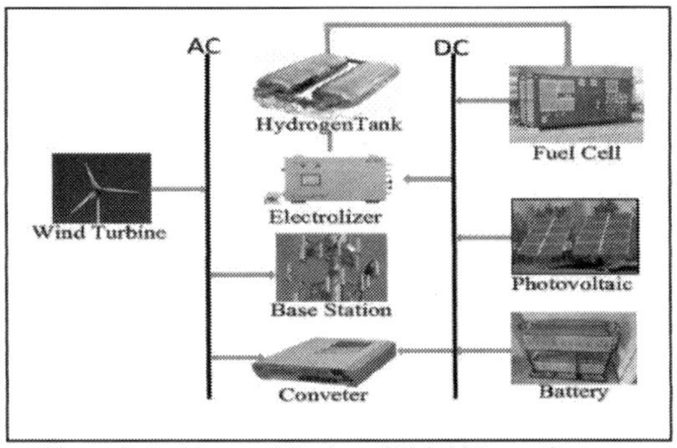

Figure 3. Architectural Design of Hybrid Power System.

2.5 Cost formulation and calculations

The Optimization by HOMER software gives detailed reports which include a range of scopes from the renewable penetration to the present net capital cost of the RES. It permits the inputs of RES (such as sun irradiances, wind speeds), battery information, load demand information, capital and value related to in operation and maintaining costs(O&MC) prices among others, again analytic sensitivity modelling the impact on the RES to varieties in any input. For the generator, the O&MC is entered as an hourly value while for other components, the annual value of O&MC is entered in the HOMER software and that value is then multiplied by the yearly operation hours of the components. This helps to get annual O&MC of the RES [14].

The total operating cost is the annualized value of all costs and revenues other than initial capital costs of the RES is determined by using the following equation [15]:

$$C_{opt,tt} = C_{anl,tt} - C_{anl,cc} \quad (1)$$

where:

$C_{anl,tt}$ = the total annual cost $[R/yr]$

$C_{anl,cc}$ = the total capital cost annually $[R/yr]$

The COE is determined using the following formula below:

$$COE = \frac{C_{anl,tt}}{E_{prim}E_{dfr}E_{grid,sls}} \quad (2)$$

where:

$C_{anl,tt}$ is the total annualized value, E_{drf} is the total deferrable energy quantity of the load and E_{prim} is the primary energy that the system serves annually.

$E_{grid,sls}$ is the annual quantity of energy oversubscribed to the grid.

$C_{NPC,tot}$ (Total present net cost) is calculated by utilizing the subsequent equation:

$$C_{NPC,tot} = \frac{C_{anl,tt}}{CRF(i_{dr}, Rpjt)} \quad (3)$$

i_{dr} = actual discounted rate annually [%]

R_{pjt} = lifetime of the project [yr]

CRF = is the capital recovery factor given by the equation:

$$CRF(i, N) = \frac{i(1+i)^N}{(1+i)^N - 1} \quad (4)$$

where i is the annual real interest rate and N is the number of years.

2.6 Component cost breakdown

The details of the components utilized to design the hybrid power system are shown in Table 2.Table 3 shows the costs of the components for the hybrid power system.

Table 2. Component details of Hybrid Power System.

Component	O&MC (R/yr)	Lifetime
PV	0	25 yrs
Wind Turbine	360	25 yrs
Batteries	7.92	10 yrs
Fuel cell	0.24	47,000 hrs
Electrolyzer	240	25 yrs
Hydrogen tank	0	25 yrs
AC/DC Converter	0	15 yrs

Table 3. Component cost details of Hybrid Power System.

Component	Size	Capital Cost (R)	Replacement cost (R)
PV	40 kW	23 098.80	0
Wind Turbine	25 kW	250 000.00	180 000.00
Batteries	1156 Ah	300 000.00	250 000.00
Fuel cell	10 kW	115 560.00	115 560.00
Electrolyzer	10 kW	30 720.00	30 720.00
Hydrogen tank	10 kg	16 024.00	16 024.00
AC/DC Converter	10 kW	21 080.00	19 800.00

3 Hybrid system composition

3.1 Solar components

To model the PV system for the selected hybrid system, the clearness index and the average solar irradiation of the data center location are considered for the HOMER simulation. The Standard Conditions (STC) test of solar panels where average solar irradiation is 6.197 $kWh/m^2/day$ and clearness index is 0.623 by data taken from Southern African Universities Radiometric Network in table 1.

The number of panels also have an important role in the cost and feasibleness analysis. The determining of the number of panels needed for the hybrid system, equation 5 is used and for this case, it is considered to be 12 panels:

$$N = \frac{P_{pv/day}}{E} \quad (5)$$

where P_{pv} is the energy produced daily and E is the power supply for one PV panel[15].

3.2 Wind turbine components

Any style of a turbine includes a theoretical maximum power potency of 58% (extracted energy carried out by a wind turbine cannot be more than 59% of wind passing through)[16]. Power coefficient is the phenomenon of the above statement and is defined as $C_{Pmax} = 0.58$. Wind turbines is not able to operate at its peak C_p. Each wind turbine type has a unique C_p and it represents the wind speed function that the turbine is operating in.

Furthermore, the components inside the wind turbine like the bearing and gearbox also have losses, which means the actual converted wind power into usable electricity is about 10-40%. The factoring of the power coefficient in equation 6 and power from the wind that can be extracted is given by[17]:

$$P = \frac{1}{2}\rho A v^3 C_p \quad (6)$$

where:

ρ is the Air density in kg/m^3, A is the rotor swept area in m^2, C_P is the Coefficient of performance and v is the wind velocity in m/s.

3.3 Fuel Cell components

A fuel cell consists of an electrolyte membrane sandwiched between two catalyst-coated electrodes the anode and the cathode. By redox chemical reaction a fuel cell acts like an electrochemical engine which produces electricity [18]. When hydrogen is passed through the anode electrode and oxygen (air) passed through the cathode electrode, electricity is generated with water, and heat being the byproducts [18].

Hydrogen fuel cells are very efficient but are also expensive to build. On a wide scale, hydrogen will not be used to produce electricity due to safety and huge costs [18]. On the anode side, the hydrogen gas is oxidized then releases electrons and creates H^+ ions as seen in equation 7. As seen in equation 8 on the cathode side, a reaction between the oxygen, electrons from the electrodes and H^+ ions from the electrolyte and water. In this process, water is the waste product and removed from the cell.

Anode side: $H_2 \rightarrow 2H^+ + 2e^-$ (7)

Cathode side: $\frac{1}{2}O_2 + 2H^+ + 2e^- \rightarrow 1H_2O$ (8)

3.4 Electrolyzer components

An electrolyzer uses an electrochemical process called electrolysis which converts electrical energy into chemical energy to produces hydrogen by using electricity to split water into hydrogen and oxygen. In this project, the photovoltaic system generates electricity to power the electrolyzer to create a chemical reaction to produce hydrogen and supply the hydrogen tanks. The equations 9 and 10 show the anode and cathode reactions.

Anode side: $1H_2O \rightarrow \frac{1}{2}O_2 + 2H^+ + 2e^-$ (9)

Cathode side: $2H^+ + 2e^- \rightarrow H_2$ (10)

3.5 Battery storage components

For this system, the Surrette 6CS25P 1156Ah with a 6V Deep Cycle is chosen and it should be noted that this battery is a new generation series 5000 which has a very high capacity with features advantages like dual case, exchangeable cells, tiny footprint and an expected lifetime of ten years. The parameters that are associated with the mathematical model of the chosen battery are presented below.

B_{sc} = Battery storage capacity N_B = Number of batteries

E_d = Energy demand D_a = Autonomous operation days

V_S = System voltage Ah_R = Battery Amp-hour rating

$B_{sc_{total}}$ = Total battery storage capacity

The daily watt-hour of the load is multiplied by three (autonomous days) to allow for bad weather when the sun doesn't shine. This provides a buffer, in case the batteries are not recharged on a daily basis.

$$B_{sc} = E_d * D_a \quad (11)$$

The batteries should not discharge below 50%, B_{sc} is multiplied by 2. This will give the total battery capacity the system needs to store to run for three days.

$$B_{sc_{total}} = B_{sc} * 2 \quad (12)$$

The required total amount of battery storage capacity for the chosen system requires 1290000 watt-hours.

$$N_B = \frac{B_{sc}/V_S}{Ah_R} \quad (13)$$

By using the nominal capacity of the battery which is 1156 Ah, the number of batteries required is approximately 186 batteries to support the load of 9.96 kW on autonomous days (days that wind and solar cannot support the system due to lack of wind and sun).

4 System design results and discussion

Figure 6 indicates the total energy required to drive the data center and this is amount to 175601 kWh/yr. In this total, PV takes 51%, Wind is 46% and Fuel Cell (FC) is allocated with 2% of the total renewable energy for the year. In the case of the emission concentration, Figure 7 presents the environmental emissions such as Nitrogen oxides, Sulfur dioxide, Particulate matter, unburned hydrocarbons, carbon monoxide and carbon dioxide using the metric scale respectively. Figure 8 presents the levelized and operating cost of the hybrid system annually for the data center in the chosen region (Cape Town) respectively. The systems performance indicators using levelized and operating cost are cost effective when compared to a diesel generator driven data center which amounts to R2.82/kWh levelized cost and operating cost of R129357/yr.

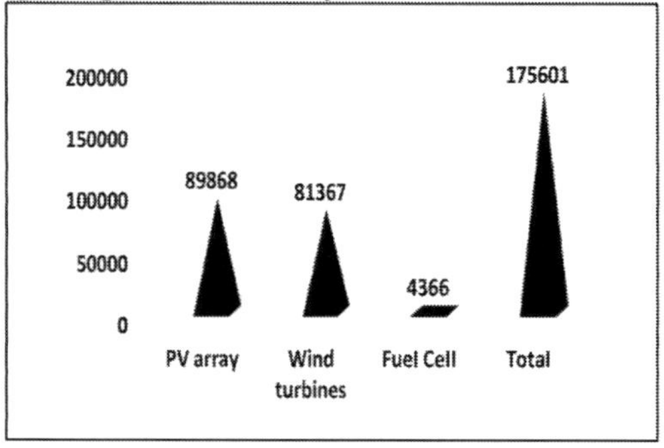

Fig. 6. Total yearly Electricity Production in kWh.

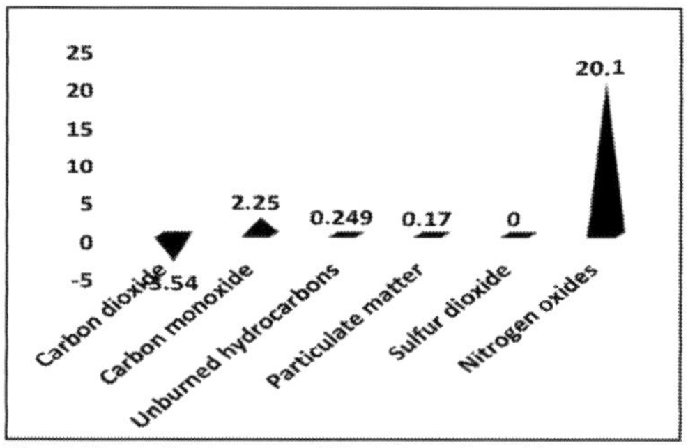

Fig. 7. Yearly Environmental Emissions Quantities

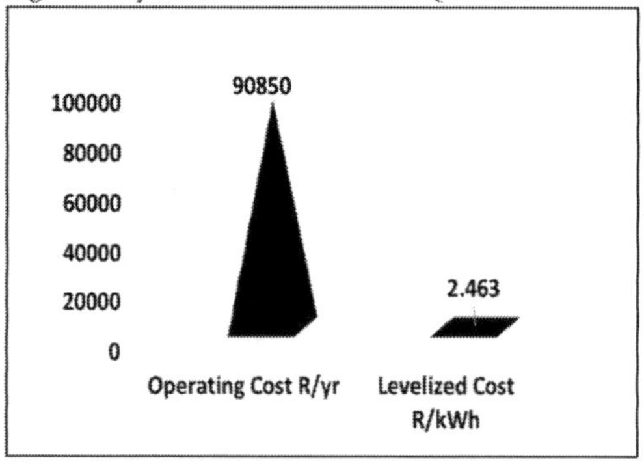

Fig. 8. Annualized Levelized and Operating Costs

5 Conclusion

Rising electricity prices and aging data centers are inflicting several businesses to require an awfully serious look into the energy potency of their computing infrastructure. The alternative sources of energy become the center of the business for the data center. Contained in the paper, is a micro-grid renewable energy system that consists of PV-Wind-FC is chosen. Based on the analysis of the system, the maintenance cost will be far less because the RES system requires less maintenance when compared to diesel generators. As a result of the aforementioned, fuel cell is chosen to acts as a backup power supply for the data center. Moreover, if the hybrid system fails to support the data center, the battery bank will provide the data center load autonomously for three days and twenty-three hours. From the results, the data center could save (885kg/yr) of CO2 emissions annually in Cape Town by removing backup diesel generators used by data centers, and this advantage will reduce the impact of greenhouse gases produced by data centers. Therefore, when RES are utilized in a hybrid system to power data centers, it will decrease the operating cost and will bring down the greenhouse gas emission that damage the environment. The simulation results showed that the solar power system can save 49.6% of the total Net Present Cost (NPC). These results indicate that the solar power system is a good choice for data centers, followed by the wind energy in order to reduce both the operational expenditures and greenhouse gases for any data center applications in other remote locations in the country.

References

1. C. Nazir. Coastal power plant: a hybrid solar-hydro renewable energy technology. Clean Energy, 2018.
2. B. Ristic, K. Madani and Z. Makuch, "The Water Footprint of Data Centers", Sustainability, vol. 7, no. 8, pp. 11260-11284, 2015.
3. S. Mishra., N. Mathur.: 'Load Balancing Optimization in /-A Cellular Networks: A Review', arXiv preprint arXiv:1412.7273, 2014.
4. M.H Alsharif., , R. Nordin., and M. Ismail.: 'A review on intelligent Data center cooperation management techniques for greener cellular networks', Journal of Communications, 2014, 9, (12), pp. 937-945.
5. A. Comninos., S. Esselaar., A. Ndiwalana., and C. Stork.: 'Airtime to cash: unlocking the potential of Africa's mobile phones for banking the unbanked', 2009.
6. G. Fettweis., E. Zimmermann.: 'ICT energy consumption-trends and challenges', in Editor (Ed.)^(Eds.): 'Book ICT energy consumption-trends and challenges' ((Lapland, 2008, edn.), pp. 6.
7. R. Gasch., J. Twele.: 'Wind power plants: fundamentals, design, construction and operation' (Springer Science & Business Media, 2011. 2011).
8. M. Khodayar, "Rural electrification and expansion planning of off-grid microgrids", The Electricity Journal, vol. 30, no. 4, pp. 68-74, 2017.
9. S. Sinha., Chandel, S.: 'Review of software tools for hybrid renewable energy systems', Renewable and Sustainable Energy Reviews, 2014, 32, pp. 192-205.
10. C.P. Castillo., B.F. e Silva. and C. Lavalle. (2016). An assessment of the regional potential for solar power generation in EU-28. Energy Policy, 88, pp.86-99.
11. S.B. Darling., F. You., T. Veselka., and A. Velosa.: 'Assumptions and the levelized cost of energy for photovoltaics', Energy & Environmental Science, 2011, 4, (9), pp. 3133-3139.
12. B. Zakeri., and S. Syri.: 'Electrical energy storage systems: A comparative life cycle cost analysis', Renewable and Sustainable Energy Reviews, 2015, 42, pp. 569-596.
13. T. Lambert., P. Gilman., and P. Lilienthal.: 'Micropower system modeling with HOMER', Integration of alternative sources of energy, 2006, pp. 379-418.
14. V. Khare, S. Nema. & P. Baredar. (2015) Optimisation of the hybrid renewable energy system by HOMER, PSO, and CPSO for the study area, International Journal of Sustainable Energy, 36:4, 326-343, DOI: 10.1080/14786451.2015.1017500.
15. C. Wang., M.H. Nehrir., and S.R. Shaw.: 'Dynamic models and model validation for PEM fuel cells using electrical circuits', IEEE transactions on energy conversion, 2005, 20, (2), pp. 442-451.
16. D.W.R. Thompson.: 'Incorporating renewable energy in a desalination plant: case study in El Paso, Texas', 2017.
17. M. Şenol., S. Abbasoğlu., O. Kükrer., and A. Babatunde.: 'A guide in installing large-scale PV power plant for self-consumption mechanism', Solar Energy, 2016, 132, pp. 518-537.
18. Z. Xia, S. Wang, L. Jiang, H. Sun, S. Liu, X. Fu, B. Zhang, D. Sheng Su, J. Wang and G. Sun, "Bio-inspired Construction of Advanced Fuel Cell Cathode with Pt Anchored in Ordered Hybrid Polymer Matrix", Scientific Reports, vol. 5, no. 1, 2015.

Design Optimization and Demand Side Management of a Solar-Assisted Industrial Heating Using Agent-Based Modelling (ABM): Methodology and Case Study

Fitsum Bekele Tilahun[1], *Ramchandra* Bhandari[1], and *Menegesha* Mamo[2]

[1]Renewable Energy Systems, ITT, TH Köln (Cologne University of Applied Science), Köln, Germany
[2]Electrical & Computer Engineering, Addis Ababa University Institute of Technology (AAIT), Addis Ababa, Ethiopia

Abstract. Heavy dependence on traditional fuels as well as connection to a significantly unreliable and inefficient grid (with about 20% energy loss) characterizes the Ethiopian industry. On the contrary, the country has many solar-suited industries for process heat augmentation in the low-to-medium temperature ranges. Thus, if properly designed and operated, solar thermal could give an opportunity for an efficient and cleaner alternative in these industries. This paper presents agent-based modeling (ABM) for an integrated optimization and demand side management (DSM) framework for solar-assisted industrial heating under varying load and weather condition. To demonstrate the validity and practicality of the proposed solution, a case study was conducted on Ethiopian textile industry. A payback period of 5.3 years and solar fraction of 66.7% was identified for an optimized system. Further with the implementation of DSM to guide the production policy of the industry, a 7.5% and 8.4% improvement in payback and solar fraction was achieved.

1 Introduction

Global estimate for industrial energy use indicates that more than 66.7% of the total energy consumption is for process heat, out of which 50% is for low-to-medium temperature range [1]. According to the same study, about two-fifths of primary energy use in industries comes from natural gas or petroleum indicating a solar replication potential that is estimated to be 15EJ which would cover about 10% of industrial energy requirement by 2030.

Identified promising industrial sectors for solar integration are: food, beverages, paper, textile, brick and blocks, chemical, plastic and flour by products [1-2]. Similar to other developing countries, these industrial sectors dominate the Ethiopian industry [3].

According to existing forecasts, the electricity consumption in Ethiopian industrial sector is expected to grow by approximately 9 folds of the 2015 consumption by 2030[4]. However, these industries are connected to the significantly unreliable national grid [5]. Consequently, the supply disruptions during the production process result in huge loss to the industries. The said unreliability of the electricity supply in the country has compelled many industries to set up their own emergency backup systems and often the diesel generator is their preferred choice. The employment of alternative local energy resources (for example, the use of solar thermal technologies to deliver industrial heat requirement, etcetera) as well as demand side management (DSM) could solve this problem.

If properly designed and operated, renewable energy resources could give an opportunity for an efficient and cleaner alternative together with the aspired reliability as is in the power industry.

The work of this paper is an attempt to address three related issues that still need a research attention. First of all, the number and fragmented nature of solar promising industrial processes have resulted in a need for a sector specific solution [1-2, 6-12]. Then more importantly, gaining the benefits of these specific solutions requires a holistic optimization framework; an approach that considers the daily performance based on weather prediction, hot water consumption variation and temperature dynamics of the thermal system. Finally, investigating the performance reliability of these optimized solutions in the presence of uncertainties in energy generation and consumption is necessary. However, scarcely few studies have answered these research questions mainly due to the inability of existing tools to handle such task [13-15]. On the other hand, agent-based models, as derivatives of complexity science, has been shown to be well suited for a natural description of systems where there is a dynamic interaction to result in complex collective patterns [15]. The primary aim of this paper is to demonstrate the capability of agent-based hybrid models for optimization and DSM analysis of an industrial scale solar thermal system. These agents have continuous thermal and consumption models that are embedded and run in parallel. Data for modelling are from measurements taken at the chosen industry and some are from manufacturer's datasheet. An optimization experiment with the total annualized cost (operating and investment) was carried out to find the optimal collector area and storage volume. These optimal parameters are then exported back to the simulation environment to investigate the possibility for improved energy use by implementing DSM to guide the industry's production policy.

2 Case study

The investigated dyeing textile process case study is found at Bahir Dar textile factory, in northern Ethiopia (11°36′N 37°24′E). The textile factory manufactures 100% cotton products, including yarns and fabrics. Electricity, diesel and furnace oil (FO) are the major energy source to the plant with annual consumptionn of 27.54million kWh, 166kL and 615kL respectively [17]. The textile plant has installed FO fired steam boiler for dying section. This boiler operates only during production of dying section and consumes an average of 250 litters FO per day running with efficiency below 70%. Increase price and poor reliability of FO are the two main problems that challenge the dyeing section of the textile factory.

On the other hand, the textile site receives an average direct and diffused solar radiation of 6.8kWh/m^2 and 1.7kWh/m^2 [18].Thus it is easy to see that a solar thermal system would have a higher replication potential for replacing the existing inefficient FO based boiler for the dyeing section of the factory. The primary aim of this study is for optimal design of a solar thermal system for the dyeing textile process and demand side management (DSM) for further energy efficiency improvement opportunity. The dyeing machine parameter is depicted in Table 1.

3 Methodology

The structure of the implemented methodology is depicted in Fig -1. The implementation is modelled in a computational framework based on Anylogic 8.1 and embedded optimizer OptQuest. The proposed work is divided into four basic procedures: i) Agent definition and communication, ii) Optimization, iii) demand side management (DSM), and iv) performance evaluation.

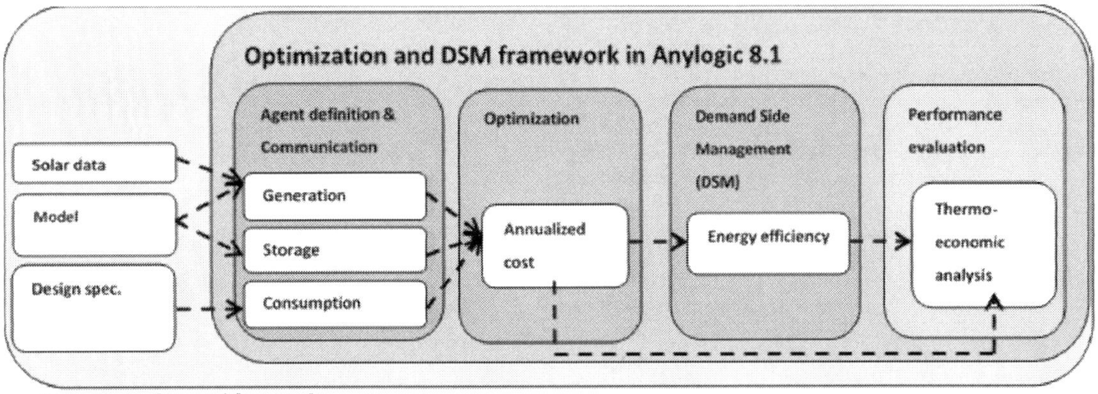

Fig. 1. The implemented framework.

3. 1 Agent definition and communication

This task involves identifying which parts of the solar-assisted industrial heating is important and creating these objects in the model as agents known as generation, storage, and consumption agents. The agent definition task creates these agents having a continuous thermal and/or demand models with embedded discrete events that run in parallel.

3.1.1 Generation agent

The generation agent used is Evacuated Tube Collector (ETC). ETC is the most efficient and convenient solar collector for low-to-medium temperature ranges [19]. The quasi-dynamic equation for ETC can be determined from [20, 21].

$$Q(t) = \eta_0 k_{\theta b}(\theta) I_b + \eta_0 k_{\theta d}(\theta) I_d - a_1(T_{coll} - T_{amb}) - a_2(T_{coll} - T_{amb})^2 - c_{eff}\frac{dT_{coll}}{dt} \quad (1)$$

The collector parameters for the modeling of the generation agent are taken from [16].

Table 1. Dyeing textile machine parameters

Parameters	Values
Mean daily production, m^2	9,406
Hot water cons, L/Kg	8 -12
Process temp, °C	40 -60
Minimum fabric weight, g/m^2	265
Maximum fabric width, m	2.45
Average washing machine speed, m/min	60
Machine efficiency, %	75

3.1.2 Storage agent

As reported in [11] a stratified storage tank helps to store solar heat efficiently and ensure lowest possible inlet temperature to collector. Thus, the storage agent is modeled as N equally segmented storage tank layers with thermal dynamics, T_n, given as

$$m_n c_p \frac{dT_n}{dt} = \dot{Q}_{conv,n} + \dot{Q}_{cond,n} - \dot{Q}_{loss,n} \quad (2)$$

Where the convective, conductive, and loss stated in Eqn. (2) can be determined from temperature dynamics of neighboring segments (T_{n-1} and T_{n+1}) given by Eqns.(3-5).

$$\dot{Q}_{conv,n} = \dot{m}_c c_p(T_{n+1} - T_n) - \dot{m}_w c_p(T_n - T_{n-1}) \quad (3)$$

$$\dot{Q}_{cond,n} = \frac{A_{str,c}k}{d}(T_{n-1} + T_{n+1} - 2T_n) \quad (4)$$

$$\dot{Q}_{loss,n} = UA_{str,s}(T_n - T_{amb}) \quad (5)$$

This storage agent is composed of N segment chain where each segment's temperature state is derived from its own as well as the neighboring segments temperature dynamics. The first and last segment has the generation and consumption agent as the connecting Chain.

3.1.3 Consumption agent

The consumption agent has three states namely idle, not available, and processing. The idle states represents the waiting mode while the dyeing machine waits inputs from other textile processes proceeding it (desizing, bleaching, washing and drying). The not available state results when the dyeing machine is periodically maintained, occasionally breakdown or randomly stopped when the factory supply power is off. The final state i.e. processing state executes continuous dyeing in accordance input parameters listed in Table 1. While in this state, the thermal energy demand of the dyeing machine is updated in the agent. Once the input fabric is entirely processed, an event will be triggered to transit the dyeing machine to the idle state. The consumption agent estimates the dyeing machines availability and hence determines its thermal energy demand.

Once the agents are defined, the remaining step is to establish the agents' communication. This communication aims at smooth interaction among agents to make them able to react to external events such as thermal energy demand from consumption agent to storage agent. In order to accomplish this task, the agent communications generates and share these patterns as well as register their status

3.2 Optimization

The optimization tool used is OptQuest optimizer which is embedded into Anylogic 8.1. This optimizer is metaheuristics and is able to generalize under uncertainty during the optimization task.

An optimization experiment with total annual cost (annualized capital cost plus operating cost) of the thermal system is formulated. Both operating and investment cost for heat exchanger network, collector area, storage volume, including the control system and piping costs are considered in the objective function of the optimization experiment. The cost of investment for heat exchanger, Cex, is calculated using [21] as

$$C_{ex} = e + f * \left(\frac{A_{tot}}{N_{min}}\right)^g * N_{min} \quad (6)$$

Whereas the cost for the collector and storage, C_i, can be approximated using the power low equation that relate the base size cost, C_b, and cost exponent factor, s, given by Eqn. (7) [11].

$$C_i = C_b \left(\frac{S_i}{S_b}\right)^n \quad (7)$$

The annualized capital cost, C_A, of a new installation over a fixed period, m, at a fixed rate of interest, i, is given by

$$C_A = C_i \frac{i(1+i)^m}{(1+i)^m - 1} \quad (8)$$

On the other hand, the operating cost, COP, is composed of the energy cost of auxiliary heat demand from boiler, C_{aux}, and maintenance cost, C_{main}, of the thermal system determined as

$$C_{OP} = C_{aux} + C_{main} \quad (9)$$

$$C_{energy} = C_{FO}\frac{Q_{FO}}{\eta_b} \quad (10)$$

The cost of maintenance is assumed to be 1.5% of the fixed capital investment.

3.3 Demand Side Management (DSM)

The aim of DSM implementation is to reformulate the production policy of the dyeing textile process in order to use the available solar thermal energy efficiently. This is done by implementing a flexible production strategy that is in line with the seasonal and daily variation of the available solar energy. Thus, the DSM effort attempts to dynamically shift in time and/or vary the daily production rate. The allowable production rate is indirectly determined from the consumption agent that is described in section A.3. This DSM effort will result in efficient utilization of the generated solar energy thereby decreasing the associated loss.

3.4 Performance Evaluation

The optimal solar thermal configuration for the considered case study is determined at the optimization step as discussed in section 3.2. Afterwards the economic analysis, solar performance and carbon dioxide saving potential of the identified optimal configuration are calculated. The payback period, the total analyzed cost and the investment cost are indicators of the economic performance whereas solar gain and solar fraction points to the extent of solar energy utilization. On the other hand, the annual carbon dioxide emission reduction indicates the GHG mitigation potential of the proposed solar thermal system. The final step involves investigating the improvement in the aforementioned performance parameters when a DSM strategy is implemented.

4 Results and discussion

The first part of this section discusses the optimization result that gives the most economically feasible solution. Secondly, the demand side energy management is elaborated that leads to energy efficiency improvement potential in the considered dyeing textile process.

The optimization approach as discussed in section B was used to determine the most economical values for collector area and solar tank volume. The procedure was applied for ETC coupled to stratified storage tank with demand generated from the consumption agent.

The data shown in Table 2 gives an output of this optimization procedure.

Table 2. Summary of results.

	Ref. system	**With DSM**
Optimized values		
Collector area (m^2)	323	323
Solar tank volume (m^3)	18.1	18.1
Solar system performance		
Solar heat gain(MWh/p)	142.1	167.6
Solar fraction (%)	66.7	75
Economic analysis		
Investment cost (€)	176838	176838
Annualized total cost (€/p)	52095	52095
Payback period (p)	5.3	4.9
Saving		
Fossil fuel (MWh/p)	203.4	141.5
Cost(€/p)	9899	9830
Co_2 (t/p)	52.0	47.0

Before investigating DSM in the dyeing textile process, a flexible production policy needed to be formulated. This policy results from the knowledge of the dyeing machine's daily and monthly availability as estimated by the consumption agent. This consumption agent considers all stochastic events such as shortage of input fabric for the dyeing machine and represents them in probability distributions. The agent also takes deterministic events such periodic maintenance and represents them as occurrences with a given rate. The monthly dyeing machine availability is portrayed in Fig. 2.

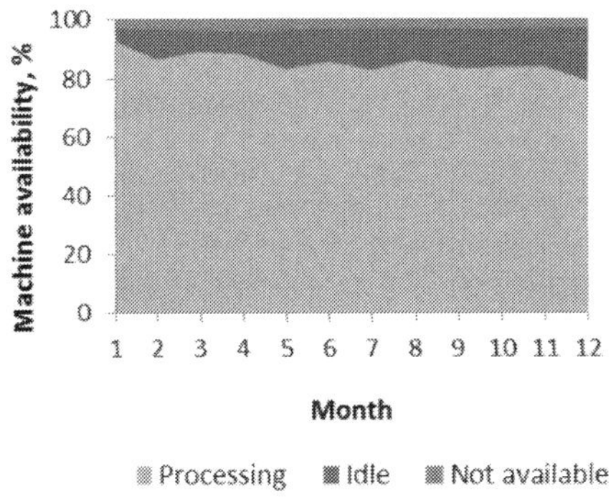

Fig. 2. Dyeing machine availability.

After estimating the machine's availability, the DSM procedure is implemented as follow: First, the consumption agent generates thermal demand based on the maximum daily dyeing machine's availability and tries to satisfy its demand from the storage agent (coupled to the generation agent). The excess of demand over supply is backlogged. On the succeeding days, the DSM procedure reviews the demand and supply inventory levels and decides on the daily production rate of the dyeing machine. This procedure goes on trying to match solar energy supply with production to arrive at a flexible production policy that results in maximum use of the generated solar energy. Thus, implementing the DSM strategy to guide the dyeing textile process, a maximum of 8.4% solar fraction with about 65.4MWh solar gain improvement could be achieved. Fig. 3 shows the monthly possible solar gain of the DSM implementation.

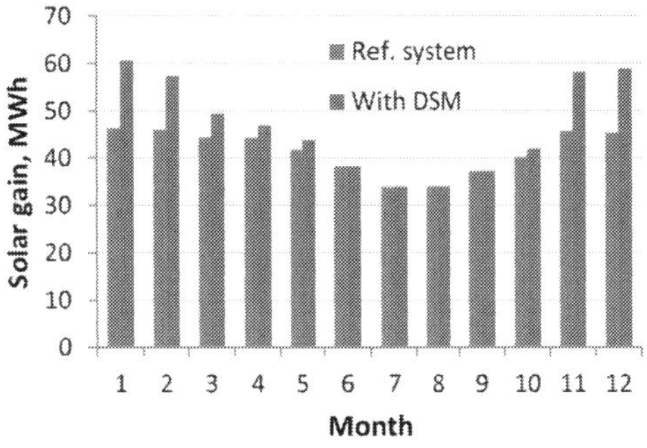

Fig. 3. DSM implementation for solar gain improvement.

5 Conclusions

The overall aim of this work was to arrive at an integrated optimization and demand side management approach, for decision support, in solar-assisted industrial heating. Demonstrated by a case study, it was found out that a payback period of 5.3 years and solar fraction of 66.7% was identified for an optimized system. With the implementation of DSM, to guide the production policy of the industry, a 7.5% and 8.4% improvement in payback and solar fraction was achieved.

Acknowledgment

The authors acknowledge the scholarship help from German Academic Exchange Service (DAAD) in supporting the PhD candidate.

Nomenclature

A	Area (m^2)
a_1, a_2	Thermal loss coefficients ($W/(m^2\,K)$, $W/(m^2\,K^2)$)
e, f, g	Fixed, scaling and, non-linear cost parameter
C	Cost (€)
C_{FO}	Specific cost of natural gas (€/J)
c_p	Specific heat capacity (J/(kg K))
c_{eff}	Effective thermal capacity KJ/(m^2K)
I	Solar irradiance (W/m^2)
i	Rate of interest
k	Thermal conductivity (W/(m K))

 https://doi.org/10.1051/e3sconf/20186402001

$k_{\theta b}, k_{\theta d}$	Direct and diffused incidence angle modifier
d	Sstorage segment distance (m)
n	Cost exponent factor
N	Number of heat exchanger
$\dot{m}$	Mass flow (kg/s)
m	Number of years
Q	Heat (J)
$\dot{Q}$	Heat flow (W)
s	Cost scaling factor
T	Temperature (°C)
U	Heat transfer coefficient (W/(m^2 K))

References

1. Solar Heat for Industrial Processes, technology brief, International Renewable Energy Agency; January 2015.
2. Ashish K. Sharmaa, Chandan Sharmab, Subhash C. Mullicka, Tara C. Kandpala. Solar industrial process heating: A review. Renewable and Sustainable Energy Reviews, 2017; 78: 124-137
3. Growth and Transformation Plan II (GTP II) (2015/16-2019/20), national planning commission, the Federal Democratic Republic of Ethiopia; May 2016.
4. The Ethiopian power sector: a renewable future, Berlin Energy Transition Dialogue, Ministry of Water Irrigation and Electricity; March 2017
5. Study on the Energy Sector in Ethiopia, Embassy of Japan in Ethiopia, report available at http://www.et.emb-japan.go.jp/electric_report_english.pdf. [Accessed 16.08.17].
6. Ashish K. Sharma, Chandan Sharma, Subhash C. Mullick, Tara C. Kandpal. Carbon mitigation potential of solar industrial process heating: paper industry in India. Cleaner Production, 2016; 112: 1683-1691
7. Mokhtar Ghazouani, Mohsine Bouya, Mohammed Benaissa. A new methodology to select the thermal solar collectors by localizations and applications. IEEE International Renewable and Sustainable Energy Conference (IRSEC), 2015:10-13.
8. M. Absi Halabi, A. Al-Qattan,n , A. Al-Otaibi. Application of solar energy in the oil industry-Current status and future prospects. Renewable and Sustainable Energy Reviews 2015; 43:296-314.
9. Badreddine Boutaghriout, Amina Bouakaz, and Chaâbane Hamouda. Investigation on the use of solar thermal energy in the agro food industry in Algeria. IEEE Renewable and Sustainable Energy Conference (IRSEC), 2013: 7-9.
10. Chaimaa El Mkadmi and Arifeen Wahed. Optimization of a solar thermal system for low temperature industrial heating process. IEEE Renewable and Sustainable Energy Conference (IRSEC), 2016: 14-17.
11. Tobias Eiholzera, Donald Olsena, Sebastian Hoffmanna, Barbara Sturmb, Beat Wellig. Integration of a solar thermal system in a medium-sized brewery using pinch analysis: methodology and case study. Applied Thermal Engineering, 2017; 113:1558-1568
12. Ilyes Ben Hassinea, Mariela Cotrado Sehgelmeble, Robert Söllb, Dirk Pietruschkaa. Control Optimization through Simulations of Large Scale Solar Plants for Industrial Heat Applications. Energy Procedia, 2015; 70: 595-604.
13. Krummenacher, P., Muster-Slawitsch, B. Methodologies and software tools for integrating solar heat into industrial processes. 13th International Conference on Sustainable Energy technologies, 2014.
14. Catherine S.E. Bale, Liz Varga, Timothy J. Foxon. Energy and complexity: New ways forward. Applied Energy, 2015; 138:150-159.
15. R. Pasupathy, S.-H. Kim, A. Tolk, R. Hill, and M. E. Kuhl. Multi-method modeling. Proceedings of the 2013 Winter Simulation Conference, 2013. P. 4089-4100.
16. The Solar Keymark. Database of certified products for collectors, http://www.solarkeymark.org/; 2017 [accessed 03.08.17].
17. Energy Audit Report of Bahir Dar Textile Share Company, Project Report No. 2013IB22, The Energy and Resources Institute, 2014
18. Measurement data from the WebelSolar(Grid) 3960Wp project, Bahir Dar university, Ethiopia, [Accessed 16.08.17].
19. M.A. Sabiha, R. Saidur, Saad Mekhilef, Omid Mahian. Progress and latest developments of evacuated tube solar collectors. Renewable and Sustainable Energy Reviews, 2015; 51:1038-1054.
20. Jie Deng, Xudong Yang, Pengsu Wang. Study on the second-order transfer function models for dynamic tests of flat-plate solar collectors Part I: A proposed new model and a fitting methodology. Solar Energy, 2015; 114:418-426.
21. Marcel Taal, Igor Bulatov, Jiri Klemes, and Petr Stehlik. Cost estimation and energy price forecasts for economic evaluation of retrofit projects. Applied Thermal Engineering, 23(14):1819-1835, 2003.
22. Vives, M., Energy production cost of solar thermal energy for industrial processes in spain. Catalonia institute for energy research (IREC), International conference on solar heating and cooling, June25-27, 2012.

Evaluation of Solar Ponds and Aplication Area

Haci Sogukpinar[1], *Ismail* Bozkurt[2], and *Serkan* Cag[3]

[1]Department of Electric and Energy, Vocational School, University of Adiyaman, Adiyaman 02040, Turkey
[2]Department of Mechanical Engineering, Faculty of Engineering, University of Adiyaman Adiyaman 02040, Turkey
[3]Department of Machinery and Metal Technology, Vocational School, University of Adiyaman, Adiyaman 02040, Turkey

Abstract. Solar ponds are heat storage systems where solar energy is collected and stored thermally. Solar ponds were discovered during the temperature variations in the lower regions of existing saltwater pond in the area is found to be higher than their surface. Later, it was constructed artificially and started to be used. These systems have heat storage capacity at moderate temperatures. Solar pons are used in many areas such as electricity generation, heating the environment, meeting the need of hot water, drying food and obtaining fresh water from salty water. In this study, the studies about solar ponds were summarized, the construction of solar pond was explained, and the application areas were examined.

1 Introduction

The ever-increasing energy demand open up the importance of the use of renewable energy resources. One of the most important renewable energy sources is the sun, although it varies according to the climatic conditions. There are many technologies that generate energy by using solar system. Solar ponds are one of the systems that convert solar energy into useable energy. Solar ponds can also be constructed artificially, like the ones in the nature. Heat energy collected and stored by solar ponds can be used for various purposes. There have been many studies on solar ponds. Bozkurt et al. [1-6] investigated the performance of the pond according to the cases of closed and open top surface with a transparent glass lid in order to reduce the heat losses from the surface of the solar ponds. In addition to this, the performance of the integrated system consisting of solar pond and flat plate solar collectors has been studied. The effect of solar collectors on the performance of the solar pond was determined. A numerical study was carried out on the effect of side wall shading on the performance of the solar pond. In the mentioned study, modeling was carried out for eight different situations, sunshine rates and solar pond performance values were calculated. The temperature distribution of the pond was determined according to the different insulation materials of the solar ponds.

Khodabandeh et al. [7] investigated the hydrothermal attributes and energy efficacy of the watere graphene nanoplatelet/platinum hybrid nanofluid flow in a horizontal spiral tube with four cross-sections used in bottom of solar ponds to determine the effect of geometry, Reynolds number and nanoparticles concentration on solar ponds. Alcaraz et al. [8] studied the design, construction, and operation of a 500 m^2 industrial solar pond in Granada (Spain). The system was used to deliver a heat stream of up to 60 °C to minimize the fuel oil consumption at the mineral processing facility. Also, the efficiency and economic analysis of the solar pond was determined during two years. Sayer et al. [9] determined to the influence of the thickness variation of all three zones on the performance of the solar pond. Furthermore, the impact of heat extraction from the heat storage zone on its temperature was investigated to examine the suitability of the deep of the solar pond for continuous power supply. Amigo et al. [10], studied the heat storage capacity of the ground beneath a salt gradient solar pond to demand a constant heat. For this purpose, a one-dimensional transient model was recommended to calculate the thermal behavior of the solar pond. Furthermore, as a case study, the solar pond located in Copiapo, Chile was modeled to evaluate the influence of the soil type and the depth to ground water the on the thermal performance of the heat storage zone temperatures. Assari et al. [11] investigated the temperature gradient, temperature behavior, and performance of two salt gradient solar ponds with thin plastic with mixed medium and conventional in the heat storage zone to determine the effect of porous medium on the efficiency of solar pond. Monjezi et al. [12] presented the mathematical model which is used to determine the temperature profile of a solar pond for the heat extraction in batch or continuous, modes for salinity gradient solar ponds. Furthermore, the performance of a solar pond in Middle Eastern and Mediterranean weather conditions was compared. Erden et al. [13] investigated to point out the performance of hydrogen production of the flat-plate collectors assisted by a solar pond system. The system was design to produce electric and hydrogen from the electrolysis and ORC to obtain the performance increase of the components. Assari et al. [14] studied phase change material (PCM) in solar ponds in order to control the stability of solar ponds during the heat extraction and improving thermal energy storage or total performance of the solar ponds. An internal heat exchanger was used in the heat storage zone. The temperature change of the system was measured and analyzed. It was shown the assisted phase change material pond decreases the temperature difference between night and day.

In this study, the recent studies on solar ponds were summarized, the structure and construction of solar ponds was discussed and the usage areas were investigated.

2 Solar pond

Solar ponds are also artificially constructed considering similarities in the nature. As seen in Figure 1, the solar pond consists of three zones. The area located on the upper surface of the pond which contains tap water is called the Upper Convective Zone (UCZ). Below UCZ there is Non- Convective Zone (NCZ). NCZ is made up of layers of increasing density. This density difference prevents heat transfer between the heat storage zone and the surface of the pond, only heat transfer occurs by conduction. Thus, the NCZ acts as a transparent insulating material. Finally, under the NCZ, there is the Heat Storage Zone (HSZ). HSZ is made up of very dense brine and is the place where solar energy is stored as heat energy. At the same time, the heat exchanger is installed in this region for heat extraction from the system. When solar pools are first built, a very intensive salt water solution is prepared and filled up to a certain height from the bottom of the pond and the HSZ is created. Then, for each layer forming the NCZ, saline solutions are prepared separately at different densities. Layers are formed at certain thicknesses to reduce the density toward the surface of the pond. During this process a plate is used which floats on salt water so that the layers do not mix.

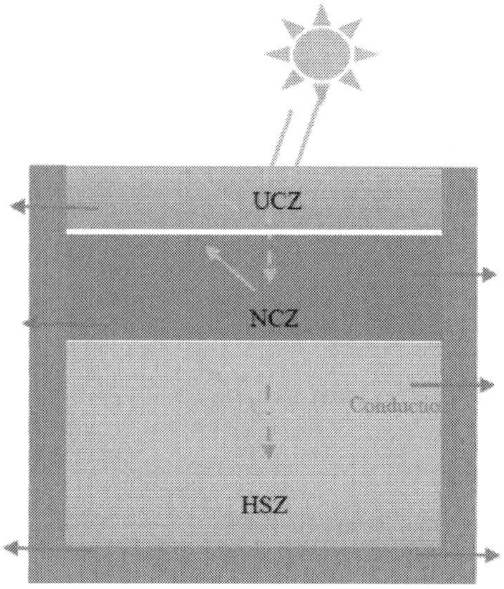

Fig. 1. The zones of solar pond

Thus, the desired salinity gradient is obtained. The prepared solutions are slowly poured onto the plate to form salt water layers at desired thicknesses. Finally, the UCZ is created using tap water and the solar pond becomes ready to store heat. Temperature sensors are placed at certain intervals within the pool to monitor the temperature changes of the solar pond. At the same time, the density of the saline solutions drawn through the thin hoses placed at regular intervals is measured by hydrometers, ensuring density.

3 Aplication area

Solar ponds are low temperature solar energy applications that can achieve temperatures between 70-100 °C. The stored heat is used in many areas such as electric generation, drying of food, heating, domestic hot water, desalination etc.

Ding et al. [15] used a small scale passive electric power generation unit to generate electricity from the heat available in the solar pond. The power generation system which consists of 120 commercially available thermoelectric cells was fabricated and tested experimentally. The system produced the maximum electric power of 40.8 W under the condition of 99 °C.

Ding et al. [16], investigated the capability of solar pond in generating electricity experimentally using a plate type power generation unit. The heat stored energy in the heat storage zone of the solar pond was used as the heat source needed for generating electricity by utilizing thermoelectric cells. The system was tested with different hot water temperatures and flow rates to determine the performance enhancement utilizing copper mesh as insertion. The system is able to generate 35.9W of electricity at the flow rate as low as 5.1 liter per minute at the hot water temperature of 81 °C.

Singh et al. [17] investigated the combined system of thermosyphon and thermoelectric modules to generate electricity from solar ponds. The temperature difference of the solar pond in the range 40–60 °C is available between the heat storage zone and upper convective zone which can be applied across the hot and cold surfaces of the thermoelectric modules to make it work as a power generator. Figure 2. shows the concept of generating electric power from salinity-gradient solar pond using a combination of thermosyphon and thermoelectric cells.

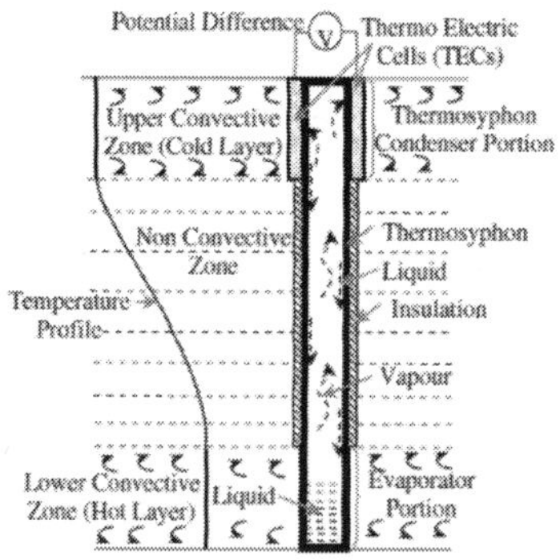

Fig. 2. The concept of generating electric power from salinity-gradient solar pond using a combination of thermosyphon and thermoelectric cells [17].

Another method of the generating electric from the solar pond is the use of organic rankine cycle. The system is configured the salt gradient solar pond matches the sources of an organic rankine cycle and the surface layer serves as cold source to cool the condenser while the bottom layer supplies the evaporator with heat. Fig. 3 shows the principle of a solar pond power plant [18].
The power plants of the solar ponds were constructed and operated at different sites around the world; 5 MW–250,000 m^2 at Beith Ha'avara, Israel; Ein Boqek, Israel 150 kW/6250 m^2; Yavne, Isreal 6 kW/1500 m^2; Alice Springs, Australia 15 kW/1600 m^2); El Paso, USA 70 kW/3350 m^2 [18]

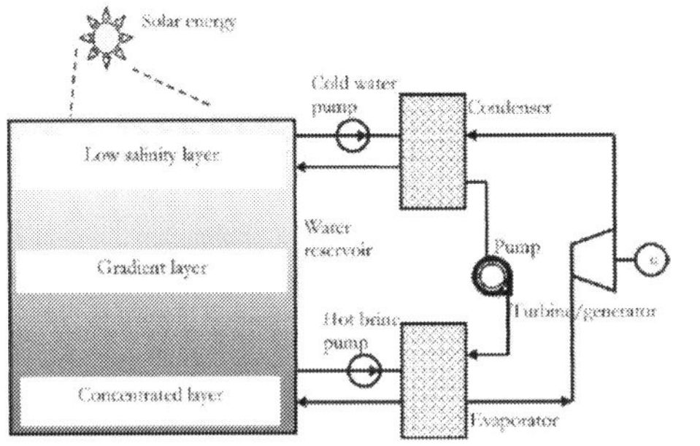

Fig. 3. The principle of a solar pond power plant [18].

Ranjan et al. [19] investigated the active solar distillation systems integrated with solar ponds. The capacity of heat energy in solar ponds storage zone available at about 50–100 °C to power desalination units even during cloudy days and off-sunshine hours. Fig. 4 shows the solar distillation system integrated with salinity gradient solar pond.

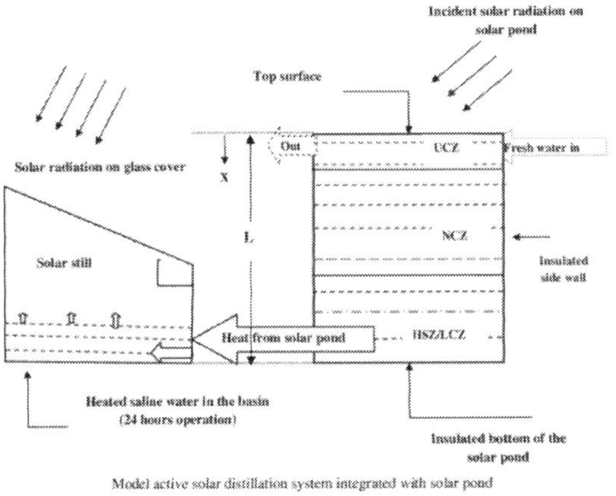

Fig. 4. The solar distillation system integrated with salinity gradient solar pond [19].

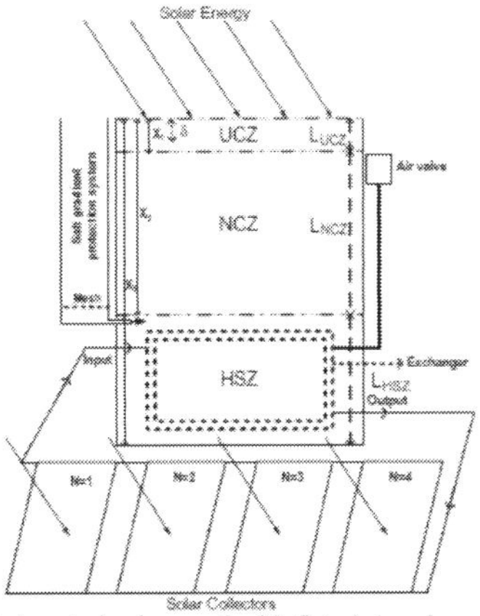

Fig. 5. The integrated solar pond with flat plate solar collectors [20].

Bozkurt et al. [20] studied heat storage performance investigation of integrated solar pond and collector system. Experimental studies were performed using 1, 2, 3 and 4 collectors integrated with solar pond. The integrated solar pond efficiencies were determined experimentally and theoretically according to the number of collectors. Figure 5 shows the integrated solar pond with flat plate solar collectors.

4 Conclusion

Solar ponds are systems that can collect and store solar energy at low temperatures (70-100 °C). High temperatures are undesirable because they negatively affect the gradient structure of the system. For this reason, the heat obtained in the solar ponds can be either used directly or transferred to other systems as heat sources. Solar pond is used directly for applications such as drying foods, domestic heating, and the distillation of the salty water. In cases where higher temperatures are needed, it is possible to use the solar ponds with the systems such as solar collectors, heat pumps. Electricity generation from solar ponds can be carried out using thermoelectric cells or organic rankine cycles.

References

1. I. Bozkurt, M. Karakilcik, I. Dincer, A. Atiz, "Transparent covers effect on the performance of a cylindrical solar pond," International Journal of Green Energy, vol. 11, 2014, pp. 404–416.
2. H. Sogukpinar, I. Bozkurt, M. Karakilcik, S. Cag, "Numerical Evaluation of the Performance Increase for a Solar Pond With Glazed and Unglazed," 2016 IEEE International Conference on Power and Renewable Energy, pp. 598-601.
3. I. Bozkurt, S. Deniz, M. Karakilcik, I. Dincer, "Performance Assessment of a Magnesium Chloride Saturated Solar Pond," Renewable Energy, vol. 78, 2015, pp. 35-41.
4. I. Bozkurt and M. Karakilcik, "The Effect of Sunny Area Ratios on the Thermal Performance of Solar

Ponds," Energy Conversion and Management, vol. 91, 2015, pp. 323–332.

5. M. Karakilcik, I. Bozkurt, I. Dincer, "Dynamic Exergetic Performance Assessment of an Integrated Solar Pond," International Journal of Exergy, vol. 12, 2013, pp. 70–86.
6. I. Bozkurt, H. Sogukpinar, M. Karakilcik, "Modeling of a Solar Pond for Different Insulation Materials to Calculate Temperature Distribution," Journal of Multidisciplinary Engineering Science and Technology, vol. 2, 2015, pp. 1378-1382.
7. E. Khodabandeh, M.R. Safaei, S. Akbari, O.A. Akbari, A.A.A.A. Alrashed, "Application of nanofluid to improve the thermal performance of horizontal spiral coil utilized in solar ponds: Geometric study," Renewable Energy, vol. 122, 2018, pp. 1-16.
8. A. Alcaraz, M. Montalà, J.L. Cortina, A. Akbarzadeh, C. Aladjem, A. Farran, C. Valderrama, "Design, construction, and operation of the first industrial salinity-gradient solar pond in Europe: An efficiency analysis perspective," Solar Energy, vol. 164, 2018, pp. 316-326.
9. A.H. Sayer, A.A. Monjezi, A.N. Campbell, "Behaviour of a salinity gradient solar pond during two years and the impact of zonal thickness variation on its performance," Applied Thermal Engineering, vol. 130, 2018, pp. 1191–1198.
10. J. Amigo, F. Suárez, "Ground heat storage beneath salt-gradient solar ponds under constant heat demand," Energy, vol. 144, 2018, pp. 657-668.
11. M.R. Assari, H.B. Tabrizi, M. Parvar, A.K. Nejad, A.J.G. Beik, "Experiment and optimization of mixed medium effect on small-scale salt gradient solar pond," Solar Energy, vol. 151, 2017, pp. 102-109.
12. A.A. Monjezi, A.N. Campbell, "A comparative study of the performance of solar ponds under Middle Eastern and Mediterranean conditions with batch and continuous heat extraction," Applied Thermal Engineering, vol. 120, 2017, pp. 728-740.
13. M. Erden, M. Karakilcik, I. Dincer, "Performance investigation of hydrogen production by the flat-plate collectors assisted by a solar pond," International Journal of Hydrogen Energy, vol. 42, 2017, pp. 2522-2529.
14. M.R. Assari, H.B. Tabrizi, A.J.G. Beik, "Experimental studies on the effect of using phase change material in salinity-gradient solar pond," Solar Energy, vol. 122, 2015, pp. 204–214.
15. L.C. Ding, A. Akbarzadeh, Abhijit Date, D.J. Frawley, "Passive small scale electric power generation using thermoelectric cells in solar pond," Energy, vol. 117, 2016, pp. 149-165.
16. L.C. Ding , A. Akbarzadeh, A. Date, "Electric power generation via plate type power generation unit from solar pond using thermoelectric cells," Applied Energy, vol. 183, 2016, pp. 61–76.
17. R. Singh, S. Tundee, A. Akbarzadeh, "Electric power generation from solar pond using combined thermosyphon and thermoelectric modules," Solar Energy, vol. 85, 2011, pp. 371–378.
18. B.F. Tchanche, Gr. Lambrinos, A. Frangoudakis, G. Papadakis, "Low-grade heat conversion into power using organic Rankine cycles – A review of various applications," Renewable and Sustainable Energy Reviews, vol. 15, 2011, pp. 3963–3979.
19. K.R. Ranjan, S.C. Kaushik, "Exergy analysis of the active solar distillation systems integrated with solar ponds," Clean Techn Environ Policy, vol. 16, 2014, pp. 791–805.
20. I. Bozkurt and M. Karakilcik, "The Daily Performance of a Solar Pond Integrated with Solar Collectors," Solar Energy, vol. 86, 2012, pp. 1611-1620.

Solar Energy Technology Choice Development

Mohammed Alsumiri[1] and *Hisham* El Khashab[2]

[1]Electrical and Electronics Engineering Technology Department, Yanbu Industrial College, Yanbu, Saudi Arabia
[2]Power Electronics and Energy Conversion Department, Electronics Research Institute, Dokki, Cairo, Egypt

Abstract. During the last decade with the modern scientific development, different solar energy (SE) technologies were introduced under vast cost variations. The huge need of energy for the use of economic development with the rise of fossil fuel price represents an urgent problem worldwide. The geographical location of Arab countries makes them subject to high average solar irradiance. So that, sustainable energy generation especially solar energy generation can be suitable alternative to conventional energy generation. To implement a solar energy application, the suitable technology choice represents a huge challenge due to high risk against budget cost. In this paper, different solar energy technologies are discussed with the suitable application to obtain the optimal energy conversion system efficiency. Both photovoltaic (PV) and Solar thermal technologies were investigated.

1 Introduction

The continuous increase in demand of electrical energy is one of the major issues worldwide. The threat of decrease in natural resources from fossil fuel makes this a more severe problem which is facing humanity. To overcome this issue, the industries, researchers and consumers have to reorganize the way how they can generate and consume energy from non-conventional sources. Employment of RE sources and improved energy savings techniques are major key elements in which this challenge can be overcome permanently. In the Arabian Gulf region, where a lot of solar radiation is available, renewable sources such as solar energy should play an important role. Renewable sources have other advantages like reduced global warming and reduced environmental pollution since there is no Carbon Dioxide emission to the environment.

Fossil fuel resources such as petrol and coal are expected to be exhausted by the end of the 21st century from the entire world according to the UNDP (United Nations Developing Program) expectations. It is clear that, renewable power generation, i.e. wind, solar and biomass have an important role in meeting the continuously increasing power demand of the world, especially during this era of technological advancements.

In this paper, the most important solar energy techniques development is discussed. Huge energy is exhausted during winter by water heating; therefore, it might become necessary to introduce solar water heaters through governmental laws to encourage the usage of such technique. Different solar systems such as Photovoltaic (PV) and solar thermal applications are considered as highly beneficial applications of RE. Concerning PV, most development regarding this area is focused on PV system optimization. Also, development took place on the technology of cell fabrications to enhance overall efficiencies of the cells. These enhancements can reduce the cost to a reasonable value among different types, for example, mono-crystalline, poly-crystalline amorphous silicon, copper indium, cadmium telluride, gallium arsenide cells and others. To ensure continuous supply from RE systems, storage techniques must be used in remote area applications such as fuel cell techniques, Nickel Cadmium gel batteries, and others. The issue of storage cost can be reduced by connecting to the grid. But this is possible only for the systems near to the electricity grid.

The annual solar radiation which has an average of 4 - 7.5 KW/m^2 is available in Middle East and North Africa (MENA) countries such as Egypt and Saudi Arabia. Moreover, the above average solar radiation is approximately five times higher than the average solar radiation in Europe which is around 1 KW/m^2 and in Greece is around 1.7 KW/m^2. Hence, solar energy can be nominated as the promising source of renewable energy in this region [1].

The goal of this investigation is to study the proper solar energy technology and to match it with the applications. SE technologies discussed [1-10] may be classified into:

I. Solar thermal energy conversion is to convert solar radiation to:
   1- Heat water. This application type has two main technologies:
   (a) Flat plate collector.
   (b) Evacuated tubes collector.
   2- Electricity via converting water to steam.
II. Photovoltaic applications, which means the direct conversion of solar irradiance to electricity.

The cost of each technology represents the main obstacle facing the spread of such technology. The efficiency of different technologies varies between 6 - 44.4% for amorphous silicon-based solar cells with multiple-junction production cells with enormous cost variation. Some of these technologies are already commercialized in the market and others are in R&D phase. This makes SE technology choice a real problem related to the application [2].

2 Solar thermal energy

Solar energy cost has fluctuated during the last decades. In 1970s oil crises began which caused a resurgence of interest in other forms of energy such as solar. After recovery, solar energy technology had got less investment. However, scientists with insignificant R&D budgets sustained the improvement in solar technologies [2].
In 1990s and especially in Japan and Germany, the solar energy has been restored due to many factors, i.e. global warming and environmental awareness, availability of energy resources and industrial countries policy. The energy supply has been forecasted to the year 2100 as presented in fig.1.

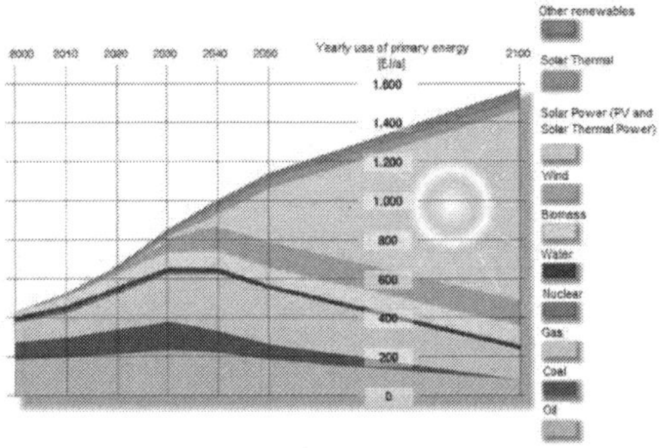

Fig. 1. Energy Supply from 2000 to 2100.
(Source: International Energy Agency / IEA).

Until 2030, the global energy demand is expected to grow at a rate of 2.5%. Also, it has been estimated that this growth in electricity demand requires cumulative investments of USD 22 trillion which means that approximately USD 1.1 trillion of investment is required per annum. The projected capacity of power generation is 4,800 GW by the end of 2030. Since one-third of the global greenhouse gas (GHG) emissions is produced by power sector, it becomes necessary to direct these investments to renewable energy technologies

2.1 Water heating technologies

Water heating represents huge energy consumption, especially in the winter season. Moreover, some applications such as swimming pools water heating consume high energy which needs further technology improvement. The well-known main water heating technologies are:

1- The Flat-Plate Collectors (FPC) technology, and
2- The Evacuated tubes collector (ETC) technology.

Both technologies are discussed and compared from the point of view of efficiency and cost.

2.1.1 Flat-Plate Collectors (FPC) technology

The major component of a typical flat-plate collector system is a glazed dark-colored or particularly coated solar heat-absorber. A network of copper tubes is connected with the plate. The network of tubes is placed in a glazed glass-covered insulator box. The glazed glass is used to protect the collection panels from dust, moisture, and other contaminants. Also, by using glazed glass a greenhouse effect is provided which can minimize energy dissipation. When light radiation falls on it, water or glycol solution that is in the copper tubes is heated up by absorption of thermal energy in the glass and tube. Then solar radiation can be converted into usable heat energy. For the domestic applications, normally one or two collection panels of a size of about 1.2 x 2.4 meters (4 x 8 feet), similar in size to a sheet of plywood, are used.

Unglazed flat-plate collectors don't have the glazed glass covering and insulation. These types are mainly used for low temperature outdoor applications like pre-heating water, winter pool heating, etc. which consumes a considerable amount of energy. Similar to the glazed type, the conversion from solar to heat energy occurs when the absorber plate is heated up by sunlight. Liquid carrying pipes are attached to the absorber plate that is to allow the heat exchange between the plates and pipes. The collector plates are universally made with black plastic or painted with "selective coatings," which absorb ultra-violet light and preserve heat better than normal black paint. The average efficiency of the mentioned technology is up to 20% with the price of (100 US$) of 2 m^2 area.

2.1.2 Evacuated tubes collector (ETC) technology

This technology is considered as the modification of the flat-plate technology. This type can be used in domestic applications. The main working part is the Evacuated tubes. The thermal energy of sun light is absorbed then converted to heat that can be used in water heating applications. Solar thermal collectors use quite a few types of evacuated tubes, i.e. "twin-glass tube", which is used by Apercus collectors. This type of tube is popular because of many advantages like reliability, performance, and cost-effectiveness.

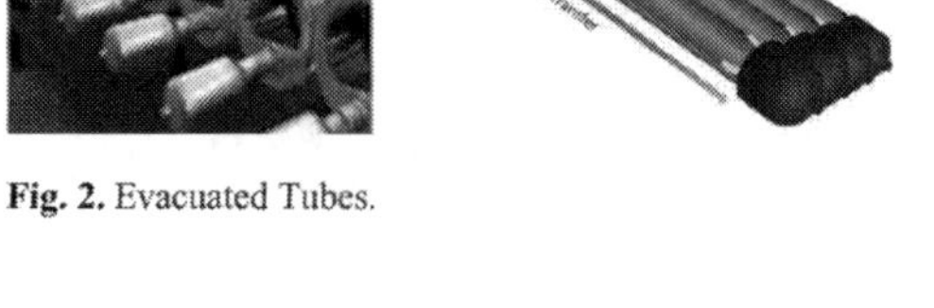

Fig. 2. Evacuated Tubes.

https://doi.org/10.1051/e3sconf/20186402003

Figure 2 shows the operation of evacuated tubes that contain dual glass tubes manufactured from extremely strong borosilicate glass. The outer tube is transparent and inner tube is coated with an aluminum nitride (Al-N/Al) coating. The sunlight is allowed to pass through the outer layer with negligible reflection and the light is collected by the inner tube. Hence, the reflection losses are minimized by improving the absorption of solar radiation which is achieved by using such arrangement of selective surfaces.

Pumping out the air which is normally present between the glass layers and heating up the top surface of the tube take place during the process of manufacturing. This process allows the two tubes to fuse together in order to create a single evacuated tube without gases. The vacuum is formed by this evacuation process, which helps in improving the performance of the evacuated tubes. The vacuum prevents the leakage of heat energy between the tubes' layers, which means that there will be no exhausted energy. Solar radiation is absorbed and converted to heat by the evacuated tube. Moreover, the vacuum influences the thermal insulation. As a result, the outer tube remains at a temperature just above the ambient temperature while the inner area of the tube reaches 150°C.

Based on the above, the performance of the evacuated tube water heaters is satisfactory even in cold weather, while the performance of the flat plate collectors is inadequate in these circumstances because of heat losses. The system efficiency may reach up to 65% with the price of (150 US$) for 15 tubes of around 2 m^2 area [3].

2.2 Solar thermal electricity generation

There are two main technologies for solar thermal electricity generation, i.e. Photovoltaic (PV) technology in which sunlight is used to generate electricity, and solar thermal technology in which the solar heat is used for electricity generation or for supporting steam and hot water to supply a turbine to produce electrical power. The overall efficiency of a solar thermal system can be narrowed by a sum of 20% as a result of boundaries of turbine efficiency. Figure 3, demonstrates the solar thermal conversion diagram to produce electricity.

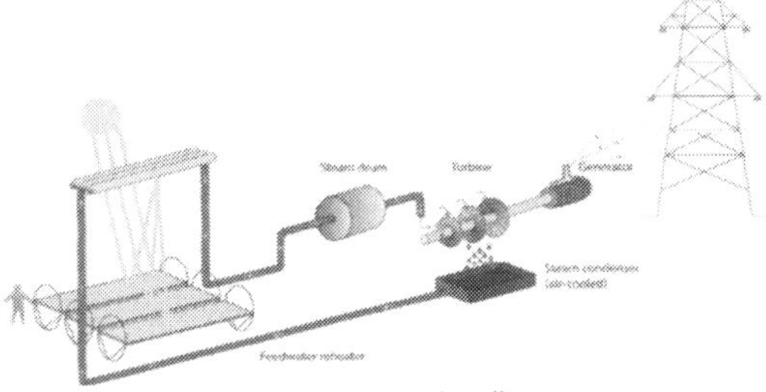

Fig. 3. Solar thermal energy conversion diagram to generate electricity [12].

3 Photovoltaic technologies

PV panel crystal technologies are of many types as follows:

1. Mono-crystalline solar cells - Made from a large crystal of silicon (efficiency around 18%).
2. Poly-crystalline Solar cells – these types of solar cell can be considered as the most widely used solar cell to make solar panels. Their efficiency is slightly less than the other types (typical efficiency between 13% - 15%). In the other hand, it cost slightly cheaper than other technology.
3. Amorphous solar panels (Thin film) – This type is made of a thin film of molten silicon. The sheet is spread directly across large plates to make the panel. The main drawback of this type is the poor overall efficiency as compared to the other two types (efficiency between 6% - 10%).
4. CIGS thin-film –This type of thin-film technology is promising due to its highest potential in efficiency enhancement. This type has higher efficiency (efficiency up to 22%) and very impressive opportunity of future application especially in building-integrated photovoltaic (BIPV) application.
5. High Concentrated PV (HCPV) cells – Up to 29%, HCPV considered being the most efficient technology. In places subjected to high sun irradiation, HCPV can ensure minimum cost of energy.
6. Third generation PV - This type make use of latest nanotechnology and organic materials. The name is used to represent solar array technologies that are far and wide considered as one of the most talented technology and that is because it has a potential for cost reduction in future.

Figure 4 demonstrates the different PV and solar thermal technologies both in the market and in R&D phase. Fig. 5 reports the conversion efficiencies of different PV cell technologies with all their developments since 1976 up to 2015 as indicated by NERL (National Renewable Energy Laboratory).

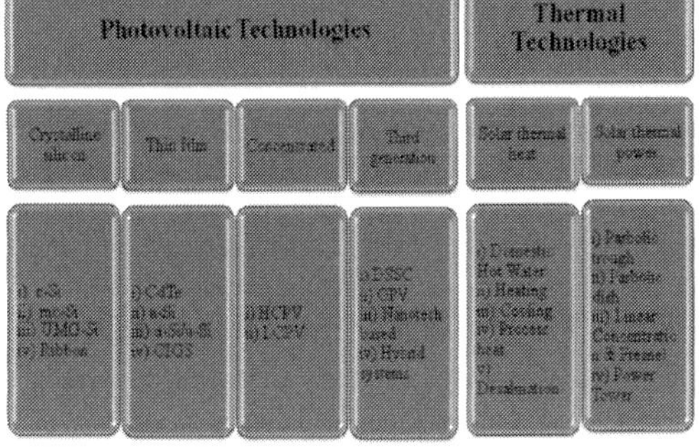

Fig. 4. Selected solar technologies for R&D activities [13].

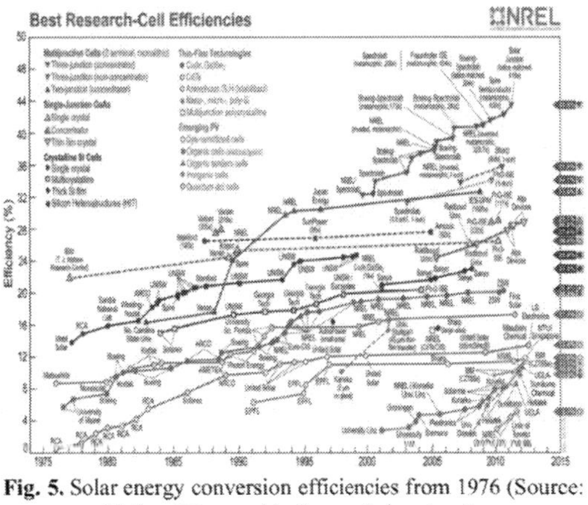

Fig. 5. Solar energy conversion efficiencies from 1976 (Source: National Renewable Energy Laboratory).

4 Technology cost

Figure 6 shows the typical efficiency of PV system and the investment costs over time in the US for three different types of PV cells: It is clear that the overall efficiency for almost all types of PV cells have been improved and still it is subject to further improvement in 2020. The major developments will be in the concentrator efficiency and its efficiency is expected to reach 25% by 2020.

From figure 6, it is clear that thin film technology has the cheaper cost, while the concentrator techniques have the higher efficiency and the suitable price. This technique can be considered as one of the promising technology for its attractive cost of energy especially in places that are exposed to a relatively high solar irradiance like the Middle East and North Africa region. The highest efficiency cells may not be economical always, for example a multi-junction cell with 30% efficiency based on exotic materials such as gallium arsenide or indium selenide produced at low volume might cost one hundred times more than an 8% efficient amorphous silicon cell in mass production, and able to deliver only about four times the output [11].

Another important factor is heating of the solar cell. As the cell gets heated up the efficiency will decrease. Actually, a 1°C temperature increase in solar cell will result in 0.45% of efficiency decrease. One of the methods used to solve this issue is to use a visibly transparent silica crystal layer over the solar panel. The silica layer emits heat into space in form of infrared radiation allowing the cell to be cooled by approximately 13 °C, since this layer behave like a thermal black body [10].

Third generation technologies is underway to be commercialized. A part from the fact that their efficiency is low, they benefit from low cost and light weight. Concentrating PV technology has the potential to have the highest efficiency among PV module. Other organic or hybrid organic/conventional technologies such as Dye-Sensitized Solar Cells (DSSC) offer low efficiency as well, however their advantages are low cost, light weight and free-form shaping. Therefore, they could fill some market place such as mobile applications, where these features are required [2].

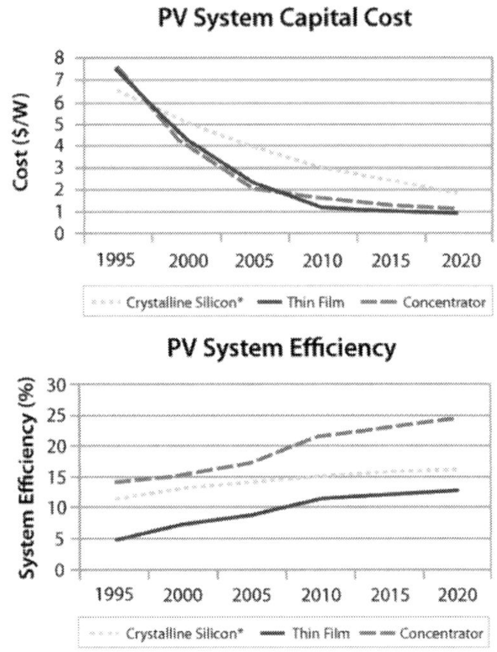

Fig. 6. PV System capital cost and efficiency up to 2020 [3].

5 Conclusion

The reasonable application with the suitable prices for heating water is the evacuated tubes panel technique. This technique may be loaned by governmental laws which can make guidance to replace solar water heating instead of electric water heaters. It is expected that the use of solar thermal technologies in both water heating and electricity generation will strongly increase in the MENA region due to the facility of technology use and due to its low cost.

PV technologies have rapid development but with high- cost variations. The Third generation technologies have high operating efficiency at a reasonable cost, which is suitable for low power applications. Other laboratory PV cells such as multi-junction cells with high efficiency and high price levels may be suitable for military and space applications.

References

1. Eldin, A.H., Refaey, M. and Farghly, A.,"A Review on Photovoltaic Solar Energy Technology and its Efficiency", https://www.researchgate.net/publication/28779239, Dec. 2015.
2. Irena working paper, "Renewable Energy technologies: cost analysis series", IRENA, vol. 1, 2012, issue 4/5.

3. US Office of Indian Energy Policy program, "Efficiency Of Solar PV, Then, Now And Future", Washington DC, 2015.
4. Dena German Energy Agency, "Information about German renewable energy, industries, companies and product(Federal Ministry of Economics and Technology)", pp. 41, 2013-2014, ISIN: B002MNZE4U.
5. Mehta, S.,"PV Technology, Production, and Cost", Outlook: 2010-2015, Greentech Media Research, October 2010, Boston, MA.
6. Arno Smets, Klaus Jäger, Olindo Isabella, René van Swaaij, Miro Zeman, "Solar Energy, Fundamentals, Technology, and systems (UIT Cambridge)", 2014, pp. 109, ISBN: 9781906860325.
7. NREL," Best research-cell efficiency", internet: http://www.nrel.gov/.
8. Green, M.A. et al, " Solar Cell Efficiency Tables progress in Photovoltaics: Research and Applications", Vol. 19, 2011, pp. 84-92.
9. J. Wennerberg, J. Kessler, J. Hedstr¨ om, L. Stolt, B. Karlsson, and M. R¨ Annelid, "Thin film PV modules for low concentrating systems", Solar Energy, vol. 69, supplement 6, 2001, pp. 243–255.
10. T. T. Chow, "A Review of photovoltaic/thermal hybrid solar technology,"Applied Energy, vol. 87, no. 2, 2010, pp. 365–379.
11. Parida, B., Iniyan, S. and Goic, R., 2011. "A review of solar photovoltaic technologies". Renewable and sustainable energy reviews, 15(3), pp.1625-1636.
12. El Khashab, Hisham, and Mohammed Al Ghamedi. "Comparison between hybrid renewable energy systems in Saudi Arabia." Journal of Electrical Systems and Information Technology 2, no. 1 (2015): 111-119.
13. Al-Jarallah, Ahmed & Al Mohiameed, Elham. (2016). Integrating Solar Photovoltaic with Urban Planning process: A review to derive Conceptual framework for Saudi Cities.

Performance Improvements of the Air-Based Solar Heating System (Improvement Method of Thermal Performance)

Youngjin Choi

Kyushu University, Platform of Inter/Transdisciplinary Energy Research, 8128581 Fukuoka, Japan

Abstract. In this study, it is proposed a heating and hot water system for a house using the air-based solar collection. The performance of existing solar heating system is analysed by annual simulation and the problems and the effect of thermal characteristics of the system are investigated due to change of heat collecting surface, heat storage method, capacity, building insulation performance and so on. Furthermore, it presents an improvement plan that maximizes the effect of the system and the performance of the system.

1 Introduction

Solar energy is essentially an intermittent, time-dependent energy source. Due to the nature of the use of solar heat that only weekly heat can be collected, there is a difference between the heatable time and the time of heating load and hot water load. These offsets and intermittency complicate the use of solar energy. In other words, heat storage is necessary for effective use of solar energy. In the case of the air-based solar heat system, since the heat capacity of air is small, it is necessary to store the heat by heat exchange with the material of the large heat capacity. There are a number of studies that attempt to solve daily-scale offsets (day-night) and annual-scale offsets (summer-winter) through various regenerative methods. In order to solve the offset of the annual scale, a lot of costs (for example, a large-scale heat storage tank) is required, so in this study, a method of compensating the offset of the daily scale is examined. The daily-scale offsets are relatively easy to compensate with water tanks or other short-term storage methods (eg, utilizing the thermal mass of the building). Recently, there are many studies that use PCM to store heat in the form of latent heat. Although the effect has been proven in many studies, the possibility of applying it to real houses is unknown due to cost problems. In this study, it is proposed a heating and hot water system for a house using air-based solar collection using foundation concrete and a space between the floor and foundation (placing plastic bottles with water) as a heat storage medium. In order to maximize the performance of the solar heating system, it is necessary to consider the overall heat balance of the thermal characteristic change of the whole building by applying each elemental technology. In this research, the air-based solar heating system is modeled and the validity of simulation model examined through comparison with experiment result. In addition, the performance of existing solar heating system is analyzed by annual simulation and the problems and the effect of thermal characteristics of the system are investigated due to change of heat collecting surface, heat storage method, capacity, building insulation performance and so on. Furthermore, it presents an improvement plan that maximizes the effect of the system and the performance of the system.

2 Simulation conditions

The simulation (ExTLA, Excel-based Thermal Load Analysis) used in this study is a thermal load calculation tool developed in the MAE laboratory of the University of Tokyo [1]. It calculates the convergence of simultaneous equations by Gauss-Seidel method using circular reference and iterative calculation function of Microsoft Office Excel. It is a feature of Excel-based simulation that it is possible to input mathematical formulas to each cell and to refer to the values of other cells from users. In the calculation method of ExTLA, it is adopted that a thermal network calculation in which the indoor temperature, the room humidity, the surface temperature of the indoor, the wall body temperature and so on. The backward difference of finite-difference methods was applied for the calculation of unsteady-state thermal conduction of the wall. The calculation is made in which convection and radiation are separated in the heat balance of the indoor surface.

An annual simulation was conducted to grasp the performance of the air-based solar heating system. The target building is a standard house prescribed by "Japanese energy saving standard (next generation energy saving standard)", and the interior space was divided into a heating space and a non-heating space. In order to secure the heat collecting area, the roof surface was set as the south-facing inclined surface, and the inclination angle of the roof surface and the heat collector was set as the latitude of Tokyo (35.4°). Also, assuming the surrounding buildings, we set the solar transmittance of the window to zero on the first floor so

https://doi.org/10.1051/e3sconf/20186402004

that there is no solar radiation acquisition from the window. Fig. 1 and Table 1 show building information. In this simulation, we were targeting a standard 4-person family, and assumed 450L/day hot water consumption (corrected M1 mode at 40°C by hot water supply) [2].

Fig. 1. Target building.

Table 1. Specification of target building.

Area	Whole floor area: 121.73m^2 Heating area of 1st floor: 40.99m^2, Non heating area of 1st floor: 19.87m^2 Heating area of 2nd floor: 44.72m^2, Non heating area of 2nd floor: 16.15m^2
Volume	328.06m^3 (including underfloor space)
Insulation of each part	Ceiling: glasswool 18K t=210mm Wall: glasswool 16K t=100mm Roof: glasswool 32K t=50mm Basis concrete: Extruded polystyrene foam t=50mm
Window	Plain double-glazed glass (Uw: 4.65W/m^{2}K)
Overall coefficient of heat transfer	UA-value: 0.83W/m^{2}K
Surface heat transfer rate of hot water tank	0.70W/m^{2}K (Insulation: 50mm, Thermal conductivity: 0.036W/mK)

The area ratio of a preliminary collector and a glass collector was set to 3:1 as preliminary collector 45m2 and glass collector 15m2. The capacity of the hot water storage tank was set to 1000L. The tank surface heat transmission coefficient of 0.7 W/m2K was input assuming a thermal insulation material of 50mm (Thermal conductivity 0.267 W/mK), and the heat loss was calculated from the difference between the outside air temperature and the temperature inside the tank. In addition, the length of heat collection side piping of the hot water storage tank was set to 30m and the thermal conductivity was set to 0.267 W/mK, and the heat loss from the pipe to the outside air was taken into consideration by the previous study. Table 2 shows the simulation condition.

Table 2. Simulation condition.

Weather data	ExpandedAMeDAS standard year (2000) Tokyo
Heating setpoint	20 °C
Heating schedule	7:00-10:00, 12:00-14:00, 16:00-23:00
Collector inclination angle	Latitude of Tokyo (35.4°)
Calculation period	Pre-calculation: January 1st - April 30th Target calculation: May 1st - April 30th
Time step	1hour
Hot water consumption	450L/day (40°C)
Internal heat generation	13.26 kWh/day
Collector area, air volume	preliminary collector: 45m^2, glass collector: 15m^2 Air flow rate: 780 m^3/h

3 Understanding the performance of the solar heating system

3.1 Simulation result of the typical house (non-solar heating collection)

In order to examine the effect of load reduction by application of solar thermal system, we first conducted an annual simulation of a typical house without a heat collecting system installed. In order to confirm the time series fluctuation of room temperature and load at a typical house, we show the result of 5 days of sunny weather that the solar thermal utilization is possible during January when the annual outside temperature is low. Fig. 2 shows the direct solar radiation from January 7th to 11th (5 days) and the direct solar radiation and the sky solar radiation intensity on the heat collecting surface (south, inclination angle 35.4°). As shown in Fig. 3, the temperature fluctuation in the room during the period (January 7th to 11th) is always lower than the heating set temperature (20°C) in the first-floor heating room where there is no solar radiation from the window. The Heating load is generated by raising the room temperature to 20°C by auxiliary heating during heating time. On the contrary, in the second-floor heating room, the room temperature rises to 20°C or more due to the influence of the solar radiation from the window during the daytime, but the room temperature has dropped from the evening when the solar radiation acquisition

https://doi.org/10.1051/e3sconf/20186402004

disappears and the auxiliary heating operation is carried out.

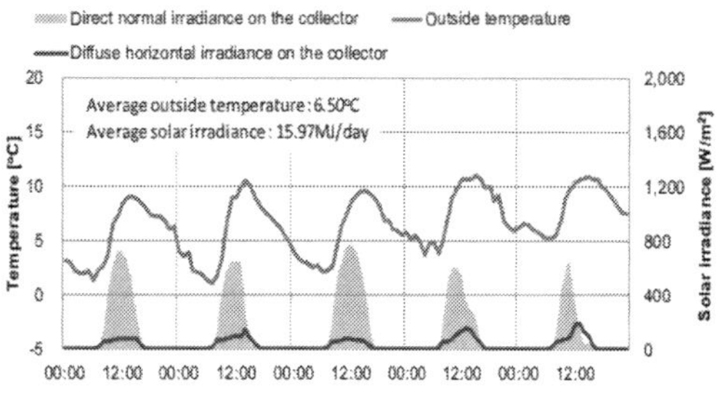

Fig. 2. Weather conditions (January 7th – 11th)

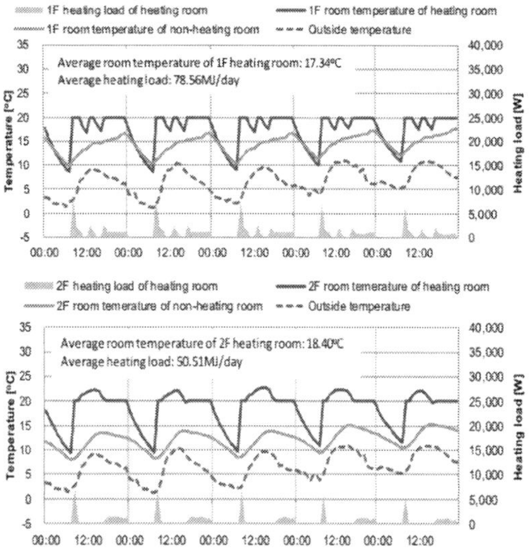

Fig. 3. Room temperature results of a typical house.

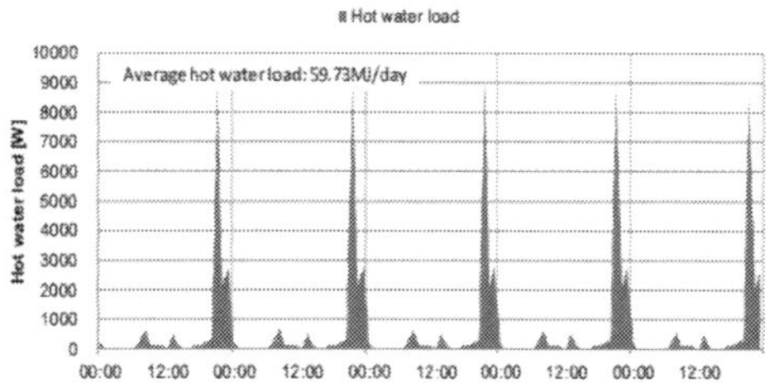

Fig. 4. Hot water supply load of typical house.

Fig. 4 shows the hot water supply load by the auxiliary heat source during the examination period. A load of 8,612W to 9,167W is generated at 21 o'clock when the usage amount becomes maximum, and a hot water supply load of about 59.73 MJ per day occurs on January 7th to 11th.

3.2 Study on air-based solar heating system

In order to improve the performance of the air-based solar heat collecting system, we examined the effect of annual heating and hot water supply load reduction by changing heat absorption/heat release of the heat collector, heat storage, and control of heat collection operation control. Regardless of the angle of the roof surface regardless of the angle of the roof surface, the area ratio of the preliminary collector to the glass heat collector is set to 3:1 based on the floor area where the minimum area is secured, and the preliminary collectors 45m2, Glass collector 15m2 was set. Fig. 5 shows the operation control of the system. Fig. 6 shows a conceptual diagram of the air-based solar heat collecting system to be studied in this research [3]. By applying the air-based solar heating system, room temperature rises by blowing the heated air during the day into the room as shown in Fig. 7, but hot air enters the room and overheats during the daytime. On the other hand, there is a problem that the room temperature drops at night. At that time, the temperature inside the hot water storage tank (Fig. 8) according to the hot water storage mode reaches a maximum of about 35°C.

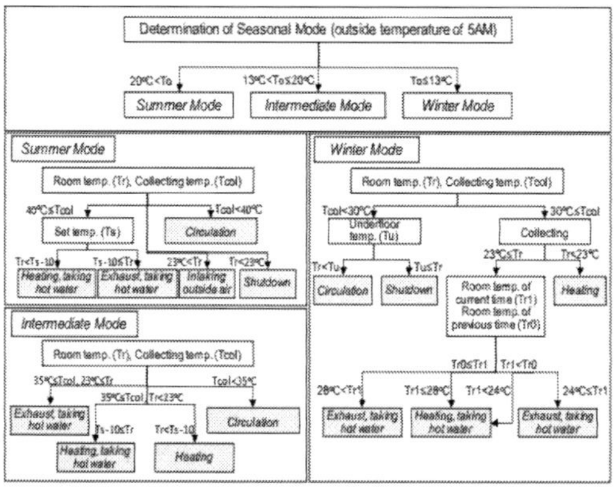

Fig. 5. Operation control of the air-based solar heat collecting system

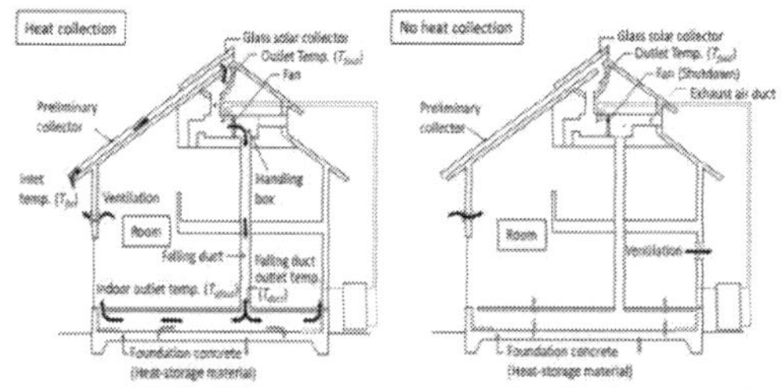

Fig. 6. Conceptual diagram of the air-based solar heat collecting system

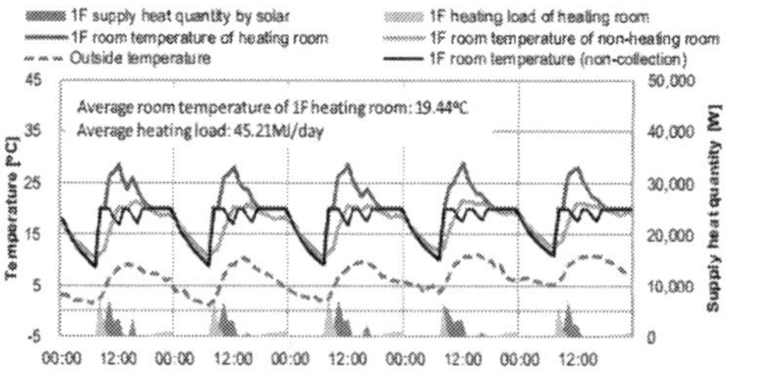

 https://doi.org/10.1051/e3sconf/20186402004

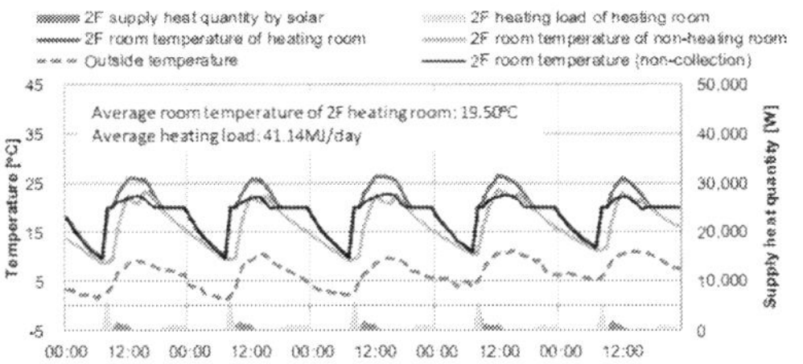

Fig. 7. Room temperature results of air-based solar system

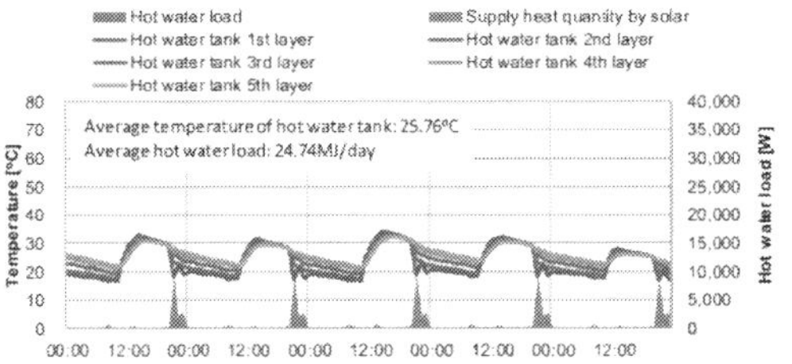

Fig. 8. Temperature inside hot water storage tank

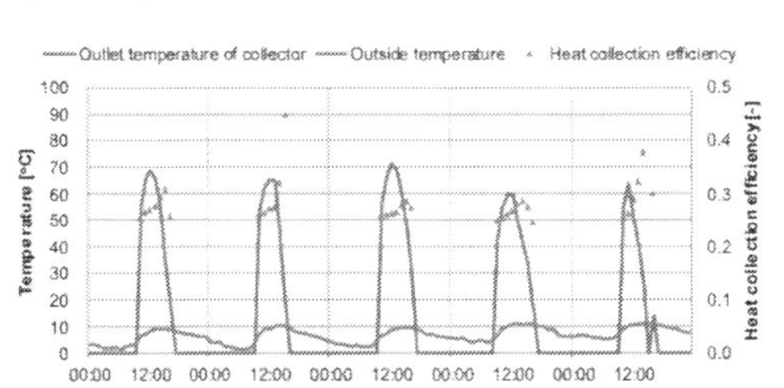

Fig. 9. Outlet temperature of collector and heat collection efficiency

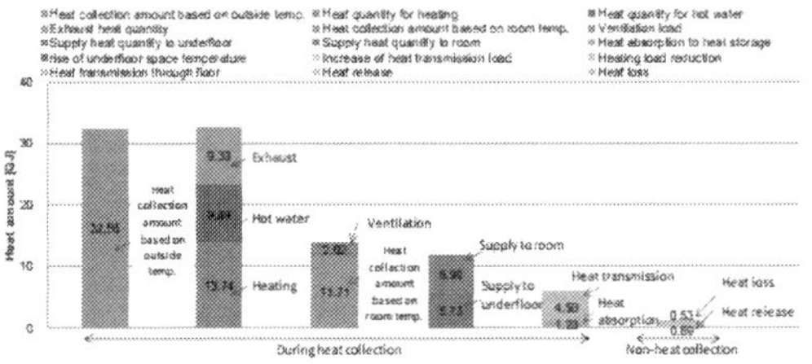

Fig. 10. Heat flow due to heat collection of air-based solar system

Table 3. Load reduction effect by application of the air-based solar heating system.

[GJ/year]	1F heating load	2F heating load	Hot water load	Reduction of heating load	Reduction of hot water load	Reduction rate of load [%]
Typical house	9.72	6.92	15.87			
Air-based solar heating system	5.67	5.58	5.55	5.39	10.32	48.3%

Fig. 9 shows the entrance and the exit temperature of the collector and the heat collection efficiency from January 7th to 11th when the air-based solar heating system is applied. Fig. 10 shows heat flux due to heat collection of air-based solar system. In order to increase the heating load reduction effect of the air-based solar heating system, it is important not only to improve the performance of the heat collector but also to increase the heat absorption performance of the heat storage. The heat storage helps that overheating in the daytime will not occur and that room temperature will not decrease by sufficiently dissipating the absorbed heat at night. In addition, it can be considered that it is possible to increase the annual heating and hot water supply load reduction effect by adjusting the hot water supply control according to the heating condition. Table 3 shows the load reduction effect by application of the air-based solar heating system.

4 Improvement plan for the air-based solar heating system

First, the glass of the collector is changed to a Low-E glass as a method of reducing the heat loss of the solar collector (case 1). It is thought that although the glass transmittance is slightly reduced due to the change of the glass, the heat resistance of the glass is increased and the heat loss to the atmosphere is reduced. Next, as a method of increasing the heat radiation amount of the thermal storage, a heat insulating material is installed below the base concrete (case 2). This is believed to reduce the loss of heat absorbed in the base concrete to the ground. In addition, as a method for releasing the heat from thermal storage, the indoor air is forcedly convected to increase the heat transfer coefficient of the surface (case 3). Since the base concrete has a limited installation area and heat capacity as a thermal storage, the additional 20L water packs are installed in the underfloor space to increase the heat capacity (case 4). In order to enlarge the surface of the additional thermal storage, the 500ml water bottles are used instead of 20L (case 5). Table 4 show the simulation cases.

Table 4. Simulation cases of air-based solar heating system.

Case1	Changing the glass of collector Existing glass: Solar radiation transmittance 0.88, U-value 6.0W/m^{2}K Low-E glass: Solar radiation transmittance 0.74, U-value 2.7W/m^{2}K
Case2	Case1 + insulation under base concrete
Case3	Case2 + indoor circulation during non-solar heat collection
Case4	Case3 + additional thermal storage (20L water packs 3000L)
Case5	Case3 + additional thermal storage (500ml plastic bottles 3000L)

The annual simulation results of the improvement of the air-based solar heat collection system are shown in Table 5. For the condition of the buildings, weather data, and system conditions of this study, 48.3% of heating and hot water load was reduced annually due to the application of the conventional air-based solar heat collection system. Also, by replacing the glass in the collector with Low-E, it was found that the heating load of the system can be reduced by about 2.5% compared to the existing system. Insulation below the base concrete resulted in a load reduction of about 0.2%. However, when the additional thermal storage is installed, the temperature of the underfloor space rises and the effect of the heat insulation is expected to increase. Lastly, it was shown that installing 3000L of 500ml plastic bottles with additional thermal storage can save the annual load up to 66.5% and keep the room temperature constant.

Table 5. Load reduction effect.

[GJ]	Heating load of 1st floor	Heating load of 2nd floor	Hot water load	Load Reduction rate [%]	1st floor	
					Avg. Temp[°C]	Daily difference [°C]
Non heat collection	9.72	6.92	15.87			
Existing system	5.67	5.58	5.55	48.3%	20.2	14.6
Case1	5.31	5.49	5.20	50.8%	20.2	14.8
Case2	5.29	5.47	5.18	51.0%	20.2	14.8
Case3	3.62	4.77	5.59	57.0%	20.7	11.1
Case4	3.32	4.59	5.81	57.8%	20.8	10.0
Case5	0.93	2.76	7.30	66.2%	21.8	4.1

5 Conclusion

As a result, the annual load reduction rate has increased by about 17.9% over the conventional air-based solar collection system. In this study, the improvement plan including various improvement factors was examined as a method to maximize the performance of the solar thermal system. However, the improvement of the indoor thermal environment by each factor and the examination of the load reduction effect are examined in the future. In this study, the performance of existing air-based solar heat collection system was evaluated. In addition, an annual simulation was conducted to examine the improvement effect for reducing the annual heating load and the hot water supply load. In order to improve the air-based solar collecting system, the glass of the solar collector was changed from ordinary glass to Low-E glass with less heat loss, and thus the effect of increasing the amount of solar heat collection and reducing the annual load was examined. In order to increase the amount of heat releasing by the thermal storage, it was examined to install insulation at the bottom of the base concrete. It was investigated the effects of forced circulation (indoor circulation) of room air and underfloor air to increase the amount of heat dissipation from the thermal storage during non-solar heat collection, and additional heat storage to increase the heat capacity of the floor space.

References

1. Fukumoto M. Research on performance comparison and optimal operation in solar heating for different types of detached houses. MS thesis. Department of Architecture Engineering, The University of Tokyo. Tokyo, Japan (2011)
2. Institute for Building Environment and Energy Conservation. Design Guidelines for Low Energy Housing with Validated Effectiveness (2007)
3. Ugagawa M, Simulation model of solar energy-based building sheath. *JSES / JWEA research presentation* (2009)
4. Mae M, Akamine Y, Kono R, Amano N, Matsueda N, Kawashima N, Kaihoh K. Development of heating and cooling system utilizing solar and earth heat for detached houses Part1: Concept of the energy saving heating and cooling system and condition of measurement in experimental house. Summaries of technical papers of annual meeting, *Architectural Institute of Japan*, 1077-1078 (2011)
5. Matsueda N, Mae M, Akamine Y, Kono R, Amano N, Kawashima N, Kaihoh K. Development of heating and cooling system utilizing solar and earth heat for detached houses: Part2: Influence on air-conditioning energy by insolation cover and earth heat of summer in residence type experiment house. Summaries of technical papers of annual meeting, *Architectural Institute of Japan*, 1079-1080 (2011)
6. Ugagawa M, Calculation method of air conditioning by personal computer. Ohmsha (1986)
7. Housing · Building Energy Conservation Organization: Building Energy Conservation Standards and Calculation Handbook (2006)
8. Ito K, Sakabe Y, Tatematsu K, Murata S, Kitadani Y, Suzuki H, Aiso K, Matsumoto T, Mae M, Iwamae A. Technological Development for Energy Generation Systems Using Roof Integrated High-Efficiency Vacuum Thermal Energy/Load Responsive Storage Etc. No.1 : Estimate for whole system performance based on individual performance tests of the developed solar hot water system parts. Summaries of technical papers of annual meeting, *Architectural Institute of Japan*, 171-172 (2010)
9. Tanaka S. The latest architectural environmental engineering, Inoueshoin; 1985
10. Tanaka S. Solar heating and cooling system. Ohmsha (1977)

Carbon-Nanotube Based Thermoelectrical Paste for Enhancing Solar Cell Effciency

Mohamed Fathi Sanad[1], *Ahmed* Shaker[1], *Sameh* O. Abdellatif[1], *Iman* Elmahallawi[1], *Hani* A. Ghali[1], and *Khaled* Kirah[2]

[1]Centre for Emerging Learning Technologies (CELT), The British University in Egypt, Cairo, Egypt
[2]Engineering Physics and Mathematics Department, Ain Shams University, Cairo, Egypt

Abstract. A super-passive cooling technique based on a thermal paste is proposed for PV efficiency enhancement in elevated temperature conditions. A mixture between carbon nanotubes and graphene having a promising Seebeck coefficient is chosen. An overall enhancement in efficiency by around 58% was reached while thermoelectrically supplying hundreds of micro-Watt per PV Watt.

1 Introduction

Nowadays, the need for energy is increasing with population and economic growth around the world. The global energy demand is expected to increase to 16 Billion Tons of Oil Equivalent (TOE) in 2030 (1 TOE ≈ 42 GJ) [1], which is a huge amount of energy that need to be provided. Most of the energy provided today comes from fossil fuel that releases huge amounts of greenhouse gases which are the primary cause of global warming. In 2013 the energy related greenhouse-gas emissions was 35.5 Gigaton, and according to the Intended Nationally Determined Contributions (INDCs), it will increase to 38.3 Gigaton in 2030 [2]. This shows the need to more green and renewable energy sources.

Thermoelectric generators (TEGs) are relatively new and promising devices that are used to harvest energy from any temperature gradient. These devices are based on the Seebeck effect in which a temperature difference between two dissimilar electrical conductors or semiconductors produces a voltage difference between the two substances. The max efficiency of a thermoelectric device is given by:

$$\eta_{max} = \frac{T_h - T_c\sqrt{1+Z(T_m)}-1}{T_h\sqrt{1+Z(T_m)}+\frac{T_c}{T_h}}$$

where T_h and T_c are respectively the hot and the cold side temperatures, $T_m = \frac{T_h+T_c}{2}$ and $Z(T_m)$ is the device figure of merit defined as:

$$Z(T_m) = \frac{S^2 T_m}{\rho k_{total}}$$

where S is the Seebeck coefficient, ρ is the electrical resistivity and k_{total} is the total thermal conductivity [4]. Therefore, researchers are working on developing new materials with higher Seebeck coefficient.

Semiconductors show superior thermoelectric behavior than metals due to their low thermal conductivity and high Seebeck coefficient. In 2004, $Z(T)$ enhancement using nanostructures was first reported. Lead telluride (PdTe) has been used in TEGs since the 1960s. Nanostructured PdTe shows a higher efficiency (11% for ΔT of 590 K) over the non-nanostructured form (5.1% for a ΔT of 417 K) by about 40-50%. This is mainly because of the lower lattice thermal conductivity due to the nanostructuring [5]. Carbon-nanotube has shown impressive thermoelectric behavior. A value of $Z(T) = 0.02$ has been reached using carbon nanotube (CNT)-filled polymer composites which is at least one order of magnitude higher that bulk Si based thermoelectric generators [6]. In addition, graphene/carbon-nanotube polymer based thermoelectric devices showed the highest electrical conductivity compared to Polymer based ternary thermoelectric composites as reported in [7].

One problem with photovoltaic panels is heating as the efficiency of the panel decreases by 0.45% for each degree rise over 25 °C [8]. Thermoelectric devices can be used as a cooling solution for photovoltaic panels. This is a two-fold advantage: increasing the efficiency and generating energy in the same time.

This paper reports the preparation and the properties of graphene/carbon-nanotube/polymer based thermoelectric paste and its potential in photovoltaic cooling and low-grade thermoelectric energy harvesting.

2 Experimental work

2.1 Thermoelectric paste preparation

Thermoelectric paste was produced using the modified Hummers method from pure graphite powder in a mixture with Multi-Walled Carbon Nanotubes (MWCNTs) [9]. In this method, 60 ml of concentrated sulfuric acid and 6 ml of phosphoric acid were stirred for 10 minutes. Then 0.5 gm of graphite was added into the solution under vigorous stirring. Around 3 gm of the potassium permanganate as strong oxidizing agent were then added slowly into the solution. This mixture was stirred for 12 hours until the mixture became green.

The green mixture is then pureed with 2 ml of hydrogen peroxide slowly while cooling in the ice bath for 30 minutes. 20 ml of hydrochloric acid and 60 ml of deionized water were added and centrifuged for 10 minutes. This was followed by decanting the residuals and then rewashing again with deionized water for 15 times until it became clear.

The solution was stirred well in a mixture of 0.6 gm of MWCNTs and 0.6 gm of graphene oxide purchased from Sigma Aldritch in presence of PEDOT in DMSO as a solvent and stirred for 3 hrs until a homogeneous solution was obtained. Finally, the solution was dried at 60 °C for 1 hr. Different samples were prepared by varying the amount of graphite and CNTs in attempt to obtain pastes with the best thermoelectric behavior according to table 1.

Table 1. Various ratios for Graphite oxide, graphene oxide and CNT in the prepared samples.

	Graphite oxide	Graphene oxide	CNT
MS-TE-001-01	1	1	1
MS-TE-002-01	2	1	1
MS-TE-003-01	1	1	2
MS-TE-004-01	1	2	1

2.2 Light source chractraization

Solar cells must be characterized under standard conditions using solar simulators. In principal, the high uniformity, temporal stability and spectrum matching are the three main parameters that distinguish solar simulators from other low-cost sources. However, commonly used low-cost light sources can still be used for characterizing PV cells, while taking the mismatching factors in consideration. Herein, a commercial H4 12V 100/90 W Halogen lamp (commonly used in most car lighting) is used [10]. Ocean Optics Red Tide USB650 FiberOptic Spectrometer [11] has been used to optically characterize the used light source.

Light is collected through a fiber terminated by a cosine corrector. This is a preconfigured, ready-to-use spectrometer in the 350-1000 nm range. Its detector has 650 active pixels, or one data point per nanometer. It uses Sony ILX511 linear silicon CCD array detector and has a sensitivity of up to 75 photons/count at 400 nm. The spectrometer reads the intensity as a number of photons/count (i.e. arbitrary units).

The spectrometer readings have been calibrated by using a reference sample to give readings in Watts/unit area. Figure 1 shows the UV-Vis-IR spectra for the used low-cost light source.

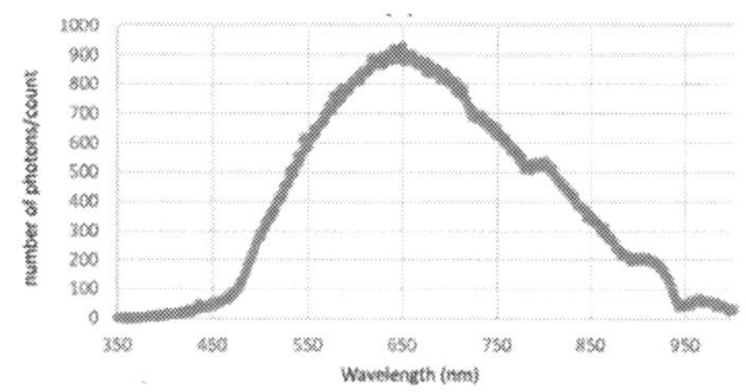

Fig. 1. UV-Vis-NIR spectrum for the selected light source.

The optical mismatching was calculated based on figure 1 with respect to the standard solar irradiance given in AM1.5G [12]. This mismatching factor will be considered in calculating the solar cells efficiency in the next section. Regarding the temporal stability, the selected source showed an acceptable stability with time reaching an overall variation less than 0.1% across five continuous hours of operation. Finally, to minimize the uniformity mismatching errors, a solar cell under test with small area (20 mm x 20 mm) is selected where the used light source is assumed to be uniform within this area.

3 Results and discussion

3.1 Thermoelectric paste optimization and chractraization

Various thermoelectric paste prepared in section 2B are characterized seeking for an optimum recipe. Current, voltage and resistance were measured with varying temperature difference for ΔT = 0 to 14 °C to investigate the optimum thermoelectric behavior among all prepared samples. This temperature range is chosen to cover the expected temperature difference due to solar cell backside where experimental measurements showed a difference of around 10 °C.

Figure 2 presents the behavior of the extracted power with variable temperature difference (ΔT) between the two sides of the sample. It was observed that sample MS-TE-004-01 achieved the highest extracted power of about 82 μW at ΔT = 14 °C. This sample was then used in the backside of the solar cell to examine its utility in enhancing the solar cell efficiency under high temperature.

On the other hand, X-ray powder diffraction (XRD) was performed for the optimum thermoelectric paste. The sample was loaded into a P-Analytical X'PERT MPD diffractometer holder using (Cu [Kα1/Kα2]) rays. Figure 3 shows the obtained pattern. X-ray peaks were observed at a diffraction angle range from 20° to 60° with 0.02° step. A broad peak at 25.9° corresponding to the (020)

https://doi.org/10.1051/e3sconf/20186402005

plane of PEDOT is observed. However, the intensity of the reduced graphene oxide (rGO) and CNT were very weak due to very low concentrations. Interaction of aggregated rGO, CNTs with PEDOT chains degrade the crystalline phase and the stack structure of PEDOT. On the other hand, there is a strong and sharp peak of graphene oxide which exhibits a (001) reflection at 9.090 corresponding to a basal spacing of d_{001} = 0.951nm.

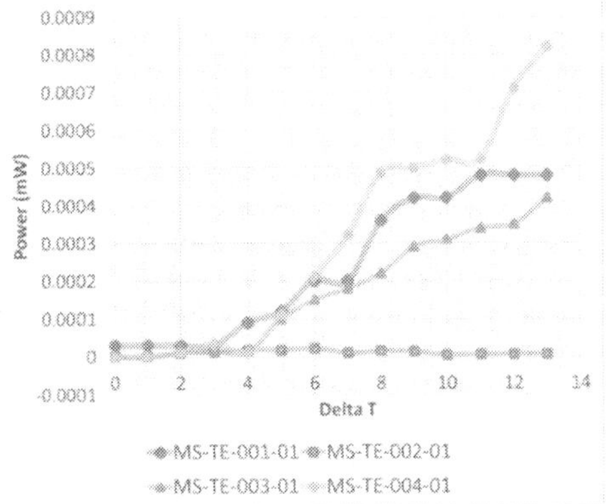

Fig. 2. Extracted power from thermoelectric generators using different CNT recipies.

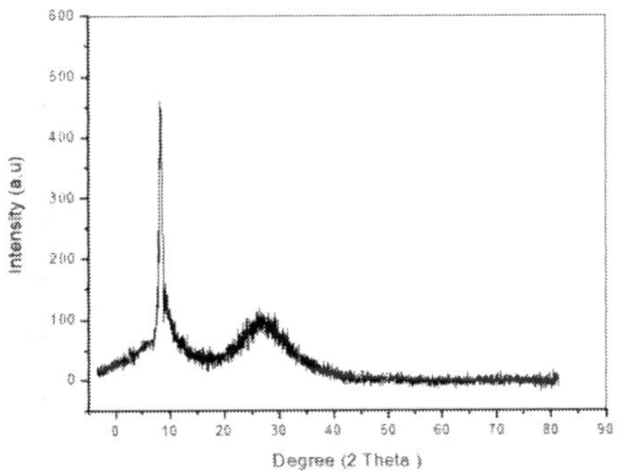

Fig. 3. XRD for the optumum thermoelecteic paste

3.2 PV module I-V chractaization

In order to investigate the utility of the thermoelectric paste on the solar cell efficiency, nanocrystalline and amorphous solar cells were used [13]. Figure 4 shows the measured *I-V* characteristic curves of amorphous silica solar cell with and without the optimized thermoelectric paste at different distances between the cell and the light source. Characterization has been achieved using Keithley 2401 source/meter was used with a four-probe station connection to vary the biased dc voltage from 0 V to V_{OC} while acquiring the light generated current.

As observed, varying the distance between the light source and the sample affects both the irradiance level and the temperature as indicated in figure 4. All measurements were made using a Keithley 2401 source/meter.

Figure 4 shows that the thermoelectric paste enhances the performance of the cell regardless of the irradiance level or the temperature. However, this enhancement varies due to the nonlinearity of the thermoelectric behavior shown in figure 2. A maximum enhancement in efficiency reaches 40% at 55 °C where the sample was placed 5 cm from the source. The effect of enhancement is captured from the short circuit current as the open circuit voltage is almost the same.

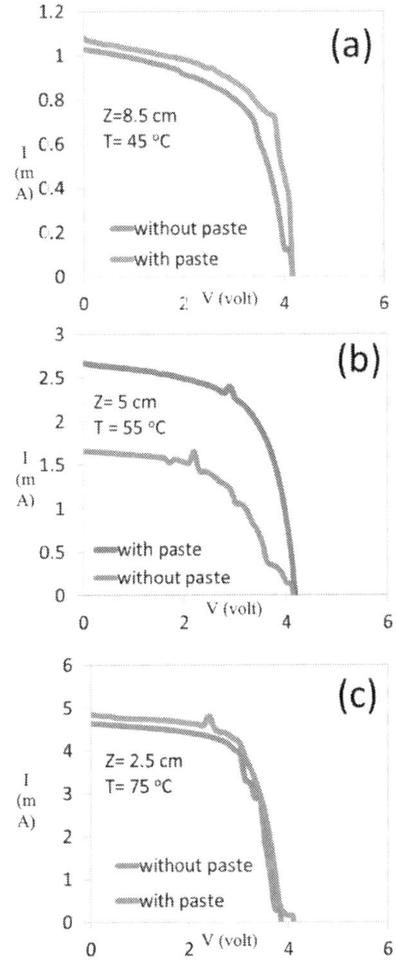

Fig. 4. I-V curve for an amourphous solar cell with and without thermoelectric paste at different distances from light source and different temperature.

The same experiment was repeated on a nanocrystalline solar cell with almost the same dimension as shown in figure 5. Similarly, the use of thermoelectric paste showed better performance on the solar cells where the non-linearity effect can still be detected. The enhancement in efficiency was found to be 58% at 55 °C where the sample was placed 5 cm from the source.

 https://doi.org/10.1051/e3sconf/20186402005

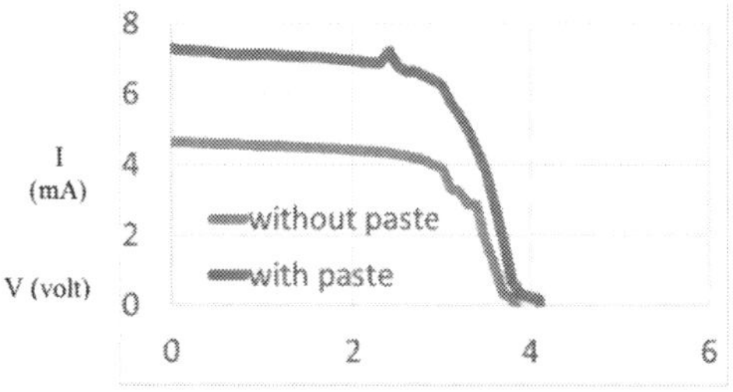

Fig. 5. I-V curves for ananocrystalline solar cell with and without thermoelectric paste for maximum effiency enhancement.

From the presented results in both Fig. 4 and Fig. 5, the utility of using the proposed CNT paste on the backside of solar cells regardless of its type, used irradiance or temperature, is clearly recognized. However, to maximize the efficiency enhancement, optimum theromelectric paste should be used. In addition, the overall performance of nanocrystalline cells is better, with respect to the thermoelectric paste utility, in comparison with amorphous solar cells.

4 Conclusion

The work presented in this paper highlighted the potential of utilizing thermoelectric paste for PV cooling especially under high temperatures. A maximum enhancement in efficiency of 58% was observed for nanocrystalline cells while the backside was pasted with a CNT thermoelectric paste. Amorphous cells also showed an overall enhancement of about 40% when using the past. Basically, the operating temperature as well as the irradiance level affects the thermal harvesting capabilities, but it is worth to mention that in all the examined cases the thermoelectric paste showed enhanced performance with respect to the reference cell.

Acknowledgment

The authors would like to acknowledge the support and contribution of the STDF in this work. This work is part of the STDF Project entitled, "Enhancement of Photovoltaic and Wind Turbine Performance for High Temperature and Low Wind Speed Environments", Project ID#26134.

References

1. "BP Energy Outlook," BP, 2018.
2. "Energy and Climate Change," INTERNATIONAL ENERGY AGENCY, 2015.
3. G. Jeffrey L, "Carbon-Nanotube-Based Thermoelectric Materials," Advanced materials, p. 35, 2018.
4. H. Jyrki Tervo, "State-of-the-art of Thermoelectric Materials," VTT WORKING PAPERS, p. 29, 2009.
5. A. Y. Xiaokai Hu, "Power generation from nanostructured PbTe-based thermoelectrics: comprehensive development from materials to modules," Royal society of chemistry , p. 13, 2015.
6. C. Y. Dasaroyong Kim, "Improved Thermoelectric Behavior of Nanotube-Filled Polymer Composites with Poly(3,4-ethylenedioxythiophene) Poly(styrenesulfonate)," American Chemical Society, p. 11., 2010.
7. G.XinLi,"Poly(3,4ethylenedioxythiophene)/graphene /carbon nanotube ternary," Elsevier, p. 5, 23 august 2016.
8. C. Cătălin George Popovicia, "Efficiency improvement of photovoltaic panels by using air cooled heat sinks," Elsevier, p. 8, 2015.
9. H. Naeem, M. Ajmal, S. Muntha, j. Ambreen, &M. Siddiq, "Synthesis and characterization of graphene oxide sheets integrated with gold nanoparticles and their applications to adsorptive removal and catalytic reduction of water contaminants," RSC Advances, 8(7), 3599-3610.
10. T. Gorkhover, M. Adolph, D. Rupp, S. Schorb, S. W. Epp, B. Erk, et al., "Nanoplasma dynamics of single large xenon clusters irradiated with superintense x-ray pulses from the linac coherent light source free-electron laser," Physical review letters, vol. 108, p. 245005, 2012.
11. P. A. Gilbert and W. Haeberli, "Experiments on subtractive color mixing with a spectrophotometer," American Journal of Physics, vol. 75, pp. 313-319, 2007.
12. S. Abdellatif and K. Kirah, "Numerical Modeling and Simulation for a Radial p-i-n Nanowire Photovoltaic Device," Energy Procedia, vol. 36, pp. 488-491, // 2013.
13. https://media.digikey.com/pdf/Data%20Sheets/Sanyo%20Energy/Amorphous_Br.pdf

Design and fabrication of a prototype solar tapping device

Anirudh N. Kulkarni[1], *B. Arjun* Bhat[1], *Chinmay D.* Sastry[1], and *Dr. K. S.* Sridhar[2]

[1] Mechanical Engineering, PES University, Bengaluru, India
[2]Department of Mechanical Engineering, PES University, Bengaluru, India

Abstract. Fossil fuel based energy sources are most commonly used because of the ease of availability and affordability, but have many long lasting negative effects, due to which a shift must be made to renewable and clean energy sources such as solar energy. Efficiency of the photovoltaic cells have increased by 1% every ten months, resulting in reduced costs and increased number of users. The project will explore the design and fabrication of an improved and efficient solar tapping device. An effort will be made to incorporate an architectural concept called "barrel vault" that is primarily used for natural lighting, over a Fresnel lens plate setup to act as a collector. The tracking mechanism incorporated is unique and can be modified to either have single axis or dual axis tracking depending on the requirements. Keeping in mind the environment and cost, an attempt has been made to fabricate the product with higher energy conversion rate, at an affordable cost using the above techniques. Upon analysing the results, the conclusion could be drawn that there was an increase in conversion rate of up to twice the traditional setup.

1 Introduction

There are primarily 2 classes of photovoltaics, or solar setups, based on concentration and solar cells, namely, conventional photovoltaics, that employ simple single junction solar cells, generally without any concentration or tracking mechanisms, and Concentrated Photovoltaics (CPV), that employ concentration setups along with multi junction solar cells [1].

Given that multi junction cells are costly and not mass produced everywhere, our setup attempts to use concentration methods on single junction PV cells that can also be a viable option for multi junction cells, essentially placing our solution between conventional PVs and CPVs.

Employing a **barrel vault** collection architecture that isn't fully used in solar devices, this allows for near-180° solar collection whilst not risking heating of the panel. The flat sides of the barrel vault have been left open allow for ventilation.

Beneath our barrel vault is a trapezoidal structure with a **Fresnel lens plate** [2] top, which is a plate with multiple small Fresnel lenses arranged on it, placed at about 80mm above the solar panel, which is the approximate focal length of the Fresnel lenses used. This allows for multiple concentration points. The trapezoidal glass panes structure allows for uniform lighting above the panel when the sun is oblique to the panel and prevents shadow formation.

Dual axis tracking [3], while being more efficient than single axis tracking [4], can cause run time issues where the tracking algorithm in one direction interferes with the other. Hence, a single axis tracking algorithm from East to West has been implemented, essentially following the sun from sunrise to sunset. The barrel vault takes care of the summer to winter sun shift and is inclined to South by default.

2 Methodology

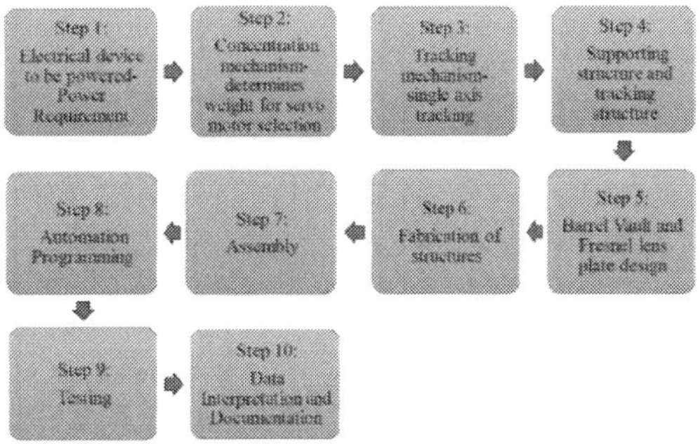

Fig. 1. Methodology Flowchart

- Power Requirement: 10 W panel for a DC bulb and a fan
- Concentration Mechanism:

a. Fresnel lens plate: 20 plano-convex lenses, 80mm focal length, producing a concentration area of 30x25 mm^2 per lens.
b. Collection setup: Barrel Vault archway consisting of a total 6 transparent panels, around 310mm diameter, of 81mm arc length per panel.
c. Trapezoidal box of transparent planes covering the distance between barrel vault and glass panel.

- Tracking: Programmed to follow the sun from 8 am to 4 pm, moving at 15° per hour

 https://doi.org/10.1051/e3sconf/20186402006

- Support Structure Analysis: Analysed with a load of 30 kgs put on the structure, corresponding to the weight of a full size solar panel and concentration mechanism. Modelled in Solidworks and analysed in ANSYS.
- Barrel vault design:
  $Diameter\ of\ barrel\ vault = 310\ mm$
  $Total\ angle\ of\ barrel\ vault = 180°$
  $Total\ number\ of\ glass\ panes = 6$
  $Angle\ of\ each\ glass\ pane = 180° \div 6 = 30°$
  $Using\ arc\ length\ formula\ l = r * \theta$
  $l = 310 * (30° * \pi \div 180) = 81.158\ mm$
- Fabrication: PLA (Polylactic acid), a biodegradable and bioactive thermoplastic aliphatic polyester, obtained from renewable sources like corn starch, cassava roots or sugar cane, was used as raw material to 3D print the base structure.
- Assembly: Structure assembled as shown-

Fig. 2. Concentration setup

- Automation programming: Done on Arduino UNO IDE; programmed to move at 15° per hour.
- Testing: Carried out for the following conditions, both with tracking and without tracking-

1. Fixed Solar Panel only
2. Panel with Fresnel Plate
3. Panel with Fresnel Plate and Barrel Vault

3 Experimentation and results

The one trial, testing was done taking, a solar panel setup, solar panel setup coupled with a Fresnel plane setup and finally the prototype that consisted of the panel coupled with both the Fresnel plane setup and the barrel vault setup. Readings were obtained without tracking the position of the sun and the data in Table 1 was obtained.

During this test, data for a basic panel, panel+ Fresnel setup and panel + Fresnel + Barrel setup, was compared. Peak current and voltage output was obtained for a basic solar panel as it works under both direct and diffuse sunlight, in contrast to the Fresnel setup that works only under direct sunlight. A steady output voltage was obtained for all the three setups. The setup involving the Fresnel setup had lower output voltage owing to shadow experienced by the panel in the current setup. The output current was maximum for the basic setup, but a low and steady value was obtained for the Fresnel setup. The current peaked at 12 noon when the sun was overhead and the sunrays were incident normally, for the final prototype. This was an indicator of the device working well with sun tracking when the rays would be incident normally.

Table 1. Test Results for complete setup without Tracking

Date	10-04-2018	SOLAR PANEL WITHOUT TRACKING		SOLAR PANEL+ FRESNEL SETUP WITHOUT TRACKING		SOLAR PANEL+ FRESNEL+BARREL WITHOUT TRACKING	
Time	Temperature (C)	Voltage 1 (V)	Current 1 (mA)	Voltage 2 (V)	Current 2 (mA)	Voltage 3 (V)	Current 3 (mA)
08:00	22	19	200	18.1	180	18.8	200
09:00	24	19.9	300	19.5	250	18.4	250
10:00	27	20.4	500	19.8	250	19.6	250
11:00	29	20	520	19.6	270	19.3	270
12:00	30	19.9	530	19.4	400	19.1	400
13:00	31	19.8	520	19.4	400	19.1	300
14:00	32	19.6	520	18.8	300	19.2	250
15:00	32	19.6	260	19.3	240	18.9	250
16:00	31	19.2	220	19	200	18.9	220

Fig. 3. Assembled Prototype

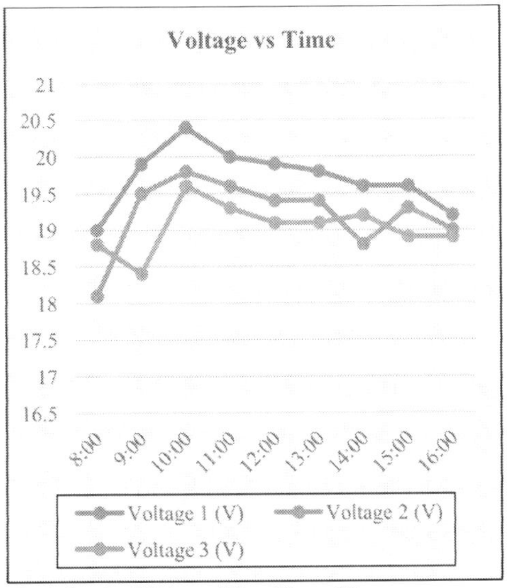

Fig 4. Graph of Voltage (V) vs Time of Day

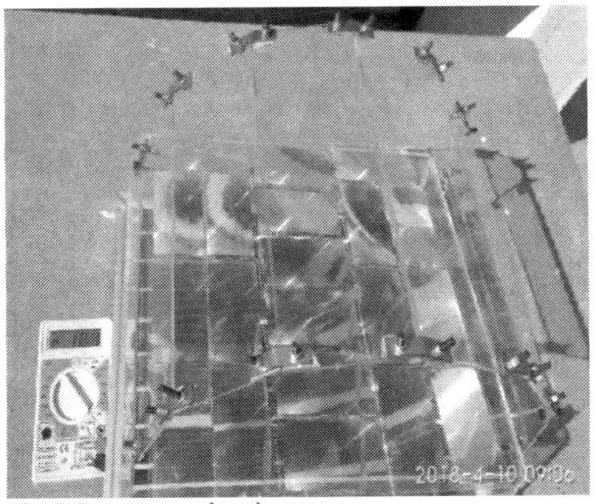
Fig. 5. Prototype and testing apparatus

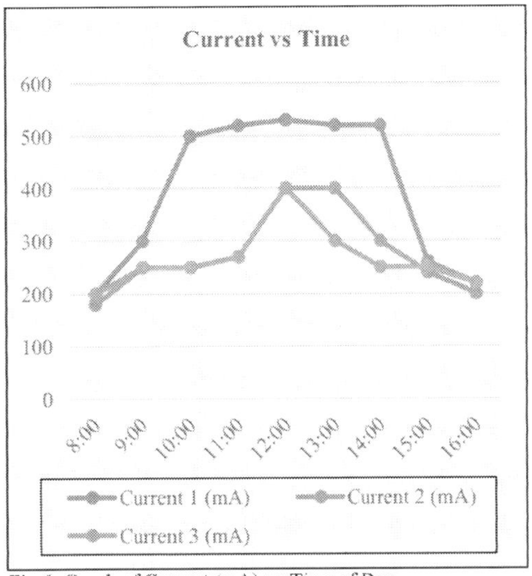

Fig 6. Graph of Current (mA) vs Time of Day

Fig. 7. Close up view of concentration zones on the panel

Fig. 5 shows concentrated sunlight being incident on the panel at 12:56 PM. Each of the concentrated sunlight rectangles have an area of 8.75 cm^2 and there is a total of 20 such rectangles, resulting in a total area of 175cm^2. Each panel rectangle area is 10 cm^2 and total area of the panel is 720cm^2 with 72 rectangles. Therefore, for just 24.3% of the panel receiving concentrated sunlight at normal incidence, the output power is 5.73W. Placing the solar cells strategically so that all of them receive concentrated sunlight could result in a power that is 4 times the previous value of 22.9W. However, a basic panel setup exposed to sunlight at the same time, resulted in a power output of 10.49W. This clearly indicates a higher energy conversion rate per cell for a combined Fresnel and barrel setup.

4 Conclusions

The prototype was successfully designed and fabricated. With our setup of barrel vault on top of the panel with the Fresnel lens to concentrate the sunlight, we could observe that only 24.3% of the panel received concentrated sunlight at normal incidence, with output power being 5.73 W. This would mean more output power per exposed area of the panel. For this setup, if the solar cells were strategically placed to receive the concentrated sunlight, the power output could rise to 4 times the current value. Following are some of the advantages of the proposed device over the basic solar setup: 1) The device has a higher energy conversion rate for the same panel area in comparison to the basic setup. 2) Lesser space required for the installation of the device for the same power output. Considering a higher energy conversion rate per cell by the incorporation of the several innovative technologies, pointed to the fact of the device being a success. Some of the aspects that can be worked on, to improve the existing device would include the following: 1) Switching to multi-junction solar cells in place of single junction cells, could result in higher output because of better temperature characteristics of the former. 2) Accurate placement of the solar cells to receive direct concentrated sunlight from the Fresnel plane. 3) Optimization and mathematical modeling of the radius of curvature of the barrel vault setup to account for the seasonal changes. 4) The current setup utilizes only single axis tracking and having dual axis tracking could improve the efficiency.

Considering a higher energy conversion rate per cell by the incorporation of the several innovative technologies, pointed to the fact of the device being a success.

Acknowledgement

The authors would like to thank the management and the Department of Mechanical Engineering of PES University for their continued support in the completion of this research.

References

1. T.S. Viana, R. Ruther, F.R. Martins and E.B. Pereria, Sol. En., vol. **85**, pp. 486-495, (2011)
2. J.-N. Juang and R. Radharamanan, *American Society for Engineering Education*, (2014)
3. T.-S. Zhan, W.-M. Lin, M.-H. Tsai and G.-S. Wang, *Computer Software and Applications Conference* (Kyoto, Japan, 2013)
4. D. S., R. R., P. A., R. Dhanabal, IJESIT, **2** (2013)

 https://doi.org/10.1051/e3sconf/20186402007

Innovation of price adjustment mechanisms to support investment in solar power in Germany

Do Thi Hiep[1], *Clemens* Hoffmann[1]

[1]Department of Integrated Energy Systems, University of Kassel, Wilhemshöher Allee 71 – 73, 34121 Kassel, Germany

Abstract. It has been widely agreed that to incentivize renewables integration into the power system, not only pricing mechanisms, but price adjustment mechanisms have played a vital role, and it has been true for the German Energiewende. This study is to carry out a detailed analysis of investment results influenced by innovative price adjustment mechanisms from an auto degression rate to a feedback system. Employing linear regression models for the historical data of investment in small-scale rooftop PV projects in Germany, we have found out a better correlation between PV system price and feed-in tariff (92.09%) under quarter feedback and monthly adjustment mechanism compared to an annual feedback system. However, the underinvestment in recent years reveals that a feedback mechanism without specific mathematical shapes was not effective enough in term of meeting the targeted volume. Therefore, further researches are to design mathematical images of feedback mechanism in order to find out the trajectory of electricity price in the future which at the same time satisfies the target of investment and economic effectiveness.

1 Introduction

Energy transition is becoming a global trend of the electricity industry. Germany is one of the leaders of this movement which is considered as a successful pattern of integration of wind and solar into the power market [1]. Compared to the rest of the world, Germany started to immigrate renewables into the power system earlier. However, it took this country long time to prepare for the integration of solar PV. The year 1990 marked the first take-off for wind technology while after 1998, German people could see a kick off solar power [2]. After almost three decades, Germany owns more than 42 GW of solar power, accounts for 21% of total installation [BMWi, 2018].

To encourage the sharing of solar power, a nation can put into use one of the three basic available pricing mechanisms including feed-in tariff, auction, CO2 pricing, or combine them. A feed-in tariff is a system that the government decides price levels (price commitment to the producer) and the market sets quantities. In contrast, an auction is a pricing mechanism in which the amount of installed capacity, electricity generation, or budget is bided. Each auction volume type has advantages and disadvantages, however, installed capacity-based volume is more convenient for designing auctions [3]. Two main kinds of auction formats consist of sealed bid auctions and dynamic descending price auctions [4]. A carbon price is an amount that must be paid for the right to emit one tone of CO2 into the atmosphere [IPCC, Glossary, 2007]. There are two main types of carbon pricing including emissions trading systems (ETS) and carbon taxes [WB]. An ETS or a cap-and-trade system caps the total CO2 emissions and allows to sell and buy. A CO2 market is established based on this mechanism. A carbon tax sets a tax rate on CO2 emission from fossil fuels use.

After the introduction of FIT for solar PV in 2000, the following years have seen the adjustment of FIT level to bring this price in line with technology cost changes [5] [6] [7] [8]. In order to reach a better control of renewables expansion, in 2015, a trial auction mechanism was introduced for ground-mounted PV [EEG 2014], and large roof-top mounted PV has been started to be applied this mechanism in 2017. For small and medium rooftop projects, project developers can choose one among three pricing mechanisms (FIT, FIP or auction) [EEG 2017] [9]. Recently, Germany has raised the idea and considered a price tag on CO2 emission with the hope that it would be an option for "the second wave Energiewende" and push this country to get the aim of Climate Action Plan 2050 [Cleanenergywire].

Under the FIT mechanism, Germany was successful in encouraging private investment in renewables [10]. However, this price system did not expose investors to a price competition. The over an increase of capacity and generation did not achieve the lowest cost. In addition, the financial burden on consumers because of a higher RE surcharge has been one of the challenges [11]. Compared to the FIT, the auction mechanism which requires competitive pricing is expected to be a better tool to control the investment. Under this system, with given targets, the low-cost projects are favored and approved.

 https://doi.org/10.1051/e3sconf/20186402007

But whatever pricing mechanism is chosen, there is a key issue that needs to be addressed: the payment levels should be based on electricity generation cost. Over the last two decades, the world has seen a significant and fast reduction of photovoltaics technology costs [Bloomberg New Energy Finance, 2015]. Furthermore, forecasting the change level of technology cost is a tough task. In case of awarding excessive profits, the overinvestment will occur. In contrast, if the set price is lower than the expectation of investors, the target of investment will not be met. Therefore, it is extremely important to define a reasonable price adjustment method.

The present study was designed to investigate the impact of different price adjustment policies on the investment in solar power as well as the electricity bill in Germany. In section 2, we outline the linear regression models, input and output variables, the research scope, and an overview of solar power integration. Section 3 is to look at the impact of policy on the investment decision in small-scale solar power projects. Conclusions, policy implications and suggestion for future researches are given in section 4.

2 Analysis framework

Using linear regression models, we examine the effect of price adjustment mechanisms on the diffusion of solar power in Germany. We investigate the relationship between PV system price and FIT, the influence of profit on investment decision. Profitability can be measured through the value of net present value (NPV), or internal rate of return (IRR).

NPV is a measurement of profit calculated by subtracting the present values of cash outflows (including initial cost) from the present values of cash inflows over a period [Investopedia]. In other words, NPV calculation is a standard method to consider whether a potential investment project should be undertaken (NPV > 0) or not (NPV ≤ 0). To compare the profitability over the years, we ignore the project scales and calculate NPV for one unit of investment.

$$NPV_t = \sum_{\tau=t}^{n} \frac{R_\tau - C_\tau}{(1+r)^\tau} - I_0 = \sum_{t=1}^{n} \frac{FLH_\tau * FIT_t - OM_\tau}{(1+r)^\tau} - I_0 \quad (1)$$

NPV_t: net present value of 1 MW put into operation in the year t (million Euro/MW).

I_0: specific investment cost of 1 MW put into operation in the year t (million Euro/MW).

R_τ, C_τ: annual revenue, cost of 1MW in the year τ (million Euro/MW/year).

OM_τ: annual operation and maintenance costs in the year τ (million Euro/MW/year).

FLH_τ: full load hours in the year τ (hours).

FIT_t: feed – in tariff of projects put into operation in the year t (cents/kWh).

r: interest rate of projects put into operation in the year t (%).

IRR is a metric used in capital budgeting to estimate the profitability of potential investments [Investopedia]. IRR is a discount rate that makes NPV of all cash flows from a particular project equal to zero. With a fixed interest rate, the higher IRR, the higher profitability. We invest in a project if IRR > r, and do not if IRR ≤ r.

We investigate the effect of price adjustment mechanisms on the investment using the following functions:

$$\widehat{FIT}_t = \hat{\beta}_1 + \hat{\beta}_2 I_o \quad (2)$$

$$\widehat{\Delta P}_t = \hat{\gamma}_1 + \hat{\gamma}_2 \pi_t \quad (3)$$

$\widehat{FIT}_t, \widehat{\Delta P}_t$: estimated feed-in tariff, new installed capacity of projects put into operation in the year t.

I_o, π_t: average PV system price, profitability (NPV or IRR) of projects put into operation in the year t.

$\hat{\beta}_1, \hat{\beta}_2, \hat{\gamma}_1, \hat{\gamma}_2$: estimated parameters of models.

In order to estimate parameters, the historical data from 2000 are collected. Although the interest rates are different for project scales, investor types, for simplicity, we take 3.5% as the interest rate for solar power projects and remains unchanged [12]. Full-load hours are assumed at 1,000 hours/year and remained unchanged. Operation and maintenance cost is assumed to equal to 1% of investment cost.

The following parts are to illustrate the investment affected by profitablity over last 28 years of solar power in Germany under the FIT mechanism.

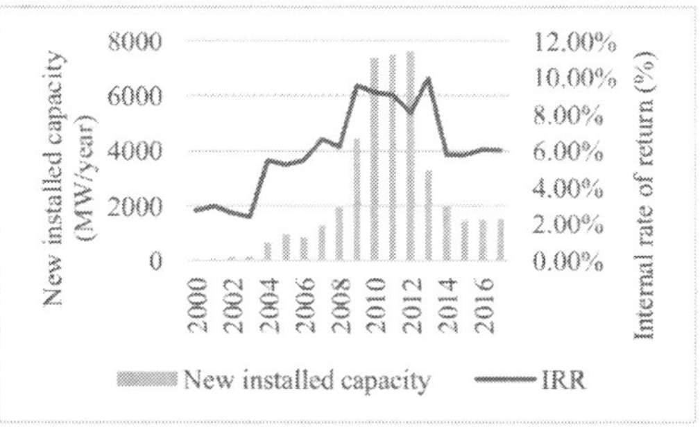

Fig 1. Internal rate of return and annual installed capacity of solar power in Germany

The investors got higher profit between 2009 and 2013 at more than 9% of IRR and a stable rate in recent years at about 6% (Fig 1).

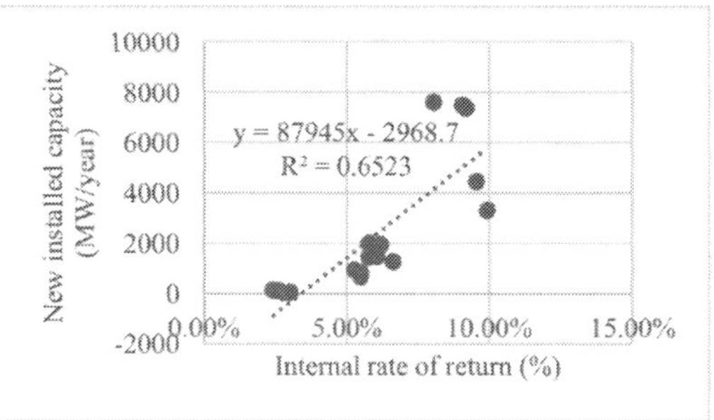

Fig 2. Correlation between internal rate of return and new investment in solar power in Germany

The linear regression function: $\widehat{\Delta P}_t = 87{,}945 IRR_t - 2{,}968$, $R^2 = 65.23\%$. Thus, 65.23% investment decision is explained by profitability (Fig 2).

3 Analysis the impact of price adjustment mechanisms on solar power investment in Germany

Currently, existing three kinds of pricing support mechanism includes FIT, FIP, and an auction in Germany. However, it is regulated that small-scale solar PV projects can choose to be paid FIT. For simplicity, this research takes the small-scale projects (up to 30 kWp until March 2012 and up to 10 since April 2012) for analyzing price adjustment mechanism under a FIT system.

In 2000, with the expected IRR at that time was approx. 5 – 7%, the FIT was set at 0.51 Euro cents/kWh for all kind of solar technology. In 2003, new rates were issued for varied sizes (small, medium and large) and technology (roof-mounted, ground-mounted). After many times of adjustment, the FIT mechanism has played a crucial role in the development of solar power. Designing the FIT mechanism, policymakers have to answer two basic questions: (1) which level the FIT should be set, and (2) how often the FIT should be adjusted. Obviously, the levels of FIT should be in line with the levelized cost of electricity (LCOE).

$$LCOE_t = \frac{I_0 + \sum_{\tau=t}^{n} \frac{C_\tau}{(1+IRR_t)^\tau}}{\sum_{\tau=t}^{n} \frac{E_\tau}{(1+IRR_t)^\tau}} = \frac{I_0 + \sum_{\tau=t}^{n} \frac{OM_\tau}{(1+IRR_t)^\tau}}{\sum_{\tau=t}^{n} \frac{FLH_\tau}{(1+IRR_t)^\tau}} \quad (4)$$

$LCOE_t$: levelized cost of electricity of 1 MW put into operation in the year t (cents/kWh).
IRR_t: internal rate of return of projects put into operation in the year t (%).

Secondly, the FIT adjustments should diminish the gap between the targeted installed capacity and actual investment. In general, the adjustment can be implemented on the time, or the quantity, or a combination of them. The following table illustrates the innovative methodologies to adjust FIT levels for solar PV in Germany.

Table 1. Alternative price adjustment mechanisms for solar power under the FIT regulation in Germany

Method	Applied Time	Frequency of adjustment	Feedback
Pre-determined	2000 – 2008	Annually	No
Feedback mechanism	2009 – 03/2012	Annually	Annual feedback
Feedback mechanism	04/2012 – now	Monthly	Quarter feedback

Beginning with a predetermined reduction rate when solar panels were new and expensive technologies. Then, the feedback mechanism which based on responsive information of actual investment was applied. Each mechanism is evaluated to be suitable for a certain phase of the diffusion. The following parts are to explain the adjustment mechanisms in a graphical way, simple mathematic equations based on the experience from the German Energiewende.

3.1 Pre-regulated degression rate

Kicking off the photovoltaics integration in 2000, Germany introduced an annual reduction scheme of FIT levels. This method was employed until 2008 [EEG 2000, 2005].

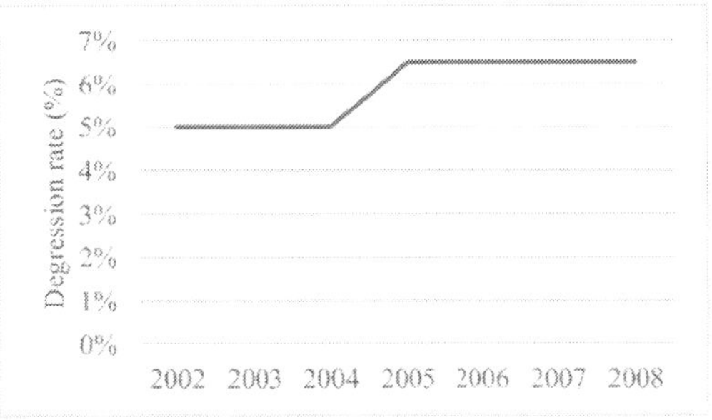

Fig 3. Annual degression rates of FIT for solar power in Germany between 2000 and 2008

The degression rate was 5%/year during the years 2002 to 2004, and 6.5%/year between 2005 and 2008. Under the pre-determined method, the adjustment FIT is defined with the following equation:

$$FIT_{t+1} = FIT_t(1 - d_{t+1}) \quad (5)$$

FIT_t, FIT_{t+1} : FIT levels in the year t and t+1, respectively.
$d_{t+1} = f(\%I_{t+1})$: degression rate of FIT in the year t+1 which is based on the prediction of PV system price reduction ($\%I_{t+1}$). The larger reduction of forecasted PV system price is, the larger degression rate is.

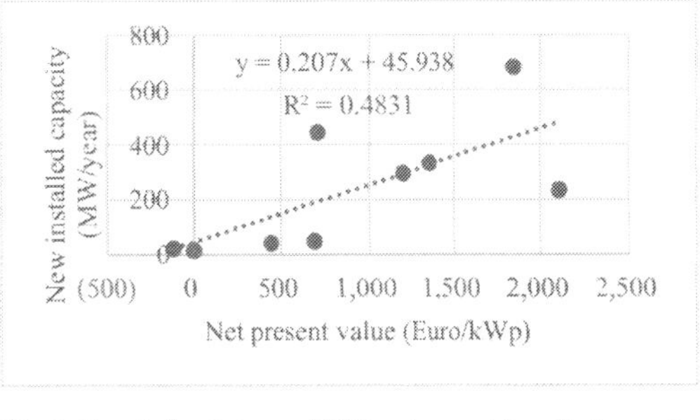

Fig 4. Correlation between NPV and annual installed capacity of small scale solar power in Germany between 2000 and 2008

The linear regression function: $\widehat{\Delta P}_t = 0.207 NPV_t + 45.938$, $R^2 = 48.31\%$.Thus, 1 Euro/kWp increase in NPV led to 0.207 MW/year increase in installation. The number 45.938 reveals that during those years, investors expected to get a higher NPV in the following years. However, during this period, only 48.31% the change of investment behavior was reflected by profitability (Fig 4).

3.2 Feedback mechanism

Because of a fast reduction of PV system cost, in 2009, a new degression scheme was adopted for photovoltaics based on feedback information of actual investment. With an annual given target of 2,500 MW/year and a prediction of PV system price reduction, a benchmark rate of degression was set at 9% in 2011 [EEG 2010]. It depends on the actual investment; the adjustment percentage can be lower or higher.

The FIT under the annual feedback mechanism is formulated as follows:

$$FIT_{t+1} = FIT_t(1 - d'_{t+1}) \qquad (6)$$

$d'_{t+1} = f(\%I_{t+1}, \Delta P^t)$: degression rate in the year t+1 not only depends on the prediction of PV system price reduction in the year t+1 ($\%I_{t+1}$), but also the gap between the target and the actual investment in the year t ($\Delta P^t = P_r^t - P_d^t$) . P_r^t, P_d^t are registered and desired installation capacity in the year t, respectively.

However, due to an unexpected faster technology cost reduction, and deficiencies of the FIT-scheme of a) being updated only annually and b) the degression being too low, a massive overinvestment of solar power between 2009 and 2012 ensured. The actual investment was more than 7,000 MW/year while the targeted installed capacity was only between 2,400 and 2,600 MW/year. During this period, the investors attained high-profit margins, and the surcharge which had to be shouldered by the majority of private consumers rose to levels that became politically unbearable.

After realizing that the high FIT levels were inappropriate, Germany switched to a more frequent adjustment mechanism afterward (monthly adjustment with quarter year feedback information). According to the new law, if the registered capacity was between 2,400 and 2,600 MW in this year, the degression rate in the next period (next three months) would be 0.5%/month. The feedback qualitatively works like: the higher the installation, the higher the degression rate and inversely.

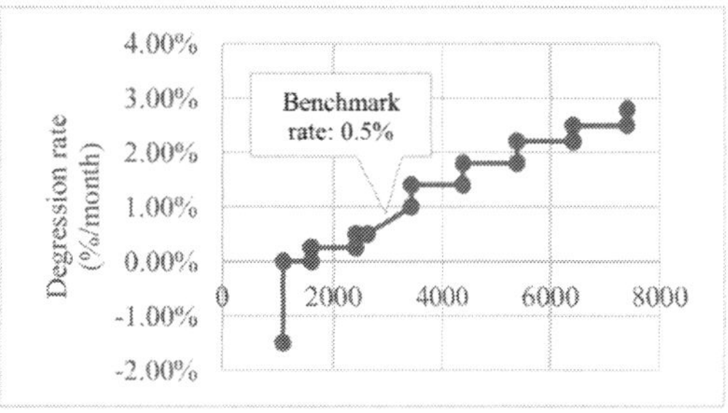

Fig 5. Monthly degression rate of FIT for solar power in Germany in 2014

The equation of the mechanism is as follows:

$$FIT_{m+1} = FIT_m(1 - d''_{m+1}) \qquad (7)$$

FIT_m, FIT_{m+1}: FIT levels at the month m, m+1 (m is the last month of the previous adjustment quarter, m+1 is the first month of the next adjustment quarter) (m = 3, 6, 9, and 12).

$d''_{t+1} = f(\%I_{m+1}, \Delta P^t)$: degression rate at the month m+1 and two following months which not only depends on the expected PV system price reduction in the month m+1 ($\%I_{m+1}$), but also the gap between the target and the actual investment in the year t ($\Delta P^t = P_r^t - P_d^t$). $P_r^t = a_m \sum_{k=1}^{m} P_r^k$, P_r^k is the registered capacity in the month k (k=1÷ 12), a_m = 4, 2, 4/3, 1 corresponds to m = 3, 6, 9, 12.

Thus, instead of depending on the time, under the feedback mechanism, the degression rate relies more upon the volume of installation. In the following part, we will investigate the effect of annual and monthly FIT adjustment on the investment in small-scale PV projects.

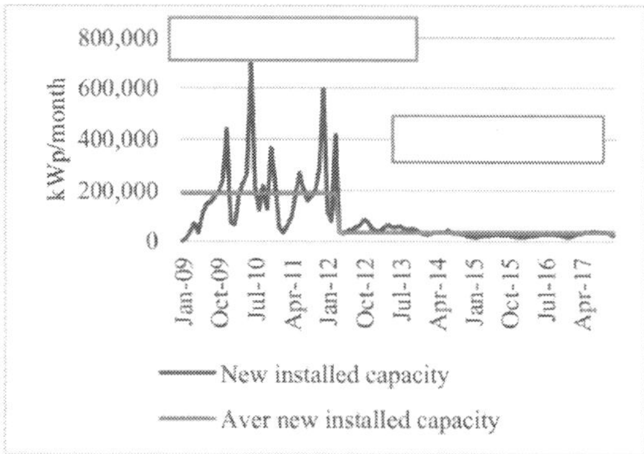

Fig 6. Investment in small-scale solar power projects in Germany between Jan 2009 and Dec 2017

A significant lower average new installed capacity at 34,669 kWp/month, and a noticeable smaller deviation at 1,743 kWp/month of investment under the quarter feedback and monthly adjustment mechanism compared to these values under the annual feedback mechanism (Fig 6).

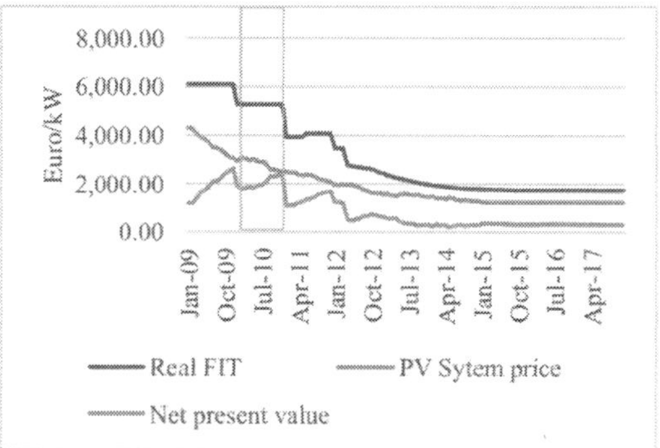

Fig 7. FIT, PV system price, NPV of small-scale solar power projects in Germany between Jan 2009 and Dec 2017

Between 2009 and 2012 the PV system price decreased continuously during the year (from around

2,950 Euro/kWp in January 2009 to 2,500 Euro/kW in December 2009, for example) while the FIT was fixed (at 5,279 Euro/kW during 2010). As a result, investors got more profit with the projects operated in the last months of the years. Since April 2014, we saw an adjustment of FIT in line with PV system price, and NPV was kept quite stable (Fig 7).

Using linear regression models, we see a better correlation between PV system price and FIT under the quarter feedback mechanism than the one of the annual responsive system.

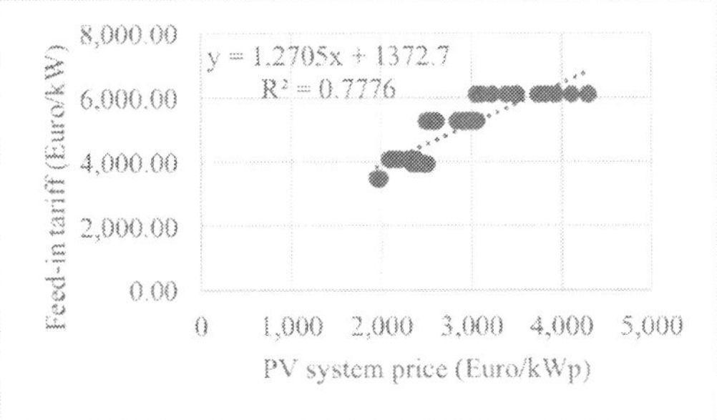

Fig 8. Correlation between PV system price and FIT in Germany between Jan 2009 and Mar 2012

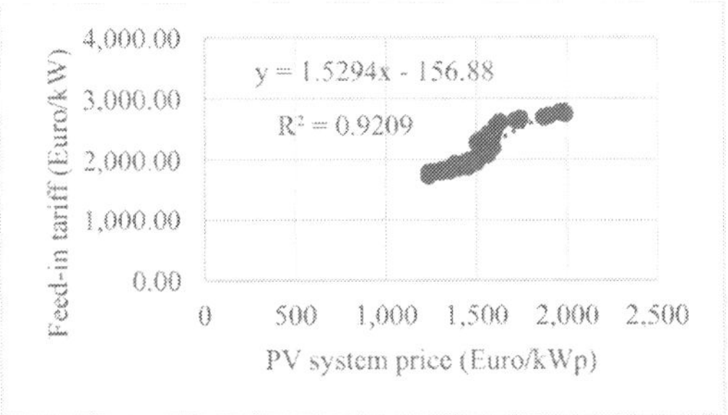

Fig 9. Correlation between PV system price and FIT in Germany between Apr 2012 and Dec 2017

For the period January 2009 – March 2012:
$\widehat{FIT}_m = 1.271 I_m + 1{,}372;\ R^2 = 77.76\%$ (Fig 8)
For the period April 2012 – December 2017:
$\widehat{FIT}_m = 1.529 I_m - 156.88;\ R^2 = 92.09\%$ (Fig 9)

A decrease of 1 Euro/kWp in PV system price caused a decline of 1.271 Euro/kW in FIT between January 2009 and March 2012, and 1.529 Euro/kW between April 2012 and December 2017. Moreover, under the annual feedback system, 77.76% change in FIT is explained by the change in PV system price while the number was 92.09% under a quarter feedback mechanism. In conclusion, a more frequent adjustment leads to a better control of FIT which reflects the cost of production.

4 Conclusion, implication, and suggestions for further works

We admit that the price adjustment mechanisms have played a crucial role in controlling the renewable energy investment and at the same time minimizing policy cost. Our findings include (1) the overinvestment in solar power led to the burden on electricity consumers because they had to pay higher EEG surcharge. The investors benefited from this event, however, the electricity users lost money. (2) The profit was the main impulse of investment decision. Net present value explained 65.23% the change in investors' decision. (3) Under the pre-regulated during the beginning years of the take-off phase, profitability demonstrated only 48.31% of investors' behavior. (4) The more frequent feedback system led to a better control of feed-in tariff. In other words, the feed-in tariff adjustment reflected better the trend of the electricity generation cost. (5) However, because there were no detailed mathematical shapes of feedback mechanisms, the quarter feedback control system was still not effective to control the volume of investment.

In this research, we have applied the linear regression models which are based on open-loop systems. However, for a mature solar power investment market like in Germany, technology costs have decreased slowly, the investment volume has been stable, an open-loop mechanism has shown some limitation that can be addressed by closed-loop systems or feedback control systems. Furthermore, the underinvestment in recent years reveals that apart from profit, existing other factors affecting investment decision such as a lower profit than investors' expectation or a rooftop, land limitation for solar power projects, etc. Therefore, further works are to design detailed mathematical shapes of feedback mechanisms which take into account other factors apart from profitability

References

1. REN21, *"Renewables 2017 global status report 2017,"* (2017).
2. S. Jacobsson and V. Lauber, *"The politics and policy of energy system transformation — explaining the German diffusion of renewable energy technology,"* vol. **34**, pp. 256–276, (2006).
3. L. Kitzing et al., *"Recommendations on the role of auctions in a new renewable energy directive,"* no. **October**, pp. 1–21, (2016).
4. L. M. Ausubel and P. Cramton, *"Dynamic Auctions in Procurement,"* Handb. Procure., no. **February**, pp. 1–21, (2006).
5. B. I. Kühn, *"International market for green electricity. Overview on German policy and opinions among German market actors. Renewable Electricity in Germany History of renewable sources of energy in Germany,"* pp. 1–14, (1999).
6. W. Gründinger, *"What drives the Energiewende? New German Politics and the Influence of Interest Groups,"* (2015).
7. Federal Republic of Germany, *"Act on the Development of Renewable Energy Sources,"* vol. 2014, no. **July**, pp. 1–74, (2014).

https://doi.org/10.1051/e3sconf/20186402007

8. Deutscher Bundestag, *"2017 revision of the Renewable Energy Sources Act."* (2016).
9. S. Tiedemann, *"Auctions for Renewable Energy Support in Germany: Instruments and Lessons Learnt,"* no. **December**, pp. 1–25, (2015).
10. Centre for Solar Energy and Hydrogen Research, *"Renewable Energy Sources in Figures,"* (2015).
11. E. Gawel, K. Korte, and K. Tews, *"Distributional Challenges of Sustainability Policies — The Case of the German Energy Transition,"* no. **September**, pp. 16599–16615, (2015).
12. C. Kost et al., *"Levelized Cost of Renewable Energy Technologies,"* no. **March**, (2018).

Thermal Stratification Performance of a Packed Bed Latent Heat Storage System during Charging

Ashmore Mawire[1], *Katlego* Lentswe[1], *Robert* Lugolole[2], *Denis* Okello[2], and *Karidewa* Nyeinga[2]

[1]Department of Physics and Electronics, Private Bag, X2046, Mmabatho 2745, South Africa
[2]Department of Physics, Makerere University, P.O. Box 7062, Kampala, Uganda

Abstract. Experimental thermal stratification evaluation of a packed bed latent heat storage is done during charging cycles. The packed bed latent heat storage system consists of adipic acid encapsulated in aluminum spheres. Sunflower oil is used as the heat transfer fluid during charging cycles. Stratification number profiles are used to evaluate thermal stratification in the storage system. Charging experiments are carried out with three different flow-rates (4 ml/s, 8 ml/s and 12 ml/s). Charging experiments are also done using the same flow-rate (8 ml/s) with three different set heater temperatures (220 °C, 240 °C and 260 °C). The lowest charging flow-rate (4 ml/s) shows the best variation of the stratification number profile since it shows the least drop from the peak value and the shortest charging interval. Different set heater temperatures show almost identical stratification number profiles. The effect of the charging flow-rate is more significant than the effect of the charging set heater temperature when evaluating thermal stratification for this particular system.

1 Introduction

The charging efficiency of thermal energy storage (TES) tanks is enhanced by thermal stratification. Stratification in the storage tank results from buoyancy effects cause by density differences in the storage. The hotter fluid rises to the top and the colder fluid falls to the bottom. A thermal gradient thus exists between the top and the bottom of the storage tank. The region between these two regions in liquid based TES systems known as thermocline should be kept at a minimal possible thickness for increased efficiency of the systems.

Studies on thermal stratification have been predominantly on water storage for domestic hot water systems [1-5]. For higher medium temperature applications (100 °C to 300 °C), the hot water storage tank is not suitable since water has to be pressurized for these applications making storage tank design complicated. Thermal oil storage tanks can be used since oil vaporizes at higher temperatures than this stipulated range. Limited recent studies have been reported on thermal performance and stratification for medium temperature applications using thermal oils [6-10], hence it necessary to carry out this study to understand the efficiency of these systems.

Latent heat storage has advantages of a larger thermal storage density and nearly iso-thermal behaviour during the storage and release of energy as compared to sensible heat storage. There have been also very few studies on oil based latent heat packed TES systems for medium temperatures [11-13]. Adipic acid has recently been proposed as a suitable PCM for medium temperature applications [14-15], but no study has ever been done using it in a packed bed storage configuration. Sunflower oil is selected as the HTF since it is cheap, food grade, and readily available, non-toxic and possesses characteristics comparable to other commercial HTFs as previously investigated in [16]. The potential application of this storage system is in solar cooking. The storage system can be used for storage of solar thermal energy which can be used at night or during cloudy periods when the sun is not available. The aim of this paper is to evaluate the charging stratification performance a Sunflower oil based packed bed latent heat storage system using adipic acid encapsulated in aluminum spheres. The objectives are to evaluate the stratification performance using different set heater charging temperatures and also using different charging flow-rates.

2 Experimental methods

2.1 Encapsulated phase change materical (PCM) spheres

Adipic acid was poured into forty spherical aluminum capsules with diameters of 0.05 m and wall thicknesses of 0.001 m. The capsules were manufactured as hollow spheres with an opening at the top to allow for pouring of PCM. The volume of the PCM in the capsules was about 80 % of the total internal volume of the capsule to allow for thermal expansion. The volume to be poured into the capsule was measured very carefully with a small measuring cylinder. Adipic acid was in a small crystalline form such that it could be poured easily in its solid form into the capsules. After pouring of adipic acid, four capsules had K-thermocouples fixed onto them to measure the PCM temperature. These thermocouples which extended into the centre of the spheres were fastened on the top of the spheres. The thermocouples acted both as sealing mechanisms and also as temperature monitoring devices. The rest of the PCM capsules were sealed with screw caps. Fig. 1 shows a PCM capsule with a screw cap and one with a thermocouple screw cap.

Fig. 1. (a) Aluminum PCM capsule with screw cap. (b) Aluminum capsule with thermocouple screw cap.

Thermo-physical properties of adipic acid obtained from open literature are shown in Table 1. The melting temperature of adipic acid is reported to be around 152 °C.

Table 1. Thermo-physical properties of adipic acid.

Property	Value
Melting Temperature (°C)	150-152 [15]
Specific Heat Capacity (kJ/kgK)	1.59 (20 °C), 2.26 (150 °C] [14]
Phase change enthalpy (kJ/kg)	213–260 [15]
Density(kg/m³)	1360 (20 °C), 1093 (163 °C) [19]
Thermal conductivity (W/mK)	0.142-0.162 [18]
Average mass of PCM in the capsules (g)	~40

2.2 Experimental setup and procedure

The experimental setup of the TES system is shown in the schematic diagram of Fig. 2.

Forty spherical PCM capsules were inserted into the storage and the top level capsules were at level A below the tank inlet port. The void between the capsules was filled with Sunflower oil up to level A. The electric unit (g) consisted of two copper spiral coils (power rating: 220V, 900 W each) to heat up the circulating oil and it was controlled by a temperature controller. The TZN4S temperature controller (h) (display accuracy: ±0.3 %) was used to set and control the required maximum temperature. An insulated stainless steel cylindrical storage tank (a) with a diameter of 0.3 m and a height of 0.54 m was used. K-type thermocouples (accuracy: ± 1 %) were placed at radial distances of 0.013 m, 0.038 m and 0.064 m to measure temperature at each axial level of the storage tank (Levels A-D). Four PCM capsules with thermocouples were placed centrally at Levels A-D, to monitor the PCM temperature at each level. The thermocouples were calibrated to an accuracy of around ±2 °C. The storage tank was filled with 4 litres of refined Sunflower oil during the experiments. A magnetic-drive pump (b) was used to circulate the Sunflower oil through the heating unit while a VLT Micro drive (c) (maximum error: 0.8% of full scale) was used to control the frequency of the pump so as to set a desired flow-rate. A flow meter (maximum operating temperature: 120 °C) (f) was used to measure the flow rate of the TES system every 10 s and a heat exchanger (e) was connected directly to the flow meter to cool it down so as not to exceed its maximum operating temperature. An HP 34970 data acquisition unit (j) was connected to the thermocouples and it was used to convert analogue signals to digital signals. A personal computer (i) was used to monitor the temperature profile of the TES system and the PCMs and it recorded data every 10 s.

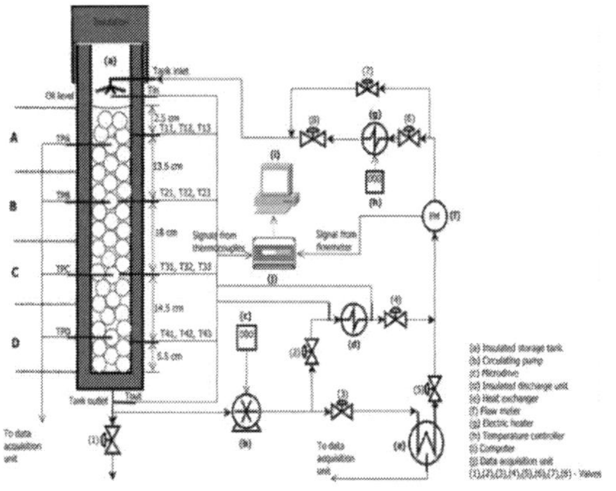

Fig. 2. Schematic diagram of experimental setup.

During the charging cycles valves (1), (2), (4) and (7) were closed while valves (3), (5), (6) and (8) were opened. Charging of the storage system was terminated when the bottom temperature of the storage tank (TD) was around 170 °C. Charging experiments were performed with flow-rates of 4 ml/s, 8 ml/s and 12 ml/s respectively at a set temperature of 260 °C to investigate the effect of the flow-rate. To investigate on the effect of the set temperature, charging experiments were performed at a flow-rate of 8 ml/s with heater set temperatures of 220 °C, 240 °C and 260 °C respectively. Charging experiments were performed from the top to the bottom of the storage tank.

3 Thermal analysis

The temperature at each level is the average of the measured radial temperatures. The stratification number is given by the ratio of the mean temperature gradients at any time to the maximum mean temperature gradient during the charging cycle [17]. The equation is expressed as:

$$Str = \frac{\left(\frac{\partial T}{\partial z}\right)_t}{\left(\frac{\partial T}{\partial z}\right)_{max}} \qquad (1)$$

where T is the average temperature at each axial level and z is axial position. The thermal gradient is expressed as;

$$\frac{\partial T}{\partial z} = \frac{1}{n-1}\left[\sum_{1}^{n-1}\left(\frac{T_{n+1}-T_{n}}{\Delta Z_{n}}\right)\right] = \frac{1}{3}\left[\sum_{1}^{3}\left(\frac{T_{n+1}-T_{n}}{\Delta Z_{n}}\right)\right] \qquad (2)$$

4 Results and discussion

4.1 Effect of flow-rate

Fluid temperatures at the four levels of the storage tank with different flow-rates at a set heater charging temperature of 260 °C are shown in Fig. 3.

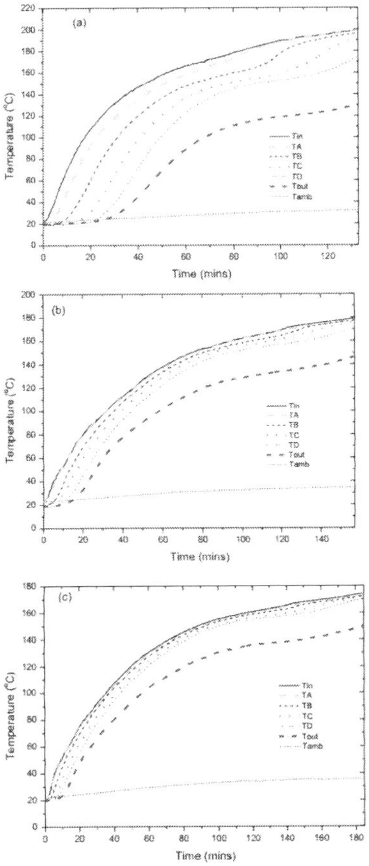

Fig. 3. Fluid temperatures at a set heater temperature of 260 °C with charging flow-rates of (a) 4 ml/s (b) 8 ml/s and (c) 12 ml/s.

An increase in the flow-rate increases the charging time. The rate of temperature rise is faster with an increase in the flow-rate which results in a faster rate of heat loss at higher flow-rates. These heat losses occurring faster with higher flow-rates result in attaining the bottom limiting charging temperature at a later time. Phase change transition commences from the top and occurs later at the bottom of the storage tank since the tank is charged from the top to the bottom. More pronounced thermal stratification and phase change transition around 150 °C is seen with the lowest flow-rate suggesting that charging at the lowest flow-rate is more beneficial for maintaining a higher degree of stratification. The phase change transition occurs earlier and is more pronounced with the lowest flow-rate. For instance, the phase change transition at the bottom of the storage tank starts at around 80 mins for 4 ml/s whereas for flow-rates of 8 ml/s and 12 ml/s it starts at around 100 mins and 110 mins respectively. The lowest charging flow-rate (4 ml/s) also results in higher temperatures at the top of the storage tank at the end of charging.

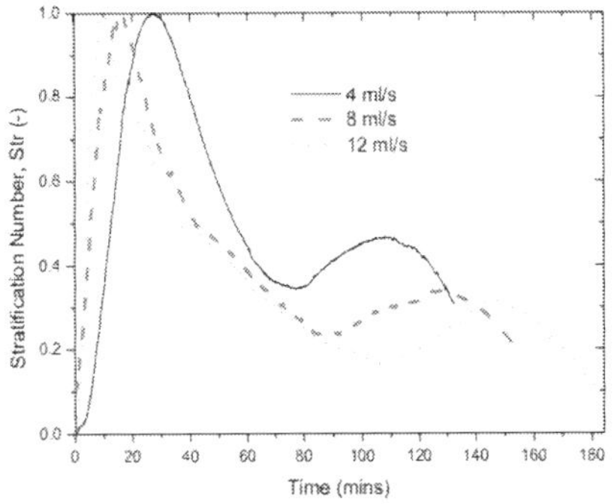

Fig. 4. Stratification number profiles at a set heater temperature of 260 °C with different charging flow-rates.

Stratification number profiles using the three flow-rates are shown in Fig. 4. The fastest rise to the peak stratification number is seen with the highest flow-rate due to the fastest rate of heat transfer. After this initial rise, the stratification number profiles begin to drop due the more pronounced and significant temperature rise at the bottom of the storage tank. The best stratification distribution is seen with the lowest flow-rate, although it has the slowest rise to the peak value. The drop of the stratification number is less with the lowest flow-rate as compared to the other flow-rates. During the phase change process, the thermal gradients of the fluid temperatures tend to decrease and flatten. This behaviour is depicted by the stratification number profiles. For the lowest flow-rate, this behaviour occurs earlier between 60 to 80 mins. For 8 ml/s and 12 ml/s it occurs later between 80 to 100 mins and between 100 to 120 mins respectively. The stratification number values for the lowest flow-rate are higher than the other flow-rates after the peak values making the best charging flow-rate. It is also important to note that there are additional secondary maxima after the peak values due to the rise in the thermal gradients of the fluid temperatures after the phase change process. The secondary peak value is higher for the lowest flow-rate and it occurs earlier at around 110 mins. Additional stratification number drops after the secondary peaks can also be explained by the drops in the thermal gradients of the fluid temperatures.

4.2 Effect of set heater charging temperature

Fluid temperature profiles with different set heater temperatures are presented in Fig. 5.

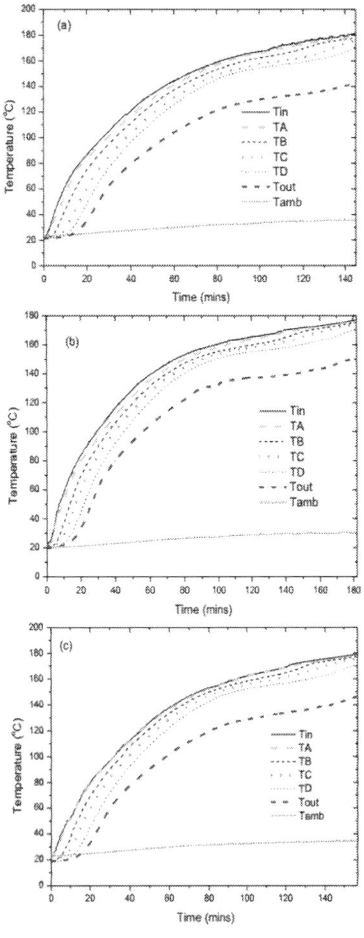

Fig. 5. Fluid temperatures at a charging flow-rate of 8 ml/s with set heater temperatures of (a) 220 °C (b) 240 °C and (c) 260 °C.

Increasing the set heater temperature seems to increase the charging time possibly due to increased heat losses associated with higher set heater temperatures. The charging time for the 240 °C (~180 mins) case is longer than that for the 260 °C (~ 156 mins) case possibly because of the lower ambient temperatures associated with 240 °C . The storage tank was charged for a longer time with 240 °C because the bottom limiting storage tank temperature was reached at a later time due to the higher corresponding heat losses at lower ambient temperature conditions. The temperature rise with the lowest set temperature (220 °C) is slightly higher at the top of the storage but it seems comparable at the bottom of the storage tank. As an illustrative example, the inlet temperature is around 160 °C at 80 mins for 220 °C whereas it is around 150 °C for 240 °C and 260 °C. The outlet temperatures are comparable and they are around 120 °C at 80 mins for the three set temperatures. Different set temperatures only result in a marginal variation of the phase transition times as compared to the flow-rate effect depicted in Fig. 3. The degree of stratification during the charging process seems to be comparable for the three set temperatures.

Stratification number profiles for different set heater temperatures with the same flow-rate of 8 ml/s are shown in Fig. 6. The rate of rise of the stratification number to the peak value is almost identical and the peak values occur at almost the same time of around 15 mins. The drops from the peak values are almost identical up to around 80 mins. These drops are due the bottom level temperatures rising appreciably from 20 to 80 mins thus reducing the axial thermal gradients. The lowest set temperature shows slightly higher values of the stratification from 80 mins to 140 mins, however, the phase change transition flattening occurs at the same time as the 260 °C case. Due to higher heat losses, the phase change flattening and the secondary peak for the 240 °C case occurs later. In general, the stratification profiles for the three set temperatures are comparable and only slightly better performance is seen with the lowest set temperature. The stratification number profile is more dependent on the flow-rate than on the set heater charging temperature. A lower flow-rate induces the phase change process more significantly and thermal stratification is more pronounced. The charging time is also reduced with the lower flow-rate for this particular system.

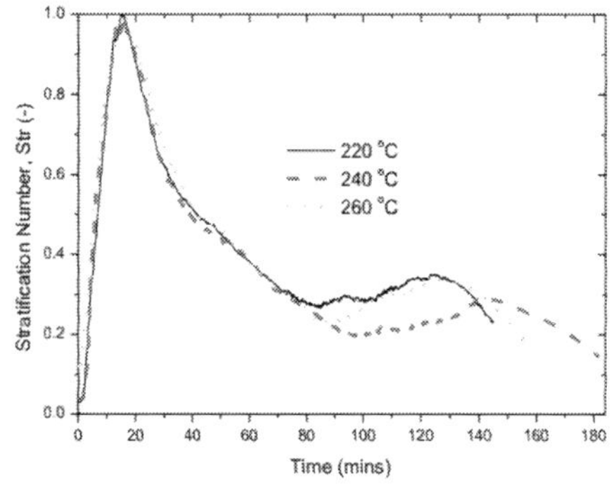

Fig. 6. Stratification number profiles at different set heater temperatures with a charging flow-rate of 8 ml/s.

4 Conclusion

An experimental setup to evaluate thermal stratification in a packed bed latent heat storage during charging cycles was designed. The packed bed latent heat storage system consisted of adipic acid encapsulated in aluminum spheres. Sunflower oil was used as the heat transfer fluid during charging cycles. Stratification number profiles were used to evaluate thermal stratification in the storage tank. Charging experiments were carried out with three different flow-rates (4 ml/s, 8 ml/s and 12 ml/s). Charging experiments were also done using the same flow-rate (8 ml/s) with three different set heater temperatures (220 °C, 240 °C, 260 °C). The lowest charging flow-rate (4 ml/s) showed the best variation in the stratification number profile since it showed the least drop from the peak value and the shortest charging interval. Different set heater temperatures showed almost identical stratification number profiles during charging. The effect of the charging flow-rate was more significant than the effect of the charging set heater temperature when evaluating thermal stratification for this particular system.

Acknowledgment

The authors wish to acknowledge NRF South Africa (Grant No: UGAG160513164998 and IFRR-Grant number: 90638) for making funds available to carry out this research.

References

1. Kong L, Zhu N. CFD simulations of thermal stratification heat storage water tank with an inside cylinder with openings (2017). Procedia Engineering 146; 394-99.
2. Wang Z, Zhang H, Dou B, Huang H, Wu W, Wang. Experimental and numerical research of thermal stratification with a novel inlet in a dynamic hot water storage tank (2017). Renewable Energy 111; 353-71.
3. Bouhal T, Fertahi S, Agrouaz Y, El Rhafiki T, Kousksou T, Jamil A. Numerical modeling and optimization of thermal stratification in solar hot water storage tanks for domestic applications: CFD study (2017). Solar Energy 157; 441-55.
4. García-Marí E, Gasque M, Gutiérrez-Colomer RP, Ibáñez F, González-Altozano P. A new inlet device that enhances thermal stratification during charging in a hot water storage tank (2013). Applied Thermal Engineering 61; 663-69.
5. Jordan U, Furbo S. Thermal stratification in small solar domestic storage tanks caused by draw-offs (2005). Solar Energy 78; 291-300
6. Mawire A, Taole S, Phori A. Performance comparison of thermal energy storage oils for solar cookers during charging (2014). Applied Thermal Engineering 73;1321-29.
7. Mawire A, Taole S. A comparison of experimental thermal stratification parameters for an oil/pebble-bed thermal energy storage (TES) system during charging (2011). Applied Energy 88; 4766-78.
8. Bruch A, Fourmigue´ JF, Couturier R. Experimental and numerical investigation of a pilot-scale thermal oil packed bed thermal storage system for CSP power plant (2014). Solar Energy 105; 116-25.
9. Bruch A, Molina S, Esence T, Fourmigue´ JF, Couturier R. Experimental investigation of cycling behaviour of pilot-scale thermal oil packed-bed thermal storage system (2017). Renewable Energy 103: 277-85.
10. Erregueragui Z, Boutammachte N, Bouatem A, Merroun O, Zemmouri E. Packed-bed Thermal Energy Storage Analysis: Quartzite and Palm-Oil Performance (2016). Energy Procedia 99; 370-79.
11. Shobo AB, Mawire A. Experimental comparison of the thermal performances of acetanilide, meso-erythritol and an In-Sn alloy in similar spherical capsules (2017). Applied Thermal Engineering 124; 871-82.
12. Shobo AB, Mawire A, Okello D. Experimental thermal stratification comparison of two storage systems (2017) Energy Procedia 142; 3295-3300.
13. Kumaresan G, Vigneswaran VS, Esakkimuthu S, Velraj R. Performance assessment of a solar domestic cooking unit integrated with thermal energy storage system (2016). Journal of Energy Storage 6; 70-79.
14. Haillot D, Bauer T, Kröner U, Tamme R. Thermal analysis of phase change materials in the temperature range 120–150 °C. (2011). Thermochimica Acta 513; 49–59.
15. Gasia J, Martin M, Solé A, Barreneche C, Cabeza LF. Phase Change Material Selection for Thermal Processes Working under Partial Load Operating Conditions in the Temperature Range between 120 and 200 °C (2017). Applied Sciences17; 772 (1–14).
16. Mawire A, Performance of Sunflower Oil as a sensible heat storage medium for domestic applications (2016). Journal of Energy Storage; 5:1–9.
17. Fernández-Seara J, Uhía FJ, Sieres J. Experimental analysis of a domestic electric hot water storage tank. Part II: dynamic mode of operation (2007). Applied Thermal Engineering 27; 137-44..
18. Adipure Datasheet (2018). http://adi-pure.invista.com/e-trolley/page_11546/index.html ,website accessed March 20018
19. Chemistry Dashboard (2018)mhttps://comptox.epa.gov/dashboard/dsstoxdb/results?search=Adipic+acid-2%2C2%2C5%2C5-d4 , website accessed March 2018

How Electrical Storage Heaters Can Reduce Wind Curtailment by Satisfying System Reserve Requirements

William D. Kerr , *David* M. Laverty, and *Robert* J. Best

School of Electronics, Electrical Engineering and Computer Science, Energy, Power and Intelligent Control (EPIC), Queen's University Belfast, Belfast, Northern Ireland

Abstract. This paper proposes how using electric storage heaters could provide a portion of the system reserve requirements which would otherwise have to be satisfied by conventional generation. With the upgrade and install of new advanced electrical storage heaters there is the potential to provide large system reserve capabilities when combined with a high-speed communications network. Together with these new more efficient heaters, not only will domestic emissions be reduced, the overall electrical grid CO2 intensity should experience an improvement due to the incorporation of the demand side management (DSM) aspect. The All-Ireland system has been used in this paper as an example of an electrical network which is undergoing a significant transition to high penetrations of renewables, while having a considerable population of potentially upgradeable electric heating sources for DSM control.

1 Introduction

As the world's electrical generation changes to incorporate large amounts of renewable energy, many new operating problems have arisen on electricity networks. The island of Ireland is such an example. Due to strict EU emissions polices both the Republic of Ireland (ROI) and Northern Ireland (NI) have set targets of 40% RES-E (Renewable Energy Sources – Electricity) by 2020 [1], [2]. In order to achieve this ambitious RES-E figure, 37% will be provide by wind generation, with the other 3% coming from various other sources [3]. The issue with this is that wind is the largest contributor to system non-synchronous penetration (SNSP), which is the percentage of total generation permitted from non-synchronous sources. This is exacerbated by the fact that the island has only two DC interconnectors of low capacity with Great Britain (GB). Consequently, the island's power system frequently operates with an SNSP of up to 65%. At this point the SNSP is curtailed due to stability concerns [4]. However, the SNSP is only part of reason wind has to be curtailed, accounting for 15.6% of the total dispatch-down [5]. Dispatch-down refers to the amount of wind energy which is available but cannot be produced due to a number of reasons, detailed in part II. Based on this, a major transition on the whole system (generation and demand) is necessary, as without this the 2020 RES-E targets would be economically impossible to reach due to the amount of wind dispatch-down. This situation will become internationally relevant as larger countries and land masses, such as GB, start to increase their renewable energy penetration to the same sort of levels. Some countries are currently running at much higher SNSP levels, for example Denmark. At the start of July 2015, Denmark was producing up to 140% of its demand from wind generation alone [6], with 42% of 2015 total electrical power generation coming from wind [7]. Denmark can currently cope with this very high wind penetration as it is heavily interconnected with Germany, Norway and Sweden. The problem with relying on interconnectors for overflow power is that wind is likely to become the major renewable source for many countries due to the maturity of the technology. This means there is going to become a point when exporting is no longer a viable option in regard to increasing SNSP, due to the receiving countries experiencing their own problems with wind and solar PV penetration. This is already under consideration from SONI, the NI transmission system operator (TSO), which limits the export capacity on the Moyle DC interconnector when modelling future constraint requirements, due to emerging network limitations in GB [8]. In light of these conceivable problems on the power networks, storage and demand side management (DSM) are considered as a solution.

In Ireland, residential space and water heating accounts for approximately 19% of the total energy consumption of the country [9], [10]. Of this heating, there is a modest amount of installed domestic electric heating on the island, with the latest census data showing Ireland having 13.5% electric central heating and Northern Ireland having 5% electric central heating [11], [12]. A report by consultancy company DNV KEMA Energy & Sustainability, which was commissioned by Glen Dimplex and SSE Plc, also claimed that Ireland has a potential electrical heating replacement of around

 https://doi.org/10.1051/e3sconf/20186403002

1 GW for distributed controllable capacity on the grid [13]. Along with updating current electric heating methods we must consider the potential replacement of typical heating methods with electric heating.

2 Reasons for dispatch-down

In previous years dispatch-down was calculated at the end of the year for a subset of windfarms and extrapolated for the rest, to give constraints and curtailment. From 2016 EirGrid now records real time wind dispatch-down into four main categories [5]:

1) Transmission Constraints:

This occurs due to resolving local network issues and is mainly due to line capacities.

2) Transmission Testing:

Occurs when wind farm testing is carried out by the TSO, e.g. for commissioning and monitoring.

3) Curtailments:

a) *High Frequency/Minimum Generation:* This occurs when attempting to alleviate an emergency high frequency event or when reserve requirements mean a minimum level of conventional generation necessary, priority dispatch or to provide ramping capabilities.

b) *SNSP Issues:* Occurs when the System Non-Synchronous Penetration must be reduced to stay within limits.

c) *ROCOF/Inertia:* Used when the Rate of Change of Frequency (ROCOF) value for the loss of the largest single infeed is unacceptably high and wind must be dispatched down as a result or when the system inertia is too low (below 20,000MWs).

4) Other Reductions:

There are a number of reductions which also fall into this category, however have not been logged due to them being outside the TSO control. The ones which are listed are:

a) Distribution Constraints: Occurs when the Distribution System Operator or the Distribution Network Operator request a dispatch.

b) Developer Outage: Occurs when a wind farm must reduce output, mainly to carry out software upgrades.

c) Developer Testing: Occurs when testing is carried out by a wind farm developer.

3 2016 dispatch-down & wind figures

In 2016 a decrease in wind generation was observed which was mainly due to the reduced wind speeds throughout the year, with the average wind speed across Ireland dropping by 9.8%. The installed wind capacity had increased by 21.3% between 2015 and 2016, with the generation dropping by 6% meaning the wind capacity factor was cut from 33% to 27%. This observation helps give an idea on the problems with reliance on wind generation and the need for a very dynamic system.

When we consider the dispatch-down of wind many previous studies have focused mainly on SNSP problems and the necessary limits needed to meet renewable energy targets based on the predicted installed capacities [3]. However, Figure 1 shows the average hourly dispatch-down energy of wind for the winter months (January, February and December). It can be observed that the three many contributors are TSO Constraints, HighFreq/MinGen and SNSP, with the other categories contributing little of significance to the overall power loss, apart from "Other Reductions" which are unavoidable/ uncontrollable.

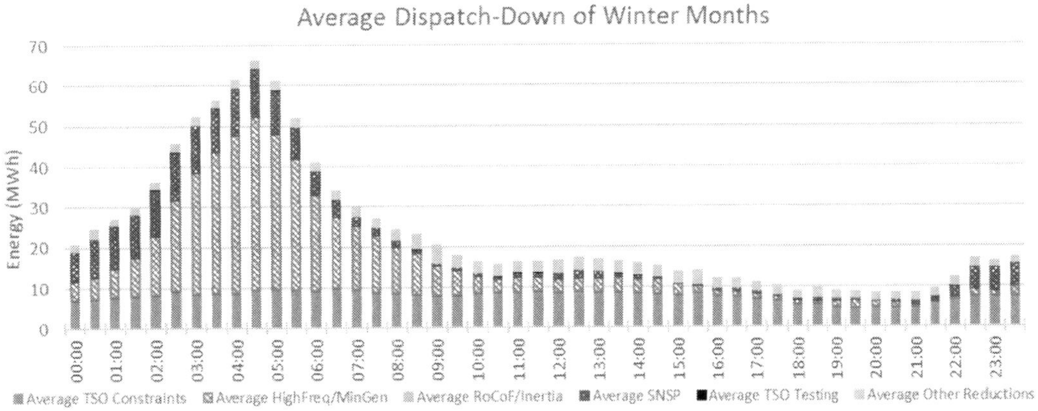

Fig. 1. Daily average dispatch-down during winter months.

The TSO constraints are very constant throughout the average day, with curtailment due to HighFreq/MinGen and SNSP spiking considerably during the slump in night time demand.

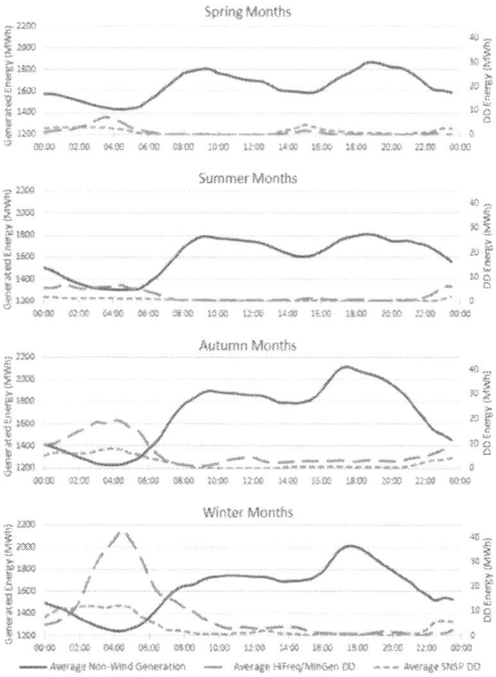

Fig. 2. Daily seasonal average of HighFreq/MinGen and SNSP against non-wind generation.

This trend is even more apparent when viewing Figure 2 which shows the relationship between the dispatch-down due to HighFreq/MinGen and SNSP against the generation of non-wind sources. In 2016 dispatch-down of wind was 2.9% (245.8 GWh), of this 33.8% (83 GWh) was due to HighFreq/MinGen and 14.4% (35.5 GWh) was due to SNSP. The fact that HighFreq/MinGen causes such a substantial amount of energy loss, shows that potentially it should be the focus of more research.

Table 1 presents the percentage breakdown of the wind dispatch-down for each season. The large percentage of TSO constraints in summer correlate with the large number of outages during this period [5]. This is mainly due to the amount of upgrading and uprating of the transmission system, as this is generally the season with the lowest wind generation and demand.

The most interesting point to be made from this table is again related to the curtailment due to the HighFreq/MinGen. From Figure 2, it can be observed that the non-wind generation has a very apparent correlation with HighFreq/MinGen. As the night time non-wind generation decreases for each season the related percentage of HighFreq/MinGen dispatch-down increases, with Autumn having both the lowest average non-wind generation point and the highest percentage of HighFreq/MinGen dispatch-down. Even though the data in this paper is from a year with lower average wind speeds, which may have skewed the impact of SNSP, there is still the premise to investigate the matter further as data from future years becomes available.

Table 1. Breakdown of seasonal dispatch-down.

	TSO Constraints	TSO Testing	HighFreq / MinGen	SNSP	RoCoF / Inertia	Other Reductions
Spring	52.8%	0.1%	16.3%	17.7%	0.0%	13.1%
Summer	72.7%	0.0%	18.8%	4.7%	0.0%	3.7%
Autumn	32.7%	0.0%	44.6%	16.6%	0.5%	5.5%
Winter	33.7%	0.3%	39.9%	17.0%	0.0%	9.1%

Table 2. Breakdown of yearly dispatch-down.

	TSO Constraints	TSO Testing	HighFreq / MinGen	SNSP	RoCoF / Inertia	Other Reductions
Spring	6.5%	0.0%	2.0%	2.2%	0.0%	1.6%
Summer	15.1%	0.0%	3.9%	1.0%	0.0%	0.8%
Autumn	8.3%	0.0%	11.4%	4.2%	0.1%	1.4%
Winter	14.0%	0.1%	16.5%	7.0%	0.0%	3.8%

The cause of HighFreq/MinGen dispatch-down is likely to be mainly due to facilitating the minimum level of conventional generation required, in order to provide dynamic stability, voltage control and system reserve [14]. The potential effects of providing reserve capabilities can be observed in Figure 2, as shown the curtailment of wind begins to increase at around the 22:00 mark for all seasons, approximately when the system demand begins to decrease below generation. However, during the winter months as the curtailment increases the non-wind generation actually increases as well. Source data also shows that around the same time interconnection on average begins to net export to GB and at 23:00 the NightSaver electrical tariff (used for electric storage heaters) comes online. With this observation and the consideration of EirGrid source data, the assumption is that conventional generators are being dispatched above their minimum base load at this time, in order to provide negative reserve (i.e. fossil generation being increase while wind generation is being curtailed) due to the potential volatility of relying on interconnection export load.

4 Benefits of controlled electric heaters on the all-Ireland grid

With the use of DSM capable electric heaters along with the incorporation of a communications network, the potential benefits for both grid and costumer are vast.

Table 2 shows the percentage breakdown of the dispatch-down for the whole year. This indicates that during the colder seasons more dispatch-down occurs, correlating with the demand of space heating. The use of electric storage heaters usually coincides with the use of an off-peak tariff. In the ROI the relevant off-peak tariff is called NightSaver (winter 23:00-08:00, summer 00:00-09:00) and in NI it is Economy 7 (winter 01:00-08:00, summer 02:00-09:00).

Off-peak tariffs are used to incentivise night time load in order to balance the demand profile. If electric storage heaters are used for DSM, their operating time will need to occur during the typical off-peak schedule as losing off-peak demand would likely create greater problems on the grid. Thankfully from Table 3 it can be observed that a huge portion of the dispatch-down occurs during these off-peak hours (NightSaver used).

Table 3. Breakdown of yearly dispatch-down during nightsaver tariff.

	TSO Constraints	TSO Testing	HighFreq / MinGen	SNSP	RoCoF / Inertia	Other Reductions
Spring	2.3%	0.0%	1.7%	1.1%	0.0%	0.5%
Summer	6.5%	0.0%	2.9%	0.7%	0.0%	0.3%
Autumn	3.0%	0.0%	7.5%	3.3%	0.0%	0.3%
Winter	5.7%	0.0%	13.2%	5.7%	0.0%	1.1%

From the DNV KEMA Energy & Sustainability report previously discussed, there are approximately 360,000 electric storage heaters on the ROI grid alone (this does not include the potential NI capacity). With this number of heaters there are huge potential benefits for using DSM with electric storage heaters for reserve capabilities, even when implemented with a basic communications network. If we considered that even just 10% of these heaters were to be DSM enabled/online at any one time, the grid impact in 2016 would have been significant. A new medium sized electric storage heater has an input of around 2.2 kW with a storage capacity of 15.6 kWh [15]. If the 10% of DSM enabled storage heaters were to be this capacity, in 2016, 91% of HighFreq/MinGen and SNSP curtailment during the winter NightSaver time period would have been potentially available (464.8 MWh of the 510.1 MWh). It is less than 100% due to the medium sized electric storage heaters only being able to provide a maximum 39.6 MWh of storage per half hour period. This is based on the assumption that providing a dispatchable load would be categorised as reserve, this would have eliminated the majority of SNSP curtailment during this period (154.8 MWh) and potentially reduce the HighFreq/MinGen considerably.

5 Communications & DSM

For the implementation of DSM incorporated electric storage heaters a low latency communication network would need to be setup. The heaters would then be aggregated together into larger loads "units" which would be of a sufficient size to be included in system scheduling for dispatch capabilities. With the implementation of this conceivable communications network the response of the storage heaters would be included in the primary operating reserve (POR) category (response time within 5 seconds). For this to work correctly, real-time information on the number of online units will be necessary as well as the available capacity, however this would need very little bandwidth in comparison to other types of data transmission. Emerging technologies in the 5G / IoT space are promising in this regard [16].

Even with a basic non-automated system, the electric storage heaters could still be used as reserve and be dispatch-down as an interruptible load similar to Turlough Hill pump storage. The install of new smarter storage heaters also brings the ability of self-contained system stability control with frequency response integrated. This however would need to have offset control in order to allow large amounts of capacity to respond in stages and prevent further large RoCoF changes.

The TSO has set the SNSP ceiling at 75% meaning their target for curtailment is 5% [17], as there will be times when potential wind generation exceeds 75% of demand and exports.

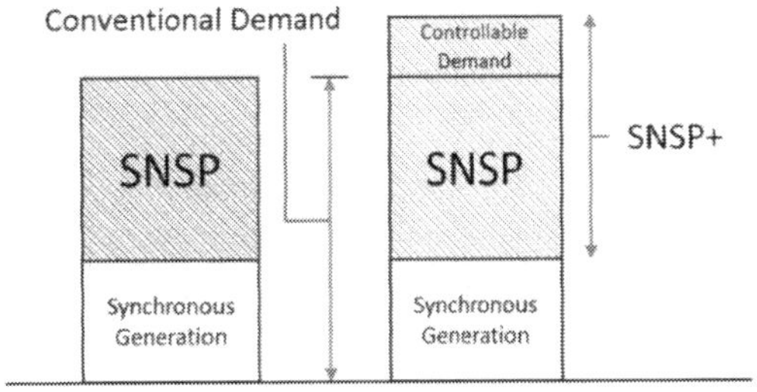

Fig. 3. Definition of "SNSP+" in the system.

Due to this 75% limit, DSM on the network would be necessary to allow for an increased amount of SNSP generation. Figure 3 helps with the understanding of how this would work describing the new generation type as SNSP+. As the controllable demand would also be essentially supplied entirely by SNSP, it can be argued to have zero carbon emission, helping to greatly reduce the system's CO_2 output.

6 Future work

This paper proposes the idea of using electric storage heaters as a means to provide system reserve capabilities using DSM. To give a full understanding for the potential benefits of this scheme future work is necessary:

- Evaluate the single electricity market operator's (SEMO) data to see the dispatch of generator units during the time of curtailment and clarify if they are supplying reserve capabilities.
- Model large scale DSM uptake to produce more realistic assumptions in terms of curtailment reduction and potential frequency response capabilities.
- Consider the potential of providing negative reserve, due to the scheduling of many units. (Having dispatchable load on the system but offline meaning it can be switched on if necessary).

7 Conclusion

The paper has considered the implementation of DSM as a means to provide system reserve capabilities. Household space heating in the form of electric storage heaters has been shown to be a good option for this application, as demand for off-peak electricity coincides precisely with a large portion of the two main sources of curtailment (HighFreq/MinGen and SNSP). The majority of this curtailment has been assumed to be due to the network having to facilitating the minimum level of conventional generation to satisfy reserve requirements. This point was furthered with Table 1 and Figure 2 showing that as night time non-wind generation decreased, the related percentage of HighFreq/MinGen dispatch-down increased.

This paper also considered the reserve category in which this DSM would fit, due to the communication schemes available. It is highly conceivable that it could be used as POR. Even with a slower less automated system it could be used in a similar way to pumped storage, being dispatched-down as an interruptible load.

Acknowledgments

This work was supported and funded by the Engineering and Physical Sciences Research Council (EPSRC) and the Centre for Advanced Sustainable Energy (CASE) project DINOSAURS.

The authors would like to thank EirGrid for providing data to enable this study as well as the CASE grant DINOSAURS. Mr. Kerr would like to thank the EPSRC for funding his PhD studentship.

References

1. Sustainable Energy Authority of Ireland, "Energy targets FAQ," [Online]. Available: http://www.seai.ie/Energy-Data-Portal/Frequently-Asked-Questions/Energy_Targets_FAQ/. [Accessed 2016 March 15].
2. M. Howley, M. Holland and K. O'Rourke, "Renewable energy in Ireland 2012," Sustainable Energy Authority of Ireland, Dublin, 2012.
3. E. McGarrigle, J. Deane and P. Leahy, "How much wind energy will be curtailed on the 2020 Irish power system?," *Renewable Energy*, vol. 55, pp. 544-553, 2013.
4. EirGrid Group, "DS3 Programme," EirGrid Group, [Online]. Available: http://www.eirgridgroup.com/how-the-grid-works/ds3-programme/. [Accessed 23 October 2017].
5. EirGrid; SONI, "Annual Renewable Energy Constraint and Curtailment Report 2016," EirGrid Group, Dublin/Belfast, 2017.
6. A. Neslen, "Wind power generates 140% of Denmark's electricity demand," The Guardian, 10 July 2015. [Online]. Available: https://www.theguardian.com/environment/2015/jul/10/denmark-wind-windfarm-power-exceed-electricity-demand. [Accessed 06 November 2016].
7. A. Neslen, "Denmark broke world record for wind power in 2015," The Guardian, 18 January 2016. [Online]. Available: https://www.theguardian.com/environment/2016/jan/18/denmark-broke-world-record-for-wind-power-in-2015. [Accessed 06 November 2016].
8. SONI, "Northern Ireland Constraints Report 2016," SONI, Belfast, 2016.
9. Sustainable Energy Authority of Ireland, "Residential energy roadmap," [Online]. Available: http://www.seai.ie/Renewables/Residential_Energy_Roadmap.pdf. [Accessed 21 January 2016].
10. M. Howley, M. Holland and D. Dineen, "Energy in Ireland," Sustainable Energy Authority of Ireland, Dublin, 2014.
11. Central Statistics Office, "Domestic building energy ratings quarter 2 2015," [Online]. Available: http://www.cso.ie/en/releasesandpublications/er/dber/domesticbuildingenergyratingsquarter22015/. [Accessed 2016 03 16].
12. Northern Ireland Neighbourhood Information Service, "Central heating: KS404NI (Settlement 2015)," [Online]. Available: http://www.ninis2.nisra.gov.uk/public/PivotGrid.aspx?ds=6665&lh=74&yn=2011&sk=136&sn=Census%202011&yearfilter=. [Accessed 2016 March 16].
13. J. Raadschelders, F. Sikkema and B. i. Groen, "Potential for Smart Electric Thermal Storage, Contributing to a low carbon energy system," DNV KEMA Energy & Sustainability, Arnhem, 2013.

14. EirGrid; SONI, "Operational Constraints Update," EirGrid Group, Dublin/Belfast, 2017.
15. Dimplex, "Quantum Heater," Glen Dimplex Group (GDC), [Online]. Available: https://www.dimplex.co.uk/product/quantum-heater. [Accessed 03 November 2017].
16. W. Ejaz, A. Anpalagan, M. Jo, M. Naeem, S. B. Qaisar and W. Wang, "Internet of Things (IoT) in 5G Wireless Communications," *IEEE Access,* vol. 4, pp. 10310 - 10314, 2016.
17. Eirgrid; SONI, "Volume Calculation Methodology and Portfolio Scenarios DECISION PAPER, DS3 System Services Implementation Project," EirGrid Group, Dublin/Belfast, 2016.

Techno-Economic Evaluation of Energy Storage Systems Built from EV Batteries – Prospective Revenues in Different Stationary Applications

Tam Thanh Nguyen[1,2], *Maik* Naumann, *Cong Nam* Truong[1], and *Andreas* Jossen[1]

[1]Institute for Electrical Energy Storage Technology at Technical University of Munich (TUM), 80333 Munich, Germany
[2]BMW Group, Petuelring 130, 80788 Munich, Germany

Abstract. Battery energy storage systems (BESSs) are already being deployed for several stationary applications in a technically and economically feasible way. This paper focuses on the revenues of industrial BESSs built from electric vehicle lithium-ion batteries with varying states of health. For this analysis, a stationary BESS simulation model is used, that is parameterised with parameters of a 22-kWh automotive battery. The comprehensive model consists of several detailed sub-models, considering battery characteristics, ageing and operating strategies, which allow technical assessment through time series simulation. Therefore, capacity fade and energy losses are considered in this techno-economic evaluation. Potential economically feasible applications of new and second-life batteries, such as photovoltaic home storage, intraday trading and frequency regulation as well as their combined operation are compared. The investigation includes different electricity price scenarios. The combined operation, followed by frequency regulation, is found to have the highest economic viability for the specified electric vehicle battery.

1 Introduction

New manufactured and used batteries from electric vehicles (EV), also called second-life batteries, have the potential to become particularly beneficial for the energy transition. Due to increasing production capacities and declining battery costs, new manufactured automotive batteries based on lithium-ion (Li-ion) are expected to be available for stationary usage in battery energy storage systems (BESSs) in the future.

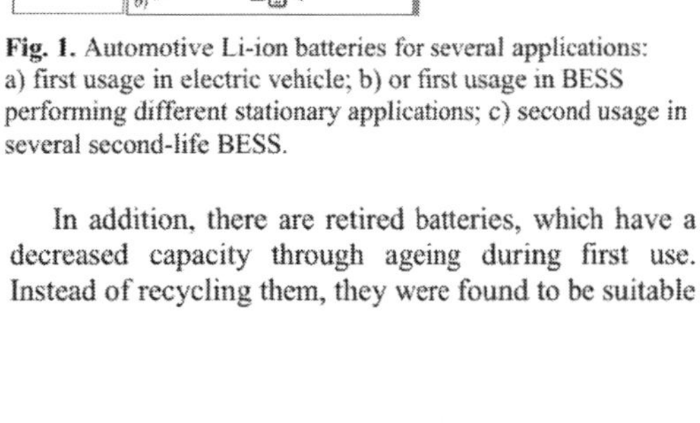

Fig. 1. Automotive Li-ion batteries for several applications: a) first usage in electric vehicle; b) or first usage in BESS performing different stationary applications; c) second usage in several second-life BESS.

In addition, there are retired batteries, which have a decreased capacity through ageing during first use. Instead of recycling them, they were found to be suitable for second use, since they still retain significant capacity. Implemented in a BESS, they are capable to fulfil diverse applications. An overview and description of the most common applications for Li-ion-based BESS is given in [1]. Figure 1 summarises and schematically shows the possible stages of deployment for batteries.

In this regard, synergies between the automotive and power generation sector arise. This paper focuses on the technical evaluation and economic benefit of BESSs built from commercial automotive battery packs of varying states of health (SOH). The research objective is to com-pare their revenues in several applications in order to determine possible revenues of EV batteries, depending on associated performance impairment and their SOH. The economic analysis is underpinned by technical simulations, which reflect the state of charge (SOC), the SOH and the storage energy over time. Technical aspects, such as system losses, cooling losses and continuous reduced capacity through ageing depend on the respective operating behaviour and strategy of the respective application and are crucial for the economic analysis. Under consideration of legal regulations and technical framework, the following applications are evaluated: Photovoltaic (PV) home storage, intraday market (IDM) and primary control reserve (PCR) as well as their combined operation (Combined) are compared through a technical and economic analysis.

For this purpose, a techno-economic model, called SimSES, depicting the above applications, is developed. The study includes the evaluation of varying SOHs,

 https://doi.org/10.1051/e3sconf/20186403003

which is an indication of different dates of removal from the EV, different driving behaviour and different mileages.

At present, there are a number of papers, which review energy storage technologies and summarise options for stationary applications. Similar model approaches in literature often solely discuss PV systems and partially neglect Li-ion ageing [3, 4]. Therefore, this research work will highlight handled cumulated energy, energy losses and degradation after a lifetime of twenty years. The simulations model the full complexity of the BESS's operational behaviour and consider technical and economic levers of four different applications.

The contribution is organised as follows. In Section 2, a brief overview about the techno-economic model and its implemented functions is given. In Section 3, technical input parameters of simulated stationary applications are presented. Section 4 displays the simulation results. Finally, Section 5 gives the conclusions to the analysis.

2 Simulation model

The utilised BESS simulation model SimSES [2] is structured in several sub-models: equivalent circuit battery model, empirical ageing model depicting calendar and cycle ageing, power electronics model and thermal model for the whole system. The sub-models are coupled in order to reflect the high interdependencies and to describe the system behaviour as accurately as possible. Figure 2 describes the simulation framework with all sub-models. Further information about the utilised power flow simulation model can be found in [2] and [5].

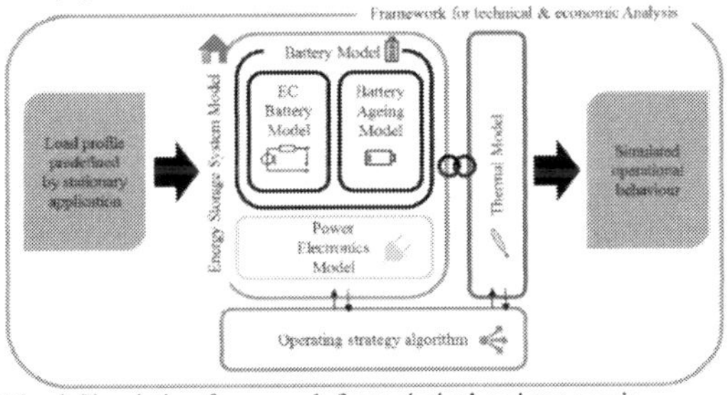

Fig. 2 Simulation framework for technical and economic analysis.

3 Technical model input of stationary applications

The simulation model is parameterised with typical performance characteristics of an automotive 22-kWh Li-ion battery pack from Table 1. The power flow is calculated and saved for each use case. The capacity loss is observed for a simulated operation of twenty years, knowing well, that the remaining lifetime of the reused Li-ion storage might not be sufficient with the little remaining SOH. It shall be investigated, which applications are more stressful to the battery pack than others and if a common End of Life (EoL) limitation at eighty percent remaining useful life is practical or not.

The model input for each of the simulated stationary applications: PV home storage, IDM, and PCR are briefly summarised in Table 2-4.

Precise simulations in an hourly time series are done with the parameter set from Table 2 matching a typical building with 4-6 rental parties with an 8-kWp rooftop PV installation. PV-generated surplus energy is charged and afterwards discharged, as soon as the household load exceeds the generated power by the PV system. Due to the highly volatile pricing in the home storage area, the economic model is executed with four electricity price scenarios for the residential use case. Thus, several possible price ranges for electricity price and feed-in rates in the future are evaluated.

An IDM strategy based on electricity prices from 2017 is implemented as a recharging strategy for scenarios where the storage needs to fulfil a load requirement in order to guarantee German primary control without any impairments. The standalone IDM application was also simulated, but as the revenues did not show to be profitable in Germany except with very precise electricity price forecasts, the use case is not further elaborated. The profitability might be different in other countries like in the U.S. where electricity prices commonly change fivefold depending on time, season and service territory.

The German PCR application is simulated in a quarter-hourly time series unlike the others due to the billing interval which is every 15 minutes.

The last investigated application constitutes a combined operation of all previous three applications. The PCR application has priority because of high remuneration and penalty reasons. PV home storage and IDM follow in this order as recharging strategies to meet the SOC band around 55% SOC for PCR. The same model input according to Table 1-4 is taken as the basis. Table 5 depicts the application hierarchy. The energy and cash flow calculations are done for a medium-sized standard battery pack.

Table 1. Model input: battery pack.

Technology	Li-ion
Nom. energy capacity	22 kWh (partially pooling)
Start SOH	50% - 100%
Lifespan simulated	20 a

Table 2. Model input: PV home storage.

Installed PV peak power	8 kWp (common size [6])
Annual Load	7 MWh (common size [6])
Electricity price scenarios	Constant increase/ linear extrapolation of historical el. price
Feed-in rate	12,20 ct./kWh (Q1 2018) 3 ct./kWh (el. market price)

Table 3. Model input: intraday market (IDM).

IDM electricity price	based on prices from 2017

Table 4. Model input: German primary control (PCR).

Power-to-energy ratio	1 : 2.5
Frequency response	based on frequency regulation from 2015 (repetitive)
PCR compensation	Based on rate from 2017 • minimum price scenario • maximum price scenario • winning bid every week

Table 5. Model input: combined operation (Combined).

Application hierarchy (implemented in control strategy)	1. primary control reserve 2. PV home storage 3. intraday market

Table 6. Technical results: photovoltaic (PV) home storage.

Start SOH	Purchased energy/ Energy fed into grid in MWh	Avg. BESS efficiency	Energy charged/ Energy discharged in MWh	Energy losses of BESS in MWh	Equivalent full cycles (EFCs)	SOH after 20 a	Calendrical ageing/ Cycle ageing share
100%	64.20/ 21.14	86.21 %	48.18/ 41.53	6.64	2,231	57.68 %*	0.55/ 0.45
90%	65.16/ 22.21	86.14%	47.10/ 40.57	6.53	2,181	51.09 %*	0.51/ 0.49
80%	67.48/ 24.83	86.00 %	44.48/ 38.25	6.23	2,059	39.27 %*	0.48/ 0.52
70%	70.82/ 28.64	85.85 %	40.68/ 34.92	5.75	1,883	27.46 %*	0.44/ 0.56
60%	75.34/ 33.86	85.73 %	35.45/ 30.39	5.06	1,641	15.63 %*	0.39/ 0.61
50%	81.40/ 40.87	85.56%	28.44/ 24.33	4.11	1,317	2.32%*	0.33/ 0.67
Solely PV, no storage	105.7349/ 69.3102						

4 Results

The results of the conducted simulations and the economic potential derived from those are presented for each of the applications in the following.

Table 6, 7 and 8 comprise cumulated power flows of purchased energy and fed-in grid energy from the consumer viewpoint as well as charged energy, discharged energy and energy losses from the energy storage viewpoint in MWh after 20 years. Furthermore, average efficiency and number of equivalent full cycles (EFC), calculated

Table 7. Technical results: German PCR.

Start SOH	Purchased energy/ Energy fed into grid in MWh	Avg. BESS system efficiency	Energy charged/ Energy discharged in MWh	Energy losses of BESS in MWh	Equivalent full cycles (EFCs)	SOH after 20 a	Calendrical ageing/ Cycle ageing share
100%	0.13/ 0.01	82.79%	47.96/ 39.71	8.25	2,220	51.66%*	0.83/ 0.17
90%	0.13/ 0.01	82.35%	46.45/ 38.25	8.20	2,150	44.83%*	0.83/ 0.17
80%	0.12/ 0.01	79.95%	39.46/ 31.55	7.91	1,826	36.91%*	0.83/ 0.17
70%	0.11/ 0.01	78.41%	34.02/ 26.36	7.66	1,574	29.66%*	0.82/ 0.18
60%	0.11/ 0.00	73.51%	27.45/ 20.18	7.27	1,270	21.86%*	0.81/ 0.19
50%	0.10/ 0.00	74.19%	21.29/ 14.52	6.7655	985	15.16%*	0.77/ 0.23

$$\eta = 1 - E_{losses} / E_{in-stored} \quad (1)$$

*Limitation: Ageing model is not validated for lower SOH. It is anticipated, that the current rates and operational voltage windows during stationary operation are much smaller than that during use in the vehicle, thus implying a linear character of the cell ageing behaviour under standard conditions.

$$N_{eq} = W_{tot} / (2\, U_N \cdot C_N) \tag{2}$$

are presented in order to evaluate the BESS performance.

The remaining degradation parameters indicate the remaining SOH after 20 years and the ratio of calendar and cycle ageing. Due to the fact that electric vehicles have not been on the market for more than four to five years, the ageing model could not be validated for SOH far below 60%. Still, there are ageing studies of single cells, which indicate, that an optimal operation in stationary BESSs may avoid or retard nonlinear aging characteristics near approximately 80% SOH and extend the life of Li-ion cells [7].

Table 8. Technical results: combined operation (combined).

Start SOH	Purchased energy/ Energy fed into grid in MWh	Avg. BESS system efficiency	Energy charged/ Energy discharged in MWh	Energy losses of BESS in	Equivalent full cycles (EFCs)	SOH after 20 a	Calendrical ageing/ Cycle ageing share
100%	85.67/ 87.11 PV Home 0.87/ 0.04 IDM	81.43%	43.27/ 34.88	8.39	2,003	34.22%*	0.89/ 0.11
90%	86.10/ 87.24 0.81/ 0.03	80.06%	41.85/ 33.51	8.35	1,938	28.10%*	0.88/ 0.12
80%	87.36/ 87.61 0.66/ 0.02	78.39%	38.06/ 29.84	8.23	1,8303	19.11%*	0.82/ 0.18
70%	89.50/88.33 0.44/ 0.01	78.39%	31.82/ 23.83	7.99	1,246	10.19%*	0.84/ 0.16
60%	91.30/ 88.94 0.28/0	71.21%	26.90/ 19.16	7.80	1,3136	01.21%*	0.84/ 0.16
50%	94.90/ 91.70 PV Home 0.01/0 IDM	65.60%	17.02/ 11.16	5.85	788	0%* from 17th month	0.84/ 0.16

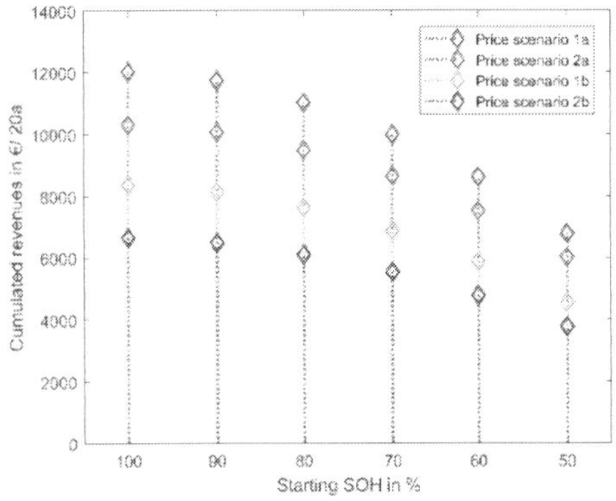

Fig. 3. Cumulated PV home storage revenues of a 22-kWh EV battery after a 20-year lifespan under several price scenarios:
1) linear extrapolation of historical electricity prices;
2) constant increase of historical electricity prices;
a) feed-in remuneration Q1 from 2018;
b) electricity market feed-in remuneration

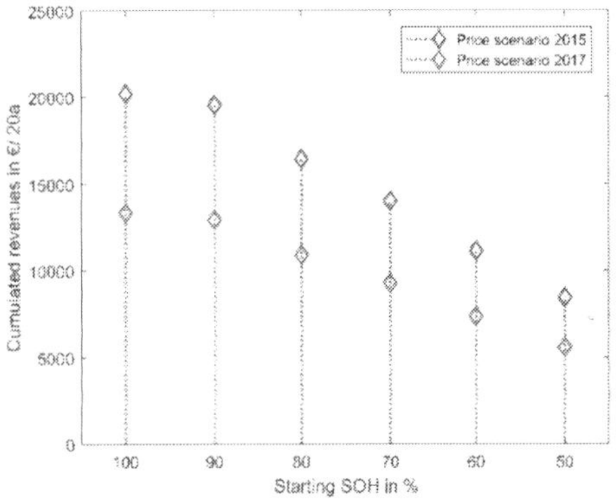

Fig. 4. Cumulated German primary control revenues of a 22-kWh EV battery after a 20-year lifespan for the historic price scenarios from 2015 and 2017.

https://doi.org/10.1051/e3sconf/20186403003

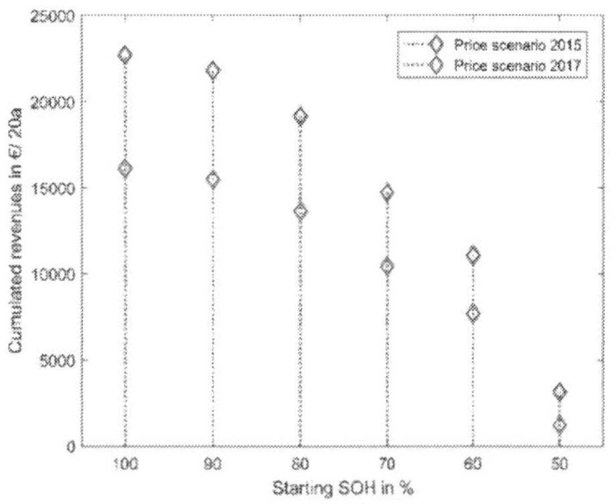

Fig. 5. Cumulated combined operation revenues of a 22-kWh EV battery after a 20-year lifespan for the historic price scenarios from 2015 and 2017; PV home storage price scenario 1a) from Fig. 3 is taken as presumption.

The difference in energy between the last line of Table 6 to the others in the same column depicts the improvement in self-consumption compared to a PV system without energy storage. Using an energy storage system requires 39% less energy purchases and 69% less energy exports from PV generation to the grid compared to solely PV.

The self-sufficiency and cumulated energy decreases with lower initial SOH. Naturally, there are the most number of EFCs and the highest system efficiencies for battery packs with more remaining capacity.

The capacity loss is the highest for the combined operation indicating the most stressful application for the storage due to the greatest utilisation rate overall. In terms of accelerated ageing, the combined operation is followed by the PV home storage application. The applications show a higher capacity loss of 39-48% and more than 59%, respectively. The PV home storage application with characteristics of high power demand and, therefore, a higher share of cycle ageing seem to have a higher impact on lifetime than the PCR application with a continuous and less demanding load profile.

Whether the more aggressively cycling works out to be more economical because more revenue is earned, is better comprehensibly when translating the cumulated energy and ageing behaviour into revenues by means of the economic model. A depreciation period of 20 years is taken because this is the standard service life of PV systems [8]. Further, interest and inflation rates are estimated as mentioned in Table 9 according to other publications [9-12]. The cumulated revenues are calculated as follows:

$$sum(revenues) = \Sigma t\, cash\, flow(t) \cdot (1 + interest\, rate)^{-t} \quad (3)$$

with t representing the number of elapsed years.

Table 9. Economic model input.

Interest rate	4% p.a.
Inflation rate	2% p.a.
Depreciation period	20 years

Figure 3, 4 and 5 present the cumulated PV home storage revenues in Euros (€) of a 22-kWh EV battery after a 20-year lifespan of uninterrupted operation in order to estimate the maximum allowable battery and installation costs for each of the applications. Different economic scenarios are considered in terms of electricity prices. Looking at the results, the combined operation is found to be the application with the highest economic viability followed by German PCR. This is in good agreement with the recent results from Shi et al. [13]. Further, taking also the operating strategy and capacity loss due to ageing into account, PCR represents the sturdier application with high profits and lesser stress on the BESS implying lower costs for maintenance or replacement of faulty battery packs. Another remarkable finding is that the difference in profit between 100% and 80% initial SOH accounts to less than 8.36-8.40%, 15.46-15.52% and 18.50-18.51% for PV home storage, combined and German PCR operations, respectively. This result reaffirms the suggested EoL found in literature at below 80% SOH, especially for the PV home storage and combined applications. The simulations with 50% initial SOH show much lower profits than the other simulations in addition to not being durable enough for a 20-year lifetime. It is advised to replace those battery packs during second use upon exhaustion in order to receive higher returns for operation.

5 Conclusion

This paper concludes that the combined application is the most economically feasible for the deployment of EV battery packs.

Contrasting to the above, the PV home storage use case only shows a positive return on invest if the overall system costs (incl. installation and maintenance) do not exceed 6,000-11,000€ for an 80% initial SOH.

When accounting for battery ageing, the German PCR application is most beneficial based on the conducted techno-economic simulations. A net present value (NPV) of 10,000-18,000€ per battery pack is estimated on the basis of an 80% initial SOH. The preceding German PCR time series simulations reveal a plausible 20-year lifespan for a 50-100% initial SOH. This conclusion can be made at this stage, but project planning for future BESSs needs to consider that the German PCR market is limited.

By means of the comprehensive embedded ageing model, the suggested, but in literature still unverified EoL of 80% SOH, can be affirmed. Further, other estimated NPVs of a new or used EV battery pack are given for all PV home storage, IDM, PCR and combined variations in SOH and aforementioned pricing scenarios. The technical simulations remain limited for the prediction of nonlinear ageing, single cell failures and unforeseeable failures due to other individual components.

References

1. Bauer, M. and Nguyen, T.T., Jossen, A., Lygeros, J.: Evaluating frequency regulation operated on two stationary energy storage systems with batteries from electric vehicles, Journ. of Energy Procedia (to be published)
2. M. Naumann, C. N. Truong, M. Schimpe, D. Kucevic, A. Jossen and H. C. Hesse. SimSES: Software for techno-economic Simulation of Stationary Energy Storage Systems. (International ETG Congress, Bonn, 2017)
3. Madlener, R., Kirmas, A.: Economic viability of second use electric vehicle batteries for energy storage in residential applications. Journ. of Energy Procedia **105**, 3806 – 3815 (2017)
4. Hoppmann, J., Volland, J., Schmidt, T.S., Hoffmann, V.H.: The economic viability of battery storage for residential solar photovoltaic systems – A review and a simulation model. Journ. of Renewable and Sustainable Energy Reviews **39**, 1101–1118 (2014)
5. Truong, C.N., Naumann, M., Karl, R.Ch., Müller, M., Jossen, A., Hesse, H.C.: Economics of Residential Photovoltatic Battery Systems in Germany: the Case of Tesla's Powerwall. Batteries **2**, 14 (2016)
6. Energie-Info: Stromverbrauch im Haushalt; (Bundesverband der Energie- und Wasserwirtschaft e.V.: Berlin, in German, 2014)
7. S.F. Schuster, T. Bach, E. Fleder, J. Müller, M. Brand, G. Sextl, A. Jossen, Nonlinear aging characteristics of lithium-ion cells under different operational conditions, Journ. of Energy Storage **1**, 44-53 (2015)
8. Bundesministerium der Finanzen. (AfA-Tabelle für die Allgemein Verwendbaren Anlagegüter ("AV"); Juris: Saarbrücken, in German, 2000)
9. Diaf, S.; Belhamel, M.; Haddadi, M.; Louche, A. Technical and economic assessment of hybrid photovoltaic/wind system with battery storage in Corsica island. Energy Policy **36**, 743–754 (2008)
10. Dufo-López, R. Optimisation of size and control of grid-connected storage under real time electricity pricing conditions. Appl. Energy **140**, 395–408 (2015)
11. Yang, H.; Wei, Z.; Lou, C. Optimal design and techno-economic analysis of a hybrid solar-wind power generation system. Appl. Energy **86**, 163–169 (2009)
12. Zucker, A.; Hinchliffe, T. Optimum sizing of PV-attached electricity storage according to power market signals—A case study for Germany and Italy. Appl. Energy **127**, 141–155 (2014)
13. Shi, Y., Xu, B., Wang, D., Zhang, B.: Using Battery Storage for Peak Shaving and Frequency Regulation: Joint Optimization for Superlinear Gains, IEEE Trans. on Power Systems **33** (2017).

E3S Web of Conferences **64**, 04001(2018) https://doi.org/10.1051/e3sconf/20186404001
ICPRE 2018

Electromagnetic Field Analysis of Operation Space of the Switch in the Substation Based on the Time Domain Integration Method

ZHANG Jian-gong[1], *Zhao* Jun[1], *GAN* zheyuan[1], *ZHANG* yemao[1] and *LU* ying[2]

[1]China Electric Power Research Institute，Wuhan，China
[2]North China Electric Power University, Beijing 102206, China

Abstract. In this paper, the approximate calculation formula is used to calculate the horizontal electric field from lightning which has arbitrary current, with the help of time domain integral equation method solving the lightning induced over voltages on overhead lines. In this paper, the time domain integral equations of horizontal electric field is established. Taking the conductor axis current as the variable, the pulse function is used as the basis function, and the discretization equation is used in space and time. Using time stepping algorithm to solve the resulting linear equations. The calculation process of this method is simple, and it is proved that the method is consistent with the conclusion of the relevant literature.

1 introduction

Air-insulated substations (AIS) are the most common type of substation. Switching operation in the station will generate a strong spatial electromagnetic field. With the promotion of smart substations, the spatial electromagnetic filed generated by the switching operation may have an impact on the smart components placed in the field. Therefore, the calculation of the space electromagnetic field generated by the switching operation in the substation is the basis for analysing whether the intelligent component can operate safely and reliably. The research of this problem has been reported in many related literatures[1-2].

From the calculation method, the time domain method has many characteristics superior to the frequency domain method in calculating the transient electromagnetic field generated by the switching operation[1]. In particular, there is no need to perform mutual transformation between the time domain and the frequency domain, and the physical quantity changes with time can be clearly expressed. Typical time domain calculation methods are the time domain finite difference (FDTD) method[2] and the multi-conductor transmission line method(MTL)[3]. Both methods are applied in the calculation of space electromagnetic fields in substations[4-6]. However, there is a are error in using this method to calculation the bending and vertical models, and these two line models are common in substations. Therefore, the Time-Domain Integration Equation (TDIE) method for analyzing the electromagnetic scattering of metal structures and surface structures of more complex structures is more suitable for the space electromagnetic field generated by switching operations in substations. However, there is a general problem of late instability in the process of applying time step for, calculation. This paper[7] proposes the method of using average elimination in both time domain and spatial domain; the literature[8] adopts a more complicated form. The wavelet-time-integral equation method has discrete space-time basis functions and has better stability. These methods can be used for time-domain integral equation method to perform spatial electromagnetic field analysis of switching operations in substations.

In the paper, for the spatial electromagnetic field analysis of switch operation in substation, firstly, the approximate calculation formula is used to derive the calculation formula of the space electric field, then time domain integral equation of the metal structure of the line conductor model is established, and in order to improve the numerical stability, the time domain integral equation is applied. The method of solving this integral equation is engineering application examples are given.

2 Calculation method

2.1 time domain calculation of space electric field

In order to establish a time-domain integral equation based on the electric field boundary condition of the conductor surface, it is first necessary to derive the time domain calculation formula of the spatial electric field. Based on the calculation formula of electric field strength and dynamic position, the calculation method of the tangential component of the electric field strength on the surface of the line conductor can be obtained, as shown in equation(1).

$$E_l(r,t)=\frac{\partial A_l(r,t)}{\partial t}+\partial\varphi(r,t)/\partial l \quad (1)$$

Among them:

$E_l(r,t)$ is the tangential component of the electric field strength at the time t of the wire segment at l at the r position; $A_l(r,t)$: Indicated along the wire l Dynamic vector of surface current; $\varphi(r,t)$ Indicated along the wire l Surface axis current dynamic scalar.

In an infinite vacuum, the relationship between $A_l(r,t)$ and the surface current is :

$$A_l(r,t)=\frac{\mu}{4\pi}\int_{l'}\frac{i(r,r',t-R/c)}{R}dl' \quad (2)$$

the relationship between $\varphi(r,t)$ and the surface charge is :

$$\varphi(r,t)=\frac{1}{4\pi\varepsilon}\int_{l'}\frac{q(l,l',t-R/c)}{R}dl' \quad (3)$$

Among them:

ε is the dielectric constant

$R=|r-r'|$ indicates the distance from the field point to the source point.

$I(l,l',t-R/c)$ and $q(l,l',t-R/c)$ separately is located r' line current and line charge at the conductor segment in the $t=R/c$

According to the law of conservation of charge:

$$i(r,r',t-R/c)=-\partial q(r,r',t-R/c)/\partial t \quad (4)$$

Equation(3)can be re-displayed:

$$\varphi(r,t)=\int_t\left(-\frac{1}{4\pi\varepsilon}\int_{l'}\frac{i(r,r',t-R/c)}{R}dl'\right)dt \quad (5)$$

To simplify the representation,

$$\psi(r,t)=-\frac{1}{4\pi\varepsilon}\int_{l'}\frac{i(r,r',t-R/c)}{R}dl' \quad (6)$$

$\psi(r,t)=\partial\varphi(r,t)/\partial t$ 。

By substituting equation(5)and (2) into equation(1), a time domain calculation form of the spatial electric field can be obtained.

2.2 Establishment of time domain electric field integral equation

The incident field and the scattered field on the surface of the thin-line structure conductor have the tangential component of the electric field strength equal[4], namely:

$$\boldsymbol{l}\times(\boldsymbol{E}^i+\boldsymbol{E}^s)=0 \quad (7)$$

In order to solve the equations of the time domain integral equation method, discrete solution are needed in the spatial and temporal domains respectively.

When the airspace is discrete, the conductor axis is divided into N segments, an d the current is selected as the basis function of the piecewise uniform function, which is denoted as ik，k=1，2...N+1. The integral along the wire acts as the inner product of the weight function. Therefore, when only spatial dispersion is considered, equation(1) is transformed into:

$$\frac{\partial\Phi(r,t)}{\partial t}+\varphi(l_k,t)-\varphi(l_{k-1},t)=0 \quad (8)$$

Among them，

$\Phi(r,t)$ is scalar potential, $A_l(r,t)$ is the result after the weighted integral operation; the scond term and the third term are integral results of the potential derivative along the direction of the wire segment l .which is:

$$\int_{l_k}\frac{\partial\varphi(r,t)}{\partial l}dl=\varphi(l_k,t)-\varphi(l_{k-1},t) \quad (9)$$

When the time domain is discrete, the time stepping method is applied, and the weight function is the $\omega_j(t)=\delta(t-t_j)$ angle j is the number of time steps calculated.

Equation(8)can form the following differential relationship:

$$\frac{\Phi_{k,j}-\Phi_{k,j-1}}{\Delta t}+\varphi_{k,j}-\varphi_{k-1,j}=0 \quad (10)$$

Equation(6)can form the following differential relationship:

$$\varphi_{k,j}=\varphi_{k,j-1}+\psi_{k,j}\bullet\Delta t \quad (11)$$

Substituting equation (11) into equation(10):

$$\frac{\Phi_{k,j}-\Phi_{k,j-1}}{\Delta t}+\varphi_{k,j-1}+\psi_{k,j}\bullet\Delta t-\varphi_{k-1,j}=0 \quad (12)$$

Combining equation(2)and (6)to get the following:

$$\Phi_{k,j}=\sum_{i=1}^{N}\frac{\mu}{4\pi}\int_{l_k}\int_{l_i}\frac{i_{i,j-n}}{|r_k-r_i|}dl'dl \quad (13)$$

$$\psi_{k,j}=-\sum_{i=1}^{N}\frac{1}{4\pi\varepsilon}\int_{l_i}\frac{i_{i,j-n}}{|r_k-r_i|}dl' \quad (14)$$

Among them:

$$j-n\triangleq(j-n)\bullet\Delta t\approx t_j-|r_k-r_i|/c.$$

n is a value indicating a time step difference;

n_{max} describe the ratio of distance between the conductor segment k away from the conductor segment and the k segment with c

$n_{min}=0$ is the distance between the axis of section k and the surface field point.。

Define two time-independent variables:

$$K^a_{k,j}=\frac{\mu_0}{4\pi}\int_{l_k}\int_{l_i}\frac{1}{|r_k-r_i|}dl'dl \quad (15)$$

$$K^b_{k,j}=-\frac{1}{4\pi\varepsilon}\int_{l_i}\frac{1}{|r_k-r_i|}dl' \quad (16)$$

Equation(13)and(14)can be re-displayed as follows::

https://doi.org/10.1051/e3sconf/20186404001

$$\Phi_{k,j} = \sum_{i=1}^{N} i_{i,j-n} K_{k,j}^{a} \quad (17)$$

$$\psi_{i,j} = \sum_{i=1}^{N} i_{i,j-n} K_{k,j}^{b} \quad (18)$$

Substituting equation (17)and (18)into equation (13),The time domain integral equation can be solved.

2.3 solution of time domain electric field integral equation

To solve equation (13) the following two aspects need to be considered. First, calculation $\Phi_{k,j}$ and $\psi_{k,j}$ when the current j moment is unknown, the processing method needs to be considered. Second, due to the existence of time integral. In order to obtain a stable numerical solution, the numerical oscillation of the late response must be considered.

The processing of the first problem is relatively simple, and the equation (13) is arranged to obtain a current at time j represented by the current at j-n(n>0), thereby solving the problem.

The second problem is the key to the numerical stability, and the current obtained is approximated, $t = t_j$, the current on the k segment can be expressed as:

$$\tilde{I}_{k,j} = \frac{1}{4}(\tilde{I}_{k,j-1} + 2I_{k,j} + I_{k,j+1}) \quad (19)$$

After calculating the current at time j+1, the current at time j is re-corrected and the used for subsequent iterative solution, the post-time instability can be eliminated to some extent with using it.

3. Algorithm verification

By comparing the induced lightning overvoltage of overhead lines in the same situation, the method is compared with the method provided in [9], verify the calculation method provided in this article

Assume that the length of the overhead line is 1000m, the diameter of single wire is 1cm, the height from ground is 10m, the lightning strike point hits the middle of the line 60m from the ground directly above the line, the lighting current propagation speed is $1.3\times10^{8}\,m/s$ .the maximum time derivative is $40kA/\mu s$ the peak is $12kA$. The lightning overvoltage waveform of the line calculated by simulation is shown in figure 1:

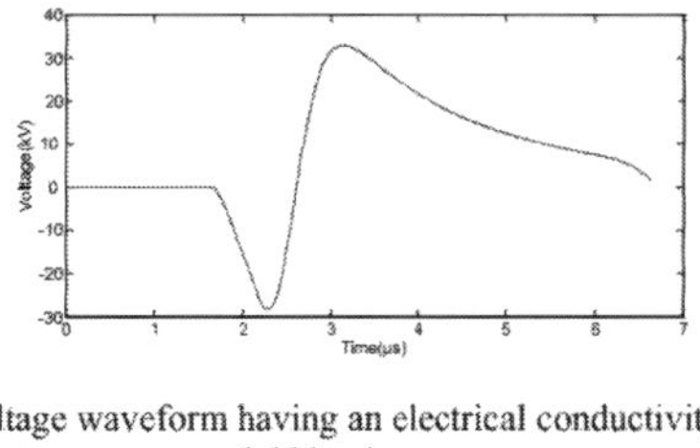

(a) A voltage waveform having an electrical conductivity of 0.001 m/s

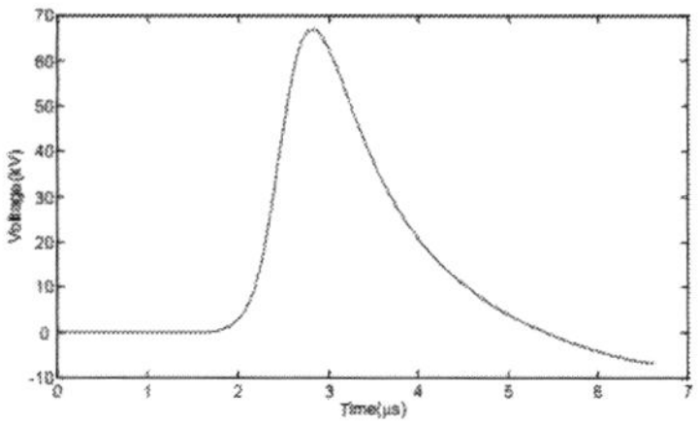

(b) Voltage waveform at infinite conductivity

Fig.1 Lightning induced overvoltage waveforms at different earth electrical conductivity

The calculations show that the difference between the method and the method provided by the literature[10]is very small, which verifies that the method is correct and feasible.

4 Engineering applications

The above calculation method can be used to calculate the spatial electromagnetic field generated in the intelligent component cabinet area during the switch operation of the AIS type intelligent substation. taking a 110kV intelligent substation as a example, two different arrangement of overhead lines are shown in figure 2.

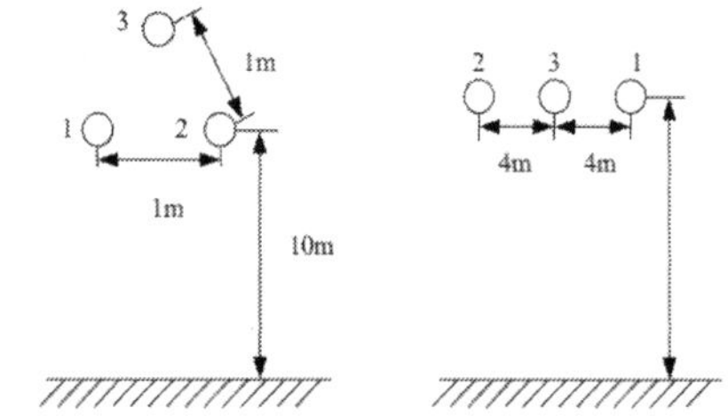

Fig. 2 Two permutations of three conductor overhead lines

Assume that the length of the overhead line is 1000m, the diameter of single wire is 1cm, the height from ground is 10m, the lightning strike point hits the middle of the line 60m from the ground directly above the line, the lighting current propagation speed is $1.3\times10^{8}\,m/s$ decaying Changshu is 0.43. the corresponding matching load is connected at both ends of the line, and the soil is considered to be well conductive. The simulated terminal lightning overvoltage waveform obtained by simulation is shown in figure 3 and 4:

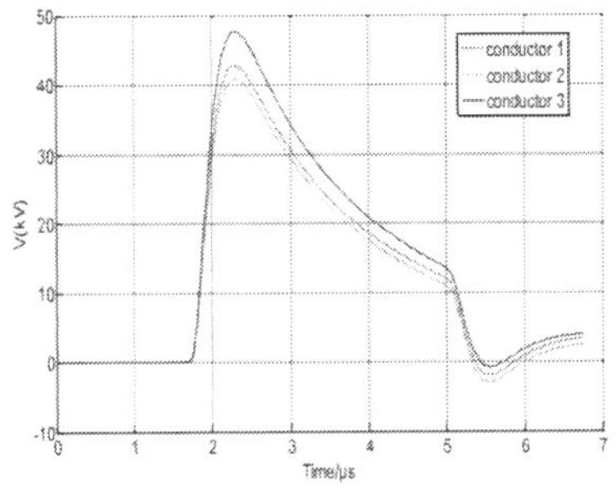

Fig.3 Sensing overvoltage of triangular arrangement overhead line end

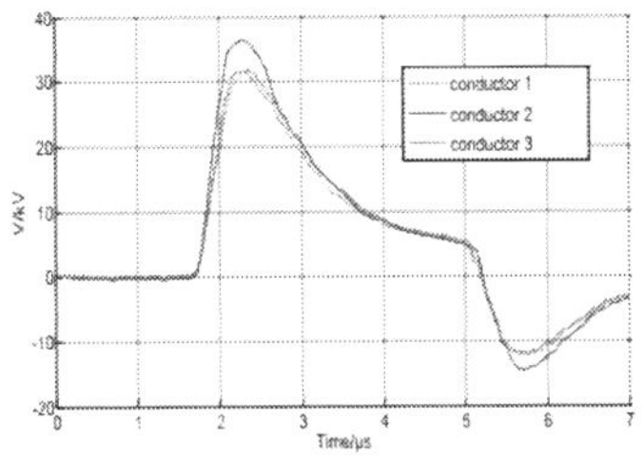

Fig. 4 Sensing overvoltage of Horizontal arrangement overhead line end

5 Conclusion

If a certain point in the transmission line is struck by lightning, considering the influence of soil –related physical parameters and long-line horizontal parameters, the derived line-sensing horizontal electric field calculation formula is applicable to any form of lightning. Solving the integral equation to obtain a lightning induced overvoltage by stablishing the time domain equation of the calculation formula.

Acknowledgement

This work was supported by the Project of SGCC: Research on the influence of VFTO on On-site protection equipment and test technology(JBB17201700082).

References

1. Wang wenbing. Transient electromagnetic field[M]. Xi'an jiaotong University Press, 1991.
2. Taflove A. Computational electromagnetics: The finite-difference time-domain method[J]. Lmn.pub.ro, 1995.
3. Paul C. Analysis of Multiconductor Transmission Lines[M]// Analysis of multiconductor transmission lines /. Wiley, 1994:45 - 54.
4. Gan zheyuan, Zhao Jun, Zhang Jiangong, et al. Calculation of transient electromagnetic radiation in shell space during operation of GIS isolation switch[J]. High Voltage Technology, 2016, 42(3):980-986.
5. Musa B U, Siew W H, Judd M D. Computation of Transient Electromagnetic Fields Due to Switching in High-Voltage Substations[J]. IEEE Transactions on Power Delivery, 2010, 25(2):1154-1161.
6. Liu L, Cui X, Qi L. Simulation of Electromagnetic Transients of the Bus Bar in Substation by the Time-Domain Finite-Element Method[J]. Electromagnetic Compatibility IEEE Transactions on, 2009, 51(4):1017-1025.
7. P. J. Davies. On the stability of time-marching schemes for the general surface electric-field integral equation[J]. IEEE Trans. Antennas Propagat., 1996, 44:1467-1473
8. Yair Shifman, Yehuda Leviatan. Transient Analysis of EM Pulse Penetration Into a Conducting Layer Using Wavelet-Based Implicit TDIE and Iterative IMC Technique[J]. IEEE Transactions on Antennas and Propagation, 2004, 52(4):1105-1111
9. Farhad Rachidi, Carlo Alberto Nucci, Michel Ianoz, Carlo Mazzetti. Influence of a lossy ground on lightning-induced voltages on overhead lines[J]. IEEE Transactions on Electromagnetic Compatibility, 1996, 38(4): 250-264.

A research based on testing method for the whole substation cascaded in digital substation

Yuekang Wu[1], *Shuheng* Chen[1], *Shi* Jing[1] and *Xueting* Shi[2]

[1]Sichuang Provincial Key Lab of Power System Wide-Area Measurement and Control, School of Mechanical and Electrical Engineering, University of Electronic Science and Technology of China (UESTC), Chengdu 611713, China
[2]Sichuang Provincial Key Lab of Power System Wide-Area Measurement and Control, School of Mechanical and Electrical Engineering, University of Electronic Science and Technology of China (UESTC), Chengdu 611713, China

Abstract. Substation as an important component of the power grid, its safe and steady operation is crucial to the power grid. After building new substation or renovating the old ones, they needs to be tested to ensure its safe operation. The method, using relay protection tester to test the substation equipment nowadays, is not available to the whole cascade related connecting equipment, and the test work need a lot of manpower and material resources. To make up for all kinds of shortcomings when testing, this paper proposes a digital substation-oriented testing method for whole station concatenation, and established a testing software to test the whole cascaded substation, especially the method can realize closed-loop simulation test for whole station. In this method, the potential faults of digital substations can be predicted by using electromagnetic transient simulation to simulate various faults of substations, and the state of primary equipment can also be tested. The testing time can be shorten and the accuracy of testing can be improved, as well as, the safe and steady operation of substation equipment can be ensured by using the method proposed in this paper.

1 Introduction

With the development of economy, the power grid plays an important role in the whole society. In the six departments of power generation, transmission, transformation, distribution, power consumption and dispatching, transformation department occupies a very critical position when smart grid is becoming a trend of China's power grid. As the important part of the construction of strong smart grid, digital substation has been greatly developed in recent years [1]. According to the smart grid planning report of the State Grid Company: "equipment information, operation and maintaining strategy are fully interactive with power dispatching, realizing the state- based, comprehensive and optimized management of life cycle. The hub and central substations are built or transformed into digital, intelligent substations, which realize the unified data collection, real-time information sharing, real-time control and intelligent regulation of the whole network.

They also support the safe and stable operation of all levels of power grid and various advanced applications" [2]. So it is particularly important to carry out correct and complete testing of digital substations.

At present, few of the methods, used in conventional substation testing with relay protection device, have realized the complete test of the organization, distribution and transmission of the input signal of the protective device, as well as the output signal [3-4]. The testing method for relay protection device has some defects which needs to be improved, and compared with [5-6], almost none of method uses the closed-loop testing method to do the test for digital substation. The drawbacks of testing ways may lead to technical problems in the construction step of digital substations, which threaten the operation safety of digital substations, and lead to major equipment accidents in some serious cases. Therefore, simulating and predicting different variety of possible and potential faults can ensure the safety and steady operation of digital substation, at the same time the check of digital substation equipment is also the key to ensure the substation operate normally. This paper proposed a method and established a software platform which can be simulated and tested in steady and transient state focusing on whole station cascaded digital substation. The platform, acting as the main station of the simulation system, carries on the data communication with each wireless sub-station, simulates the data uploading of collector simulator of the primary system, and realizes support of the switch configuration and on-off state reading synchronously, so that it can realize the integrity of the digital substation testing include secondary system.

2 System design

2.1 System construction

The main idea of the testing method, proposed in this paper, is to treat the secondary system of the digital substation which is tested as a whole, and also to bring the merge unit (MU) and the network switch into the detection scope for maintain the integrity of the secondary system connection while testing. The testing method will utilize testing simulation platform which established by C++ language and the algorithm is based on method of Electro-Magnetic Transient Program (EMTP) to simulate the running status of the connection line in digital substation and transmit the simulation result to the collector simulator through wireless transmission. Also, synchronous injection of experimental signal to the MU of tested substation by multiple collector simulators with distributed layout is adopted in this method. This method also will use the intelligent switch box to return the statues of breakers to collectors. Then the status will be backed to the platform in order to check on/off statues of relay protection device under fault condition. Eventually this method can check the correctness of organized information, the distributed function of MU and network switch by using the network analyser, fault oscillography and the monitoring system of substation secondary system.

The testing system shown in Fig. 1 can be used to test the whole performance of the secondary system of digital substation and test the behaviour of relay protection devices.

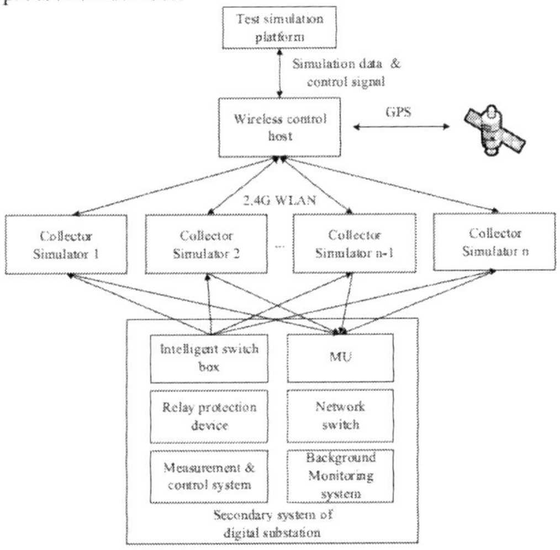

Fig. 1. System synchronization and overall test diagram.

The testing system of digital substation showed above can realize the accuracy of testing. The key factor of this process is to transmit the time domain simulation data to secondary system simultaneously. In order to realize the experimental scene close to the actual operation scene of digital substation, the data of any time period obtained by each MU must be synchronized, and the time interval between the two segments must also comply with the sampling rate requirement of the MU. At the same time, the accuracy of switching behaviour which intelligent switch boxes act depends on the synchronous clock. In this paper, the wireless control host provides synchronous clock to each collector simulator through Global Position System (GPS) timing [7-9]. This paper makes a long-distance experiment of the synchronous clock so that it can simulate the distance between the real substation monitoring room and the testing position. The test indicates that the time difference is within 10 minutes between the distance, which is wireless host to each sub-terminal, of 50-70 and 5-10 meters, while the speed of date uploading are the same.

2.2 Introduction to testing method

The whole station cascaded testing method of digital substation is a digital substation integrated performance testing technology, based on power network simulation, wireless synchronization, transmission and decentralized injection. The main function of the method is to treat the secondary system of digital substation as a whole. The method not only test the function of digital substation equipment, but also make the MU and network switch whose function can take the organization and distribution of information in digital substation into the testing scope completely. In order to test the overall performance of the digital substation, it's necessary to maintain the integrity of the connection and the input information while testing the secondary system of digital substation. This test can realize the electromagnetic transient simulation of digital substation and the result can be sent to each sub-terminal in wireless form. Then the terminal transmits the information to the MU through optical fibre, thus completing the integrity performance test of the secondary system of digital substation. The test system includes the notebook computer, the substation simulation platform, the wireless control host and the collector simulators.

The flow chart of the test method is shown in Fig. 2.

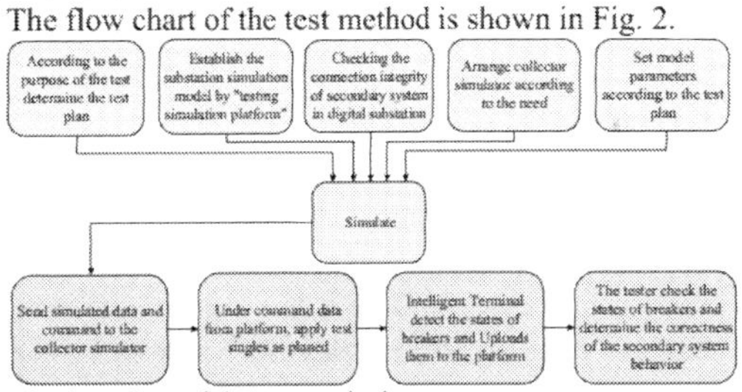

Fig. 2. Flow chart of the test method.

The method can improve the efficiency of the system testing as well as ensure the integrity of the system testing. The purpose of the test is designed and the method is determined in advance, according to the scheme, the corresponding model and parameters are set up on the testing simulation platform. The collector simulators will be reasonably arranged depending on the actual needs of the scheme, and then the simulation will be initiated. The data will be injected into each simulators synchronously after simulation. When the operations mentioned above have been finished, the test will start, the operation will be carried out rely on the presetting control instruction, and then the parameters and states will be collected and transmitted back to the testing simulation platform. The tester will determine the validity of the secondary system behaviour according to the action behaviour of the intelligent terminal.

2.3 Closed-loop simulation

At present, most of the relay protection testing devices used in China are protective tester [10]. It still stays in the test stage of open loop on single device, and can only complete the basic function test of protection device. However, it is hard to test the whole performance of power system equipment, to carry out the closed-loop test, as well as the system-level test of relay protection device. Hence, in order to improve the stability and reliability of secondary equipment in substation, it is necessary to study the closed-loop system test for secondary equipment in substation. Especially for the digital substation based on IEC 61850 [11], It's vital that the test for the closed-loop system of relay protection device must be done. The method proposed in this paper are fit for system-level, which can do the basic function test, on single unit, and systematic test simultaneously. It can also ensure that the software and the hardware of whole substation work well. The closed-loop for simulation is shown in Figure. 3:

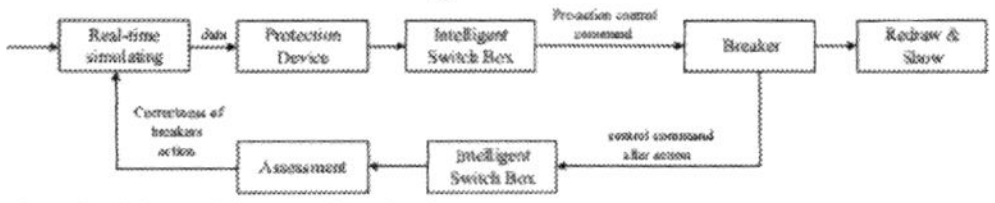

Fig. 3. Closed-loop simulation.

The simulation platform carry out simulation by the given parameters and setting, after that it will upload the result to protection device, then the protection device then transfers the switch action instructions to the intelligent switch box, the breaker acts according to the pre-action control command which is uploaded by the intelligent switch box. When the action completes, the platform will redraw the breaker state, meanwhile, the action statues will be returned back to the intelligent switch box, box will continually send the data back to assess. In assessment part, people will judge whether the acting of breakers is right. If the breakers have done incorrectly, the assessment segment will give the alarm to tell people which breaker is wrong and find out the causes. Then it will send the right statues of breakers to the platform for simulation again in order to do the correct experiment. Hence the closed loop have been achieved.

Through closed-loop simulation, it is possible to test the whole system and avoid under the condition of only testing the performance on a single device. At the same time, the whole information transmission loop should also be involved in the test scope, which can quickly find out the secondary equipment fault and feedback to the testers. For example, data transmission problems like damaging optical fibre which cause faults like device protection locking and unusual action and so on, it will be quickly found and warning message will be informed instantly. Optical cables have been used for signals transmitting in digital substation instead of the power cables, hence it's crucial to test the secondary equipment and secondary circuits, including optical fibre links. By adopting the testing method provided in this paper, the electric devices can be quickly tested according to the feedback result as well as information transmission loop. Meanwhile, the efficiency can be improved, the system testing becomes more comprehensive.

3 Transient model establishment

With the gradually increase of power grid voltage level and the capacity, the transient process decay time increases, and the influence of transient components on the protection operation becomes more and more serious. Because the conventional protection inspection instrument can't simulate the transient components, it can only be limited to the conventional test of the protection device, and can't analyse the results of the protection operation well. In the past, the test of the action behaviour of the protective device is also done by using the dynamic simulation, which can only be done indoors and can't be applied to the real substation, so the expansibility and flexibility are relatively low. But the computer technology develops quickly, it is possible to get transient simulation of power system by computer, and the transient simulation results can be output smoothly, therefore, in this paper it will use numerical method to calculate the electromagnetic transient.

In view of the fast changing speed of electromagnetic transient process, it is generally necessary to calculate and analyse the transient values of voltage and current in milliseconds. Numerical integration algorithm is the basic method of power system transient stability analysis. In this paper, the element model is analysed and the element transient equivalent model is established by numerical integration method.

The main numerical methods for power system stability calculation are Euler method, Runge-Kutta method, predictive correction and trapezoid method for solving differential equations, Newton method for solving algebraic equations and iterative solution for triangular decomposition, etc. The numerical integration method of differential equation can be divided into two kinds: explicit integration algorithm and implicit integration algorithm.

For a set of equations

$$\begin{cases} \dfrac{dY}{dt} = F(Y,t) \\ Y(t_0) = Y_0 \end{cases} \tag{1}$$

When we use explicit integration algorithm to calculate, we can calculate the algebraic equations and differential equations separately, only use the value at $t-\Delta t$ when we calculate the value at t ,this method can easy programming and convenient expanding, but because of the existence of interfacing error between the differential equation and algebraic equation, the stability of integral calculation is not high. Implicit integration method can maintain the stability of integral calculation very well. Using ordinary differential equation

https://doi.org/10.1051/e3sconf/20186404002

$y' = -y + x + 1$ to show the difference of calculation stability, this method can also be improved in nonliner ordinary differential equation, but in this paper will not describe in details.

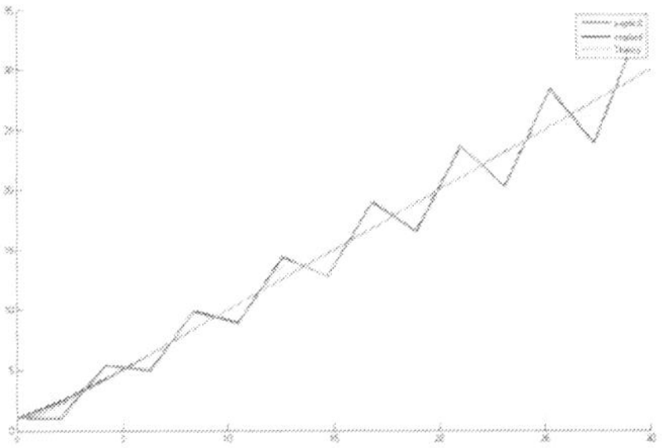

Fig. 4. Calculation stability of explicit and implicit integration algorithm.

In Figure.4 we can see that explicit integration become deviate the theoretical value with increase, and on the contrary implicit integration algorithm become more and more close to the theoretical value. Figure.4 can verify the result that using implicit integration algorithm can get well calculation stability and get good performance.

At the same time, the truncation error caused by Euler method and trapezoidal method is different, Euler method has the first order precision and the trapezoidal method has the second order precision. Therefore, the trapezoidal formula of implicit integral is used to differentiate the differential equation, and the differential equation and network equation are used to get the calculation result.

For example, the open loop transfer function like:

$$\frac{7s+100}{s^4+7s^3+28.44s^2+65.11s} \tag{2}$$

Given unit step signal to this system, getting the output curve of closed-loop transfer function.

Figure.5 is the precision difference of Euler and trapezoid method:

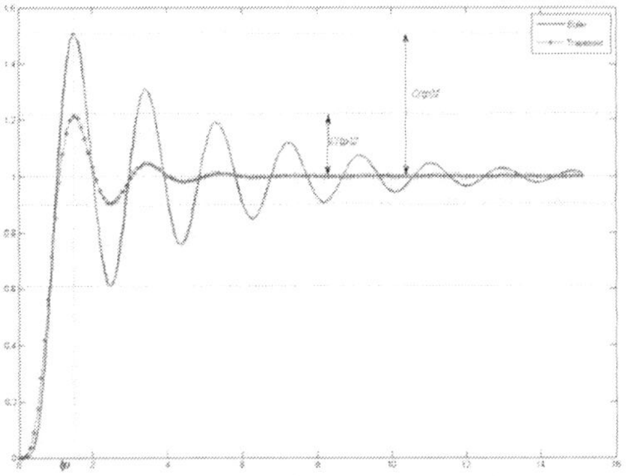

Fig. 5. Precision difference of Euler and trapezoid.

From Figure.5 we can see that the maximum overshoot value of Euler method is higher than Trapezoid method, and the response speed of trapezoidal method to steady state is much more faster than that of Euler method. The oscillation time and amplitude of Euler method are larger than that of trapezoidal method. In the numerical calculation process, there are other algorithms with higher accuracy like RK algorithm and Simpson algorithm, but compared with implicit trapezoidal integration method, there are not easy to program and the implicit trapezoidal integration method can meet the accuracy in real situation. Therefore, the implicit trapezoidal integration method is used to calculate the electromagnetic transient process, which can meet the speed of computation and meet higher accuracy.

Based on the core idea of implicit trapezoidal integral method, the algebraic difference equations of the common components of each circuit are derived one by one, and the corresponding transient equivalent model is obtained, which provides the model basis for subsequent calculation. In this paper, implicit trapezoidal integral method had been used to get the three-phase resistance, inductor series coupling circuit as an example:

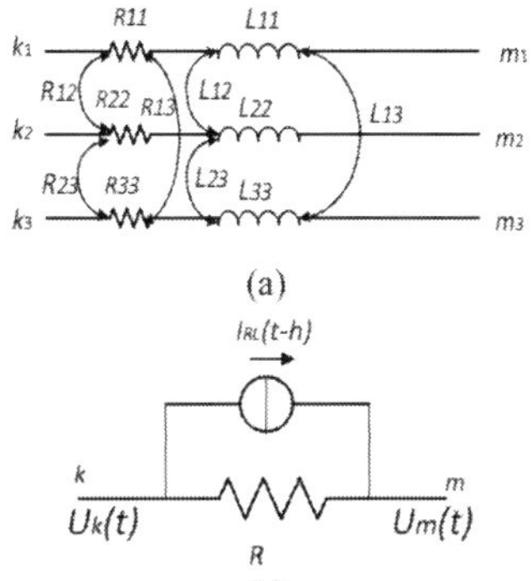

Fig.6. (a) is the three-phase resistance, inductor series coupling circuit.(b) is the transient equivalent model of (a).

If the three-phase circuit is balanced, the differential equations are listed according to the law of electromagnetic induction.

$$Ri_{km}(t) + L\frac{di_{km}(t)}{dt} = u_k(t) - u_m(t) \tag{3}$$

In this equations k is represented k_1, k_2, k_3 and m is represented m_1, m_2, m_3.

Using the trapezoidal integral formula to differencing the differential equation can get:

$$i_{km}(t+h) = S^{-1}\left[u_k(t+h) - u_m(t+h)\right] + I_{km}(t) \tag{4}$$

In equation (4)

$$S = R + \frac{2L}{h} \tag{5}$$

$$I_{km}(t) = S^{-1}[S-2R]i_{km}(t) + S^{-1}[u_k(t) - u_m(t)] \tag{6}$$

Also according to formula (6) can get

$$I_{km}(t) = H[u_k(t) - u_m(t) + SI_{km}(t-h)] - I_{km}(t-h) \tag{7}$$

In equation (7)

$$H = 2\left[S^{-1} - S^{-1}RS^{-1}\right] \tag{8}$$

In the formula (8) showed above, S is a resistance matrix

$$S=\begin{bmatrix} R_s & R_m & R_m \\ R_m & R_s & R_m \\ R_m & R_m & R_s \end{bmatrix}+\frac{2}{h}\begin{bmatrix} L_s & L_m & L_m \\ L_m & L_s & L_m \\ L_m & L_m & L_s \end{bmatrix} \quad (9)$$

In the above

$$\begin{cases} R_s=\frac{1}{3}(R_0+2R_1) & R_m=\frac{1}{3}(R_0-R_1) \\ L_s=\frac{1}{3}(L_0+2L_1) & L_m=\frac{1}{3}(L_0-L_1) \end{cases} \quad (10)$$

L_s is the self-inductance and L_m is the mutual-inductance, R_s is the self-resistance and R_m is the mutual-resistance, L_1 and R_1 are the positive sequence inductor and resistance , L_0 and R_0 are the zero sequence inductor and resistance, and H is admittance matrix, due to the existence of coupling, they are all full array. If there is no resistance in the coupling branch, that is, R is a zero matrix, in this moment $S=\frac{2L}{h}, H=\frac{h}{2L}$, it's a three-phase balanced coupling inductor circuit.

According the implicit trapezoidal integral method can get all transient equivalent model of common electrical components, and then get the admittance matrix , according the admittance matrix iteration can do the transient calculation and get the calculate result.

4 Result

The digital substation-oriented testing method, mentioned in the thesis, and the established system can obtain an electrical logic model. This model is able to do the numerical simulation and obtain the data which come from the steady-state operation and fault transient system. Moreover, it can offer the instant feedback to the platform operators as well while simulates operating-state of digital substation. It's obvious that the model could lay the foundation for providing more humanized man-machine interface.

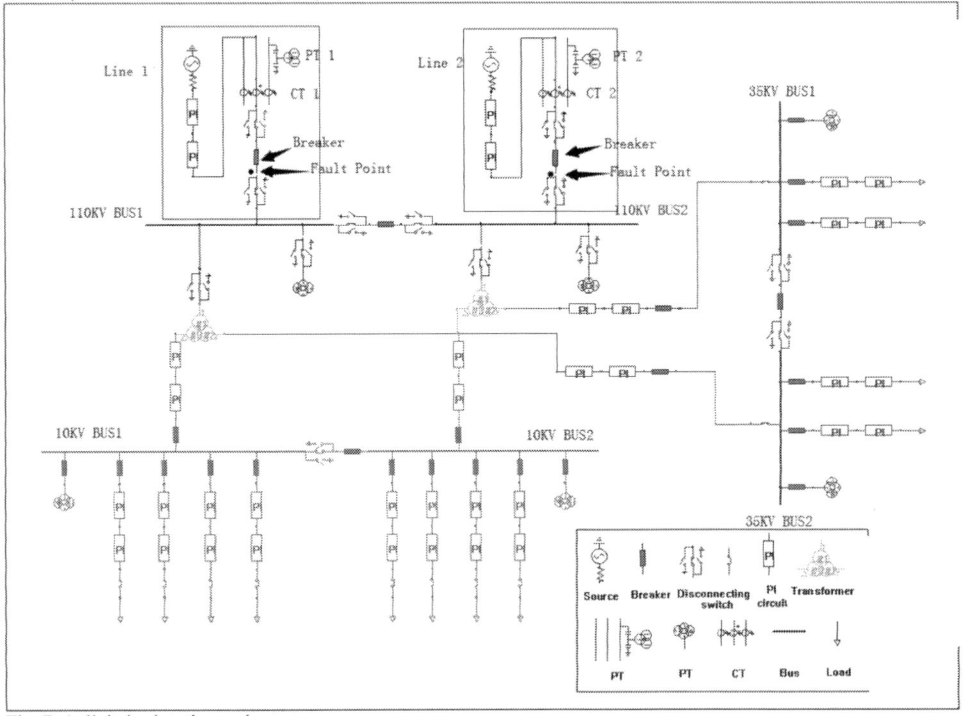

Fig. 7. A digital substation under test.

The system based on the method which is proposed in this paper, have been used in a substation of China Southern Power Grid. For instance, the traditional fault of substation such as single-phase transient grounding fault and three-phase transient grounding fault can be simulated. The single-phase transient fault and the three-phase transient grounding fault are set to a 110KV substation line 1 and line 2 respectively in Figure. 7. In Figure. 7, the Fault Point is the point where set fault and t1 is the time when fault initiated, and t3 was the ending time, the simulating result can be seen in Figure. 8 and Figure. 9 respectively.

4.1 Single-phase transient grounding fault

At time t1, a fault is applied to the line 1 of phase A and ends in time t3. Owing to the delay of control command, the breaker will work till t2, and end in t4. During the time of fault, Phase A is grounded, which causes the voltage becoming zero. The current amplitude of phase A increase slightly as well as the other two phase voltages. At the same time, the statue of breakers also show in the Figure. 8.

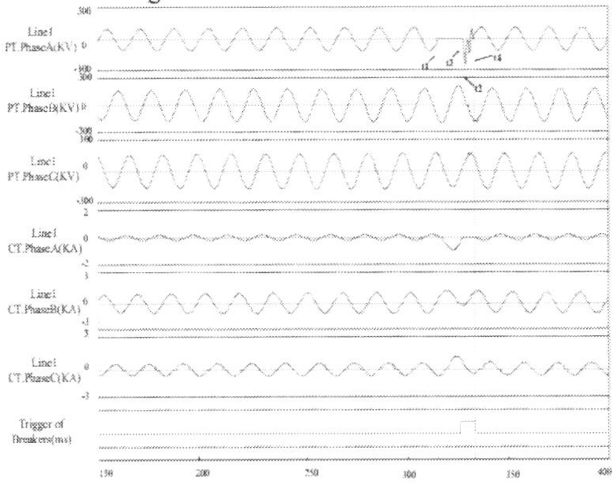

Fig. 8. Single-phase Transient Fault of Line 1 Phase A.

4.2 Three-phase transient grounding fault

At time t1, a three-phase transient grounding fault is applied to the three phase of line 1 and fault ends in time t3 .Owing to the delay of control command, the breaker will act till t2. During the time of fault, the voltages of phase A, B, C all become zero because of grounding. The statue of breakers is shown in Figure. 9.

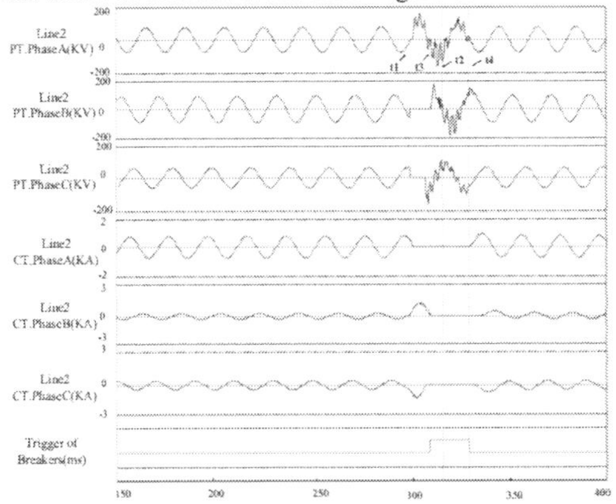

Fig. 9. Single-phase Transient Fault of Line 1 Phase A, B, C.

Through the two common faults mentioned above, it is obvious that the system established, using this method, can truly simulate the real substation faults and fulfil the requirements of the real substation transient simulation.

4.3 Field application

The method proposed in this paper and the established software has applied in a digital substation in China Southern Power Grid, and has verified the correctness of function. Figure.10 and Table.1 are the result when the simulate platform give the simulate result to a line of 10KV capacitor which its current is 35.0A and the time of duration is 650ms, the protection measurement and control device displays the over current I and II segment action, and showing the information of breakers (the I, II and III segment protective setting is 33.3A 0ms, 11.5A 200ms and 3.47A 500ms) trip operation.

Table. 1 Results of protection test in field application

Overcurrent action / Analog && duration	Seg.I	Seg.II	Seg.III
3.0A 650ms	0	0	0
7.3A 650ms	0	0	1
15.0A 250ms	0	1	0
34.0A 10ms	1	0	0
34.0A 250ms	1	1	0
34.0A 650ms	1	1	1

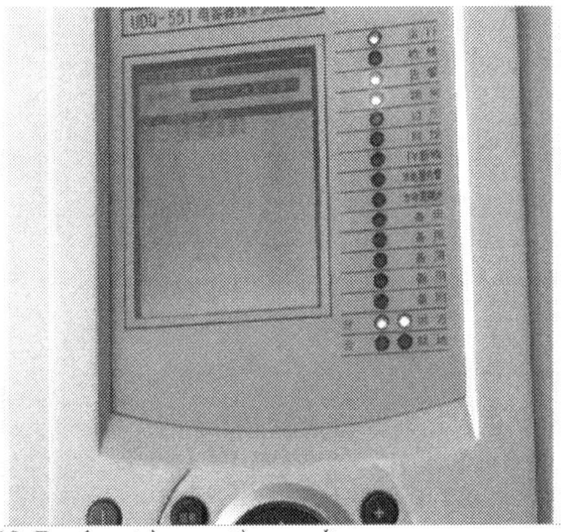

Fig.10. Breakers trip operation result.

4.4 Time of synchronization

Through the wireless synchronization function, the substation injecting data synchronously can be realized, the simulation system can send the same sets of data to the collector simulators.

By sending the same sets of data to the collectors and then gather the data to MU, the phase difference of data collected by different collector terminals is about 0.08 °, while the time is about 4us. That is to say, the time difference between wireless host and each terminal is 4us, within a reasonable range and can be termed as injected synchronously. The whole station cascaded testing method proposed in this paper and the system established by using the unified clock and data synchronous upload in each department can perform well. It represents the real situation for simulating substation.

5 Conclusion

The traditional substations do a function test on the primary electrical equipment only though the giving analog value [12] which is unavailable to the substation equipment connected wholly. This method ignores the correctness of interconnection between devices and communication equipment such as optical fibre and network switch in substation. All devices only do conformance testing, in other words, this method only do the test for the inputting and outputting data and specification aspect testing, seldom do the functional tests under the whole station cascaded statues.

This paper proposes a cascade station method in digital substation test. Problems, like incomplete function test, potential fault detection and so on, can be avoided by this method. It can find out the potential faults of digital substation effectively, not only limited to single device performance test, but also to the whole digital substation for system-level testing. At the same time, the system-level test results can also ensure that the functions of the secondary equipment like the relay protection devices, the measurement and control device are normal. The problems of traditional substation test have been solved. The method mentioned in this thesis avoids the separation of communication and devices and improves the test efficiency dramatically.

Announcement

Supported by "The Fundamental Research Funds for the Central Universities" (ZYGX2015J105) for sponsoring this research.

References

1. Dechang Yang, Li Yong, C.Rehtanz, et al. Research on the composition and development planning of Chinese smart grid. J. power grid technology,(2009), **33 (20)**: 13-20.
2. Jun Feng. Principle and Test Technology of Digital Substation. M. *China Electric Power Press* (2011).
3. Gao Xiang, Zhang Peichao. Main features and key technologies of digital substation. J. power grid technology (2006) **30 (23)**: 67-71.
4. Li Xianmei, Huang Jiadong, Tang Baofeng. Analysis and research of relay protection testing technology in digital substation. J. power system protection and control (2012) **40 (3)**: 105-108.
5. J. Bowen, "Substation commissioning and turnover planning," *IEEE Ind. Appl. Mag.*, vol. **6**, no. **2**, pp. 8–22, Mar./Apr(2000).
6. Wu Zai-jun, Hu Min-qiang. Analysis of IEC61850-communication networks and system in substations. J. Electric Power Automation Equipment (2002).
7. Yu Pengfei, Yu Qiang, Deng Hui, et al. Application scheme of IEEE 1588 precise time synchronization protocol. J. automation of electric power system (2009), **33 (13)**: 99-103.
8. Zhao Shanglin, Hu Minqiang, Dou Xiaobo, et al. Research on clock synchronization technology in digital substation based on IEEE1588. J. power grid technology, (2008) **32 (21)**: 97-102.
9. Jing S, Huang Q, Wu J, et al. Wireless clock synchronization method for whole-view test of Intelligent Substation. J. Automation of Electric Power System, (2013) **37(9)**:103-109.
10. Daboul M, Orsagova J, Bajanek T, et al. Testing protection relays based on IEC 61850 in Substation Automation Systems. C. *International Scientific Conference on Electric Power Engineering*. IEEE, (2015):335-340.
11. Kaneda K, Tamura S, Fujiyama N, et al. IEC61850 based Substation Automation System. C. *Joint International Conference on Power System Technology and IEEE Power India Conference*, 2008. Powercon. IEEE, (2008):1-8.
12. Zhou Chunxia, Zhan Rongrong, Jiang Jianning, et al. 500kV digital substation dynamic model test research. J. Power grid technology, (2010) **34 (6)**: 193-197.

A Novel Submodule Applied in HVDC Hybrid DC Breaker Topology

Junmin Wu[1],*Xizi* Zhang[1],*Cui* Li[1] , *Jun* Zhang[2], *Zhibin* Zhao[2]

[1]State Key Laboratory of Advanced Power Transmission Technology (Global Energy Interconnection Research Institute of State Grid Corporation of China), Beijing 102209, China
[2]State Key Laboratory of Alternate Electrical Power System with Renewable Energy Sources (North China Electric Power University), Beijing 102206, China)

Abstract. In recent years, with the vigorous development of power electronic technology, renewable energy sources such as wind power and solar energy utilization efficiency continuing to improve, and the application scale expanding unceasingly, DC power grid technology and the development of multiterminal HVDC has become a key link of transmission technology in our country. As the key equipment of biulding DC system, The direct current breaker need to be optimized on the aspect of topology,to promote the development of dc transmission. This article puts forward a new type of submodule which can be applicated in hybrid DC breaker. After a brief review of DC circuit breaker research status, the paper will give the topology structure, and analyzes its working principle. Then, verification and electrical stress calculation will be done in Saber Sketch simulation. Based on this, two other present scheme will be introduced, and finally a comparison will be made between the diffent scheme. As is concluded, the novel submodule can save the power elctronic device cost effectively

1 Introduction

Large-scale renewable energy access, transportation, and consumption request interconnectino and complementary with load, energy storage, thermal power, hydropower and other energy in Internet more quickly and widely, this means that the traditional power grid structure, electric equipment and operation mode will be subject to severe challenges. Constructing cross-regional dc grid, is the inevitable development trend in today's power transmission system.[1-2]

In the recent double side DC transmission system, when failure occurs in certain transmission line, the protection strategy is cutting off protection switch in the converter valve, or cutting off nearby AC breaker, which will stop the entire transmission system of power transmission, so as to achieve the aim of protecting grid system. But in multiterminal HVDC system, the strategy will greatly reduce the transmission efficiency and the stability of the system. We hope that when failure occurs on the transmission line, the faults will be cut off quickly and specifically, to ensure the normal operation of other transmission line. Therefore, DC circuit breaker is important equipment in the future dc power grid.

On the aspect of high voltage hybrid DC circuit breaker, there are corperations and research institudes have made development in the recent years[3-11].In 2012,ABB Corp has developed a 320kV hybrid DC breaker full-scale prototype based on press-pack IGBT, the turn off current is 9kA,and fault duration time is 2ms[6-8].In 2013,Alstom Corp has developed a 120kV hybrid DC breaker full-scale prototype which can turn off 7.5kA fault current[9].At the end of 2013,State Grid Corperation of China successfully developed a 200kV hybrid DC breaker ,which can turn off 15kA fault within 3ms[10-11].

In this paper, a novel high voltage hybrid DC circuit breaker topology will be proposed. This topology is based on Diode and IGBT which is called "Diode Full-bridge submodule, DFS". The auxiliary branch can be made by many DFS in series. This novel topology can not only maintain existing performance, but also reduce the cost by 20%~30%.In the first part of paper, the structure and working principle will be introduced, and then verificate the circuit breaker working process in simulation soft. In the second part of paper, two other present scheme will be introduced, and finally a comparison will be made between the DFS scheme, Anti-series scheme and IGBT full-bridge scheme.

2 The Structure and working prinple of novel topology

2.1 Topology structure

The novel high voltage hybrid DC circuit breaker topology structure is shown in Fig.1.Main branch is composed of

https://doi.org/10.1051/e3sconf/20186404003

fast mechanical switch and little DFS, which will flow the normal working current. Auxiliary branch is composed of series DFS. the DFS structure is shown in Fig.2

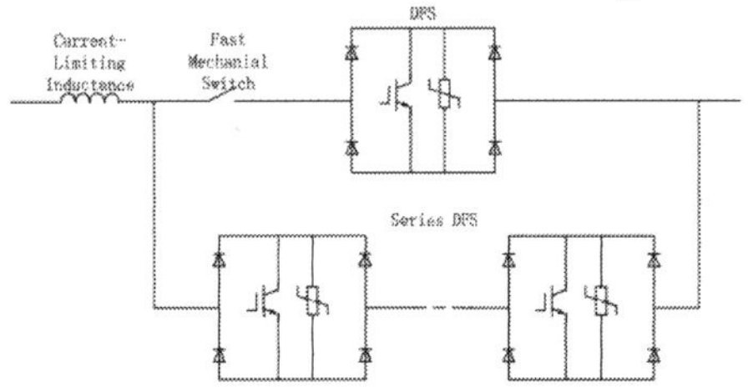

Fig.1 New Type Hybrid DC Breaker Topology

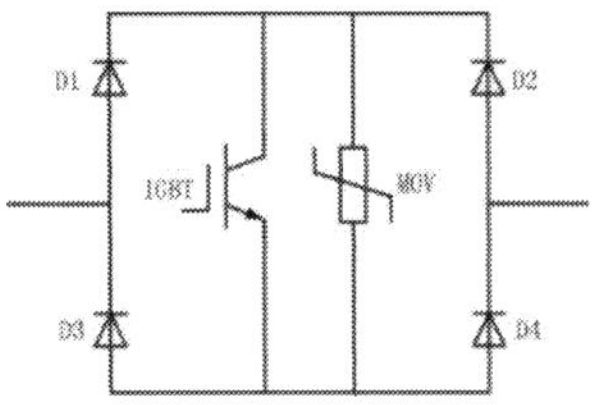

Fig.2 Diode Full-bridge Submodule Topology

In the DFS topology,D1,D2,D3,D4 is Diode. When DFS begins to conduct fault current from left to right, the current path is D1-IGBT-D4.When DFS begins to turn off, the current path is D1-MOV-D4.In turn, when DFS begins to conduct fault current from right to left, the current path is D2-IGBT-D3.And when DFS begins to turn off, the current path is D2-MOV-D3.So the double-side fault current can be turned off in DFS topology. The Fig.3 is waveforms of DC circuit breaker based on the DFS. Ia means normal working current value, Ib means equipment protection start value, Ic means circuit breaker turn off current value.i_main means main branch current,i_zy means auxiliary branch current, and i_arr means arrestor branch current.

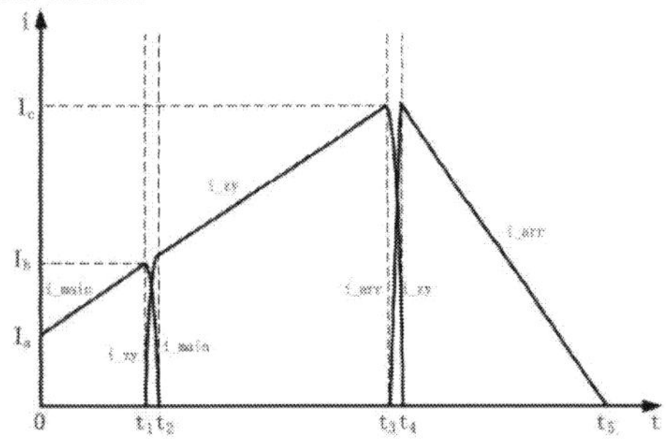

Fig.3 The DC breaker waveform based on DFS

2.2 Working principle

Suppose the current flow left to right: When DC transmission line is on normal working state, the current flow through the main branch. The number of DFS on main branch could not be too much because DFS here only endure the total IGBT Vce(sat) on the auxiliary branch, which is several hundreds of volt, and fast mechanial switch has been open when IGBT on the auxiliary turn off.

When the fault happens in the transmission line and fault current rise to the set value, the DFS on main branch should be turned off. Meanwhile, the DFS on auxiliary branch are turned on, so the current transfer from main branch to auxiliary branch, as is shown in Fig.4.

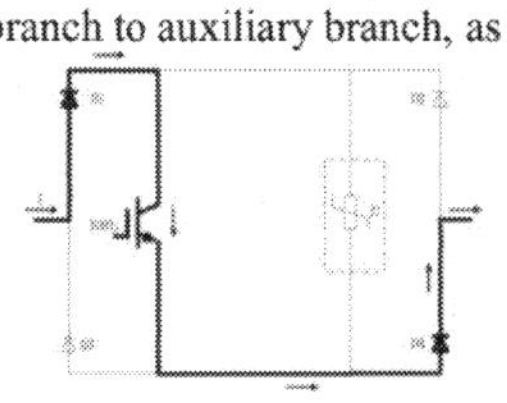

Fig.4 Turn-on process of DFS

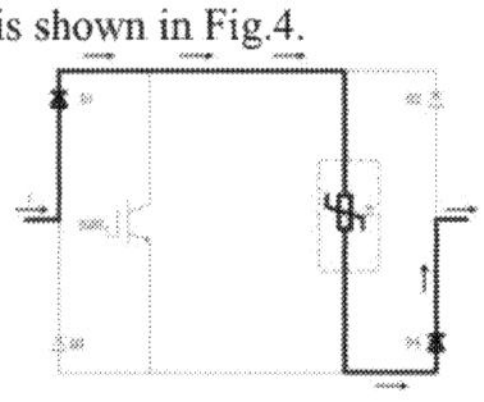

Fig.5 Turn-off process of DFS

After several millsecond, the fast mechanical switch has restored insulating ability. After that, DFS on the auxiliary can be turned off, which results in the voltage on current limiting inductance rise rapidly, until the operate voltage of arrestor. Then the fault current will flow through the arrestor, which absorb the fault energy. Until the fault energy and fault current fall down to zero, the disconneter open, and working process is over, as is shown in Fig.5.

When the fault current flow in anti-direction, the working principle is similar, as is shown if Fig.6.

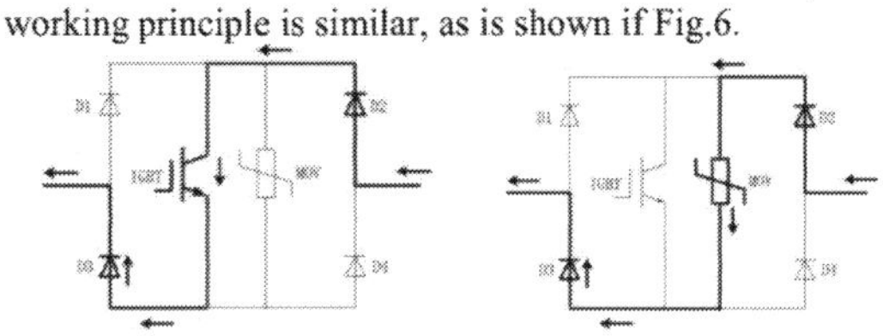

Fig.6 Turn-on and Turn-off process of DFS when current anti flow.

2.3 Simulation analysis

In order to verify the working principle and performance, and calculate the electric stress in IGBT in auxiliary branch, the simulation work has been done in Saber Sketch. First of all, ABB 4.5kV/2kA press-pack IGBT has been modeled in Saber Sketch according to the datasheet. Then, the simulation has run in the situation of 35kV transimission line.

Tab.1 The related simulation parameter in Saber

Parameter	Value
Udc/kV	35
Ls/mH	18
Rs/Ω	35
Irating/kA	1
Iset/kA	1.3
Isc/kA	5
tsc/ms	2
IGBT numbers	18

According to the simulation results, the electric stress of IGBT in DFS on auxiliary branch has been studied as follows:

(1)Commutation process: When fault current reach the set value, the commutation process begins. Actually, the IGBT on auxiliary branch will turn on first, after several us delay time, the IGBT on the main branch will turn off. This is because if IGBT on auxiliary branch hasn't been turned

 https://doi.org/10.1051/e3sconf/20186404003

on in time when IGBT on main branch has been turned off, there will be a pulse voltage across the IGBT on both branch, which will bring damage on power electronic devices. In the commutation process, the max di/dt rate is 0.8kA/us, IGBT device turn on in the situation of load shorted, so this process could be also called the "Over Current Turn-on process", as is shown in Fig.7.During this process, the current distribution of IGBT chips will be diverse because of the inconsistency of press distribution inside press-pack IGBT and stray inductance between chips and thus, the reliability of device will be reduced[16-17].

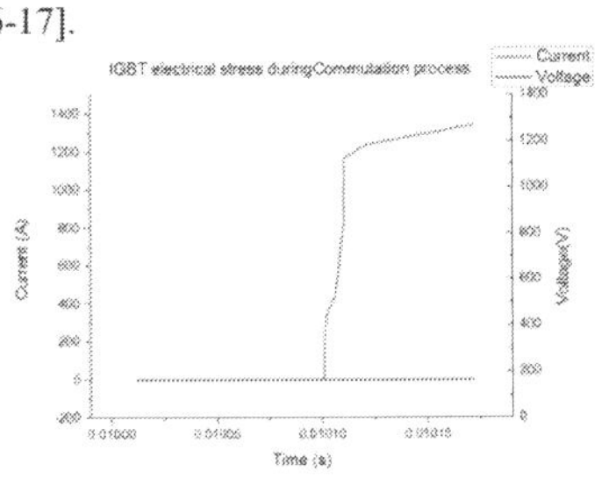

Fig.7 The current and voltage waveform in single IGBT device during commutation process.

(2)Current rise process: When the commutation process is over, the fault current will keep rising cause fast mechanical switch need several millseconds to restore insulating ability. The current rise rate depends on current-limiting inductance, and the voltage across IGBT on auxiliary branch is IGBT saturated voltage Vce(sat).In this process, for the purpose of utilize IGBT current turn off capability as much as possible, the fault current will rise from set value to near saturated current Ice(sat),which is 5-6 times the rating current. Therefore, the temperature inside IGBT device will rise because of the conduction loss. So the demand for IGBT device work in this situation and maintain reliability is the important request of device design[18].

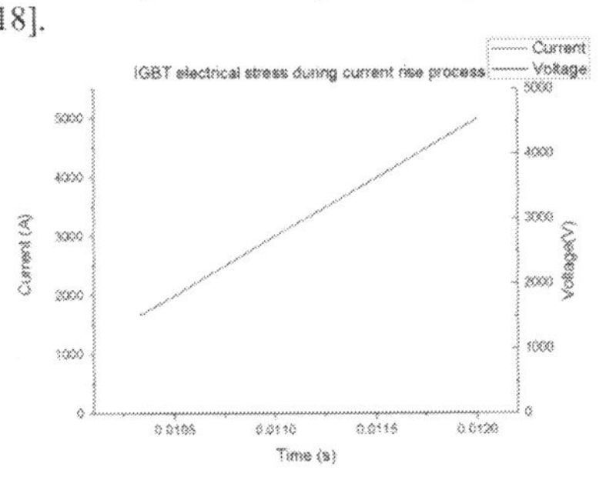

Fig.8 The current and voltage waveform in single IGBT device during current rising process.

(3)Turn-off process: After fast mechanical switch restored insulating ability, IGBT on auxiliary branch can be triggered to turn off. The turn-off current valus is 5-6 times rating curent, the current change rate is 0.5kA/us, voltage change rate is 0.2kV/us. During the turn-off process, IGBT device will suffer large electrical and thermal stress, and transient temperature will rise, which will raise the harsh demand of IGBT device application reliability[19-22].

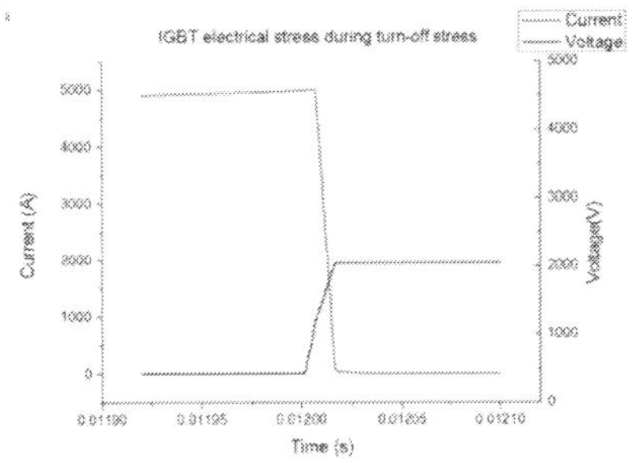

Fig.9 The current and voltage waveform in single IGBT device during interrupted process

3 The Comparison of Different Hybrid DC Breaker

3.1 Other two present main topology

(1)Anti-series Submodule Scheme: The topology based on anti-series submodule is shown in Fig.10 and Fig.11,which is applied in ABB Corp's scheme in 320kV/9kA Hybrid DC breaker. This topology is composed of three paralelled branch: branch one is main branch, branch two is auxiliary branch and branch three is arrestor branch. Main branch is composed of fast mechanical switch and little anti-series submodule, which is used for conduct normal working current. Auxiliary branch is composed of many series anti-series submodule, which is used for turn off fault current. Arrestor branch is composed of many paralelled arrestor, which is used for absorbing the short energy.

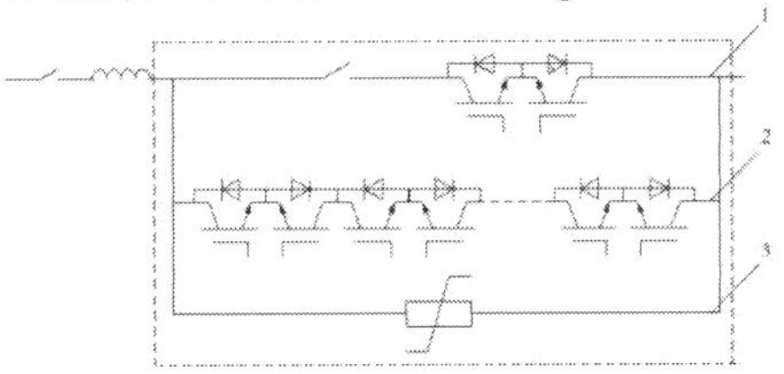

Fig.10 Hybrid DC Breaker Topology based on anti-series submodule

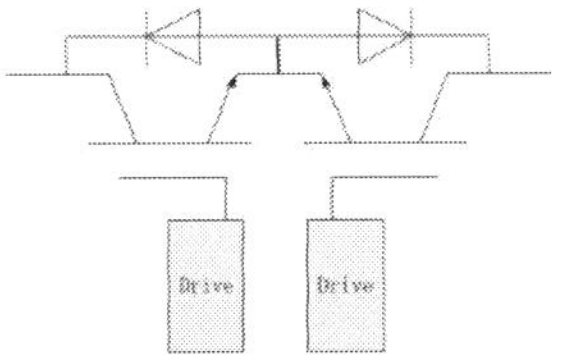

Fig.11 Anti-series submodule structure

(2)IGBT Full-Bridge Submodule Scheme(FBS):The topology based on FBS is shown in Fig.12 and Fig.13,which is applied in State Grid Corp in China in 2013 for 200kV/15kA Hybrid DC breaker scheme.This topology is also composed of three paralelled branch,the structure and working principle is similar to anti-series submodule scheme, and the difference is anti-series submodule is replaced by FBS,and it can turn off doubled fault current.

3.2 Comparison of three submodule scheme

As is verified in Saber Sketch simulation, there is little difference of electrical and thermal stress in single IGBT between three diffent submodule. The main difference exists in cost. The Diode Full-Bridge Submodule(DFS) contains one IGBT device,one IGBT drive and four power diodes; The Anti-series Submodule(AS) contains two IGBT devices, two IGBT drives and two power diodes;The IGBT Full-Bridge Submodule(FBS) contains four IGBT devices, four IGBT drives and four power diodes. Because the FBS scheme can be simply equivalent to two paralelled IGBT branch, so we reduce half the IGBT device, drive and power diode in FBS module for comparison in the same performance.

Due to the different price of IGBT device, drive and power diode between different model and manufacturers, we make the comparison using the unit of "Diode number", as is shown in Tab.2.

Tab.2 The price comparison between three submodule using the unit of "Diode number".

"Diode" price	DFS	AS	FBS
1IGBT+1Drive=2Diode	6	6	6
1IGBT+1Drive=3Diode	7	8	8
1IGBT+1Drive=4Diode	8	10	10
......	......	......	......
1IGBT+1Drive=+1Diode	+1	+2	+2

We can infer from the Tab.2 that, if the price of 1 IGBT device and 1drive is higher, or, the price of 1 Diode is lower then the hybrid DC circuit breaker scheme used DFS will save more cost. According the present market price, we estimate that 1 IGBT + 1 Drive = 4 Diode, that is, DFS will save around 20% cost.

4 Conclusion

DC power grid based on flexible HVDC technology application prospect became bright and clear. as one of the technical bottleneck, DC circuit breaker research has developed rapidly. Based on this ,this article first bring a novel Diode Full-Bridge submodule, which is composed of four power diode and 1IGBT.Then we verify the working principle of the hybrid DC circuit breaker using FBS in 35kV transmission line in Saber Sketch. Finally, we compare DFS with Anti-series Submodule(AS,ABB Corp)and IGBT Full-Bridge Submodule(FBS,SGCC).On the one hand, the performance in the three submodule is around the same; On the other hand ,DFS will save 20%~30% cost from the aspect of power electronic device.

Acknowledgement

Project supported by National Science and Technology Major Project(2015ZX02301)

References

1. Wen Jialiang, Ge jun , Pan yan, et al. Development and prospect of power electronic devices for DC power network [J]. Power System Technology, 2016(3).
2. Wen Jialiang, Wu rui, Peng chang et al. Analysis of application prospect of DC power network in China [J]. Proceedings of the CSEE, 2012, 32(13):7-12.
3. Xu zheng, Tu ruiqing, Qiu peng. Looking at the development direction of DC power transmission technology from the 2010 international power grid conference [J]. High Voltage Engineering, 2010, 36:3070-3077.
4. Wen jun, Zhang yigong, Han minxiao, et al. A new generation of HVDC technology [J]. Power System Technology, 2003, 01:47-51.
5. Huang A, Peng C, Song X. Design and development of a 7.2 kV/200A hybrid circuit breaker based on 15 kV SiC emitter turn-off (ETO) thyristor[C]// Electric Ship Technologies Symposium. IEEE, 2015.
6. Franck C M. HVDC circuit breakers：A review identifying future research needs[J] . Power Delivery，IEEE Transactions on，2011，26(2)：998-1007.
7. JÜRGEN HÄFNER, BJÖRN JACOBSON " Proactive Hybrid HVDC Breakers - A key innovation for reliable HVDC grids" [C] (Cigré session 2012)
8. Eicher S, Rahimo M, Tsyplakov E. 4.5kV Press Pack IGBT Designed for Ruggedness and Reliability[J]. Power Electronics, 2004, 3:1534-1539 vol.3.
9. W. GRIESHABER，L. VIOLLEAU" Development and test of a 120 kV direct current circuit breaker" [C] (Cigré session 2014)
10. Wei xioaguang, Gao chong, Luo xiang, et al. Design scheme of new high voltage DC circuit breaker for flexible DC transmission network [J]. Automation of Electric Power Systems, 2013, 37(15):95-102.
11. DC research institute of China netcom research institute: " the 200 kv high voltage DC circuit breaker developed by China netcom research institute has reached the international leading level." [EB/OL].
12. http://www.sgcc.com.cn/xwzx/gsxw/2015/05/325722 .shtml
13. Yao tao, Scheme design and principle verification of high voltage DC circuit breaker [D]. China electric power research institute，2014.
14. Zhang xuena, Zhao chengyong, Pang hui et al.. Mmc - based DC side fault control and protection strategy for multi-terminal DC transmission system [J]. Automation of Electric Power Systems, 2013, 37(15):140-145.
15. Wang shanshan, Zhou xiaoxin, Tang gaungfu, et al. Analysis of over-current of modular multilevel converter HVDC module with double extremely short path [J]. Proceedings of the CSEE, 2011(1):1-7.
16. Hassanpoor A, Hafner J, Jacobson B. Technical assessment of load commutation switch in hybrid HVDC breaker[J]. IEEE Transactions on Power Electronics, 2014, 30(10):3667-3673.

17. Cwikowski O, Barnes M, Shuttleworth R, et al. Analysis and simulation of the proactive hybrid circuit breaker[C]// IEEE, International Conference on Power Electronics and Drive Systems. IEEE, 2015.
18. Basler T, Lutz J, Jakob R, et al. The influence of asymmetries on the parallel connection of IGBT chips under short-circuit condition[C]// - European Conference on Power Electronics and Applications. IEEE, 2011:1-8.
19. Musing A, Ortiz G, Kolar J W. Optimization of the current distribution in press-pack high power IGBT modules[C]// Power Electronics Conference. 2010:1139-1146.
20. Lutz J. Packaging and Reliability of Power Modules[C]// International Conference on Integrated Power Systems. VDE, 2014:1-8.
21. Hong T, Pfirsch F, Reinhold B, et al. Transient avalanche oscillation of IGBTs under high current[C]// Ispsd. 2014:43-46.
22. Basler T, Lutz J, Jakob R, et al. Surge current capability of IGBTs[C]// International Multi-Conference on Systems, Signals and Devices, Ssd. 2012:1 - 6.

Inductance extraction of Press-Pack IGBT by Considering Displacement Current

Jinyuan Li[1],Zhongyuan Chen[1],Lei Zhang[1], Chouwei NI[2] and Zhibin ZHAO[2]

[1]State Key Laboratory of Advanced Power Transmission Technology (Global Energy Interconnection Research Institute of State Grid Corporation of China), Beijing 102209, China
[2]State Key Laboratory of Alternate Electrical Power System with Renewable Energy Sources (North China Electric Power University), Beijing 102206, China)

Abstract. Taking the displacement current generated by charges accumulation into consideration, this paper presents a inductance extraction method for press-pack IGBT where conductor might not form a closed loop in simulation. Charge accumulating on ends of conductor is considered and then Ampere's circuital law (ACL) could be meet. Vector potential A also could be affected by displacement current, which leads to a different formulation of partial inductance. A simplified model based on practical structure are used in numerical experiment for comparison between existing method and proposed method.

1 Introduction

The press-pack IGBT has become much more attractive than ever due to its high voltage and high current ratings. In the switching process, the maximum current overshoot of paralleled chips inside PPI decides the application limit of the device, which is directly affected by stray parameters of the package module.

The inductance in package is important for current distribution between chips, which mainly exists in the conductors of package. The magnetic coupling effect between conductor segments that don't form a closed loop needs to be considered because the return current path might be distant.

Therefore, Ruehli [1] proposed the definition of partial inductance based on flux linkage method and gave a series of formulas to establish the relationship between incomplete loops and closed loops. In recent years, Paul [2] and Holloway [3] made some detailed analysis and summary. Afterwards, the partial element equivalent circuit (PEEC) method that based on the partial inductance theory is widely applied [4-6].

To obtain partial inductance, magnetic vector potential A generated by each conductor segment needs to be calculated at first. Kalhor's study [7] showed that two charges accumulating with time need to be placed at both ends of a conductor, which generate displacement current in space, to meet Ampere's circuital law (ACL) in incomplete loop. Kalhor [8] pointed out that Biot-Savart's Law is still applicable in this situation, but effect on A is not analyzed. According to (1), differential form of ACL is applied in the calculation of A, which means total current that includes displacement current is used.

$$A(r)=\frac{1}{4\pi}\int_{V'}\frac{\nabla'\times B(r')}{|r'-r|}\mathrm{d}V' \qquad (1)$$

Hence, this paper presents an inductance extraction method of partial inductance calculation by considering both conduction current and displacement current. In section II, A is derived when charges accumulating at both ends of a conductor. Section III gives a method of partial inductance calculation based on the result of A. Section IV makes a numerical experiment for a simplified structure and show difference of results between methods before and after considering displacement current.

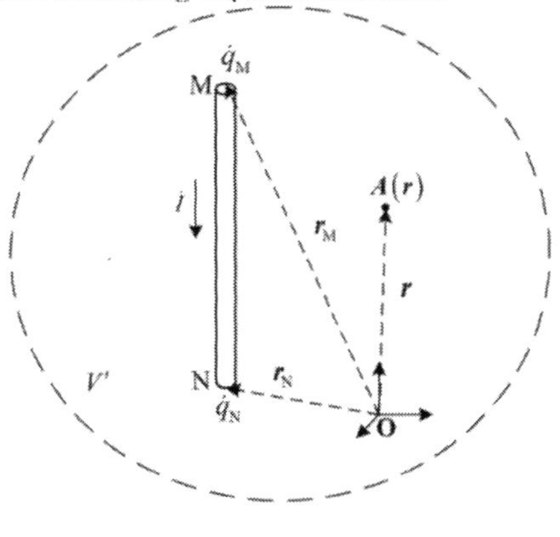

Fig. 1. Conductor segment carrying current in space.

2 A Generated by Displacement Current

As shown in Fig. 1, when a conductor segment carries alternating current $\dot{I}$, opposite charges accumulate respectively at both ends of the conductor, noted as points M and N, and generate displacement current in space. The opposite charges are noted as $\dot{q}_{\mathrm{M}}$ and $\dot{q}_{\mathrm{N}}$. According to current continuity equation, (2) is given.

$$-j\omega\dot{q}_{\mathrm{A}} = j\omega\dot{q}_{\mathrm{B}} = \dot{I} \tag{2}$$

As mentioned above, ACL is satisfied in this model. So (4) is obtained by substitute (3) into (1).

$$\nabla\times\dot{\boldsymbol{B}} = \dot{\boldsymbol{J}}_{\mathrm{C}} + \dot{\boldsymbol{J}}_{\mathrm{D}} \tag{3}$$

$$\dot{\boldsymbol{A}}(\boldsymbol{r}) = \frac{\mu_0}{4\pi}\int_{V'}\frac{\dot{\boldsymbol{J}}_{\mathrm{C}}(\boldsymbol{r}') + \dot{\boldsymbol{J}}_{\mathrm{D}}(\boldsymbol{r}')}{|\boldsymbol{r}'-\boldsymbol{r}|}\mathrm{d}V' \tag{4}$$

where V' represents the whole area of source, $\boldsymbol{r}'$ and $\boldsymbol{r}$ are space vectors pointing to source point and field point, respectively. $\dot{\boldsymbol{J}}_{\mathrm{C}}$ is current density of conduction current $\dot{I}_{\mathrm{C}}$. $\dot{\boldsymbol{J}}_{\mathrm{D}}$ is sum current density of displacement current generated by $\dot{q}_{\mathrm{M}}$ and $\dot{q}_{\mathrm{N}}$, and can be calculated by (5). $\dot{\boldsymbol{D}}$ represents electric displacement vector.

$$\dot{\boldsymbol{J}}_{\mathrm{D}}(\boldsymbol{r}') = j\omega\dot{\boldsymbol{D}}(\boldsymbol{r}') \tag{5}$$

$$\dot{\boldsymbol{D}} = \dot{q}\boldsymbol{r}_{\mathrm{q}}/\left(4\pi r_{\mathrm{q}}^3\right) \tag{6}$$

According to Gauss' law, $\dot{\boldsymbol{D}}$ generated by a charge in space is given in (6), where $\boldsymbol{r}_{\mathrm{q}}$ is space vector pointing from charge to field point. $\dot{\boldsymbol{J}}_{\mathrm{D}}$ consists of two parts, $\dot{\boldsymbol{J}}_{\mathrm{DM}}$ and $\dot{\boldsymbol{J}}_{\mathrm{DN}}$, due to different charges $\dot{q}_{\mathrm{M}}$ and $\dot{q}_{\mathrm{N}}$. Combine (5) and (6),

$$\begin{aligned}\dot{\boldsymbol{J}}_{\mathrm{D}}(\boldsymbol{r}') &= \dot{\boldsymbol{J}}_{\mathrm{DA}}(\boldsymbol{r}') + \dot{\boldsymbol{J}}_{\mathrm{DB}}(\boldsymbol{r}') \\ &= \frac{-\dot{I}}{4\pi}\left(\frac{\boldsymbol{r}'-\boldsymbol{r}_{\mathrm{M}}}{|\boldsymbol{r}'-\boldsymbol{r}_{\mathrm{M}}|^3} - \frac{\boldsymbol{r}'-\boldsymbol{r}_{\mathrm{N}}}{|\boldsymbol{r}'-\boldsymbol{r}_{\mathrm{N}}|^3}\right)\end{aligned} \tag{7}$$

where $\boldsymbol{r}_{\mathrm{M}}$ and $\boldsymbol{r}_{\mathrm{N}}$ are space vectors pointing to M and N, respectively. Thus, sum $\dot{\boldsymbol{A}}$ in (4) can be also spited into two items $\dot{\boldsymbol{A}}_{\mathrm{C}}$ and $\dot{\boldsymbol{A}}_{\mathrm{D}}$, corresponding to current density. Therefore, $\dot{\boldsymbol{A}}_{\mathrm{D}}$ can be calculated as followed.

$$\begin{aligned}\dot{\boldsymbol{A}}_{\mathrm{D}}(\boldsymbol{r}) &= \frac{\mu_0}{4\pi}\int_{V'}\frac{\dot{\boldsymbol{J}}_{\mathrm{D}}(\boldsymbol{r}')dV'}{|\boldsymbol{r}'-\boldsymbol{r}|} \\ &= \frac{\mu_0\dot{I}}{16\pi^2}\int_{V'}\frac{-1}{|\boldsymbol{r}'-\boldsymbol{r}|}\left[\frac{(\boldsymbol{r}'-\boldsymbol{r}_{\mathrm{M}})}{|\boldsymbol{r}'-\boldsymbol{r}_{\mathrm{M}}|^3} - \frac{(\boldsymbol{r}'-\boldsymbol{r}_{\mathrm{N}})}{|\boldsymbol{r}'-\boldsymbol{r}_{\mathrm{N}}|^3}\right]dV' \\ &= \dot{\boldsymbol{A}}_{\mathrm{DM}}(\boldsymbol{r}) + \dot{\boldsymbol{A}}_{\mathrm{DN}}(\boldsymbol{r})\end{aligned} \tag{8}$$

Similarly, $\dot{\boldsymbol{A}}_{\mathrm{D}}$ has two parts $\dot{\boldsymbol{A}}_{\mathrm{DM}}$ and $\dot{\boldsymbol{A}}_{\mathrm{DN}}$ generated by $\dot{q}_{\mathrm{M}}$ and $\dot{q}_{\mathrm{N}}$. Hence formulation after derivation is shown.

$$\dot{\boldsymbol{A}}_{\mathrm{DM}}(\boldsymbol{r}_1) = -\mu_0\dot{I}\boldsymbol{e}_{r_1}/(8\pi) \tag{9}$$

Equation (12) indicates magnitude of $\dot{\boldsymbol{A}}_{\mathrm{DM}}$ is constant in space, and direction is in accordance with $\boldsymbol{r}_1$. We can easily derive that $\dot{\boldsymbol{A}}_{\mathrm{DM}}$ is irrotational.

$$\nabla\times\dot{\boldsymbol{A}}_{\mathrm{DM}} = 0 \tag{10}$$

In the same way, $\dot{\boldsymbol{A}}_{\mathrm{DN}}$ generated by another charge $\dot{q}_{\mathrm{N}}$ can be calculated by (14), where $\boldsymbol{r}_2$ is space vector pointing to field point from point N.

$$\dot{\boldsymbol{A}}_{\mathrm{DN}}(\boldsymbol{r}_2) = \mu_0\dot{I}\boldsymbol{e}_{r_2}/(8\pi) \tag{11}$$

Consequently, sum $\dot{\boldsymbol{A}}$ consists of three parts, $\dot{\boldsymbol{A}}_{\mathrm{C}}$, $\dot{\boldsymbol{A}}_{\mathrm{DM}}$, and $\dot{\boldsymbol{A}}_{\mathrm{DN}}$, and has a different direction with conduction current.

$$\begin{aligned}\dot{\boldsymbol{A}} &= \dot{\boldsymbol{A}}_{\mathrm{C}} + \dot{\boldsymbol{A}}_{\mathrm{D}} = \dot{\boldsymbol{A}}_{\mathrm{C}} + \dot{\boldsymbol{A}}_{\mathrm{DM}} + \dot{\boldsymbol{A}}_{\mathrm{DN}} \\ &= \dot{\boldsymbol{A}}_{\mathrm{C}} + \mu_0\dot{I}\left(-\boldsymbol{e}_{r-r_{\mathrm{M}}} + \boldsymbol{e}_{r-r_{\mathrm{N}}}\right)/(8\pi)\end{aligned} \tag{12}$$

When any two conductor segments with same current $\dot{I}$ are connected at the end, as shown in Fig. 2,

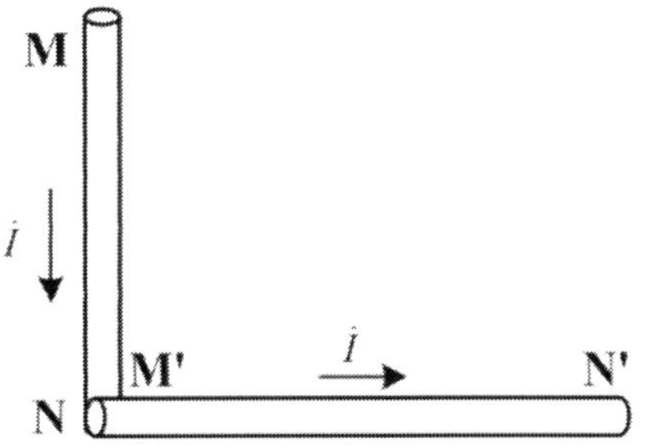

Fig. 2. Connection of two conductors.

The four end points are noted as M, N, M', and N', where N and M' is connected. On the joint, because opposite charges accumulate in the same location and generate opposite displacement current, they generate opposite A in space as (16).

$$\dot{\boldsymbol{A}}_{\mathrm{DN}} = -\dot{\boldsymbol{A}}_{\mathrm{DM'}} \tag{13}$$

where $\dot{\boldsymbol{A}}_{\mathrm{DN}}$ and $\dot{\boldsymbol{A}}_{\mathrm{DM'}}$ are generated respectively by charges accumulating on point N and M'. Therefore effect of them is compensated. Sum A is

$$\begin{aligned}\dot{\boldsymbol{A}} &= \dot{\boldsymbol{A}}_{\mathrm{C}} + \dot{\boldsymbol{A}}_{\mathrm{D}} \\ &= \dot{\boldsymbol{A}}_{\mathrm{C}} + \dot{\boldsymbol{A}}_{\mathrm{DM}} + \dot{\boldsymbol{A}}_{\mathrm{DN}} + \dot{\boldsymbol{A}}_{\mathrm{DM'}} + \dot{\boldsymbol{A}}_{\mathrm{DN'}} \\ &= \dot{\boldsymbol{A}}_{\mathrm{C}} + \dot{\boldsymbol{A}}_{\mathrm{DM}} + \dot{\boldsymbol{A}}_{\mathrm{DN'}}\end{aligned} \tag{14}$$

Similarly, when conductor segments form a closed loop, all effect of displacement can be compensated based on current continuity equation. That means the introduced item $\dot{\boldsymbol{A}}_{\mathrm{D}}$ will not influence any results of closed loops by existing method, so it can be also applied in a closed loop. But $\dot{\boldsymbol{A}}_{\mathrm{D}}$ may modify result of partial inductance for conductor segments.

https://doi.org/10.1051/e3sconf/20186404004

3 Partial Inductance Calculation

Existing method of partial inductance in [4] assumed that conductor segment S1 as a portion of half-infinite loop in Fig. 3,

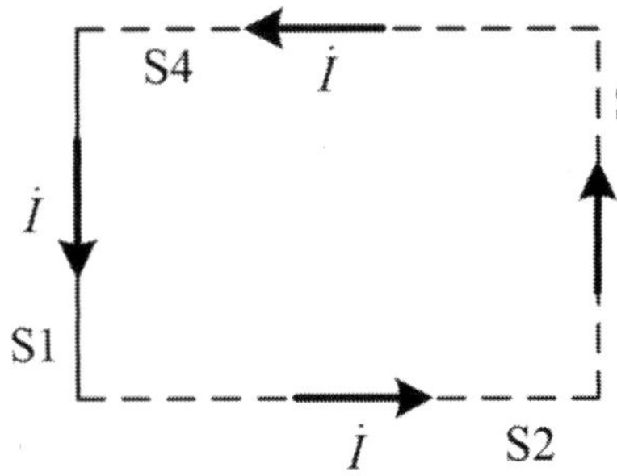

Fig. 3. Conductor segment S1 in a half-infinite loop.

where segments 2 and 4 have infinite length. Partial inductance can be calculated by (18) based on existing method. $\boldsymbol{A}_{Ci}$ is generated by conduction current in segment S*i*.

$$L_{p11}=\frac{1}{\dot{I}}\oint\dot{\boldsymbol{A}}_{C1}\cdot d\boldsymbol{l}=\frac{1}{\dot{I}}\sum_{i=1}^{4}\int_{Si}\dot{\boldsymbol{A}}_{C1}\cdot d\boldsymbol{l}$$
$$=\frac{1}{I}\int_{S1}\boldsymbol{A}_{C1}\cdot d\boldsymbol{l}+0+0+0 \quad (15)$$

Considering displacement current, because direction of $\boldsymbol{A}_D$ is different with current, last three parts of integral in (18) is not all of zero. So partial inductance of conductor segments is redefined as followed.

$$\frac{1}{I}\oint(\boldsymbol{A}_{C1}+\boldsymbol{A}_{D1})\cdot d\boldsymbol{l}$$
$$=L'_{p11}+L'_{p12}+L'_{p13}+L'_{p14} \quad (16)$$
$$=\frac{1}{I}\sum_{i=1}^{4}\int_{Si}(\boldsymbol{A}_{C1}+\boldsymbol{A}_{D1})\cdot d\boldsymbol{l}$$

Where $\boldsymbol{A}_{C1}$ and $\boldsymbol{A}_{D1}$ are generated by segment S1. L'_{pij} represents mutual partial inductance between segment S*i* and S*j*. When *i*=*j*, L'_{pij} means self partial inductance of segment S*i*. Therefore, inductance of segment S1 can be calculated by

$$L'_{p11}=\frac{1}{I}\int_{S1}\boldsymbol{A}_{C1}\cdot d\boldsymbol{l}+\frac{1}{I}\int_{S1}\boldsymbol{A}_{D1}\cdot d\boldsymbol{l} \quad (17)$$

As mentioned above, $\boldsymbol{A}_{D1}$ is irrotational, i.e., the second integral is merely related to locations of head and end points in path S1, leading to a convenient calculation.

Similarly, mutual partial inductance between two segments S*i* and S*j* is defined by

$$L'_{pij}=\frac{1}{I}\int_{Sj}\boldsymbol{A}_{Cij}\cdot d\boldsymbol{l}+\frac{1}{I}\int_{Sj}\boldsymbol{A}_{Dij}\cdot d\boldsymbol{l} \quad (18)$$

where $\boldsymbol{A}_{Cij}$ and $\boldsymbol{A}_{Dij}$ are along path S*j* and are generated by segment S*i*.

4 Numerical Experiment for Comparison

A simplified structure based on practical module, as shown in Fig. 4, is used for numerical experiment and comparison of proposed method and existing method.

Fig. 4. Copper emitter pillars layout of press-pack IGBT

Inductance due to pillars is important part of stray inductance of package and has obvious influence on current distribution [9], hence we compare the partial inductance of pillars using both proposed method and existing method. The calculated model is shown in Fig.5.

Fig. 5. Calculated model

We make one simplification that the notches of pedestals are neglected, so all of pedestals are cuboid. Self-inductance and mutual-inductance of pillars are calculated and shown followed.

Table. 1. Comparison of partial inductance calculation using existing method and proposed method

Partial Inductance	Existing method	Proposed method
Self-L of 1	1.83 nH	1.24 nH
Mutual-L of 1&2	0.665 nH	0.362 nH
Mutual-L of 1&3	0.337 nH	0.172 nH
Mutual-L of 1&4	0.225 nH	0.114 nH

Such 16 pillars lead to 256 results of inductance, so typical results are picked and shown in Table. 1. Results show that displacement current obviously affect calculation of partial inductance in practical situation.

The influence caused by inductance extraction is analyzed by simulation of IGBT current calculation. The electrical structures of press-pack IGBT in Fig.4 is shown followed.

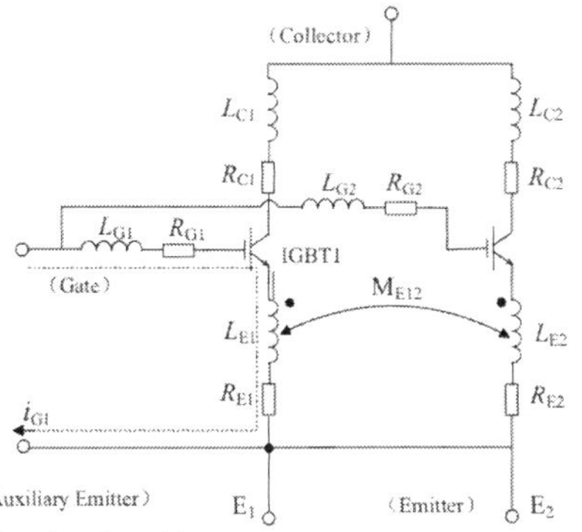

Fig. 5. Calculated model

where L_{C1}, L_{C2}, R_{C1}, R_{C2} represent collector inductance and collector resistance for IGBT1 and IGBT2 separately, L_{E1}, L_{E2}, R_{E1}, R_{E2} are emitter inductance and emitter resistance. L_{G1}, L_{G2}, R_{G1}, R_{G2} represent inductance and collector resistance of gate side. M_{E12} represents mutual inductance of emitter side.

Simulation and experiment are made for comparing above two methods. Result of current is shown followed.

Table. 2. Result of current

Current peak	Existing method	Proposed method	Experiment Measurement
Chip 1	7.49	7.58	8.10
Chip 2	21.4	21.6	22.6

Results show that proposed method could improve the calculation accuracy of simulation. The influence will be obvious whereas IGBT module become larger.

5 Conclusion

Precious inductance extraction of conductor is important in analysis of electrical performance. This paper proposes a method of inductance extraction for press-pack IGBT where conductor segments might not form a closed loop. By considering displacement current, the conductor segment model could meet ACL. Vector potential $\boldsymbol{A}$ also could be affected by displacement current, which leads to a different formulation of partial inductance. A simplified model based on practical structure are used in numerical experiment, and obtains different results of existing method and proposed method.

Acknowledgement

This work was supported by the 02 Support Project: Research on key technology of IGBT with high short circuit shutdown capability (SGRI-GB-71-16-003)

References

1. A. E. Ruehli, "Inductance Calculations in a Complex Integrated Circuit Environment," *IBM Journal of Research and Development*, vol. 16, no. 5, pp. 470-481, Sept. 1972.
2. Paul C R, "The Concept of Partial Inductance," in *Inductance: loop and partial*, New Jersey: John Wiley & Sons, 2011, pp. 195–245.
3. Holloway C L, Kuester E F, Ruehli A E, et al. "Partial and internal inductance: Two of Clayton R. Paul's many passions," *Electromagnetic Compatibility, IEEE Transactions on*, vol. 55, no. 4, pp. 600-613, 2013.
4. Cao Y S, Jiang L J, Ruehli A E. "Distributive radiation and transfer characterization based on the PEEC method," *Electromagnetic Compatibility, IEEE Transactions on*, vol. 57, no. 4, pp. 734-742, 2015.
5. Antonini G, Romano D. "Adaptive-cross-approximation-based acceleration of transient analysis of quasi-static partial element equivalent circuits," *Microwaves, Antennas & Propagation, IET*, vol. 9, no. 7, pp. 700-709, 2015.
6. Hackl Y, Scholz P, Ackermann W, et al. "Multifunction Approach and Specialized Numerical Integration Algorithms for Fast Inductance Evaluations in Nonorthogonal PEEC Systems," *Electromagnetic Compatibility, IEEE Transactions on*, vol. 57, no. 5, pp. 1155-1163, 2015.
7. Kalhor H. "Comparison of Ampere's circuital law (ACL) and the law of Biot-Savart (LBS)," *Education, IEEE Transactions on*, vol. 31, no. 3, pp. 236-238, 1988.
8. Kalhor H. "The degree of intelligence of the law of Biot-Savart," *Education, IEEE Transactions on*, vol. 33, no. 4, pp. 365-366, 1990.
9. TANG Xinling, CUI Xiang, ZHAO Zhibin, et al, "Analysis of Transient Current Distribution Characteristics of Parallel Chips in Press Pack IGBT," *Proceedings of the CSEE*, vol. 36, pp. 1-11, 2016 (in Chinese).

Influence of Parasitic Inductances on Switching Performance of SiC MOSFET

Jinyuan Li[1], *Meiting* Cui[1], *Yujie* Du[1], *Junji* Ke[2] *and Zhibin* Zhao[2]

[1]State Key Laboratory of Advanced Power Transmission Technology (Global Energy Interconnection Research Institute of State Grid Corporation of China), Beijing 102209, China
[2]State Key Laboratory of Alternate Electrical Power System with Renewable Energy Sources (North China Electric Power University), Beijing 102206, China)

Abstract. Compared to the silicon power devices, silicon carbide device has shorter switch time. Hence, as a result of the faster transition of voltage (dv/dt) and current (di/dt) in SiC MOSFET, the influence of parasitic parameters on SiC MOSFET's switching transient is more serious. This paper gives an experimental study of the influence of parasitic inductance on SiC MOSFET's switching characteristics. Most significance parameters are the parasitic inductances of gate driver loop and power switching loop. These include the SiC MOSFET package's parasitic inductance, interconnect inductance and the parasitic inductance of dc link PCB trace. This paper therefore focuses on analysis and comparison of different parasitic parameters under various operation conditions in terms of their effect on overvoltage, overcurrent and switching power loss.

1 Introduction

Power semiconductor devices are attracting increasing attention as key components in a variety of power electronic systems. Silicon power devices have improved significantly over the past several decades, but they are now approaching performance limits imposed by the fundamental material properties of silicon, and further progress can only be made by migrating to more robust semiconductors [1]. As a wide-bandgap semiconductor, Silicon carbide device has superior physical and electrical properties such as the high blocking voltage, high-frequency capability, high-temperature tolerance and low-loss.

An increase in switching frequency has always been in great quest to push up the power density and facilitate the miniaturization of power electronics converters [2]. Due to SiC MOSFET's fast switching speed, it is gradually obtaining the widespread application in high-frequency power converters. However, in the traditional hard switching converter topology, high speed silicon carbide MOSFET device results in a very fast switching transient thereby subjecting the converter to bear higher du/dt and di/dt. The larger slopes of the voltage and current will induce larger current and voltage respectively in stray capacitances and inductance in the converter circuit. This leads to larger overshoots in voltage and current profiles with the potential effect of damaging SiC MOSFET device. In addition, electromagnetic oscillation may also occur due to the interaction between the voltage and current transitions with the circuit stray inductance and capacitance. This may increase switch losses as well as contributing to electromagnetic interference problems.

There has been a significant effort on the study of SiC MOSFET, but the majority of research have focused on the design, packaging, modeling and application of SiC MOSFET power device [3-6]. Only a few studies on the influence of parasitic parameters on SiC power MOSFET switching characteristics have been mentioned. Investigations into the effect of parasitic elements on switching performance of power device can be classified into three categories. The first method focuses on the extraction and modeling of these parasitic parameters with ANSYS Q3D to measure their effect using circuit simulations [7-8]. This, however, needs a precise device model. The second method involves developing an analytical model of the MOSFET [9-11]. But it is very difficult to obtain the accurate analytical expressions of the whole switching transient characterization in case of taking account of all parasitic elements. Therefore, certain parasitic elements such as gate inductance and the change of terminal parasitic capacitance of MOSFET may be neglected. The last approach is the experimental method. Though the method still cannot suffice to explain the physical meaning behind those observations, it can intuitively display the relationship between parasitic parameters and switching performance, and show the level of influence of each parasitic parameter [12-13].

This paper analyzes circuit operation and

systematically studies the effects of individual parasitic inductance under various operation conditions by conducting lots of experiments. All the gate loop and power switching loop parameters have been independently controlled and studied based on the SiC MOSFET switching characterization test bench. Their influences on SiC MOSFET switching characteristics can be shown from the experimental waveforms. Moreover, the degree of influence of all parameters have been evaluated in terms of overvoltage, overcurrent and switching loss.

2 Test Circuit Operation Simply Analysis

Most of the power electric systems usually operate under clamped inductive load, and the load inductance is relatively large. After SiC MOSFET switches off, the load inductance current generally will not be discontinuous as it continues flowing through the freewheeling diode. When SiC MOSFET turns on, there will be a diode reverse recovery process. The single pulse experiments cannot reflect the diode reverse recovery process. Hence, a double pulse test (DPT) has been used for evaluating SiC MOSFET's switching characteristics in this work. Double pulse test schematic is shown in Fig. 1. The gate driver test signal is a double pulse waveform as shown in Fig. 2.

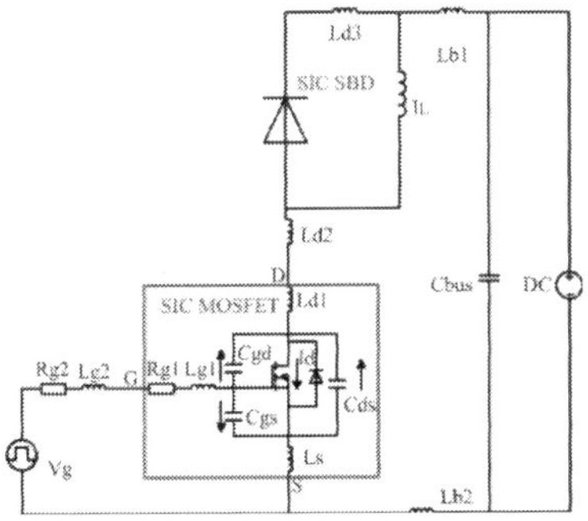

Fig.1 Double pulse test equivalent circuit with parasitic parameters

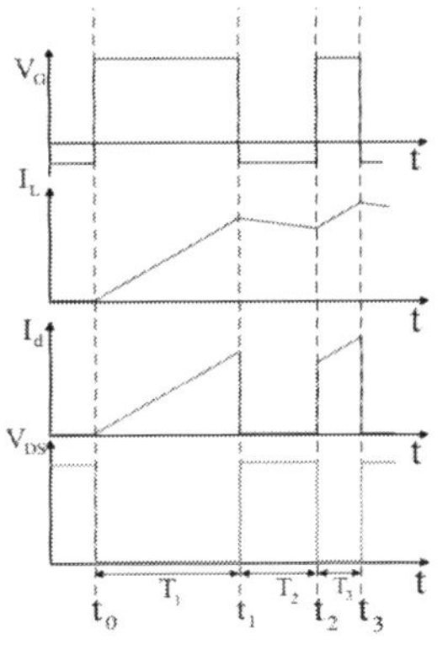

Fig.2 Double pulse test waveforms

At time t_0, the SiC MOSFET which is the device under test (DUT) is turned on. It is held on until current reaches the desired level. At time t_1, the gate is forced to negative level and the turn-off switching characteristics of the SiC MOSFET and the turn-on characteristics of the diode are recorded. During t_1-t_2, the SiC MOSFET is kept off and the inductor current freewheels through the diode. At time t_2, the device is turned back on. The turn-on characteristics of SiC MOSFET and the turn-off characteristics of the diode are then captured. At time t_3, the SiC MOSFET is turned off, the measurement is completed and then the inductor current falls to zero through the freewheeling diode.

The schematic of this circuit taking consideration of all parasitic parameter is shown in Fig. 1. Based on derivation of Faraday's Law, it is clear that inductance is only defined for a closed contour. According to the loop concept, for the stray inductances' parameters to be studied, the network's stray inductance can be narrowed down to three primary lumped inductances according to the current paths. 1) the gate loop inductance, L_G, formed by the gate current path, where $L_g = L_{g1}+L_{g2}$; 2) the main switching loop inductance, L_d, formed by the drain current path, which includes the stray inductances of the MOSFET and Diode packages as well as PCB interconnections, $L_d=L_{d1}+L_{d2}+L_{d3}+L_{b1}+L_{b2}$; 3) the common source inductance L_S that exists in both loops, which can be treated as a mutual inductance between the gate loop and the switching loop.

3 Experimental Study Of The Parasitic Parameters

3.1 Hardware setup

In order to evaluate the parasitic parameters impact on the device switching performance, the DPT bench needs to be designed and optimized. The implemented prototype is designed to accommodate a TO-247 for the switch and the diode. The test circuit is constructed using a 1200V/36A SiC MOSFET C2M0080120D and a 1200V/33A SiC Schottky diode C4D20120D from Cree Semiconductor in TO-247 package. As shown in Fig.3, the test platform includes 4 parts of PCB including the digital signal processor (DSP) evaluation board 00IC TOP2812, gate drive board, isolation type power supply board and dc bus board. The DSP system board is used to generate the required double pulse control signal. The gate drive board uses the ACPL-W343 driver chip which is internally integrated with an optical coupling isolation to magnify the control signal. At the same time, the isolation type power supply board based on MORNSUN QA01C is used to provide positive and negative bias voltage for gate driver. In addition, a + 24 V and -5 V transient suppression diodes are connected between the gate and source terminals to protect gate from breakdown.

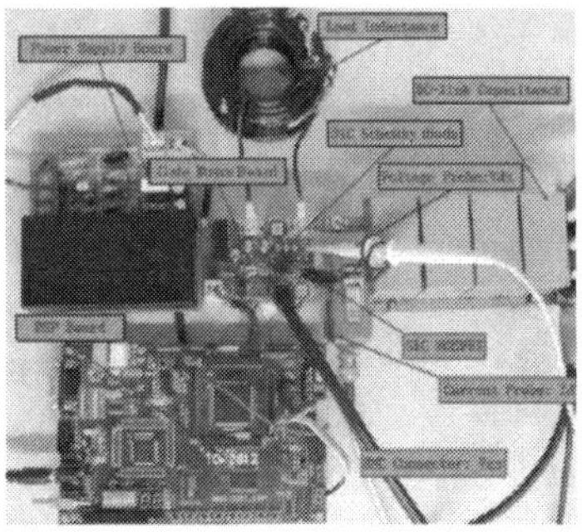

Fig.3. Double pulse test bench

3.2 Analysis of switching characteristics

Due to the parasitic inductances and a d*i*/d*t* during switching time, there are inductive voltage drops or overshoot across the parasitic inductances. These voltages together cause a voltage drop on the drain-source voltage:

$$\Delta V_1 \approx L_{\text{loop}} \frac{\mathrm{d}i_f}{\mathrm{d}t} \tag{1}$$

$$\Delta V_2 \approx L_{\text{loop}} \frac{\mathrm{d}i_r}{\mathrm{d}t} \tag{2}$$

Where:

$$L_{\text{loop}} = L_d + L_s \tag{3}$$

The MOSFET switching waveforms under the influence of this inductance is shown in Fig. 4. The device is switched under 500 V and 18 A condition with a gate resistance R_G of 37 Ω. In the experiment, total power loop parasitic inductances of the test bench are estimated and compared by Eq.1 and Eq.2 in table I. The result shows that the average of total power loop parasitic inductances is about 136nH.

Table 1. Total Power Loop Parasitic Inductance Computation

Compute by Eq.1			Compute by Eq.2		
ΔV_2	$\mathrm{d}i_r/\mathrm{d}t$	L_{loop}	ΔV_1	$\mathrm{d}i_f/\mathrm{d}t$	L_{loop}
60V	434A/μs	138nH	56V	418 A/μs	134nH

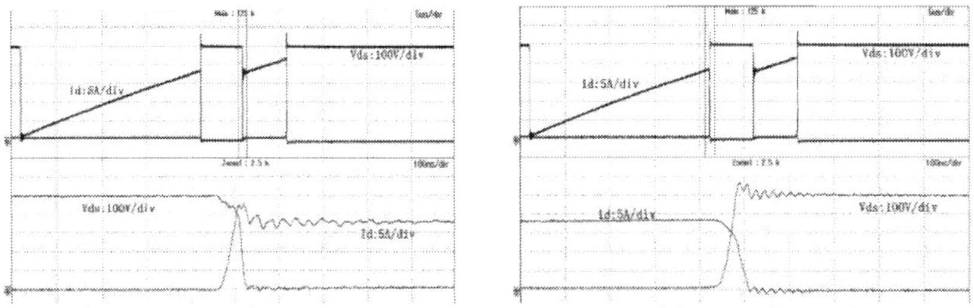

a) turn-on waveform b) turn-off waveform

Fig.4. Experimental drain current and drain-source voltage waveforms

3.3 Influence study of different parasitic inductances

To study the influence of these parasitic parameters, the current paths in the tester have been broken at several points, in which way small different inductances can be inserted into the loop to mimic the change of parasitic ones. The different inductances are made of several air-core coils with the corresponding inductances of 20nH, 34nH, 48nH, 64nH, and 77nH. In the experiment, each type of stray parameter is controlled with the rest of the parameters kept constant, such that its effect can be studied independently. The influence on the device stress is shown in Fig. 5, which indicate that the gate loop parasitic inductance has only a minor effect on the SiC MOSFET switching waveforms. The drain-source voltage overshoot increases with the increase of power loop parasitic inductance L_d. Compared to the voltage overshoot without insertion of an additional inductor, the voltage overshoot with insertion of a 77nH inductor increases by approximately 8%. However, drain current overshoot decreases with the increase of inductance L_d. The common source inductance L_s relieves the device of switching stresses by slowing down the switching process due to its negative feedback nature. Due to the fast switching characteristic of SiC devices, the change rate of voltage and current is faster than Si devices. The same L_d with a lager current rate will result in larger voltage overshoot, similarly, L_s will cause the slower switching speed. So L_d and L_s have a greater impact on the switching process of SiC than Si devices. In addition, Fig.6 shows the effect of parasitic inductances on the switching energy E_{on}, E_{off} and E_{total}. According to IEC standard, they are defined as follows.

$$E = \int_{t_1}^{t_2} V_{ds} \cdot I_d \mathrm{d}t \tag{4}$$

$$E_{\text{total}} = E_{\text{on}} + E_{\text{off}} \tag{5}$$

As is shown in Fig.6, the increase in L_g almost did not affect the switching energy. L_d reduces the turn-on energy a little bit due to its snubber effect, and increases

 https://doi.org/10.1051/e3sconf/20186404005

the turn-off energy a little bit due to the higher V_{ds} overshoot. However, the total loss of both L_g and L_d are still almost unchanged. Fig.6 also indicate that L_s has more influence on switching energy. The switching energy at turn-on and turn-off transients significantly increases with greater L_s. This is because the inductance L_s forms negative feedback in gate drive loop, and increases the switching time. The total switching loss when an additional 77nH inductance is inserted is approximately three times bigger than E_{total} without insertion of the inductance. For Si devices, larger L_s will also leads to increased switching loss, but the effect is relatively small.

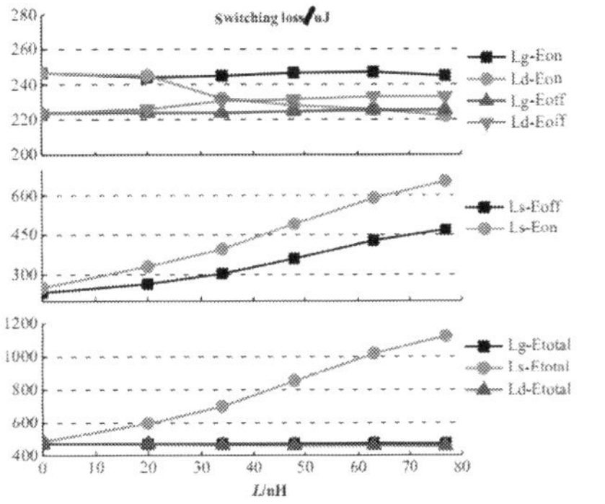

Fig.6. SiC MOSFET switch loss vs. various parasitic inductance

3.4 Influence of parasitic inductances under different load current

Fig.7 shows the influence of parasitic inductances on voltage overshoot and current overshoot under different values of load current. The experimental bus voltage is kept constant at 500V. It is observed that the relationship between voltage overshoot and load current is almost linear, which proves the correctness of the Eq.1. On the basis of the rated current, the current overshoot percentage increases a little bit due to the increase of slope of current with greater load current.

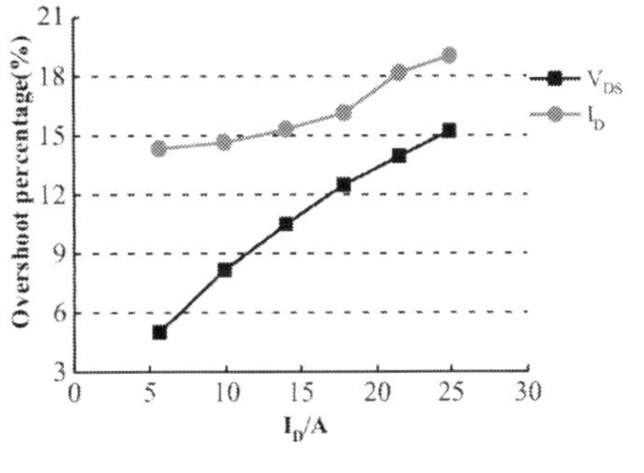

Fig.7. SiC MOSFET stress under load current

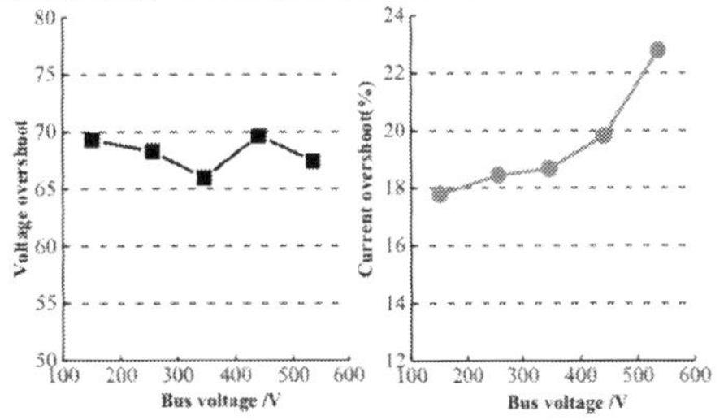

Fig.8. SiC MOSFET stress under various bus voltage

3.5 Influence of parasitic inductances under different bus Voltage

In the practical application, the device maybe work under different voltage levels. In order to estimate the influence of parasitic parameter on SiC MOSFET electrical stress, many experiments are conducted in this paper under a variety of bus voltages. As shown in Fig.8, the drain-source voltage overshoot almost remains constant, which implies that the overshoot is not dependent on the bus voltage level. However, for the higher voltage levels, the overshoot is relatively smaller. Due to the constant length of the switching period, the higher the bus voltage the greater the slope of drain-source voltage. Hence, the voltage transition would induce a displacement current on the output capacitance of the device, thereby increasing the current overshoot in turn-on transient.

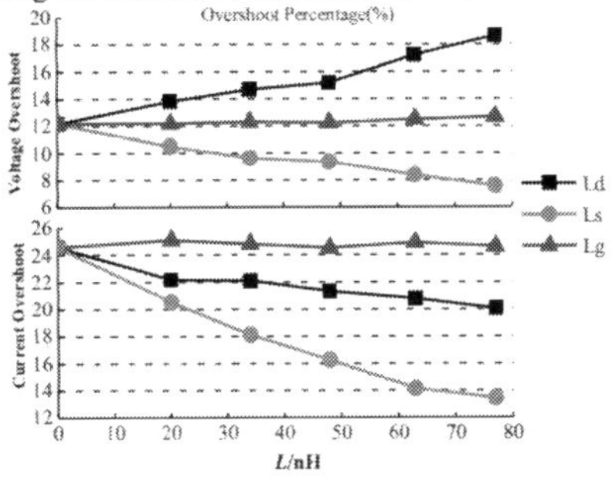

Fig.5. SiC MOSFET stress vs. various parasitic inductance

4 Conclusion

This paper has systematically studied the influence of the parasitic inductances under various operating conditions on the SiC MOSFET's switching characteristics, electrical stress and switching loss. It is shown that a larger power loop inductance will result in more voltage overshoot but less current overshoot. The larger inductance will also lead to more turn-off losses but less turn-on losses while the total loss remains unchanged. It's found that the common source inductance has the most significant effect on the switching performance because it is included not only in the main power circuit but also in the gate drive circuit. The experimental results show that the inductance, Ls, slows down the turn-on and turn-off processes, thus increasing the turn-on and turn-off switching power losses. The conclusion is useful for switching loss measurement system design, and the

https://doi.org/10.1051/e3sconf/20186404005

common source inductance should be minimized for realizing a good switching performance.

Acknowledgement

This work was supported by the National Key Research and Development Program (2016YFE0102800)

References

1. Kimoto T, Cooper J A. Fundamentals of silicon carbide technology [J]. 2014.
2. B. J. Baliga, Fundamentals of Power Semiconductor Devices. New York: Springer, 2008.
3. Lee J S, Chun D H, Park J H, et al. Design of a novel SiC MOSFET structure for EV inverter efficiency improvement[C]//IEEE, International Symposium on Power Semiconductor Devices & Ic's. 2014:281 - 284.
4. Miao Z, Mao Y, Ngo K, et al. Package influence on the simulated performance of 1.2 kV SiC modules[C]// Wide Bandgap Power Devices and Applications. IEEE, 2015.
5. Xu Y, Li H, Zheng T Q, et al. Study on the Pspice Simulation Model of SiC MOSFET base on its Datasheet[C]// Ifeec. 2015.
6. Sasagawa M, Nakamura T, Inoue H, et al. A study on the high frequency operation of DC-DC converter with SiC DMOSFET[C]// Power Electronics Conference (IPEC), 2010 International. IEEE, 2010:1946-1949.
7. Xiang xiang F. Characterization and Modeling of SiC Power MOSFETs [J]. 2012.
8. Anthon A, Hernandez J C, Zhang Z, et al. Switching investigations on a SiC MOSFET in a TO-247 package[C]// Industrial Electronics Society, IECON 2014-,Conference of the IEEE. IEEE, 2014.
9. Xiao Y, Shah H, Chow T P, et al. Analytical modeling and experimental evaluation of interconnect parasitic inductance on MOSFET switching characteristics[C]// Applied Power Electronics Conference and Exposition, 2004. Apec '04. Nineteenth IEEE. 2004:516-521 Vol.1.
10. Wang J, Chung S H, Li T H. Characterization and Experimental Assessment of the Effects of Parasitic Elements on the MOSFET Switching Performance [J]. IEEE Transactions on Power Electronics, 2013, 28(1):573-590.
11. Ren Y, Xu M, Zhou J, et al. Analytical loss model of power MOSFET [J]. IEEE Transactions on Power Electronics, 2006, 21(2):310-319.
12. Z. Chen, D. Boroyevich, and R. Burgos, "Experimental parametric study of the parasitic inductance influence on MOSFET switching characteristics," in Proc. Int. Power Electron. Conf., 2010, pp. 164–169.
13. Y. Q. Shen, J. Jiang, Y. Xiong, Y. Deng, and X. N. He, "Parasitic inductance effects on the switching loss measurement of power semiconductor devices," in IEEE Int. Symp. Ind. Electron. Jul. 2006, vol. 2, pp. 847–852.

Electric Field Analysis of Press-Pack IGBTs

XinlingTang[1], *Yan* Pan[1], *Yanfang* Chen[1], *Pengyu* Fu[2] and *Zhibin* Zhao[2]

[1]State Key Laboratory of Advanced Power Transmission Technology (Global Energy Interconnection Research Institute of State Grid Corporation of China), Beijing 102209, China
[2]State Key Laboratory of Alternate Electric Power System with Renewable Energy Sources (North China Electric Power University), Beijing 102206, China

Abstract. High voltage IGBT module is the ideal option for the VSC-HVDC power transmission application. At present, wire-bonded technology and press-pack technology are available packaging technologies for high voltage IGBT. The press-pack IGBTs have such advantages as low inductance, low thermal impedance and short circuit failure mode than the wire-bonded IGBT module, which especially suit for high voltage power transmission application by series connection. However, the electrical insulation failure modes of press-pack IGBTs are much less known with limited literature published. In this paper, we presented the electric field analysis of a 3D press-pack IGBT model under DC rating voltage test condition. The electric field distribution of the press-pack IGBT stack was solved as an electrostatic problem by employing the finite element method. The results revealed the potential electrical insulation failure modes of the press-pack IGBTs: corona discharge at the edge of silver plate, partial discharge at the micro gap between die and PEEK frame and creeping discharge at the surface of PEEK frame.

1 Introduction

The demand to achieve a balance between environment and energy is urgent. The fossil energy supports the industrialization of human society but also causes the greenhouse effect. Searching for clean and renewable energy alternatives is key importance to the development of human society. However, the bulk distribution of available alternatives (such as wind, solar etc.) is usually far from the load center (especially in China). Achieving bulk and reliable transmission of the clean and renewable energy is a technological challenge.

HVDC power transmission technology is a key technology to realize the integration of clean and renewable energy to the power grid [1]. Power electronics is an enable technology to achieve the conversion and control of power. Power electronics based converter is the key component of HVDC power transmission system. In fact, HVDC power transmission technology is mainly subdivided into two categories: LCC-HVDC and VSC-HVDC according to the power semiconductor devices used in the converter. The half-controlled thyristors are used in the converters of LCC-HVDC technology. However, the full-controlled devices (such as IGBT, IEGT) are used in the converters of VSC-HVDC technology. Comparing with LCC-HVDC technology, VSC-HVDC technology is more flexible and is especially suitable for integrating the versatile renewable energy into the power grid.

Continuous improvements in the device design and in their manufacturing process have gradually increased the ratings of the IGBT. The VSC-HVDC converter is a main application of the high voltage and high power IGBT device. In the power transmission application, a high level of reliability of the device is required to prevent costly or catastrophic failure. In this context, the device packaging is assumed to play a critical role. At present two packaging technologies for IGBT devices are available for application in the MW power range: the wire-bonded IGBT module and press-pack IGBTs. Power modules have been object of extensive investigation including their characterization under variable operating conditions and the analysis of failure mechanisms [2]. The press-pack packaging eliminates bonding wires and solder layers, and is claimed to offer improved reliability [3]-[5]. However, the knowledge on press-pack IGBTs is much less mature with only limited literature published. In addition, High voltage blocking capability is one of the most important performance for the press-pack IGBTs. Therefore, the analysis of the electrical insulation failure modes of press-pack IGBTs needs to be performed.

2 Insulation Failure of Press-Pack IGBT

In Fig.1, the explosive view and schematic of the cross section of a conceptional press-pack IGBTs are shown. The packaging technology evolves from press-pack thyristors and GTOs, which have proven to be excellent in ruggedness and reliability [2]. Press-pack thyristors and

GTOs use a single large die in the capsule. However, it is still not feasible to fabricate a single large IGBT die for high voltage application from a single wafer due to the fine pattern of the IGBT cell structure. To overcome the technological challenge, press-pack IGBT is developed as a multi-die device, as shown in Fig. 1. The device consists of multiple die stacks, and each stack is composed of an IGBT or FRD die, two molybdenum plates, and a die frame. The emitter electrode is multi-stamp and collector electrode is single-stamp. Both electrodes transfer the external clamping force to the individual IGBT or FRD stack.

High operating voltage (⩾3300V) of IGBT device results in higher demands on electrical insulation capability as well as partial discharge resistance [6]. Unlike the power module in which silicone gel is used to provide the dielectric strength necessary to guarantee the required insulation capability [7], [8], press-pack IGBT uses protective gas to provide electrical insulation. When studying the insulation failure mechanism of high voltage press-pack IGBT, the dies, the protective gas and the plastic die frame should be the main concerns.

In the press-pack IGBTs, the main insulation aging processes are partial discharge and creeping discharge. Partial discharge occurs in the protective gas when the electric field exceeds the critical electric field, which depends on the pressure and the gas gap. Creeping discharge occurs at the interface of gas and dielectric when the tangent electric field exceeds the critical electric field, which depends on the dielectrics and the condition of the gas. To improve the electrical insulation performance of press-pack IGBTs, it is essential to analyze the electric field distribution in the structure.

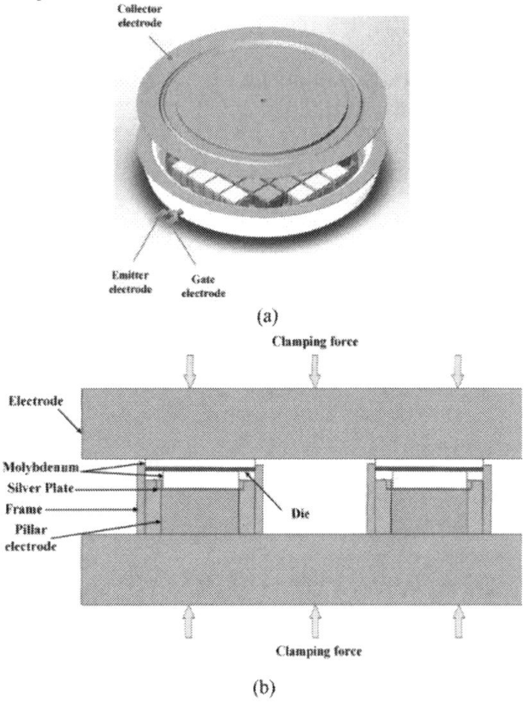

Fig. 1. Press-pack IGBTs. (a) explosive view. (b) Schematic of its cross section view

3 Electric Field Simulation

3.1 Modelling of a Press-Pack IGBT Stack

Press-pack IGBT is a mulit-die device. When analyzing the electric field of the press-pack IGBT under rating voltage test, it only needs to analyze the electric field of a single IGBT stack. In Fig. 2, a 3-D model for the analysis of a press-pack IGBT stack is presented. In the model, the IGBT stack is located in the center of the solving region. The outer box is the equivalence of the package housing. The upper surface is an analogy to the collector side and the lower surface is an analogy to the emitter side. The other 4 surfaces are an analogy to the ceramics of the package housing.

The model simulates the condition of press-pack IGBTs under V_{CES} type test [9]. The V_{CES} (3300V) is applied to the collector of IGBT die, the molybdenum plate attached to the collector of the IGBT die, and the collector side of the outer box. Zero voltage is applied to the emitter of the IGBT die, the silver plate, the molybdenum plate attached to the emitter of the IGBT die and the emitter side of the outer box. The other 4 surfaces are set to be Neumann boundary condition of the electric potential.

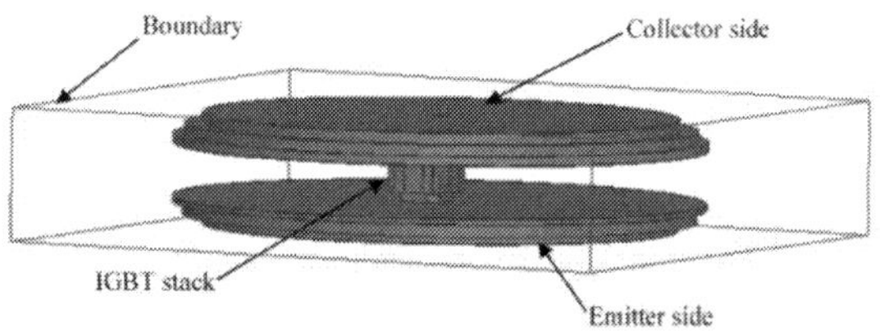

Fig. 2. 3-D model for the analysis of a press-pack IGBT stack

3.2 Analysis Method

The fundamental description of all macroscopic electromagnetic phenomenon is Maxwell equations:

$$\begin{cases} \nabla \cdot \vec{D} = \rho \\ \nabla \times \vec{E} = -\dfrac{\partial \vec{B}}{\partial t} \\ \nabla \cdot \vec{B} = 0 \\ \nabla \times \vec{H} = \vec{J} + \dfrac{\partial \vec{D}}{\partial t} \end{cases} . \quad (1)$$

When the rate of change of the system is low, the electric filed induced by the changing magnetic field could be neglected. If the electric field of the system only has Coulomb field component, then this kind of electric field is called quasi-static electric field. In this context, the electric field of the press-pack IGBT stack could be assumed to be quasi-static electric field due to the low change rate of magnetic field. Under this assumption, the electric field could be represented as

$$\vec{E} = -\nabla \varphi . \quad (2)$$

In equation (2), φ represents the electric potential. Do the divergence operation to the both side of the fourth equation in (1), and we obtain

$$\nabla\cdot(\nabla\times\vec{H})=\nabla\cdot(\vec{J}+\frac{\partial\vec{D}}{\partial t})=0\ . \quad (3)$$

Substituting $\vec{J}$ and $\vec{D}$ in (3) with $\vec{J}=\gamma\vec{E}$, $\vec{D}=\varepsilon\vec{E}$ and $\vec{E}=-\nabla\varphi$, the equation (3) is changed into

$$\nabla\cdot(\gamma\nabla\varphi)+\nabla\cdot(\frac{\partial}{\partial t}(\varepsilon\nabla\varphi))=0\ . \quad (4)$$

The phasor form of equation (4) is

$$\nabla\cdot(\gamma\nabla\dot{\varphi}+j\omega\varepsilon\nabla\dot{\varphi})=\dot{0}\ . \quad (5)$$

In equation (5), $|\dot{J}_c|=|\gamma\nabla\dot{\varphi}|$ represents the conduction current and $|\dot{J}_d|=|j\omega\varepsilon\nabla\dot{\varphi}|$ represents the displacement current.

When deciding the electric field control equation of a region filled with material, the electric property of the material is vital. The ratio of $|\dot{J}_d|$ to $|\dot{J}_c|$ is

$$\frac{|\dot{J}_d|}{|\dot{J}_c|}=\frac{|j\omega\varepsilon\nabla\dot{\varphi}|}{|\gamma\nabla\dot{\varphi}|}=\frac{\omega\varepsilon}{\gamma} \quad (6)$$

If the ratio is much smaller than 1, the electric field of the region could be assumed to be controlled by electrostatic field. Otherwise, if the ratio is much larger than 1, the electric field of the region could be assumed to be controlled by DC electric field. The two electric field types mentioned above are constant. Equation (4) descripts the time-varying process of electric field. From Equation (4), the transition process of the electric field can be characterized by the relaxation time

$$\tau_r=\frac{\varepsilon}{\gamma} \quad (7)$$

If the test duration is much shorter than the relaxation time, then the electric field should be analyzed by using the electrostatic field. Otherwise, DC electric field should be used.

The electric property of important materials involved in the model is shown in Table I [10]-[11]. The relaxation time of metals in the model is much lower than 1s, and the metals are considered to be electrostatic. The relaxation time of PEEK is 47.2 min, and the relaxation time of N_2 is 24.6 h. During the short DC test, the test time is much lower than the relaxation time of dielectric materials. Therefore, the electric field distribution is governed by electrostatic field. Under the power frequency voltage test, the ratio of $|\dot{J}_d|$ to $|\dot{J}_c|$ of the dielectric material region is much higher than 1. Therefore, the electric field distribution is governed by electrostatic field.

Table 1 Electric parameters of dielectric materials in the model

Material	Electric conductivity γ ($\Omega\cdot$m) (25°C)	Relative dielectric constant ε_r
PEEK	10^{-14}	3.2
N_2	10^{-16}	1.00058

The analysis is solved by employing Finite Element Method (FEM). However, the typical electromagnetic field solver does not solve the semiconductor equations. Many TCAD softwares are developed to fulfill this purpose (such as Silvaco). But for electric field analysis in the package, they are complex to use. In this simulation, IGBT die is modelled as silicon, and the dielectric constant of silicon was employed. The surface exception the one including emitter of the die is set to be 3300 V. This assumption is an ideal behavior for the IGBT die. In fact, due to the development of the die edge termination technique (such as guard rings or Junction Termination Extension), the manufacture of the die is toward to this goal. But the local electric field enhancement should be concerned, it could result in the partial discharge in the protective gas on the passivation layers of the IGBT dies. This situation is complex and is not considered in the simulation.

4 Results

The model of a press-pack IGBT stack is solved as an electrostatic problem by using the finite element method. Fig. 3 shows the electric potential lines around the metals and dies. The electric potential lines around the metallization edge of the die emitter and the silver plate are highly uneven distributed. The electric field at these regions could be highly reinforced, therefore weaken the electrical insulation of the device. The reinforcement of the electric field on the surface of die should be considered during the die design process.

Various edge termination techniques and passivation layers should be adopted to avoid the occurrence of partial discharge on the die surface under the rating voltage test. If the edge termination and passivation of the chip is good enough to block the rating voltage, then the chip surface should not be the weak point in this packaging structure. Another weak area is the edge of the silver plate. Comparing with other metals used in the structure, the silver plate is much thinner and has a relatively sharper corner. The analysis result in Fig. 3 shows that the electric potential lines are condense around the corner of the silver plate. It is highly risky to occur corona discharge at the corner of silver plate in the protective gas.

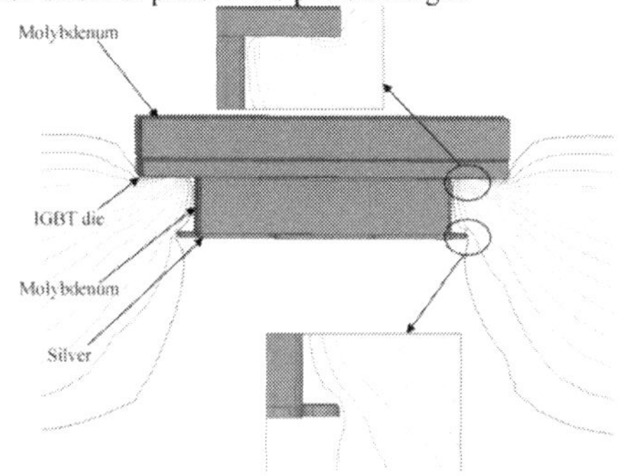

Fig. 3. Electric potential lines around the metals and dies

https://doi.org/10.1051/e3sconf/20186404006

On the other hand, the creeping discharge of dielectric material should be considered. Fig. 4 shows the electric potential distribution in the PEEK frame. The upper branch of the PEEK frame has a dramatic voltage drop, and the tangent component of the electric field is high. As the trace P1 to P2 shown in Fig. 4, it is possible to occur creeping discharge at the surface of the PEEK frame. Another region should be considered is the micro gap between the chip and PEEK. If the gap between the chip and PEEK is too narrow, the potential difference between the two may be high enough to cause micro-arcing in the gap. In this model, the gap between the PEEK and chip is wide enough to avoid this failure mode.

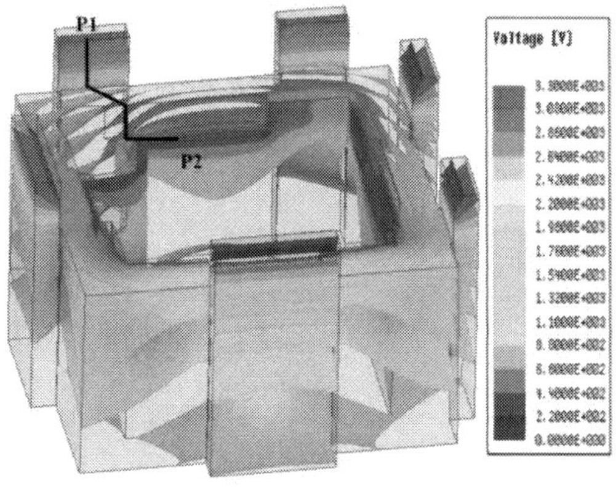

Fig. 4. Electric potential distribution in the PEEK frame

Fig. 5 shows the electric field distribution of the PEEK frame and the surface of the silver plate. The maximum electric field intensity is 9.16 kV/mm, which is located at the corner of the silver plate. The dielectric strength of the nitrogen gas is 3.3 kV/mm at 25°C and 1 atm[11]. The maximum electric field intensity on the silver plate is far beyond the dielectric strength and the corona discharge at the corner of the silver plate is expected. The result of electric field analysis shows that the die surface has a maximum field strength of 85 kV/mm. However, this unexpected high value is caused by the assumption made during the modelling. In fact, the die used in the device passes the rating voltage test for the bare die. The edge termination and passivation of the chip should be well-designed to avoid the breakdown on the surface of die. The electric field analysis result about the die could be taken as a reference, but the electric field analysis results about the other components are reliable.

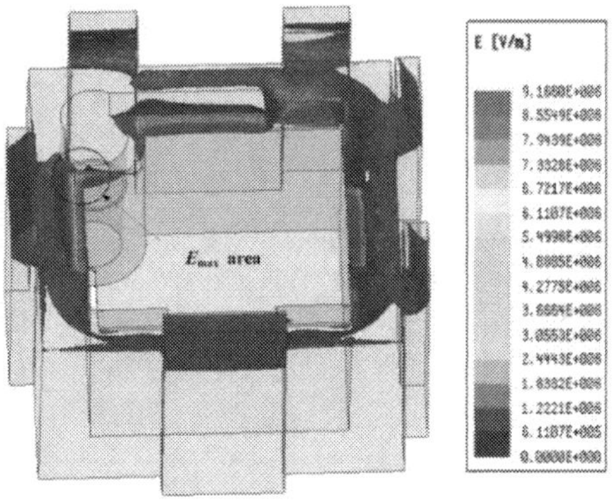

Fig. 5. Electric field distribution of PEEK frame and silver plate

5 Conclusions and Outlook

The electric field analysis of a press-pack IGBT stack was performed by using an FEM based electromagnetic field solver. The electric potential distribution and the electric field intensity were calculated during the V_{CES} test condition for the press-pack IGBTs. During the analysis process, we mainly concern three kinds of insulation failure modes: corona discharge, partial discharge and creeping discharge. For the present packaging structure, the calculation results show that:

1) The maximum electric field intensity at the edge of the silver plate is much higher than the electric strength of the protective gas N_2. It is highly risky to occur corona discharge at this region.

2) The chip region is an electrical insulation weak point of the assembly. The potential partial discharge at the surface of the chip should be considered during the chip design process. The chip used in the device should firstly pass the rating voltage test for the bare die. The micro gap between the chip and PEEK frame should be avoid to occur the micro-arcing at this region.

3) The potential creeping discharge at the surface of PEEK should be concerned. The tangent component of the electric field at the PEEK surface may be high enough to result in the creeping discharge.

The analysis presented in the paper shows the potential electrical insulation failure mode of the device under the rating voltage test. Further experiments should be done to verify the conclusions made here. The partial discharge characteristics of the protective gas and the creeping discharge characteristics of the insulation material in the protective gas should be investigated in these experiments. Another work should be done is to study the electrical insulation failure modes of the devices under other test conditions (such as reliability test).

Acknowledgement

The work presented in this paper has been supported by the National Science and Technology Major Project (2015ZX02301).

References

1. D. V. Hertem, O. G. Bellmunt and J. Liang, *HVDC grids for transmission of electric energy: offshore grids and a future supergrid.* New Jersey: Wiley-IEEE Press, 2016, pp.45-78.
2. T. Poller, T. Basler, M. Hernes, S. D'Arco, and J. Lutz, "Mechanical analysis of press-pack IGBTs," *Microelectronics Reliability*, vol. 52, pp. 2397-2402, 2012.
3. K. T., Y. K., W. H., T. Y., K. H., and S. Y., "Ruggedness and reliability of the 2.5 kV-1.8 kA power pack IGBT with a novel multi-collector structure," in *Proc. ISPSD*, 1998, pp. 437-440.
4. F. Wakeman, D. Hemmings, W. Findlay, and G. Lockwood, "Pressure contact IGBT, testing for reliability," *PCIM*, Nuremberg Germany, 2000.
5. S. Eicher, M. Rahimo, E. Tsyplakov, D. Schneider, A. Kopta, U. Schlapbach, and E. Carroll, "4.5kV press pack IGBT designed for ruggedness and reliability," in *Proc. IAS*, 2004, pp. 1534-1539.
6. T. Lebey, D. Malec, S. Dinculescu, V. Costan, F. Breit, and E. Dutarde, "Partial discharges phenomenon in high voltage power modules," *IEEE transactions on dielectrics and electric insulation*, vol. 4, pp. 810-819, 2006.
7. D. Frey, J. L. Schanen, J. L. Auge, and O. Lesaint, "Electric field investigation in IGBT power modules," in *Proc. Conference on solid dielectrics*, 2004.
8. J. Fabian, S. Hartmann and A. Hamidi, "Analysis of insulation failure modes in high power IGBT modules," in *Proc. IAS*, 2005, pp. 799-805.
9. IEC, "IEC 60747-9 International Standard, Part 9: Insulated-gate bipolar transistors (IGBTs)," Switzerland, 2003.
10. Victrex PEEK datasheet, available at: https://www.victrex.com/~/media/datasheets/victrex_tds_450gl30_black.pdf
11. M. S. Naidu and V. Kamaraju, *High voltage engineering*: Tata McGraw-Hill Education, 1995, pp. 35-39.

Optimal Reconstruction of Power Distribution Network based on Reliability and Economy

Zhang Kaiyu[1], *Feng* Yuyao[1], *Ren* Lijia[2], *Yuan* Quanning[2]

[1]State Grid Shanghai Electric Power Research Institute, Shanghai, China
[2]Shanghai Puhai Qiushi Electric Power High Technology Co.,Ltd., Shanghai,China

Abstract. Great significance shall be attached to the research on how to improve reliability of user-end power supply with limited expense by coordinating reliability and economic efficiency of power grid, so that to meet the increasing demand on reliability of electric power, and achieve considerable economic effectiveness. In this paper, process of power distribution network reliability calculation based on minimum-path failure mode and effect analysis mode is expounded, and a demonstration on calculation of power distribution network reliability is provided. Furthermore, cost-benefit analysis, which includes calculations of cost and benefit of reliability, is introduced to reflect reliability efficiency through power interruption cost calculation. At the end of this paper the cost-benefit Analyses is applied on the optimal reconstructed power distribution network.

1 Introduction

Power distribution network is one of the most important link in the chain of electric system, which has immediate influence on user-end reliability and quality of power supply. Higher benchmark for power supply reliability has been set by users. Power companies' attempts on elevating reliability level of their power distribution networks are inevitably restrained by the cost. Yet, users always prefer power supply with a low price and a high reliability, an incompatible contradiction between reliability and economy is thereby created. Alternative perspectives toward the contradiction shall be considered. From the perspective of economy, optimization measures shall be selected to improve reliability by analyzing cost-benefit coefficient of proposed reliability-improving method. From the perspective of reliability-economy relation, maximizing power reliability with limited funds has always been a challenge for analysis on reliability of power distribution networks.

Relevant researches have been carried out by domestic and abroad experts for an effective resolution of the contradiction. Reference [1] proposed a new comprehensive assessment method for the coordination of reliability and economy by integrating the concepts of satisfaction degree and inkling; Reference [2] introduced the concept of safety economic equivalent, and indicated a direction for future researches on reliability-economy relation; Reference [3] presented resolutions proposed domestically and abroad, as well as pros and cons of such resolutions based on mathematical models, matching current situation of reliability-economy coordination; Reference [4] started from power interruption cost, rendered reliability with a status of economic value, and thereby assessed various means for improving reliability. Level of reliability has been more frequently considered as an economic value, and integrated with cost for a comprehensive analysis on reliability and economy of power distribution networks. Possibility of power interruption is high in a low-reliability municipal power distribution network. Damage for social users caused by power interruptions is called power interruption cost, which is a mirror for the value of reliability.

This paper consists of analysis on reliability and economy of power distribution system, assessment and prediction on reliability, introduction on cost-benefit method for comprehensive assessment on reliability of a network by integrating economy and reliability, and calculation of cost and benefit of reliability for the optimal scheme based on comprehensive assessment.

2 Assessment on reliability of power distribution network

2.1 Reliability assessment based on minimal path method

Various methods for assessing reliability of power distribution network are specified in the current situation of power distribution network research theories. The Minimal Path Method is used herein for reliability assessment [5].

Definition of Minimal Path: If any section in a path connecting two points is missing, the whole path is interrupted, then, the path is the minimal path between the two points. Figure 1 shows a simplified circuit of points and arc. Arcs connecting nodes in fixed directions are called directed arcs, and the arcs connecting nodes in unfixed direction, such as Arc c, are called undirected arcs. In the Figure, it is presumed that the nodes are completely reliable, while the arcs are with potential failures. As shown in the Figure, there are 4 minimal paths between Node 1 and 2, namely ad, ace, be, bcd. Failure of any arc between the nodes would lead to disconnection between Node 1 and 2, and missing of minimal path between the nodes.

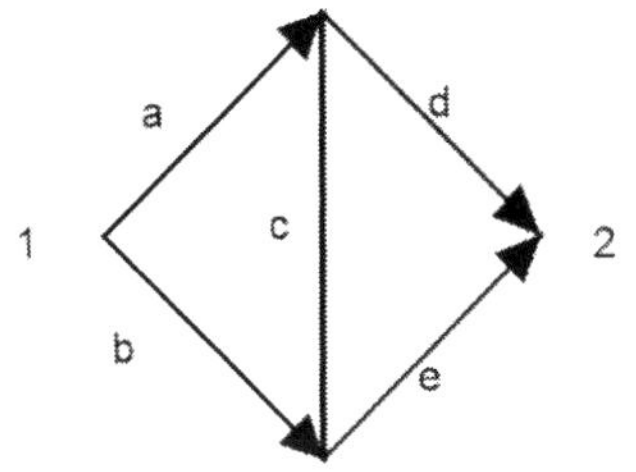

Fig. 1. Simplified circuit of node and arc

In radiation power networks which are commonly used in practice, the minimal path is started from load point, and ended at power supply in a reverse direction, paths below the load point are non-minimal paths. For any single load point, its components may be minimal path component or non-minimal path component, effect of component failure on non-minimal path on power supply to load node is shifted to corresponding node on the minimal path where calculations of components are conducted to obtain reliability index of load point and the system. Steps of calculation are as described below.

(1) Simplify circuit diagram, depict minimal path and non-minimal path of the desired load node, and locate component on its corresponding path.

(2) Shift all components on non-minimal path to minimal path.

(3) Perform failure analysis calculation for every component on minimal path, and summarize the results into a FMEA statement.

(4) Calculate reliability index of load point and power distribution network respectively.

2.2 Reliability assessment calculation

Reliability and reliability index of a system are calculated in a 4-node network. Reliability of the power distribution network is assessed by building a consequence analysis table of failure mode with the minimal path method. Data of node network table are as shown in Table 1 and 2. In Figure 2, QF refers to circuit breaker, QS refers to disconnecting switch or interconnecting switch, Fa, Fb, Fc and Fd refer to fuse protectors on load points respectively, while Ta, Tb, Tc and Td refer to transformers.

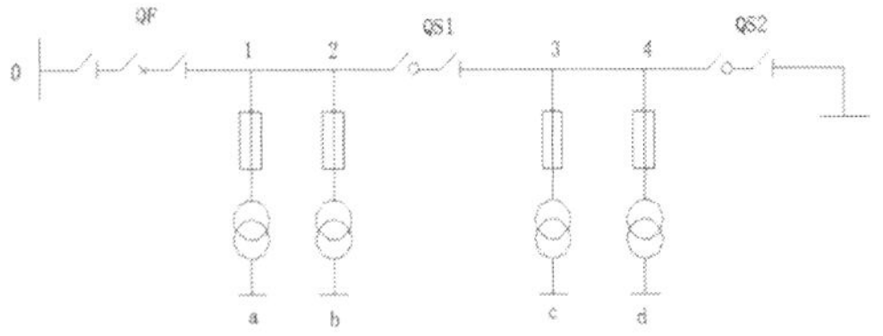

Fig.2. Structure diagram of power distribution network frame

Table 1. Data from Practical Example

Component	Failure location and isolation Time (h)	Interconnection Switch Time (h)	Pre-arranged Interruption Ratio (times/100km*year)	Pre-arranged average interruption duration Time (h)
Feeder	0.5	1	6	7

Table 2. Data of Network Frame Example

Component	Circuit Length (km)	Interruption Ratio (times/100km (set)*year)	Repairing Time (h)	Users supplied by the load point	Connected Load(kW)
Main Feeder 1	2	0.25	3		
Main Feeder 2	2	0.2	2		
Main Feeder 3	1	0.25	3		
Main Feeder 4	2	0.2	2		
Branch Line a	1.5	0.2	2		
Branch Line b	1	0.1	2		
Branch Line c	1	0.1	2		
Branch Line d	1.5	0.2	2		
Breaker		0.25	3		
Disconnecting Switch		0.25	2.5		
Fuse		0.2	2		
Transformer		0.35	4		

Load Point a				6	1200
Load Point b				5	1000
Load Point c				5	1000
Load Point d				6	1200

Taking the calculation of Load Point C for example. Following the above-mentioned steps of minimal path method, both minimal path and non-minimal path for Load Point C shall be drawn first, and then components on both paths shall be confirmed. Components on non-minimal path shall be shifted into components on minimal path. For Load Point C, any failure on Brach a and b may be shifted to corresponding nodes on main feeder, and any failure on Main Feeder 4 and Branch d may be shifted to main feeder node corresponding to Branch c, and components on non-minimal path could be shifted to minimal path thereby.

Prepare a FMEA statement with data of the network, and perform analysis on Load Point C as shown in Table 3.

Table 3. Analysis Table for Load Point C

Load Point c					
	Component		**λ (times/year)**	**r (h/times)**	**u (h/year)**
		Main feeder 0-1	0.005	1	0.005
		Main feeder 1-2	0.004	1	0.004
		Tc	0.0035	4	0.014
	Component on minimal	**Fc**	0.002	2	0.004
	path	**Circuit breaker**	0.0025	1	0.0025
		Disconnecting switch	0.0025	1	0.0025
		Main feeder 2-3	0.0025	3	0.0075
		Branch 3-c	0.001	2	0.002
Failure		**Branch 1-a**	0.003	1	0.003
		Branch 2-b	0.001	1	0.001
	Component on	**Main feeder 3-4**	0.004	2	0.008
	non-minimal path	**Branch 4-d**	0.003	0.5	0.0015
		Fd	0.002	0.5	0.0025
		Fb	0.002	1	0.002
		Main feeder 0-1	0.12	1	0.12
	Component on minimal	**Main feeder 2-3**	0.12	1	0.12
	path	**Main feeder 1-2**	0.12	1	0.12
		Branch 3-c	0.06	7	0.42
Pre-arranged interruption		**Main feeder 3-4**	0.12	7	0.84
	Component on non-minimal path	**Branch 1-a**	0.09	1	0.09
		Branch 2-b	0.06	1	0.06
	Total		0.668	3.19	2.132

We can also get minimal path analysis tables for Load Point a, b and d, respectively, and eventually obtain reliability indexes of the load points and system, as shown in Table 4 and 5.

Table 4. Reliability Indexes of Load Points

Index of Load Point	**Load a**	**Load b**	**Load c**	**Load d**
Annual interruption frequency - load point(SAIFI-LP)	0.37	0.424	0.668	0.76
Average interruption duration - load point (SAIDI-LP)	2.388	2.799	2.132	2.7675
Average supply availability - load point (ASAI-LP)	99.973%	99.968%	99.976%	99.968%
Energy not supplied - load point(ENS-LP)	0.3834	0.3195	0.2434	0.3791

Table 5. Reliability Index of System

System Index	**Index Value**
Average interruption duration of system (SAIDI)	2.526727273

Average supply availability (ASAI)	99.9702%
Average interruption frequency of system(SAIFI)	0.556
Energy not supplied (ENS)	11100.7
Average Energy not supplied (AENS)	504.576

3 Calculation method of cost-benefit analysis on power supply reliability

Cost-benefit analysis method is commonly adopted for comprehensive assessment on reliability and economy, which consists of analyses on cost of reliability and benefit of reliability. Users prefer highly reliable and economic electrical power to avoid economic interruption costs. Power companies always seek for the possibility of improving reliability of power supply with the minimum cost. Large investment can improve reliability of power supply from a comparatively low level. Yet, funds for further improvement of reliability from a certain level could be increased exponentially.

A price has to be paid to see a higher reliability of power supply. Cost-benefit analysis is required to achieve high reliability with a reasonable expense. Cost-benefit analysis is important for the theoretical basis it could provide the decision makers with.

$f(R)$ represents increment of investments on equipment and construction for improving reliability; $g(R)$ represents loss of sales and compensations caused by insufficient power supply and interruption due to reliability of network; $C(R)$ represents annual total cost of power supply. Axis R represents current reliability level of power distribution network.

$$C(R)=f(R)+g(R) \quad (1)$$

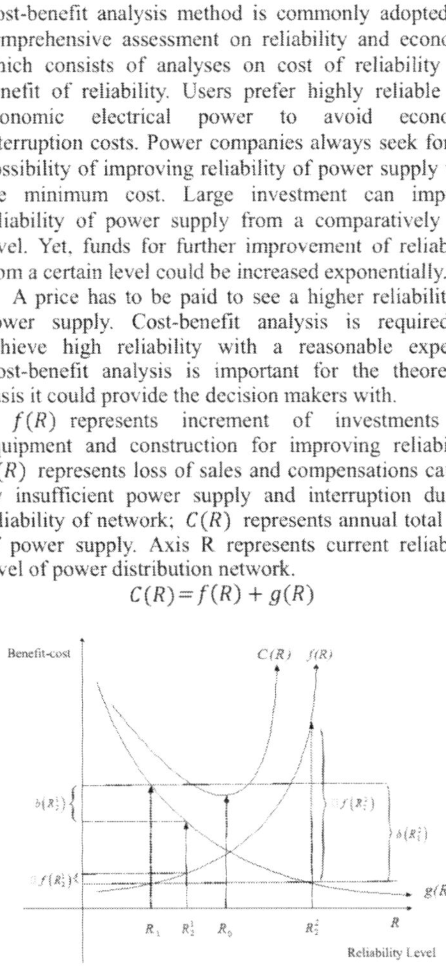

Fig.3. Benefit-cost relation at various reliability level

In Figure 3, R_0 represents the optimal reliability level, corresponding to $C(R_0)$, the lowest total cost of power distribution network. On both sides of the equation (1) , take the derivative of to get the following,

$$\frac{dC(R)}{dR} = \frac{df(R)}{dR} + \frac{dg(R)}{dR} \quad (2)$$

because, $\frac{dC(R)}{dR}|_{R-R_0}=0$

$$\frac{df(R)}{dR}|_{R-R_0} + \frac{dg(R)}{dR}|_{R-R_0}=0 \quad (3)$$

$df(R)/d(R)$ represents amount of investment required for lifting reliability by one level; $dg(R)/d(R)$ represents foreseeable benefit from lifting reliability by one level.

Physical significance of Equation (3): when construction investment required for lifting reliability by one level equals to benefit, the optimal balance of reliability and economy, the optimal reliability, and the optimal investment plan are obtained.

However, due to multiple influencing factors, practical construction could hardly be consistent with expectation to achieve the optimal reliability level [7]. Therefore, $b(R)$, the concept of post-reconstruction benefit of network is introduced:

$$b(R) = g(R_{initial}) - g(R_{reconstruction}) \quad (4)$$

Suppose R_1 is the current reliability level of network, R_2^1 and R_2^2 are two potential reconstruction schemes. The two schemes shall be compared as below:

Scheme 1: Expense for reconstruction is

$$*f(R_2^1) = f(R_2^1) - f(R_1) \quad (5)$$

Post-reconstruction benefit is

$$b(R_2^1) = g(R_1) - g(R_2^1) \quad (6)$$

because $(R_2^1) >* f(R_2^1)$, so Scheme 1 is feasible.

Scheme 2: Expense for reconstruction is

$$*f(R_2^2) = f(R_2^2) - f(R_1) \quad (7)$$

Post-reconstruction benefit is

$$b(R_2^2) = g(R_1) - g(R_2^2) \quad (8)$$

because $b(R_2^2) <* f(R_2^2)$, so Scheme 2 is unfeasible.

The concept of cost/benefit ratio [8-9] has to be introduced for determining the more favorable scheme:

$$\frac{\text{Cost}}{\text{Benefit}} = \frac{*f(R)}{b(R)} \quad (9)$$

As shown in the formula above, the scheme with the lowest known cost/benefit ratio is the optimal one. Therefore, the formula for network planning scheme shall be:

$$\frac{\text{Cost}}{\text{Benefit}} \leq 1 \quad (10)$$

Criterion for the optimal scheme shall be:

$$min\left(\frac{Cost}{Benefit}\right) \quad (11)$$

Computer-aided calculation shall be used for the formula, and cost-benefit of reliability shall be added in the computerized algorithm for reliability of power distribution network to solve equation (11).

4 Analysis on an example of power distribution network optimization

In this paper, a power network is optimized based on the cost-benefit analysis, economic cost and benefit of reliability improvement are calculated respectively. Economic benefit is represented by the interruption cost. When the benefit is greater than the cost, the plan is feasible, and a larger difference value represents a more favorable plan. The average electricity price converting method is adopted herein for calculating interruption cost, average electricity price and converting coefficient are set to be 0.65 Yuan and 25, respectively.

A 4-node power distribution network is optimized herein with schemes as shown in Figure 4, 5 and 6:

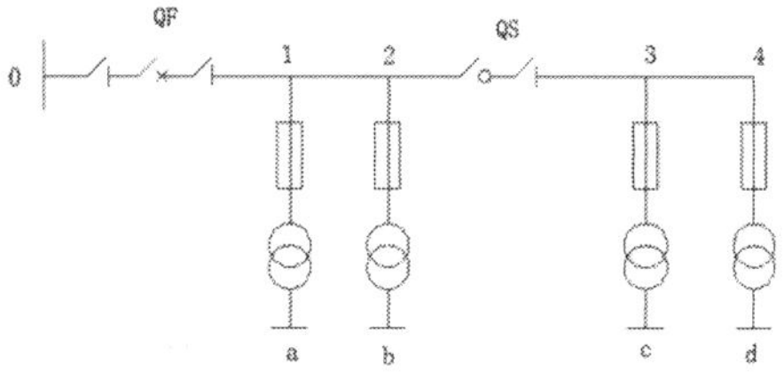

Fig. 4. Network Structure of Scheme 1

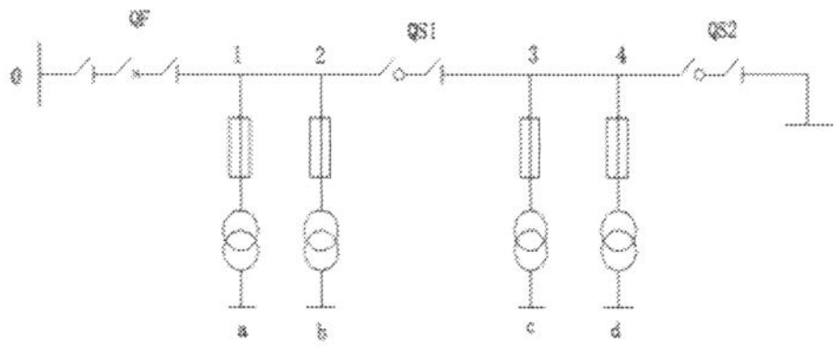

Fig. 5. Network Structure of Scheme 2

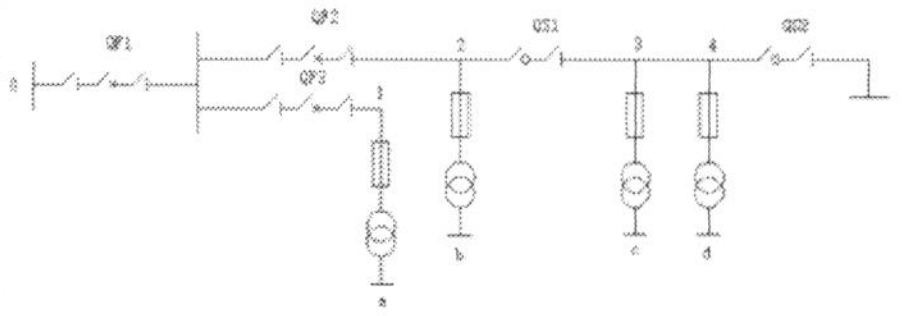

Fig. 6. Network Structure of Scheme 3

Minimal path failure mode and consequence analysis as mentioned above are adopted for reliability assessment in the cost-benefit analysis. Reliability indexes of 3 schemes are calculated as shown in Table 6. Interruption costs of 3 schemes are calculated with average electricity price converting method based on quantity of energy not supplied, as shown in Table 7.

One-time investment on main feeder applied on network frame is 150,000 Yuan/km, unit price of fuse is 600 Yuan, unit price of disconnecting switch is 1000 Yuan, expense for installation of a 1,250MVA transformer is 80,000 Yuan, unit price of circuit breaker is 4,000 Yuan, investment on branches is 30,000 Yuan/km, expense for installation of interconnecting ties and interconnecting switches is 170,000 Yuan. Economic costs of the 3 schemes are calculated based on the prices above, respectively, with the results shown in Table 8.

Table 6. Comparison of Reliability Indexes of Optimal Systems

System Index	Index Value 1	Index Value 2	Index Value 3
System average interruption duration (SAIDI)	3.62439772	2.5267273	2.363784091
Average supply availability (ASAI)	99.9586%	99.9702%	99.9730%
Average interruption frequency (SAIFI)	0.554	0.556	0.550
Energy not supplied (ENS)	15957.216	11100.7	10400.65
Average energy not supplied (AENS)	725.328	504.576	472.75682

With data of energy not supplied as shown in the table above, interruption costs are calculated with electricity price converting method based on the 3 reliability indexes, respectively, and results of which are shown in Table 7, 8 and 9:

Table 7. Estimation of system interruption cost

	Scheme 1	Scheme 2	Scheme 3
Interruption cost	259304.6625	180386.375	169010.5625

Table 8. Total cost of system

	Scheme 1	Scheme 2	Scheme 3
Economic cost (10,000 Yuan)	153.04	170．04	201.24

Table 9. Analysis of Cost-Benefit

	Optimized Scheme 1	Optimized Scheme 2
Difference in interruption cost (Yuan)	78918	11376

https://doi.org/10.1051/e3sconf/20186404007

Difference in economic cost (Yuan)	170000	312000

As shown in the tables above, improvement of reliability can be achieved by both of the schemes. Based on cost-benefit analysis, a larger difference between economic benefit and cost of reliability improvement corresponds to a more favorable scheme. Benefit from reliability improvement may be represented by reduced interruption cost. As shown in Table 9, difference in interruption cost before and after the reconstruction is less than economic cost of reliability improvement in both of the schemes. Such results are obtained based on a low supply capacity and a small user group. If supply capacity and user group were expanded, quantity of energy not supplied would be increased dramatically based on the reliability index shown in the table above, and the difference in interruption cost before and after the reconstruction would be greater than the difference in cost, and Scheme 1 would be the optimal.

5 Conclusion

Assessment and estimation of reliability of power distribution system is drawing more attention of power system researchers. Level of reliability determines economic loss borne by users and power companies, even influences economic development and social improvement of the country. While blindly pursuing the highest reliability level is impracticable, reliability and economy must be analyzed as a whole. In this paper, a reliability assessment is carried out on a simple power distribution network based on the minimal path failure mode and consequence analytical method, and levels of reliability and economy before and after reconstruction are compared based on cost-benefit analysis. Interruption cost is calculated with average electricity price converting method. The best optimization scheme is selected.

Reference

1. Ramirez-Rosado I J, Bernal-Agustin J L, Reliability and Costs Optimization for Distribution Networks Expansion using an Evolutionary Algorithm [J], Power System, IEEE Transactions on, 2001,16(1):111-118.
2. Wang Kun, Economic Equivalent of Power System Safety and its Optimization [J] Modern Electronical Technology, 2006 (16): 49-51.
3. Niu Hui, Cheng Haozhong, Zhang Yan, Chen Chen, Literature Review on Reliability and Economy of Power Network Expansion Plan [J], Automation of Power System, 2000(01):51-56.
4. Zhou Limei, Fan Mingtian, Estimation and Assessment of Interruption Cost for Users of Municipal Power Network [J], China Power, 2006 (07) :70-73.
5. Xiang Yingjie, Reliability Assessment and Economic Research of Municipal Power Distribution [D], North China Electric Power University, 2015.
6. Wan Dong, Assessment Method for Reliability and Economy of Power Distribution Network [D], Nanchang University, 2015.
7. Qiu Sheng, Zhang Yan, Sun Jiansheng, Wang Jingliang, Wang Zhipei, Teng Letian, Luo Min, Measures for Improving Reliability of Medium Voltage Distribution Network and Relevant Cost-benefit analysis [J], Relay, 2005 (13):34-38+49.
8. Feng Huajun, Network Reliability in Asset Entire Life-cycle Management [D], North China Electric Power University, 2016.
9. Xu Xianglei, Calculation and Improvement of User-end Supply Reliability of Power Distribution Network [J], Science Information, 2009 (32) :78-79.

Study on Reliability Evaluation Method Based on Improved Monte Carlo Method

Liu Jun[1], *Yang* Fan[1] , *Ren* Lijia[2]

[1]State Grid Shanghai Electric Power Research Institute, Shanghai, China
[2]Shanghai University of Engineering Science, Shanghai, China

Abstract. The advancement in science and technology comes with continuously expanding power system scale, increasingly complex system operation condition and higher requirements for accuracy and speed of power system reliability evaluation, but actual calculation methods cannot meet the needs. Therefore, there is need to improve the reliability of conventional power distribution network so that requirements of calculation speed and calculation accuracy can be met. In this paper, reliability of the power distribution network will be evaluated using improved Monte Carlo method with uniform sampling. The average value is obtained through calculation of state of multiple sub-intervals and test functions, which effectively improves calculation accuracy, and further increases the utilization of random numbers. By improving the uniform sampling method, the Monte Carlo simulation variance is reduced, and evaluation and calculation efficiency is improved. At the same time, unqualified power grid is selected for analysis. Based on the simulation results, qualified power distribution networks are compared to point out where the requirements are not met. Also, comparative analysis is made on the effect of power distribution network grid structure etc. on the user's power supply. Finally, suggestions for improving power distribution network reliability are given from equipment reliability, grid structure.

1 Introduction

Power supply reliability means the ability of the power supply system to provide continuous power to the user. The interruption of power supply will not only incur huge economic losses, but also affect people's lives and social stability. High power supply reliability is not only what power consumers demand, but also what power supply companies pursue in self-development. Researches on reliability of power distribution network gradually began in the 1960s, originally as qualitative analysis. As the research deepens, pure qualitative analysis could not meet the research needs. Researchers adopted quantitative indicators to measure reliability. So far, the research progress has always been concerned by experts in related fields. In developed countries such as the United States, Japan and France, relevant agencies have been set up to analyze, research and plan power distribution networks, and reliability evaluation system has been relatively complete.

Commonly used reliability assessment methods can be categorized into analytical and simulation methods. The basic idea of analytical method is to build a corresponding model according to original reliability data of the components, and then obtain power grid reliability indicators through mathematical analysis. Analytical method features clear physical concept, high model precision, and clear logic concept. However, the computation burden increases sharply when component number increases, and direct use of analytical method faces difficulties when the power grid reaches a certain scale. Failure mode and effects analysis [1] (FMEA) and network methods are commonly used analytical methods. By using fault traversal algorithm, FMEA classifies component failures, determines the impact scope of the failure in combination with preorder first traversal and postorder first traversal methods. Its algorithm computation increases rapidly with the power distribution network increase [2-3]. Hence, many optimized algorithms are introduced based on analytical method, for instance, network methods based on topological structure of the power distribution system which include minimum path method [4-5], minimum cut-set method[6], network equivalent method[7], and region separation method[8].

Simulation method is recently introduced in the field of reliability analysis, which samples according to probability distribution of components, judges the state of components, and then obtains reliability-related indicators using statistical knowledge, mainly including Monte Carlo simulation method [9-10]. Simulation method, with intuitive and convenient characteristics, can well handle dynamic load variation. However, its accuracy is directly proportional to the consumed time. It often takes time to enable high accuracy in result, which is not conducive to targeted evaluation [11].

Computation speed means a lot for a good practical model and algorithm. This paper calculates and analyzes the actual power distribution network using improved Monte Carlo method of discrete sampling method, and studies measures to improve power supply reliability of power distribution network.

2 Monte carlo simulation evaluation method

Monte Carlo method is widely used in reliability evaluation. The specific idea is: First, the network is divided into components, and the failure probability of the grid is simulated according to the failure probability of each component. The grid state is determined by these components. Such a simulation constitutes an assessment of the grid to be simulated. Then, continue with this method, and set a certain amount of simulation as a sample. When enough data is collected, analysis can be made to obtain relevant indicators. Monte Carlo method supports evaluation of complex systems regardless of power distribution network structure. The number of samples required is also independent of the structure, so it can relatively well reflect some random factors. However, a large number of samples are required to obtain results close to the truth value, which will greatly increase the simulation time and affect the overall assessment as components show similar status.

2.1 Ordinary sampling method

In the conventional Monte Carlo method [43-48], each component is divided into two states independent from each other and free from interference by other components. The state is generally set to 1 when the device is in normal operation and to 0 when in a fault state. The failure rate of the component is in the interval [0,1]. By generating a random number and comparing it with the failure rate of the component, the component state can be judged. State in which the primary grid is operated can be obtained by simulating each component and multiplying it with the probability function. Repeat the above steps constantly, and the average system reliability index can be calculated (generally estimated value due to the limitation of the sample number) when the sample reaches a certain size. The convergence criterion β is then introduced, which concerns the expected variance of the reliability index and the variance of the component state function.

In order to meet the pre-set accuracy, Monte Carlo method requires a lot of simulation, which will sacrifice the operation time. On the contrary, guaranteeing simulation time demands reduced convergence accuracy as a price.

2.2 Average and scattered sampling method

The sampling method needs improvement for better speed and accuracy of Monte Carlo simulation. There are currently many effective improvement methods available. In this experiment, average and scattered sampling method is adopted to sample the operation state of the power system according to the following specific sampling steps:

(1) The interval [0,1] is divided into h intervals according to the maximum failure rate of the component in the system;

(2)A random number a_j is generated for each element in the power system;

(3) If there is k making $(k\text{-}1)/h < a_j \leq k/n+\beta_j$ (β_j is the failure rate of component j), then component j is faulty in interval k and in normal operation in other intervals ;

(4) The operation state of each component is averaged, and then operation state of each component is normal or with $1/h$ failure probability, so the system operation status is obtained;

(5) Statistical analysis is made for the system operation status to determine whether the sample variance coefficient satisfies the convergence condition. Stop sampling if the condition is met. Otherwise, turn to step (2).

2.3 the basic steps of Monte Carlo method to assess power system reliability

According to the above theory, Monte Carlo method is mainly composed of three parts: system state sampling, system state analysis and system evaluation index statistics. The main steps of the assessment are shown in Figure 1.

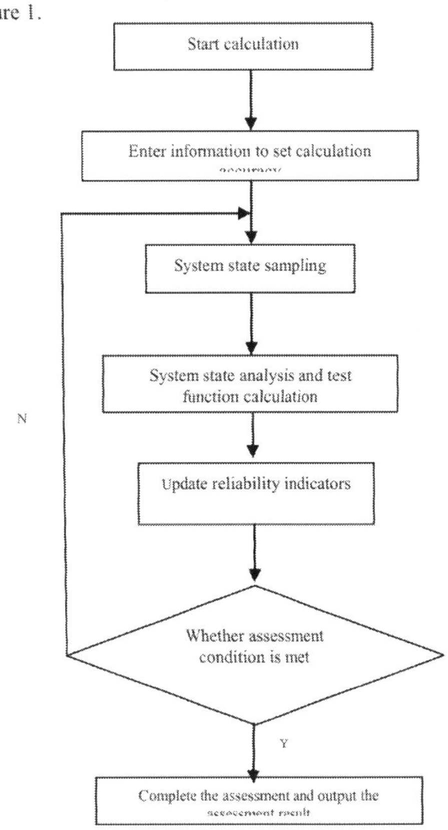

Fig. 1 Calculation Flow Chart of Monte Carlo Method

It can be known from the above figure that Monte Carlo method need proceed with random state sampling of the system, without involving the duration and transfer rate of the state. The assessment process requires a higher degree of accuracy, so sampling algorithm can be improved during the sampling process to reduce the contrast and improve the calculation accuracy.

After completing the sampling of the system state, the test function value of the state needs to be analyzed and calculated. The test function calculation method corresponds to reliability evaluation index. State analysis of the power system mainly starts from two aspects: on the one hand, the power system state demands load flow analysis to judge fault existence in the system; on the other hand, the actual system state should be optimized based on load flow analysis to eliminate the fault state in the system and guarantee safe system operation. DC load flow analysis is a commonly used method for system optimization analysis.

3 Reliability example analysis of power distribution network

Reliability of actual power distribution network in a region is analyzed using dispersed sampling method such as the improved Monte Carlo described earlier. Figure 2 shows the power distribution network of a region. The power distribution network consists of 13 load points, 8 switches, 3 incoming lines, and 25 lines. The statistical data cover failure rate, repair time, load point location, user number, average load, line length, etc., specifically as follows:

Line failure rate of 25 lines: 0.0150 (times/year • km)

Node failure rate: 0.065 (times/year • kilometers)

Line repair time: 5 (hours)

Node repair time: 20 (hours)

Load position: 2, 5, 6, 7, 11, 12, 13, 14, 15, 19, 20, 21, 24

User number at each load point: 147, 126, 1, 1, 132, 147, 1, 79, 1, 76, 79, 76, 1

Line length: 0.8, 2.5, 1.6, 0.9, 1.6, 2.5, 0.6, 1.6, 0.75, 0.9, 3.2, 2.8, 0.6, 3.5, 1.6, 2.8, 3.2, 2.5, 3.2, 1.6, 0.8, 2.8, 2.5, 3.2 , 2.8 (km)

The system component operation state is subject to average and scattered sampling. The component working status is judged and component repair time is simulated. Then, failure times and failure time of each load point are calculated according to the sampled system status. Whether experiment objective is reached is judged after each sampling calculation. Several experimental objectives were tried in this experiment, including sampling times, result variance and sampling time etc. The result variance should be taken as objective if there is a high requirement for result accuracy. The sampling times or time should be taken as objective if there is high requirement for system operation time. Finally, the results of all sampling operations are counted to obtain the average value, i.e. the target value of the power distribution network reliability index. In this experiment, common sampling method and average and scattered sampling method were used for reliability analysis of the same network. The specific experimental data of multiple experiments is shown in Table 1.

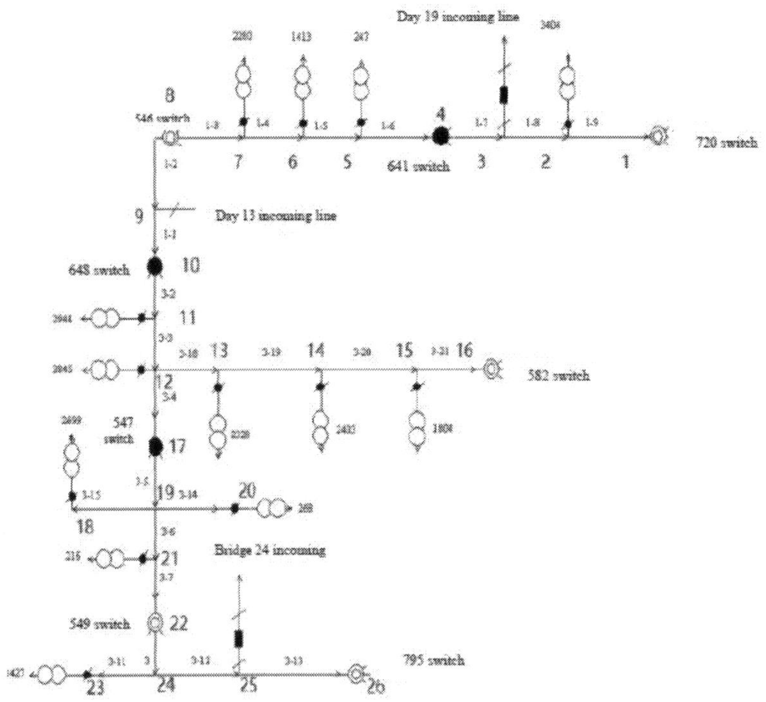

Fig. 2 Power Distribution Network Structure

Table 1 Reliability Calculation Results of Power Distribution Network

	SUM	SAIFI	SAIDI	CAIDI	ASAI	ENS	Time
Ordinary sampling	1773	0.167118535	0.880575877	5.269169432	0.999899478	2.31458926	600.2812500
Ordinary sampling	1802	0.174769591	0.884536473	5.061157757	0.999899026	2.357938961	600.2812500
Ordinary sampling	1811	0.170485244	0.859122941	5.039280361	0.999901927	2.281123884	600.2656250
Ordinary sampling	5523	0.168725837	0.835373268	4.951069048	0.999904638	2.238578926	1800.2656250
Ordinary sampling	21447	0.169801149	0.837376009	4.931509677	0.999904409	2.264007962	7200.2812500
Uniform sampling	6345277	0.166269346	0.830837862	4.996939506	0.999905156	2.255971281	1800.0156250
Uniform sampling	6390384	0.166207687	0.830577778	4.99722841	0.999905185	2.255046977	1800.1406250
Uniform sampling	198496449	0.166114804	0.830629097	5.000331562	0.999905179	2.254982628	55000.0156250

In the reliability assessment, statistics is made on average power outage frequency index of the system, average power outage time, users' average power outage time and expected power shortage indicator. SUM value means the time duration of Monte Carlo simulation usage, which is in unit of year. The last time indicates program operation time in unit of second.

By analyzing the data obtained, it can be seen that simulation speed of both Monte Carlo method using uniform sampling method and Monte Carlo method using ordinary sampling method increases as the simulation period increases. In addition, there is a large gap in the indicators between the minimum simulation period and the maximum simulation period. For closer years, the simulated indicators are also closer. Comparison reveals that the average and scattered sampling method not only performs more simulations in a shorter time, but also greatly improves result accuracy. In terms of computational efficiency, average and scattered sampling method has been improved by about two orders of magnitude compared with ordinary sampling method, and the accuracy has also been increased by an order of magnitude.

In terms of reliability, the system has extremely high reliability. The index meets the reliability guidelines for Shanghai power distribution network that the reliability of full-caliber power grid should be above 99.99%, and the reliability of city center should be above 99.999%. It also meets the national power industry requirement that the power supply reliability in the region should be above or equal to 99.990%. Also, the accuracy grade far exceeds these requirements. The average power outage time of the system is 0.16~0.17 times, which is at a relatively high level compared with that in other cities. However, the user's power outage time is as much as 5 hours per household per year, exceeding the prescribed 52 min.

4 Factors that Affect Reliability and Reliability Improvement

4.1 Factors Affecting Reliability

The factors affecting reliability of power supply can be classified as equipment factors, network structure, and external factors.

1. Equipment factors mean the power supply of the power grid is affected for the reason of the equipment itself. When new equipment is put into operation, there will be a run-in period. The failure rate will gradually decrease after the equipment operates for a period of time. Then, the equipment will enter a period of stability when the failure rate will be reduced to a minimum. However, the failure rate will gradually increase when it approaches the service life.

2. In terms of network structure, factors such as line segmentation and connection may affect line maintenance, which will greatly influence the control of the outage range. At the same time, factors such as wire diameter, power supply radius range, switch reliability and whether the cable ring network is connected in series to the overhead network will also affect power supply reliability.

3. The power outage event means the power company needs to cut off a part of the load when there is need to operate the relevant equipment according to the actual operation of the power grid. When it is scheduled in advance according to a reasonable plan, the power outage scope can be reduced and the power outage duration can be shortened so that normal use of electricity by neighboring users will not be affected as much as possible. In addition, power supply may be rationed based on the operation condition when the current power supply capacity cannot meet the requirements but outage is unnecessary.

4. External factors are mainly divided into climate factors and human factors.

4.2 Reliability improvement

The power distribution reliability of power distribution network can be improved by taking the following measures based on the specific grid conditions:

1. Improve equipment reliability: The power grid is composed of various electrical equipment. The power grid will be more reliable in case of reliable equipment, and equipment reliability is a necessary condition for reliable power grid. According to statistical values of equipment reliability parameters (such as the average

number of failures per year, the repair time required for each failure, etc.) and impact analysis of equipment failure on power grid reliability, it is possible to judge which equipment has poor reliability and which equipment greatly influences power grid reliability, so that management and maintenance of such equipment can be strengthened, or replacement can be made.

2. Change or strengthen the grid structure: a strong grid structure can effectively reduce the average time of scheduled and fault outage, which serves as the basis for improving power supply reliability. Grid structure irrationality will affect the power supply reliability of the grid to a large extent. In the power grid planning and construction, it is necessary to take into account the path which the corresponding load can be transferred to after equipment failure from the aspect of connection mode to improve power supply reliability of the power grid. At different load levels, the grid operates in different ways and its power supply reliability may not be identical. For different load levels, the power supply reliability level can be improved by changing operation mode of the power grid.

3. Change the equipment maintenance mode: The fault power outage maintenance or scheduled power outage maintenance greatly affects the power supply reliability during the maintenance period. By strengthening the maintenance efforts or taking hot line maintenance operation mode, equipment fault maintenance time and planned maintenance time can be shortened. By adopting condition based maintenance, the scheduled maintenance times that require power outages will be less.

4. Invest in distribution automation equipment: Input of distribution automation equipment can greatly shorten the fault point search and isolation time, rapidly restore the load power supply in non-fault areas, reduce users' power outage time and power outage range, and significantly improve power supply reliability of the power grid.

5 Conclusion

With the increased complexity and scale of power system grid structure, some traditional reliability calculation methods have been unable to meet the needs of modern distribution network reliability assessment due to the limitations in calculation accuracy and speed. Monte Carlo method should be used more as its accuracy is free from influence of power system scale change and its mathematical model is relatively simple. This study evaluated power distribution network reliability using improved Monte Carlo method, sampled the state of power distribution network system with analysis and calculation made. Statistical analysis was made on power system evaluation indicators, and the evaluation value was made to converge to the truth value of the value to be evaluated when there were ample samples. The results show that the improved Monte Carlo method with average and scattered sampling improves the speed and accuracy of reliability assessment, saves computation time, and provides a good direction for future reliability assessment of power distribution networks.

References

1. Li Zhimin, Li Weixing. Fault Traversal Algorithm for System Reliability Assessment Based on Radial Distribution[J]. Automation of Electric Power Systems, 2002, 26(2): 53-56.
2. Li Weixing, Li Zhimin, Liu Yingchun. Reliability Assessment of Complex Radial Distribution System[J]. Proceedings of the CSEE, 2003, 23(3): 69-73.
3. Xu Zhenxia, Zhou Jiangxin. Improved Fault Traversal Method for Reliability Assessment of Complex Distribution Networks [J]. Power System Technology, 2005, 29(14):64-67
4. Bie Zhaohong, Wang Xifan. Reliability Analysis of Power Distribution System [J]. Electric Power, 1997(5):10-13.
5. Xie K, Zhou J, Billinton R. Reliability Evaluation Algorithm for Complex Medium Voltage Electrical Distribution Networks Based on The Shortest Path[J]. IEE Proceedings Generation, Transmission and Distribution, 2004,150(6):686-690.
6. ÖZdemirA, Caglar R, Mekic F. A New Active Failure Simulation Approach for Distribution System Reliability Assessment[C].International Conference on Power. System Technology, 1998. Proceedings. Powercon. IEEE, 1998:237-240 vol. 1.
7. Wang P, Billinton R, GoelL. Unreliability Cost Assessment of an Electric Power System Using Reliability Network Equivalent Approach [J]. IEEE Power Engineering Review, 2007, 22(7):58-58.
8. Wei Zhinong, Zhou Fengwei, Xiao Chuanling, et al. Reliability Evaluation Algorithm for Complex Medium Voltage Distribution Network Analysis Based on Simplified Network Model[J]. Power System Technology, 2006, 30(15):72-75 .
9. Goel L. Monte Carlo Simulation-Based Reliability Studies of a Distribution Test System[J].Electric Power Systems Research, 2000, 54(1):55-65.
10. Huda ASN, Živanović R. Accelerated Distribution Systems Reliability Evaluation by Multilevel Monte Carlo Simulation: Implementation of Two Discretisation Schemes[J]. LET Generation Transmission & Distribution, 2017, 11(13):3397-3405.
11. Wan Guocheng, Ren Zhen, Wu Risheng, et al. Application of Hybrid Method in Reliability Assessment of Complex Distribution Network [J].Proceedings of the CSEE, 2004, 24(9):92-98.

Research on Power Knowledge Model Based on Group Wisdom

Zhao Hongying[1], *Yu* Yang[1], *Ji* Youlang[1], *Zhu* Jun[1] , *Zhang* Jinqiu[1]

[1]Electric Research Institute, Jiangsu Electric Power Company, Nanjing 210019, China

Abstract. This study described the plan of knowledge management and crowd innovation for Jiangsu electric company based on literature analysis on crowd innovation research and investigation of the company's innovation state as well as its electric service. A knowledge management system was designed for all the experts in electric field, all the workers and the customers in order to achieve crowd innovation. This system was designed based on the consideration of innovation process, evaluation and motivation, and it is implemented with the Web 2.0 technology.

1 Introduction

The public innovation model has been tested in practice and is considered to be the key to improving the efficiency of corporate innovation [1]. Under the strategic principle of "One Strong Three Excellent" (Strongly Constructed Power Grid, Excellent Assets, Excellent Service, and Outstanding Achievements) of State Grid Power, State Grid Jiangsu Electric has implemented State Grid's "Innovation, Creation, and Entrepreneurship" requirements, insisting that "The development path of innovation, strategy, and time is the first to strengthen the innovation management system and the overall innovation culture. It has established the "four-wide" innovation concept of "full-factor, full-service, full-staff, and full-scale collaboration" and has made clear innovations. The direction, through the creation of a comprehensive and collaborative innovation atmosphere, supplemented by a number of initiatives in platform support and policy management, continues to drive the company's overall innovation and development, making it a truly endogenous driving force for strong transformation.

On the basis of discussing the theory of the crowd-building mode, this article takes the innovation and development of the Jiangsu Power Company as an example, and elaborates on how the crowd-building mode is combined with information methods and has been applied and achieved results in practice. Specifically, on the basis of the overall innovation plan, Jiangsu State Power of the State Grid integrates the resources and needs of internal power experts, front-line employees, and external power customers, and creates a cross-professional, cross-regional knowledge innovation platform to share knowledge. Pass experience, apply breakthrough mode to seek breakthroughs and innovations from practice, enhance enterprise's marketing services in all directions, enhance corporate image, and improve customer satisfaction.

2 Related work

The history of human history has been inseparable from the word innovation. From drill fires to household appliances, from radio radio to VR, every step of scientific creation of technological innovation is a disruptive "unheard of", and every company that can capture disruptive technologies has therefore acquired in the business world. Great success. However, this traditional innovation model is often inefficient, and the reason for this is that it does not meet the two major competencies of innovation literacy: one is the lack of driving force for exploration, and the other is a relatively narrow transmission channel [2].

Crowd Innovation came into being. As an extension of open innovation, the "public innovation" model has the fundamental difference from the traditional innovation model in that it breaks through the barriers to the innovation category, mobilizes the enthusiasm of the internal and external public, and expands the source of creativity. Professor Chesbrough, founder of Open Innovation, believes that an open environment can better help companies integrate internal and external resources, create an interactive, intersecting, divergent ideological atmosphere, and greatly increase the probability of successful innovation [3]. The theory of complex adaptive system (CAS) mentioned that the diversity of intelligence resources caused by individual differences has provided a boost to knowledge innovation and intelligence emergence, and has become one of the important sources of innovative new transmission channels [4]. Another important source is external resources and information. In 2013, China introduced the "Enterprise Innovation Capability Evaluation Index System" draft for consultation. One of the indicators' collaborative innovation capabilities emphasizes the importance of the cooperation between industry, universities, and research institutes, and encourages the integration of resources such as collaborative innovation and technology introduction and digestion. Western innovation management practices go even further, and their understanding and application of "open style" have long since escaped the directly linked partners. For example, GE's "Ecomagination Challenge", through co-funding with VCs to fund innovative companies with innovative ideas, has brought about a low-cost, high-efficiency, GE social responsibility. The results of innovation [3]. In particular, it is worth noting that large companies with large profits and large markets often do not give up their mature products and achievements in order to try out subversive breakthroughs, but this type of cooperation in investment financing can avoid such risks. Taken together, the crowd-sourced model breaks through the conceptual boundary of the internal and external environment, and its openness can be continuously penetrated according to requirements, reducing the intangible costs to a considerable extent.

Another important feature of the crowd-sourced model is the close connection with the Internet [5]. The innovation of Internet technology has allowed explosive expansion of information volume. Through the storage, processing, sharing, and integration of huge amounts of information, companies can use the external environment more closely. In this context, Song Gang and Zhang Nan (2009) proposed the innovation 2.0 theory. They believe that through the development of information and communication technologies and their integration with people's lifestyles, working methods, and organizational methods, one is based on knowledge and information. The core society is being formed, and the innovation model based on traditional laboratories and scientific researchers is being subverted, and it has turned to the user-centered, employee-centered, mass-centered innovation 2.0 model [5]. Web 2.0 is the embodiment of innovation 2.0 in the field of Internet. The purpose of the organization's choice of Web 2.0 is to look at the "collective wisdom" it uses to integrate open innovation. Its functions range from increasing intimacy with users to expanding the internal knowledge management system. [6]. User-led is the core of the web2.0 environment. This positioning allows the leading community in crowd-sourced space to transition from the corporate government to the public, promoting the initial seeding stage of the creative platform, and even linking education and training, social media, and attracting talents. And other aspects [7]. Liu Zhiying and other scholars (2015) proposed the community-development-market (CDM) conceptual model of the crowd-building model [2]. That is to say, the web2.0 platform serves three functions: First, Knowledge community, where the masses gain knowledge sharing; Second, the development platform, the public's innovation needs and collaborative innovation can be achieved in the interactive function; the last is the technology market, promote technology flow, form an online trading market. In the final analysis, Internet technology is a human-driven knowledge management network. This is particularly important for technology-based industries.

The technological innovation in the power industry is inseparable from four key words: safety, economy, cleanliness and continuity. Due to the universality of the audience in the energy industry, the advantages of web2.0 technology are more prominent. The community interaction function brings relatively symmetrical information communication channels to scientific research work, company employees and ordinary consumers. In addition to the concept of open innovation above, we undertake the scientific knowledge of consumer's electric energy knowledge, further fulfill corporate social responsibility and expand the possibility of innovation of the people. Web2.0 technology, as a mature Internet product, is bound to be the best platform and mainstream trend for enterprise organizations with knowledge and technology as the core application of crowd-building space.

4.1 Knowledge innovation process

The knowledge creation process is divided into creative communication process, creative refinement process, and creative reward process, as shown in Figure 1.

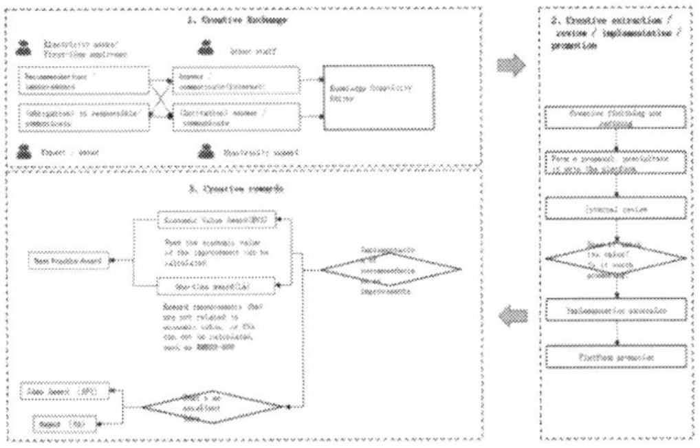

Fig. 1 China National Network Jiangsu Electric Power Knowledge Creation Process

4.2 Knowledge innovation mechanism

The State Grid Jiangsu Electric Power Knowledge Innovation Evaluation System is based on the overall innovation evaluation mechanism of Jiangsu Electric Power Company on the State Grid and is designed as follows.

4.2.1 Knowledge Organizational Structure Evaluation

The Corporate Innovation Management Committee is responsible for the leadership of innovation evaluation.

The internal evaluation organization is an organization that will be set up in the future evaluation process. It is mainly responsible for the extraction, review, implementation, and promotion of ideas, and is mainly composed of experts in various fields.

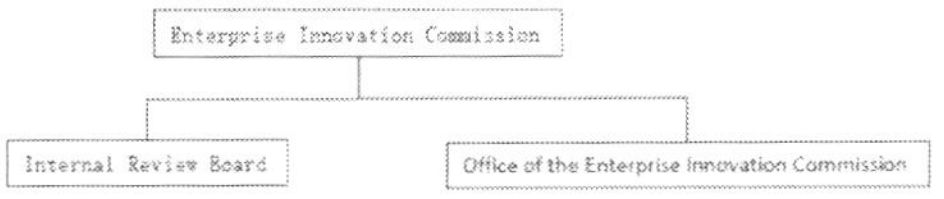

Fig. 2 Knowledge and organization structure of Jiangsu Electric Power Network

4.2.2 Knowledge Creation Evaluation Organization's Responsibility

The company's innovation management committee is responsible for the leadership of innovation evaluation. The main responsibilities include: leading the company's innovation evaluation work, coordinating and resolving major problems in the evaluation work; approving the company's innovation work evaluation plan; approving the company's innovation work evaluation index system; reviewing and approving the company's innovation work Evaluation report.

The Office of Corporate Innovation Management Committee (established in the branch of the Enterprise Association) is responsible for organizing the company's innovation evaluation. Its main responsibilities include: formulating (revising) and releasing the evaluation index system of the company's innovation work; formulating the evaluation plan for the company's innovation work and organizing the company's innovation work Evaluation; compiles and publishes evaluation reports on company innovation work; organizes the development of improvement measures for evaluation issues; and enhances the management of the company's innovation work; organizes and coordinates other work on innovation evaluation of the company.

The internal accreditation body is an organization that will be set up in the future evaluation process and is mainly responsible for the extraction, review, implementation, and promotion of ideas. After the ideas submitted by the users, employees, experts, etc., collected, are refined and refined, proposals will be formed, which will then be deposited on the platform for review by the internal review committee. After these ideas are reviewed, ideas that match the company's value and are worthy of promotion will be promoted and eventually promoted on the platform.

4.2.3 Knowledge Innovation Work Evaluation Index System

For knowledge-creating work, it can be evaluated in three dimensions: input, process, and effectiveness [9].

Innovation investment includes organizational guarantees and resource guarantees. Organizational guarantees include innovations in the organizational structure and mechanisms, including two major tasks: First, establish leadership institutions for knowledge innovation, and improve the working mechanism, including the establishment of a leading group in this unit, and the chief responsible person as the first person responsible for knowledge innovation work. The responsible person in charge shall personally participate in and guide the advancement of knowledge innovation, establish and improve the working mechanism of information notification, inspection, assessment and evaluation; second, formulate annual opinions on knowledge innovation work, and opinions shall have clear ideas, goals, tasks, and time schedules, etc. The elements and working opinions must reflect the company's decision-making and deployment. The key work tasks are clear, pertinent and operational.

Resource protection also includes two aspects: First, the level of employee participation in knowledge innovation, that is, the number of employees participating in various types of knowledge innovation in the current year/the total number of employees*100%; second, the proportion of knowledge innovation incentive investment in total cost, ie, The amount of investment in knowledge innovation awards for the current year/the total cost of the year* 100%.

The innovation process includes project management and publicity training. The project refinement management includes: the level of knowledge innovation refinement, the level of knowledge innovation review, and the level of promotion of knowledge innovation achievements. Publicity training includes two aspects: First, the effectiveness of knowledge innovation and publicity work (80%). The highlights of this unit's knowledge innovation are incorporated into the company's (local) work reports, information, and notifications, or publicity and reporting in the company's (local) newspapers, websites, etc.; the highlights of this unit's knowledge innovation are incorporated into the State Grid Corporation of China. Work reports, information, and notifications (at or above the provincial or ministerial level), or publicity and reporting in newspapers, websites, etc. of the State Grid Corporation (at or above the provincial level). The second is the rate of innovative knowledge training (20%). Number of training courses organized by this unit on new knowledge, theory, or methods / Total number of training courses organized by the unit * 50% + Employees participating in training on new knowledge, theory or methods, etc. Total staff time * 50%.

Finally, the results of knowledge innovation include awards, application of results, and results.

The entire "lead geese" talent training project is aimed at the common development of the company and its employees, and it has established an integrated employee career growth

4.3 Knowledge Innovation Mechanism

Incentive mechanism refers to the means and environment that can stimulate and guide people to produce specific behaviors. Establishing and improving incentive mechanisms is an effective means to mobilize the enthusiasm and creativity of talents and maximize their potential.

Incentive mechanism design is mainly based on positive incentives. It mainly includes Economic Value Award (EVA), One-time Award (LA), Ideas Award (AFI) and Best Practice Award.

Among them, the economic value award for the innovative economic value brought about by the improvement and can be calculated, according to the economic value of discretion to give rewards; one-time award refers to the reward and economic value is not related to the creativity or EVA can not be calculated, reward 20-8000 yuan The idea prize means that improvement or creativity is not implemented but good attention is given to the idea. The Economic Value Award (EVA) and the One-time Award (LA) were introduced by the internal review committee and the Best Practice Award was launched.

5 Electric power knowledge platform design

5.1 Platform design principles

First of all, the design of the knowledge creation platform establishes the principle of "people-centered", which is embodied in encouraging users to share, allowing more people to participate, and making it easier for users to read:

(1) Users share. Under the Web 2.0 model, power users, power line workers, and power expert workers can share ideas without time and geographical constraints. Users can get the information they need and they can post their opinions. Makes the user a viewer of both the content of the site and the content of the site. This means that the Web 2.0 site provides users with more opportunities for participation. For example, blog sites and wikis are typical guidelines for user-generated content. The tag technology (user setting tag) directly assigns the information in the traditional website to the user.

(2) Many people participate. In Web1.0, Internet content is customized by a small number of editors (or webmasters), such as portals. In Web2.0, information is provided by organizations and individuals.

(3) In Web 1.0, the Internet is "reading Internet" and Web 2.0 is "readable and readable Internet." Although everyone participates in information feeds, it is a small part of people who, on a large scale, contribute most of their content.

Second, in order to realize the effect of knowledge change from quantitative to qualitative, the design of the platform establishes the principle of "information aggregation", which allows information to accumulate on the network and facilitates users to conduct historical queries to innovate on the basis of knowledge accumulation.

Thirdly, the platform for knowledge and creativity should be one of the manifestations of Crowd-innovating space. It should have "characteristics of popularity, openness, self-organization, and interaction" [10]. The third principle of creating a platform design is "establishing a community where interest is the aggregation point." Under the Web2.0 model, the knowledge crowdfunding platform can aggregate groups interested in one or more issues. These groups have higher frequency of knowledge exchange, higher content commonality, and are more likely to produce innovations in specific areas.

Finally, the knowledge-creating platform is based on the principle of enhancing interactivity, and has also done a lot of design in human-computer interaction and human-to-human interaction. This interactivity is not only reflected in the user's interaction with the web server in the process of publishing content, but also includes the interaction between different users of the same website and the interaction of information between different websites. Web2.0 includes the services we often use. BBS and blogs are the main representatives. The "one-to-many" and "many-to-many" modes of communication coexist. Examples include RSS, blogs, podcasts, wikis, P2P downloads, social bookmarking, SNS, communities, sharing services, and more. Blog is a very important element in Web2.0, because it breaks the information monopoly of the portal, and in the future, the status of blog will be more important. Web2.0 actually expands the information source of Web 1.0 to make it diversified and personalized.

5.2 Platform technology

The power knowledge crowdfunding platform adopts Web2.0 related technologies for design, including blogs, RSS, encyclopedias, online abstracts, social networks (SNS), P2P, and instant messaging (IM).

A blog is an easy-to-use website that allows users to quickly post ideas, communicate with others, and engage in other activities.

RSS is a simple technique (also called syndicated content) that sites use to share content with other sites. The technology originally derived from the browser "news channel" is commonly used for news and other in-line websites such as blogs.

WIKI Encyclopedia, Wiki is a multi-person collaboration writing tool. Wiki sites can be maintained by multiple people (or even any visitor), and everyone can express their opinions or expand or explore common themes.

5.3 Creative scene design

Through research and interviews, we collected business requests from different users in the field of Jiangsu power marketing for the Knowledge Innovation platform. The first-line user's working environment was considered. The design of the system platform supports the convenient application of the mobile terminal. The crowded platform will be up to the enterprise management personnel, experts, business management personnel, down to the first-line power employees, new employees, etc. are included in the platform to use the object, to achieve from the front line to find problems, from the business to find a breakthrough in innovative models.

5.4 Platform Functional Architecture

https://doi.org/10.1051/e3sconf/20186404009

In order to meet the needs of different groups of people, based on the principles and positioning of platform design, we incorporate the following major functions into the platform of knowledge and creativity: creating space, creative refinement, results promotion, and system management. The public space includes: mutual aid questions and answers, expert networks, wikis, etc. System management includes: space classification management, question and answer classification management, expert management, creative extraction management, and results promotion management. The functional structure of the system is shown in Figure 4.

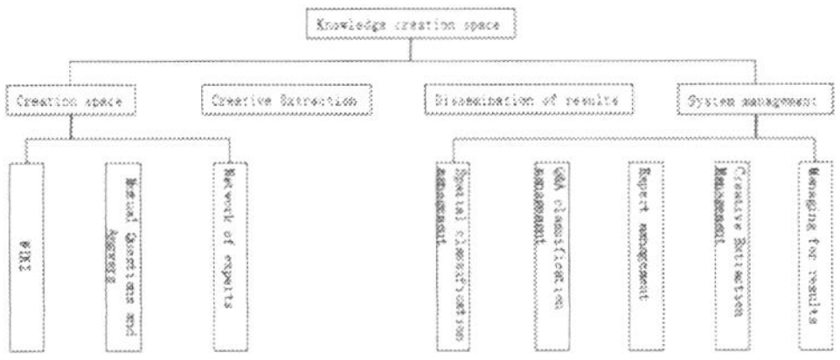

Fig. 3 China National Network Jiangsu Electric Power Knowledge Creation Platform Function

6 Conclusion and outlook

With the continuous innovation and development of information technology, the innovation model itself is also iteratively updating at a faster rate. The Zhongchuang mode is an innovative development mode that came into being in the context of the knowledge society, which effectively expanded the sources of innovation and enhanced the drive for innovation. This article takes Jiangsu State Power as the research object, specifically analyzes the development status of Jiangsu Electric Power Co., Ltd. of the State Grid and the conditions of the application of the crowd-building model, and designs the organizational structure and incentive mechanism of the crowd-sourced model applicable to power companies, and develops it in conjunction with Web 2.0. The knowledge-creating platform has added a detailed case for the application of the "Internet+Column" model. In the future, we can also analyze the application issues and solutions of the crowd-sourced model based on the operational effectiveness of the knowledge-based crowd-building platform, and form a set of perfect innovation models that meet China's national conditions and the development of the power system.

References

1. Chen Jin. Technological Innovation Studies for Super Innovative Nation [J]. Technology Economics, 2015 (34): 1-4
2. Liu Zhiying, Chen Qingxiang, Xu Yi. The Conceptual Model and Theoretical Analysis of Crowd Innovation [J]. Science of Science and Management of S. & T., 2015 (36): 52-61
3. Chesbrough, H. GE's ecomagination Challenge: An Experiment in Open Innovation, California Management Review [J], 2012 (3): 140-154
4. Wang guishan, Jin fu. The Model and Countermeasure of Intellectual Resources Management in High &New Technological Enterprise [J]. Science, Technology and Dialectics, 2006, 23(2):103-108.
5. Song gang , Zhang nan. Innovation 2.0: Democratizing Innovation in Knowledge Society [J]. China Soft Science, 2009(10): 60-66
6. Paul Griffiths and Teresita Arenas, 'ENTEL: A Case Study on Knowledge Networks and the Impact of Web 2.0 Technologies', Electronic Journal of e-Learning [J], 2014 (12): 383-393.
7. Steven De Hertogh, Stijn Viaene, And Guido Dedene, 'Governing Web 2.0', Communications of The ACM, Business Source Complete [J], 2011(54):124-130.
8. Wang shushuang. Construction of Science and technology enterprise human resources management system [J]. Scientific Management Research, 2012, 30(6):94-97.
9. Xu liping, Jiang xiangrong, Yin chong. A research on the evaluation index system about innovation ability of enterprises [J]. Science Research Management, 2015(s1):122-126
10. Fu qunying , Liu zhiying. Crowd Innovating: Connotation and Operation Model[J]. Science of Science and Management of S. & T., 2016(37): 3-10.

Research on the control system of an amphibious transmission line Inspection UAV based on fuzzy PID control

Lu Shengbiao[1], *Xia* Liangbiao[1], *Shi* ying[1], *Han* Xiping[1], *Pang* Tong[1], *Cai* Ruyi[2]

[1]Guangxi power grid limited liability company, Yulin Power Supply Bureau, Yulin, China
[2]School of Electrical Engineering, Wuhan University , Wuhan, China

Abstract. The amphibious UAV combines the advantages of multi rotor UAV and line inspection robot. It can carry out two working modes: air flight and online driving. It can observe the running state of transmission line and improve the quality of patrol. In this paper, a fuzzy PID control algorithm is proposed to make the amphibious UAV flight control more flexible and close to the transmission line, and the simulation and software control process design are carried out.

1 Introduction

UAV is an advanced unmanned self propelled vehicle, which has the ability of high altitude, long distance, fast operation and self operation. It can quickly patrol the transmission line and improve the efficiency of transmission line inspection.At present,UAV has a problem of insufficient endurance, limited vision, easy to defect, and higher quality of control personnel.The inspection robot needs to be manually installed on the conductor, which is heavy in weight and difficult to hoist. At the same time, the robot is slow in walking speed, difficult to surmount obstacles, and has long navigation time and poor practicability.

In this paper, a new design scheme of amphibious patrol unmanned air vehicle is proposed based on the advantages of the inspection robot and the multi-rotor UAV. The fuzzy PID control algorithm is used to make the flight control more flexible and close to the target.

2 Method

2.1 Amphibious UAV inspection process

The amphibious patrol UAV is improved in the basic structure of the rotorcraft, including the suspension rod, the sliding power device and so on. The amphibious UAV can be hung on the ground line for inspection. The other power unit is used to drive the U theory. The UAV's obstacle crossing operation is completed by the flight unit and the patrol equipment[1].

Amphibious type UAV in flight line below the starting unit, the motor drives the propeller to rotate the rotor lift, flying close to the power line to guide hanging on the line, when the sensor determines has hanging off the motor, each rotor, and start line patrol unit. When the UAV meets the obstacle, it stops moving on the line and starts the flight unit again for the flight winding barrier[2].

2.2 Basic action of amphibious UAV

The amphibious UAV takes the multi rotor unmanned aerial vehicle as the foundation of the basic flight unit. The flight unit controls the change of lift by adjusting the speed of the propeller with the motor, so as to control the attitude and position of the aircraft.

The lift force produced by the propeller is proportional to the square of the angular velocity of the propeller, that is, the speed of the motor is doubled and the tensile force increases to the original number of times. The flight attitude of the aircraft can be controlled by controlling the motor drive by a given PWM signal. When the rotation speed of each propeller is the same, the reverse torque produced by the two adjacent propellers counteracts each other, and the whole horizontal force is in the equilibrium state[3]. The speed of each propeller is increasing or decreasing, which can guarantee the vertical rise or decline of the aircraft unit. If the total lift of each propeller is greater than gravity, then the whole aircraft will steadily increase and decrease. Otherwise, the sum of the lift of the rotor is equal to the weight of the aircraft, and when the speed is zero, the vehicle will remain hovering[4].

Because the UAV controls the flight by adjusting the motor speed, it will cause the system not very stable, so an appropriate control method is needed to control the flight of the aircraft.

2.3 Flight control process of amphibious UAV

The control of amphibious UAV adopts closed loop control. It compares the expected value with the actual measured value, and constantly corrections the actual measured values to achieve the numerical value we need. A high flying (for example, ignoring the influence of the air resistance and the external environment), take off from the ground, the plane will be at a certain acceleration to accelerate the rise; when the speed reaches a certain value and larger and the target height difference, the drone into the uniform rising stage; when the height of the UAV and close to the target range that movement into the deceleration phase[5]. The position of flight expiration stops. As shown in the figure, the flexible approach to the specified height hover is achieved by controlling the acceleration of the deceleration.

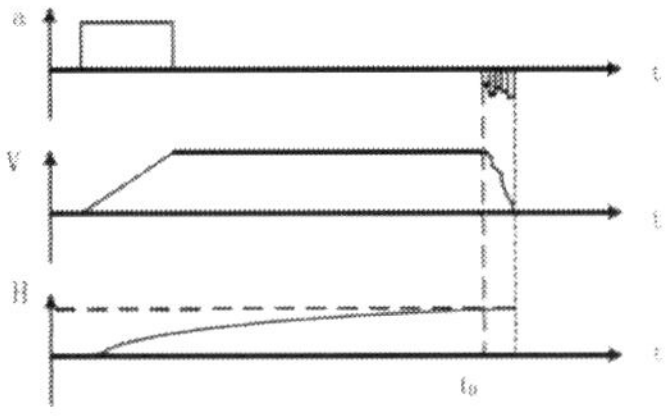

Fig. 1 Control deceleration acceleration method to fly to specified hover

The control process is adjusted by the difference between the actual height and the calibrated height as the adjustment. Considering the comprehensive factors, the fuzzy PID control principle is adopted for the handling of the control quantity. The general PID control can eliminate the static error, and can improve the control precision when the error value is relatively small. However, PID control has overshoot phenomenon, and at the same time, the reaction time is longer when there is interference. It is not suitable for direct flight control algorithm of amphibious UAV[6].

The fuzzy control has good robustness and can adapt to the change of parameters in a relatively large range. Compared with the traditional PID control, the dynamic response stability performance is high. The control of flight unit requires high real-time performance. Traditional PID can't meet the real-time requirement of controller[7]. Combining fuzzy control, it can shorten regulating time and meet real-time requirement, and also can restrain overshoot and overshoot behavior very well. Therefore, the combination of PID and fuzzy control can meet the design requirements.

3 Research on fuzzy PID control algorithm

3.1 Fuzzy PID control

Fuzzy PID control is a combination of fuzzy control and PID control. Fuzzy control is a nonlinear control. It is a numerical control based on fuzzy set theory, fuzzy language variable and fuzzy logic reasoning. It is the control method of fuzzy reasoning and decision processing of human behavior experience. Firstly, the control experience has been adapted into fuzzy rules, and then the real-time signal actually collected signal after fuzzification, fuzzy treatment as the input value input to the fuzzy inference module, output is obtained after the reasoning will be added to the controller, in order to achieve further control. The fuzzy controller is a kind of language controller, because the fuzzy control rules used are described according to the fuzzy condition statement in the fuzzy theory. The design contents of fuzzy controller include: first, select the set of all words that can describe the input and output variables, and then define the fuzzy subset of every fuzzy variable. Finally, we need to set up the corresponding fuzzy controller's control rule.

The change of the actual measurement error value and the calibration value of E and the error rate of EC as the input and processed by fuzzy controller (fuzzy parameters, fuzzy reasoning, fuzzy parameter solution) processing by the corresponding Kp, Ki, Kd values of three parameters input to the PID controller, the final output control the target object[8].

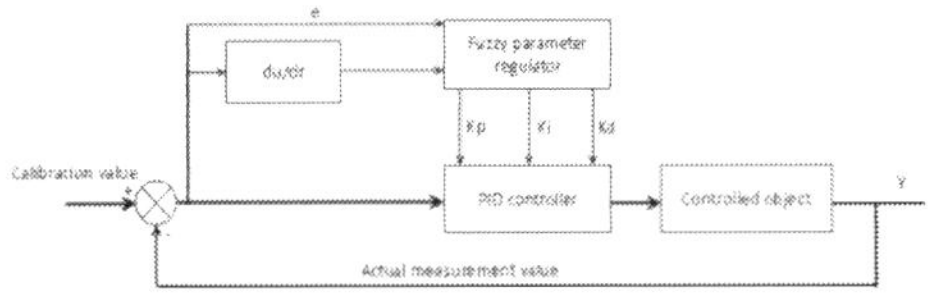

Fig. 2 control structure diagram

3.2 Construction of PID fuzzy controller for amphibious UAV

Fuzzy PID control is to find a fuzzy relationship based on PID, KD, KI, KP three parameters and E and EC, and continuously modify all parameters in real time to meet the current control requirements of E and EC, and build fuzzy PID controller under the condition of high control.

(1)Brushless DC motor transfer function

Assuming that the load of the brushless DC motor is constant, only the angular velocity of the output motor is considered, then the transfer function of the brushless DC motor is obtained.

$$\frac{\Omega(s)}{U_a(s)} = \frac{K_T}{LJs^2+(LK_D+RJ)s+K_eK_T+RK_D} \quad (1)$$

Among them, L is self inductance of three phase stator, R is the resistance of stator windings in every phase, KT is expressed as electromagnetic torque constant, Ke back EMF constant, KD is the damping coefficient of rotation part, and J moment of inertia. According to the characteristics of the motor and the system, the values of all the parameters of the motor and the transfer function of the motor can be obtained.

$$G(s)=\frac{523500}{s^2+87.35s+10470} \quad (2)$$

(2)Design of fuzzy PID controller

According to the existing control experience and the curve analysis of the measured value and the calibrated

https://doi.org/10.1051/e3sconf/20186405001

value, the adjustment is made.The size of the three parameters of the PID. Fuzzy control is used to improve the setting of three parameters of PID.

3.3 System control simulation

Based on the mathematical model of brushless motor, the traditional PID control and fuzzy PID control are simulated by Matlab/Simulink software, so as to verify the control effect of fuzzy PID.

After repeated experiments, the three proportional parameters of PID are set to (0.4, 1, 0.01). The step signal sampling time is 0.001, and the system running time is 2 seconds. The step response of the system can be obtained, as shown in Figure 3.

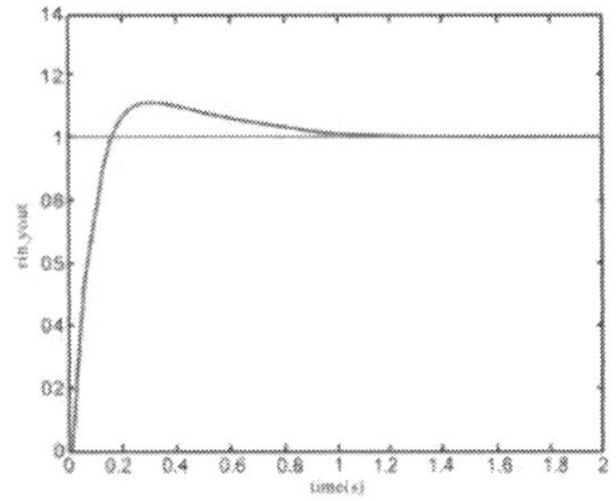

Fig. 3 Simulation results of step response

The simulation chart shows that the system can be adjusted to the specified position in a short time, but the phenomenon of overshooting occurs. This overshooting behavior is not allowed for the fly-flying patrol robot to fly to the specified height during flight. It is easy to collide with the transmission line.

Run the simulation program, the system step response shown in Figure 4. Under the step response, compared with the traditional PID in Fig. 3, the rise time and the adjustment time of the fuzzy PID control system are greatly reduced, the reduction of the overshoot is obvious, and the dynamic performance of the system is greatly reduced improve.

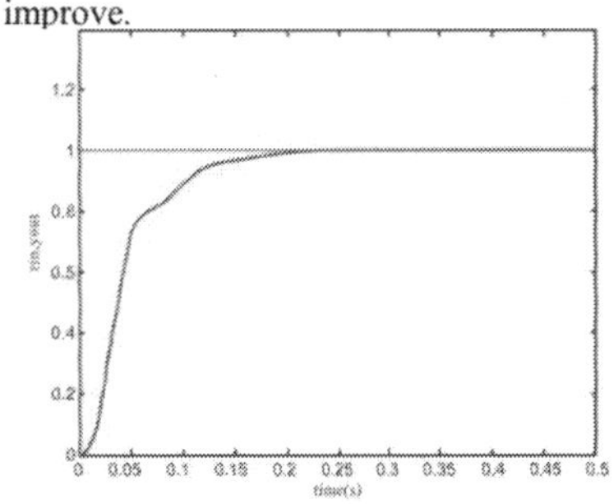

Fig. 4 Step response of fuzzy PID system

Based on this control algorithm, we also need to verify the system's response speed when the algorithm encounters interference. Therefore, at the position of the 400th sampling point of the algorithm, the amount of interference is added and the system responds to the system through Matlab, as shown in Figure 5.

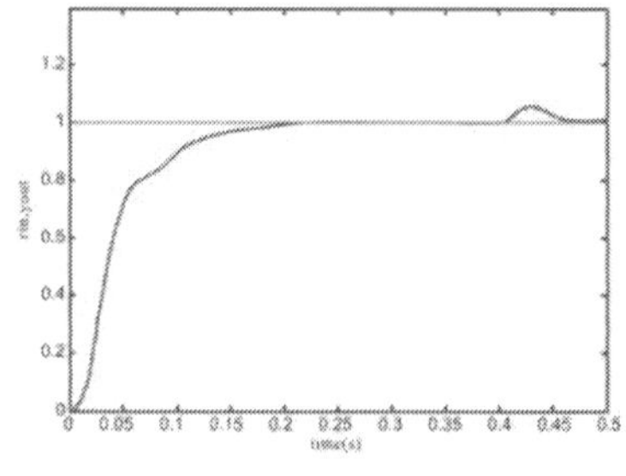

Fig. 5 Step response of system with interference

The simulation results show that overshoot occurs when 0.4s is used, but the overshoot is relatively small, and at the same time, it can quickly and smoothly overshoot. Therefore, the flight control of the flying skid line robot can also respond to the system very quickly if it interferes[9].

The above simulation results show that fuzzy PID has good control effect, and fuzzy control algorithm is applied to the flight control of the flying skid line robot, which is in line with the design requirements.

3.4 Design of flight control software

The overall goal of the software design for the control system of the amphibious patrol UAV is to start the various functional modules of the flight control system and make it work normally, and achieve stable flight according to the established plan. The flight control of the UAV common and the biggest difference is that the amphibious patrol UAV flight in a transmission line for flight target, close to the target need to slow down and flexible contact, therefore, control of selecting the appropriate amount of control software design, using the control theory of fuzzy PID control algorithm to control the system. The main control chip of flight unit to complete the action process as follows: first, system initialization, and then check the flight system is normal, normal self start motor driver, the AD will open the collection and return attitude information, then execute the control algorithm to calculate the output control volume, will control the amount back to the motor driver.

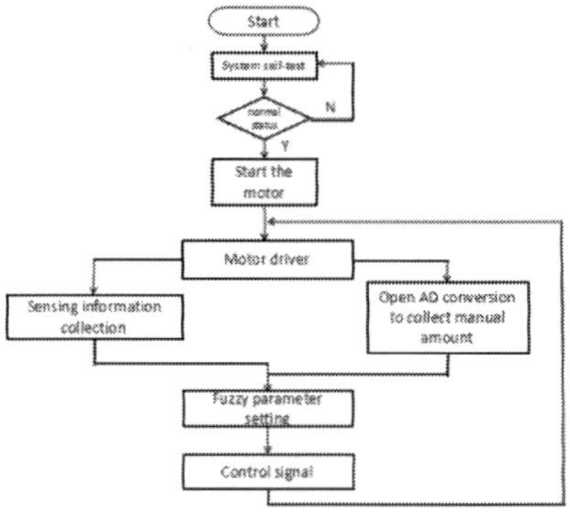

Fig. 6 system startup process

In the process of navigation, the chip samples the result of the angular velocity and the linear acceleration in the result register of the ADC. The sampling flowchart is shown in Figure 7.

In motor control, first initialize hardware, setting control signal value is initialized; after the start, the attitude signal collected into reliable speed feedback signal according to the feedback signal by the fuzzy PID control algorithm parameter tuning output control signal, the motor reaches the desired speed, in order to adjust the attitude objective. The fuzzy PID controller software flow chart is shown in Figure 8.

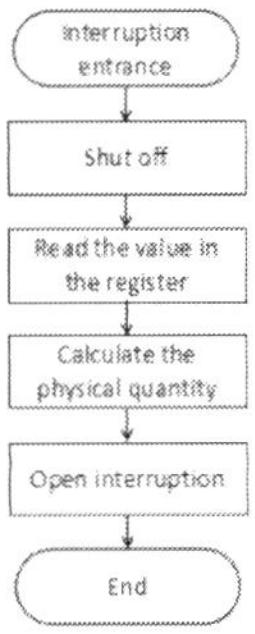

Fig. 7 sampling flow chart

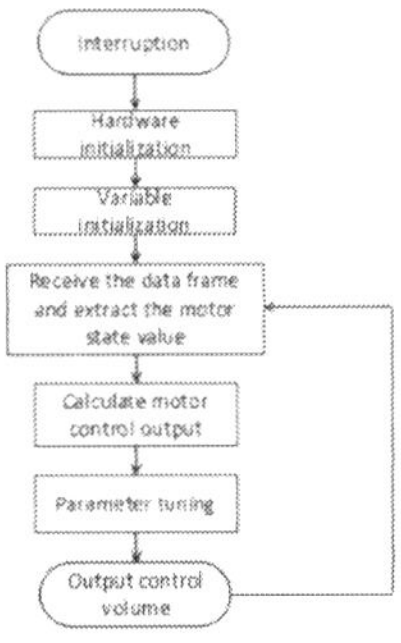

Fig. 8 software flow of fuzzy PID controller

4 Conclusion

The amphibious inspection UAV needs to fly near the transmission line and hangs on the transmission line. This paper designs the operation flow of the UAV flexible proximity transmission line, and designs and realizes the fuzzy PID control algorithm according to the operation flow. The simulation results show that the fuzzy PID control system can shorten the rise time and adjust the time, reduce the overshoot, and improve the dynamic performance of the system. However, it does not consider the impact of wind and other environmental factors. At the same time, the system intelligence is not high enough, Autonomous flight research.

References

1. Zheng Weigang. Research on distribution network inspection system base on infrared image technology for UAV[D]. Shenyang Agricultural University, (2013).
2. Xu Huadong. Research on Intelligent Obstacle Avoidance Methods for Power Lines Inspection with UAVs[D]. Nanjing University of Aeronautics and Astronautics, (2014).
3. Wu Feilong, Lin Han, Tang Mingwen. Application of Multi-relay Method in Inspection of Large Unmanned Aerial Vehicle Transmission Line[J]. Electric Power,02:104-110,(2015).
4. Liang Kun. Research on Ground Supervision and Control Technology of Power Line Inspection with Unmanned Aerial Vehicle[D]. Nanjing University of Aeronautics and Astronautics, (2014).
5. Whitworth C C,Duller A W G,Jones D I,et al. Aerial video inspection of overhead power lines [J].Power Engineering Journal ,15(1): 25-32, (2001).
6. Montambault S, Cote J, St.Louis,M.Preliminary results on the development of a teleoperated compact trolley for live-line working[C].Proceeding of the 2000 IEEE 9th International Conference on Transmission and Distribution Construction, Operation and Live-line Maintenance, Montreal, Canada:21-27(2000).
7. Wang Hongguang, Jiang Yong,Liu Aihua,et al.Research of Power Transmission Line Maintenance Robots in SIACAS[C]. 2010 1st International Conference on Applied Robotics for the Power Industry. Canada, October 5-7, (2010).
8. Jidai Wang,Aiqin Sun,Wei Su, et al.Development of an Expert Control Strategy for Controlling Obstacle Crossing of a High-voltage Transmission Line Inspection Robot[C]. Proceedings of the 18th International 195 Conference on Automation & Computing, Loughborough University, Leicestershire,UK,8 September 2012.
9. Julien Beaudry, Kristopher Toussaint , Nicolas Pouliot. On the Application of VTOL UAVs to the Inspection of Power Utility Assets. International Conference on Applied Robotics for the Power Industry,(2010).

Analysis of passive interference on radio station from AC UHV power

GAN Zheyuan[1], *ZHANG* Jiangong[1] , *ZHAO* Jun[1] and *Lu* ying[2]

[1] China Electric Power Research Institute ,Wuhan,China
[2]North China Electric Power University, Beijing 102206, China

Abstract. The problems of Electromagnetic Interference in Airborne Intelligence Radar Stations near the AC UHV Transmission Lines is researched in this paper, which is mainly passive interference. And then analyzed the passive interference of AC UHV lines. It mainly studied the influence of the tower and line on the distance detection performance of the air-to-air intelligence radar of the UHV transmission lines. and integrated the influence of the surface reflection surface, the shadowing effect and the secondary radiation echo interference, Proposed a passive interference protection distance for UHV transmission lines and airborne intelligence radar stations

1 introduction

With the rapid development of current science and technology, UHF radio stations(such as analog TV and digital TV broadcasting station, military aviation radars, etc.),which have strict requirements on the surrounding electromagnetic environment. The solution of passive interference of these stations with UHV transmission lines with large size metal structures has become a research hot spot[1,2].

The air-to-air intelligence radar is the radar which used to search, monitor, identify airborne targets and determine their coordinates and motion parameters, it also known as the air-to-air search radar[3].It is an important electronic technology for acquiring air target information in modern warfare equipment.

In this paper, it analyzed the passive interference of the AC UHV transmission lines, and mainly studied the influence of the tower and line on the distance detection performance of the air-to-air intelligence radar of the UHV transmission lines, and then integrated the influence of the surface reflection surface, the shadowing effect and the secondary radiation echo interference, Proposed a passive interference protection distance for UHV transmission lines and airborne intelligence radar stations.

2 Analysis of passive interference of transmission lines and airborne radar stations

The passive interference of transmission lines to radar is mainly manifested in three aspect[4]:

• The secondary radiation of radar electromagnetic waves by obstacles such as transmission lines and towers can generate fixed interference echoes to the radar;

• When the ground facilities such as transmission lines or substations are at a certain distance from the radar, the effective reflection surface of the radar position may be destroyed in this direction, which affects the performance of the radar detection;

• The orientation of the ground facility of the transmission line or substation may cause a loss of shielding from the electromagnetic waves emitted by the radar.

For the meter-band radar, the electric field strength radiated from the antenna to a point in space is the vector sum of the direct wave and the ground reflected wave, and its detection performance is related to the radar position. According to the requirements of Standard 2[5], for the meter wave radar, the detection power at different elevation angles can be described by the following formula:

$$r = R_0 \cdot F(\alpha) \cdot \sqrt{1 + D^2|R|^2 + 2D|R|\cos\left(\frac{4\pi h}{\lambda}\sin\alpha + \phi\right)} \quad (1)$$

Where: R_0 ——radar free space nominal distance; $F(\alpha)$——the vertical direction coefficient of the radar antenna; D、R、ϕ——respectively represent the ground diffusion coefficient, the ground reflection coefficient and the reflected wave lag phase; h——the height of radar antenna; λ ——radar operating wavelength.

When the ground is an ideal reflection condition, the sum of all items under the root number is close to 4 at a certain elevation angle, and the radar detection distance can be increased by nearly 1 times after the square root is at a certain elevation angle. It can be seen that the terrain around the radar antenna has a great influence on the radar detection performance. The size of this area is usually calculated according to the Fresnel zone required by the radar elevation angle of the first lobe.

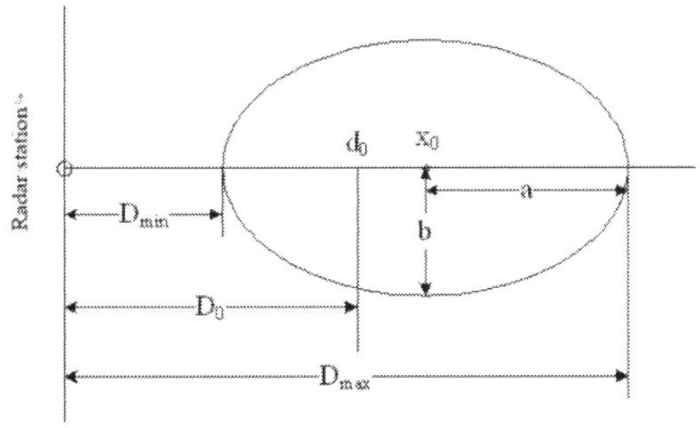

Fig. 1 The first wave of disc center elevation of Fresnel zone

This area is an ellipse, as shown in Figure 1. The relationship of its long axle a 、 short half shaft b 、 Ellipse center x_0 with radar antenna height h 、 operating wavelength λ 、 target elevation angle α is as follows:

$$x_0 = \frac{h}{\sin\alpha}\left(1+\frac{\lambda}{2h\sin\alpha}\right) \tag{2}$$

$$a = \sqrt{\frac{h\lambda}{\sin^3\alpha}\left(1+\frac{\lambda}{4h\sin\alpha}\right)} \tag{3}$$

Where: h ——the center height of the radar antenna; λ ——radar operating wavelength; α ——the elevation angle on the vertical plane of the antenna.

Then, the farthest reflection distance D_{max} required by the first lobe elevation angle, the closest reflection distance D_{min} are:

$$D_{max} = x_0 + a = \frac{h}{\sin\alpha}\left(1+\frac{\lambda}{2h\sin\alpha}\right) + \frac{2h}{\sin\alpha}\sqrt{\frac{\lambda}{4h\sin\alpha}\left(1+\frac{\lambda}{4h\sin\alpha}\right)} \tag{4}$$

$$D_{min} = x_0 - a = \frac{h}{\sin\alpha}\left(1+\frac{\lambda}{2h\sin\alpha}\right) - \frac{2h}{\sin\alpha}\sqrt{\frac{\lambda}{4h\sin\alpha}\left(1+\frac{\lambda}{4h\sin\alpha}\right)} \tag{5}$$

If the first lobe elevation expression $\alpha_1 = \arcsin\left(\frac{\lambda}{4h}\right)$ is brought in (4) and (5), it is simplified and available:

$$D_{max} = 23.3\frac{h^2}{\lambda} \tag{6}$$

$$D_{min} = 0.7\frac{h^2}{\lambda} \tag{7}$$

The distance between the Geometric reflection center and the radar d_0 is:

$$d_0 = \frac{h}{\tan\alpha} \tag{8}$$

Put α_1 the expression substituted into (8) gives:

$$d_0 = 4\frac{h^2}{\lambda} \tag{9}$$

As can be seen from the above formula, the distance between each reflection point is proportional to the square of the height of the antenna h, and is inversely proportional to the wavelength λ .According to the analysis: the electromagnetic energy density of the antennas of different distance segments is not the same, and more than 60% of the energy is concentrated near the center point of geometric reflection. For this reason, when considering the influence of ground obstacles on radar detection performance, it can be calculated according to radar's geometric reflection point. According to stricter requirements, the working frequency of the radar is taken as 300MHz, and the height of the antenna center is taken as 10m. Substituting the equation (6), (7) and (9) can be obtained: D_{max} =2330m、 D_{min} =70m、 d_0 =400m, the protection distance of meter wave radar to UHV transmission lines is 2330m.

Ground facilities such as high-voltage overhead transmission lines may have a shadowing effect on direct waves and ground reflection waves radiated from radar antennas at a certain distance outside the Fresnel zone. Even at a distance other than the farthest point of the effective reflective surface, the ground reflected wave may be shielding, so that the detection performance at a low elevation angle is greatly affected or the vector sum of direct and reflected waves at a certain point in space is affected, as shown in Figure 2. The average height of the UHV transmission line route tower is 50m, d_0 takes 400m, and the height h of the radar antenna is 10m away from the ground. The protection distance between the UHV transmission line and the radar calculated according to Figure 2 is:

$$d_0 + d = d_0\left(1+\frac{H}{h}\right) = 400\left(1+\frac{50}{10}\right) = 2400m \tag{10}$$

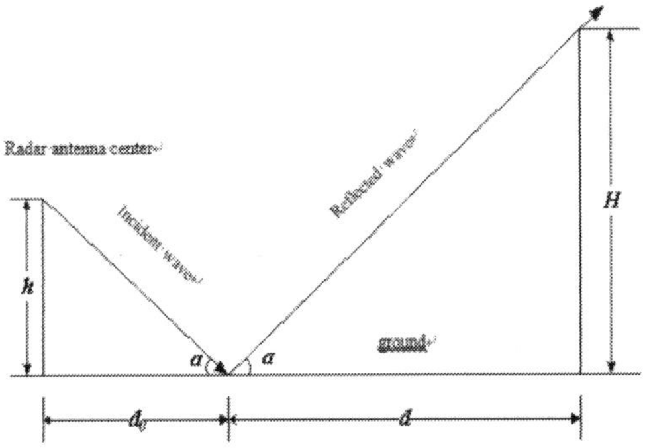

Fig. 2. Schematic diagram of transmission line blocking reflection wave

3. Simulation test of the influence of transmission lines on radar

The test of the influence of obstacles such as UHV transmission line routes and towers on radar detection performance is most ideally carried out on actual circuits. Considering that the transmission line and the tower itself are relatively large, it is difficult to carry out the physical test. Therefore, the" scaling ratio" method is used for the simulation test.

According to Wang Qi, Zhao Jie, and Zhao Zhibin's research on the "passive interference of 800kV DC transmission lines to the radio stations in the short-wave band", the parameters set in this paper are as follows. The experimental prototype is a "drum type" iron tower. The ratio is 30:1, the tower height is about 100m (100/30m in the test), the transmission line adopts 8 split double loop form, the wire splitting pitch is 400/30mm, and the wire type is 8×LGJ-500/35 aluminum cable steel reinforced. Stranded wire with a wire radius of 16/30mm and a pitch of 500/30m. The ground wire is aluminum cable steel reinforced with a grounding resistance of no more than 15Ω. During the test, a 3-pitch 4-tower model was selected for testing.

The transmitting antenna and receiving antenna adopt the double-ridged horn antenna with working frequency band of 1~18GHz, its average gain is 11.3dB, the impedance is 50Ω,the maximum power is 300W,and the peak power is 500W. The power of the transmitted signal source is E8257D with a maximum output power of 10dBm and a maximum operating frequency of 40GHz. The receiver uses the E8257D spectrum analyzer and its operating frequency range is 9kHz~26.5GHz. The signal source and the transmitting antenna and the spectrum analyzer and the receiving antenna are respectively located in the two sides of the transmission line and the tower, the signal source is connected with the transmitting antenna, and the spectrum analyzer is connected with the receiving antenna[3]. The test layout is shown in Figure 3.

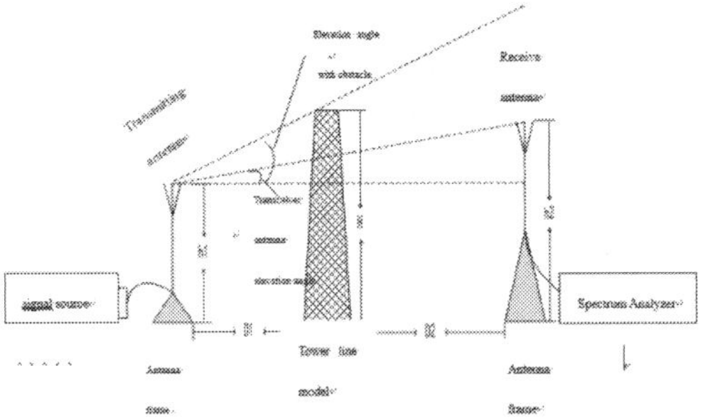

Fig.3 Layout of measurement

Where: $D1$ ——distance between the transmitting antenna and the tower line model, units: m; $D2$ —— distance between the receiving antenna and the tower line model, units: m; $h1$ ——the height of the transmitting line, units: m; H ——the height of the tower and line model, units: m; $h2$ ——the height of the receiving line, units: m; α ——the elevation angle between the transmitting antenna and the obstruction highest tangent, in degree; β ——the elevation angle between the connecting line of the transceiver antenna and the horizontal line, in degree.

The main parameters are set as follows: based on the source power and the receiving distance, the distance between the transmitting antenna and the receiving antenna is determined, that is , ($D1+D2$)is 100m, test frequency selection 3GHz, 7GHz and 12GHz. During the test, follow the steps below.

• Testing the propagation characteristics of electromagnetic waves in free space.

Select the open space, fix the transmitting antenna and receiving antenna at intervals of 100m,the transmitting antenna frame height is 1.5m; when the transmitting signal frequency is set to 3GHz, 7GHz and 12GHz respectively, The corresponding erection height of the receiving antenna is 2.1m, 2.5m, 3m and 6m respectively. Record the power indicator displayed by the spectrum analyzer;

• Testing the effect of iron tower on the propagation characteristics of electromagnetic waves.

Keep the setting parameters of the spectrum analyzer and the output power of the signal source unchanged, and set the position of the receiving antenna and transmitting antenna unchanged. The tower model is set at 70m, 50m, 30m and 10m from the transmitting antenna respectively. The corresponding erection height of the receiving antenna is 2.1m, 2.5m, 3m and 6m respectively. Read the power indicator displayed by the spectrum analyzer;

• Testing the influence of iron towers and transmission lines in the propagation characteristics of electromagnetic waves.

Keep the setting parameters of the spectrum analyzer and the output power of the signal source unchanged, and set the position of the receiving antenna and the transmitting antenna unchanged. Set the assembled iron tower and transmission line models at positions 70m,

50m, 30m, and 10m away from the transmitting antenna. Record the power indicator displayed by the analyzer.

From the test results, the mainly influence of transmission line is the line, and the tower has less influence on the signal transmission. The results of the passive interference simulation test of the transmission line to the radar station during the 1988 GB13618-92"Electromagnetic Environment Protection Requirements for Airborne Radar Stations" are given. The test principle and arrangement are the same as this test. The test results also show that the influence of high-voltage overhead transmission lines on signal propagation is mainly the influence of split conductors, and the influence of towers can be ignored. The main reason is that the actual size of the tower forms a shielding angle to the transmitting antenna smaller than the critical unshielded opening angle.

3 Conclusion

In summary, for the meter band radar, the electric field intensity radiated from the antenna to a point in space is the vector sum of the direct wave and the ground reflected wave, its detection performance is related to the radar position, and the path of the ground reflected wave should not be blocked by the ground object. Direct waves and ground reflected waves radiated from the radar antenna may also have a shadowing effect at a certain distance outside the Fresnel zone.

When using a 1000kV UHV single-circuit transmission line, the guard spacing is 2400m. If the height of the radar position is significantly higher than the average height of the transmission line conductor, the protection distance of the meter wave radar to satisfy the effective reflection surface is 2400m, and the remaining radars can satisfy the active interference protection distance of 1600m. However, before determining the protection distance between the 1000kV UHV transmission line and the air-to-air radar station, it is also necessary to negotiate with the radar station management department. In the specific implementation, if it is difficult to ensure the protection spacing, other areas can be negotiated and resolved in the case that the main responsibility area is satisfied to meet the above requirements.

Acknowledgement

This work was supported by the Project of SGCC (GY71-16-011).

References

1. Yin Hui, Zhang Xiaoming, Wang Yantao, et al. Research on 3D Visualization of Electromagnetic Interference for Ultra-high Voltage Transmission Lines[J]. High Voltage Engineering, **40**, 12, 2014.
2. Yang Jiawei, Tang Bo, Huang Li, et al. Research on Interference between UHV Power Transmission Lines and Air Defense Surveillance Radar Station[J]. Journal of China Three Gorges University, 2018.
3. Ding Lufei, Geng Fulu, Chen Jianchun, et a. Radar principles[M]. Beijing. Publishing House of Electronics Industry, 2014.
4. GB13618-92. Electromagnetic Environment Protection Requirements for Airborne Radar Stations[S]. Beijing. Standard Press of China, 1992
5. GJBZ20195-3. Selection Criteria for Military Ground Radar Positions[S]. Beijing. Standard Press of China, 1993

 https://doi.org/10.1051/e3sconf/20186405003

Analysis of active interference on radio station from AC UHV power

ZHANG Jiangong[2], GAN Zheyuan[2] and ZHAO Jun and Lu ying[2]

[1] China Electric Power Research Institute，Wuhan，China
[2]North China Electric Power University, Beijing 102206, China

Abstract. The problems of Electromagnetic Interference in Airborne Intelligence Radar Stations near the AC UHV Transmission Lines are researched in this paper, which is mainly active interference. The calculation method for the interferences is analysed, and then the proper protection distances of the AC UHV transmission line to the radar station are given. The paper analyzed the active interference of AC UHV transmission lines, and focuses on the active interference caused by the line corona on the active radar of nearby airborne intelligence when the transmission line is under normal operation. In the end, this paper proposes the recommended values of active interference protection distance for 1000kV AC UHV transmission lines and airborne intelligence radar stations.

1 Introduction

With the rapid development of current science and technology, UHF radio stations(such as analog TV and digital TV broadcasting station, military aviation radars, etc.),which have strict requirements on the surrounding electromagnetic environment. The solution of passive interference of these stations with UHV transmission lines with large size metal structures has become a research hot spot[1,2].

The air-to-air intelligence radar is the radar which used to search, monitor, identify airborne targets and determine their coordinates and motion parameters, it also known as the air-to-air search radar[3].It is an important electronic technology for acquiring air target information in modern warfare equipment.

The paper analyzed the active interference of AC UHV transmission lines, and focuses on the active interference caused by the line corona on the active radar of nearby airborne intelligence when the transmission line is under normal operation. And then, this paper proposes the recommended values of active interference protection distance for 1000kV AC UHV transmission lines and airborne intelligence radar stations.

2 Analysis of active interference of transmission lines and airborne radar stations

The operating frequency of the radar[3].The operating frequency of the applicable radar has been extended from the high frequency band(3MHz to 30MHz)to the millimeter band(30GHz to 300GHz) .The air-to-air intelligence radar studied in this paper is a commonly used in China. The frequency is 80MHz to 3000MHz, which can be divided into two frequency bands: 80MHz to 300MHz and 300MHz to 3000MHz. The definition of guard spacing and the guard spacing limit[4] are given in GB13618-1992" Electromagnetic Environment Protection Requirements for Airborne Information Radar Stations". Table 1 shows the spacing between the overhead power lines and the airborne intelligence radar stations at different voltage levels. If it cannot meet the requirements of Table 1 or special circumstances, it can be calculated according to the electromagnetic environment protection criteria of the air intelligence mine station. The interference effects of high-voltage overhead transmission lines and substations considered in the electromagnetic environment protection criteria mainly come from two aspects: First, the high-voltage overhead transmission lines, substations, etc. in the live operation, will produce radio interference; Second, the high-voltage overhead transmission lines, towers and other facilities are used as obstacles, if they are too close to the radar station, it will destroy the requirement of the reflector surface of the radar array, which will cause the deformation of the radar lobe and affect the detection performance. In addition, high-voltage overhead transmission lines, towers and other facilities will also cause a certain shielding angle or electromagnetic wave attenuation of the radar, which may cause the missing radar target or detection distance. According to GB13618-92 "Electromagnetic Environment Protection Requirements for Airborne Information Radar Stations"[4] (hereinafter referred to as "Standard 1") and GJBZ20195-3 "Selection Criteria for Military Ground Radar Positions"[5] (hereinafter referred to as "Standard 2") ,the protection spacing for

 https://doi.org/10.1051/e3sconf/20186405003

radar overhead transmission lines can be calculated separately. The larger one is the final guard spacing value.

Table 1. The protection space between overhead transmission line and intelligence radar station

		Protective spacing	
		80~300MHz	300~3000MHz
High-voltage overhead transmission lines	500kV	1.6km	1.0km
	330~330kV	1.2km	0.8km
	110kV	1.0km	0.7km

For the radio interference level of High-voltage overhead transmission lines above 30MHz, the above has mainly come from accidental factors such as sparks discharge of iron towers, insulators strings, and metal fittings. At present, there is no effective calculation and prediction method, and only through actual measurement.

The measuring instrument uses SCHWARZBECK 9k~3000MHz receiver, supporting 30MHz~300MHz broadband double-cone antenna and logarithmic period antenna of 200MHz~1000MHz frequency band. The radio interference quasi-peak of the antenna in both the vertical and horizontal polarization directions was measured in the measurement[6].

Since the radio interference level above 30MHz is very low, the method of comparative measurement under the condition that the test line segment is charged and not charged id adopted. The data measured under the condition that the line segment is not charged is the background radio noise, and by comparison, it is concluded whether the radio interference of the UHV transmission line in this frequency band is significant[7-8]. Figure 1 shows Radio interference quasi-peak spectrum in the 30 MHz to 1000MHz band .Figure 2 shows the influence of antenna polarization direction on measuring results.

Radars generally operate in ultrashort waves and above, and are therefore mainly affected by high frequency noise pulses. The mechanism of radar being reduced by noise interference[3] is as fellow: the interference noise from the transmission line enters the radar receiver and is superimposed on the internal noise, reducing the signal-to-noise ratio, causing the radar to reduce the target discovery probability. To maintain the same probability of discovery and false alarm probability, the power of the minimum detectable signal must be increased, but this will reduce the detection rang of the radar.

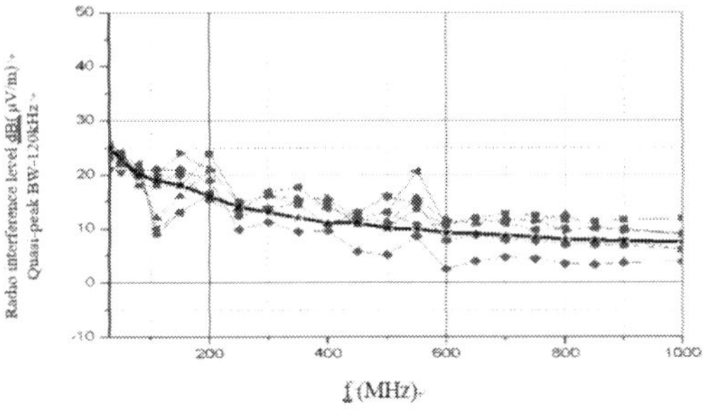

Fig.1 Radio interference quasi-peak spectrum in the 30 MHz~1000MHz band

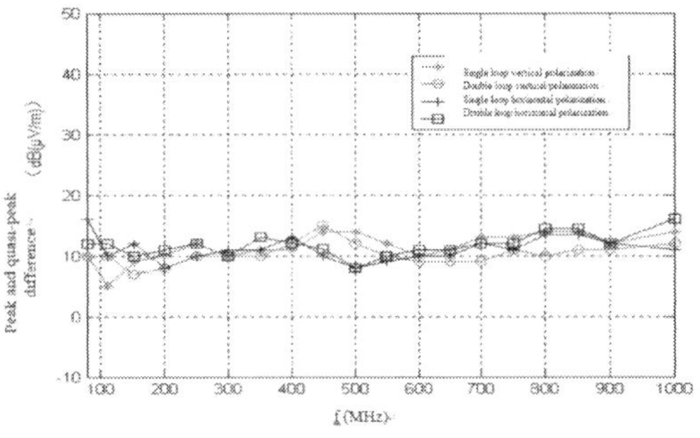

Fig. 2 The influence of antenna polarization direction on measuring results

Figure 3 and 4 summarize two typical comparative measurements. Both measurements were performed in good weather. The measurements did not distinguish between the radio interference levels before and after charging. Therefore, it can only be explained that the UHV line has low radio interference level in this frequency band and is completely submerged in background radio noise[9,10]. Background radio noise is mainly caused by unstable factors such as space radiation and meteorological activities, so that it is presented in an Irregular fluctuations with a small range.

The observation of the iron tower the insulator string, the fittings, etc, of the test line segment by the ultraviolet image also did not clearly reveal the portion where the spark discharge exists(this is an observation made for the specific measured object).

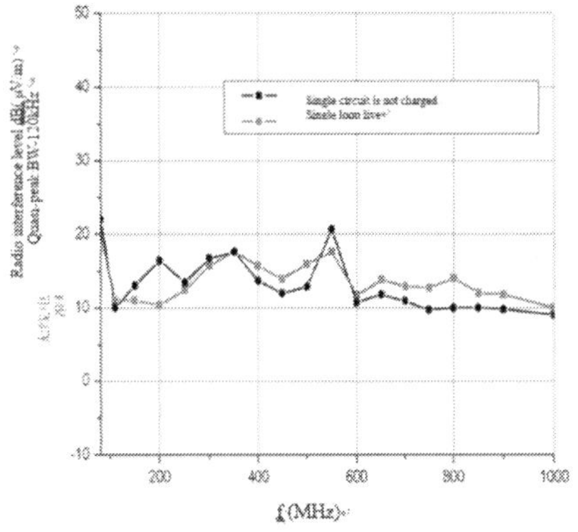

Fig . 3 Comparison I of the radio interference level before and after Charged (vertical polarization)

https://doi.org/10.1051/e3sconf/20186405003

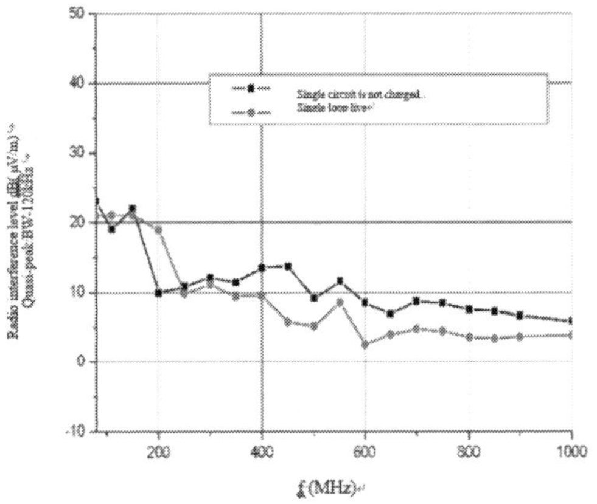

Fig. 4 Comparison II of the radio interference level before and after Charged (vertical polarization)

The measurement results show that the difference between the active radio interference generated by the UHV transmission line and the background noise is very small. Therefore, the impact on radar at working frequencies above 80MHz is very small. The guard spacing can follow the active guard spacing of the 500kV transmission line. That is, the guard spacing is 1600m when the working frequency is 80MHz~300MHz; the guard distance for the radar up to the operating frequency of 300MHz~3000MHz is 1000m.

3. Simulation test of the influence of transmission lines on radar

The test of the influence of obstacles such as UHV transmission line routes and towers on radar detection performance is most ideally carried out on actual circuits. Considering that the transmission line and the tower itself are relatively large, it is difficult to carry out the physical test. Therefore, the" scaling ratio" method is used for the simulation test.

According to Wang Qi, Zhao Jie, and Zhao Zhibin's research on the "passive interference of 800kV DC transmission lines to the radio stations in the short-wave band", the parameters set in this paper are as follows. The experimental prototype is a "drum type" iron tower. The ratio is 30:1, the tower height is about 100m (100/30m in the test), the transmission line adopts 8 split double loop form, the wire splitting pitch is 400/30mm, and the wire type is 8×LGJ-500/35 aluminum cable steel reinforced. Stranded wire with a wire radius of 16/30mm and a pitch of 500/30m. The ground wire is aluminum cable steel reinforced with a grounding resistance of no more than 15Ω. During the test, a 3-pitch 4-tower model was selected for testing.

The transmitting antenna and receiving antenna adopt the double-ridged horn antenna with working frequency band of 1~18GHz, its average gain is 11.3dB, the impedance is 50Ω,the maximum power is 300W,and the peak power is 500W. The power of the transmitted signal source is E8257D with a maximum output power of 10dBm and a maximum operating frequency of 40GHz. The receiver uses the E4408B spectrum analyzer and its operating frequency range is 9kHz~26.5GHz. The signal source and the transmitting antenna and the spectrum analyzer and the receiving antenna are respectively located in the two sides of the transmission line and the tower, the signal source is connected with the transmitting antenna, and the spectrum analyzer is connected with the receiving antenna[3]. The test layout is shown in Figure 5.

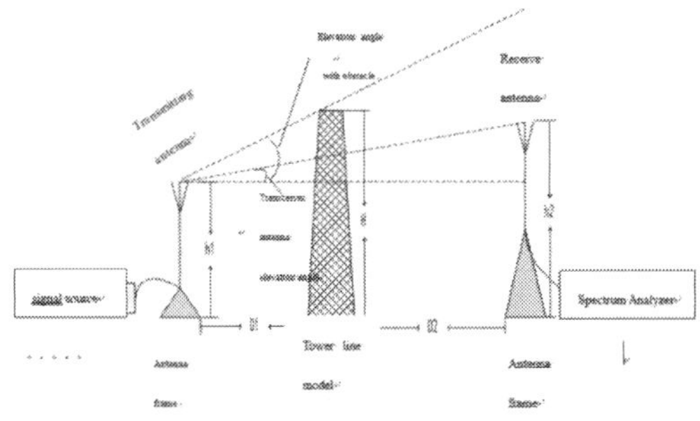

Fig.5 Layout of measurement

Where: $D1$ ——distance between the transmitting antenna and the tower line model, units: m; $D2$ —— distance between the receiving antenna and the tower line model, units: m; $h1$ ——the height of the transmitting line, units: m; H ——the height of the tower and line model, units: m; $h2$ ——the height of the receiving line, units: m; α ——the elevation angle between the transmitting antenna and the obstruction highest tangent, in degree; β ——the elevation angle between the connecting line of the transceiver antenna and the horizontal line, in degree.

The main parameters are set as follows: based on the source power and the receiving distance, the distance between the transmitting antenna and the receiving antenna is determined, that is , ($D1+D2$)is 100m, test frequency selection 3GHz, 7GHz and 12GHz. During the test, follow the steps below.

- Testing the propagation characteristics of electromagnetic waves in free space.
  Select the open space, fix the transmitting antenna and receiving antenna at intervals of 100m,the transmitting antenna frame height is 1.5m; when the transmitting signal frequency is set to 3GHz, 7GHz and 12GHz respectively, The corresponding erection height of the receiving antenna is 2.1m, 2.5m, 3m and 6m respectively. Record the power indicator displayed by the spectrum analyzer;
- Testing the effect of iron tower on the propagation characteristics of electromagnetic waves.
  Keep the setting parameters of the spectrum analyzer and the output power of the signal source unchanged, and set the position of the receiving antenna and transmitting antenna unchanged. The tower model is set at 70m, 50m, 30m and 10m from the transmitting antenna

respectively. The corresponding erection height of the receiving antenna is 2.1m, 2.5m, 3m and 6m respectively. Read the power indicator displayed by the spectrum analyzer;

- Testing the influence of iron towers and transmission lines in the propagation characteristics of electromagnetic waves.
  Keep the setting parameters of the spectrum analyzer and the output power of the signal source unchanged, and set the position of the receiving antenna and the transmitting antenna unchanged. Set the assembled iron tower and transmission line models at positions 70m, 50m, 30m, and 10m away from the transmitting antenna. Record the power indicator displayed by the analyzer.

The test results show that when the transmission frequency is 12GHz, only the tower has a larger attenuation of the received signal than when there is a tower cable. The reason is that when the tower attenuation test is performed, the front of the tower is opposite to the transmitting and receiving antennas, and when there is a tower cable test, the minimum face of the tower is opposite to the transmitting and receiving antennas, resulting in the above results. From the test results, the mainly the influence of the transmission line is the line, and the tower has less influence on the signal transmission. If there is a high-voltage transmission line between the radar and the target, the echo signal received by the radar will pass through the transmission line twice. Since it is difficult to achieve a six-fold increase in the conductivity of the model, the absorption loss of the radar wave by the steel material used in this test model is about 2-3 times larger, so the test results may be stricter.

3 Conclusion

In summary, Since the actual measurement shows that the UHV transmission line has a low radio interference level above 80MHz in the radar operation, it is basically submerged in the background radio interference noise, and even if it is strictly considered, it will not exceed the current 500kV transmission line. The active interference protection distance of the radar to the UHV transmission line can be implemented according to the requirements of the protection spacing of the 500kV transmission line in" Standard 1", and the protection distance between the UHV transmission line and the radio station is 1600m. When using a 1000kV UHV single-circuit transmission line, the guard spacing is 2400m. If the height of the radar position is significantly higher than the average height of the transmission line conductor, the protection distance of the meter wave radar to satisfy the effective reflection surface is 2400m, and the remaining radars can satisfy the active interference protection distance of 1600m. However, before determining the protection distance between the 1000kV UHV transmission line and the air-to-air radar station, it is also necessary to negotiate with the radar station management department. In the specific implementation, if it is difficult to ensure the protection spacing, other areas can be negotiated and resolved in the case that the main responsibility area is satisfied to meet the above requirements.

Acknowledgement

This work was supported by the Project of SGCC (GY71-16-011).

References

1. Yin Hui, Zhang Xiaoming, Wang Yantao, et al. Research on 3D Visualization of Electromagnetic Interference for Ultra-high Voltage Transmission Lines[J]. High Voltage Engineering, **40**, 12, 2014.
2. Yang Jiawei, Tang Bo, Huang Li, et al. Research on Interference between UHV Power Transmission Lines and Air Defense Surveillance Radar Station[J]. Journal of China Three Gorges University, 2018.
3. Ding Lufei, Geng Fulu, Chen Jianchun, et a. Radar principles[M]. Beijing. Publishing House of Electronics Industry, 2014.
4. GB13618-92. Electromagnetic Environment Protection Requirements for Airborne Radar Stations[S]. Beijing. Standard Press of China, 1992
5. GJBZ20195-3. Selection Criteria for Military Ground Radar Positions[S]. Beijing. Standard Press of China, 1993
6. Gan Zheyuan, Wang Yanhai, Zhang Jiangong, et al. Suppression of HF Passive Interference in Transmission Lines by Using Magnetic Tube[J]. high Voltage Engineering, **43**, 5, 2017.
7. Xiang Tao, Hu Yuxian. The electromagnetic Environment Test of the Navigation Station of the Airport in Dazhou, Sichuan Province[J]. China Radio, **11**,1, 2017(11)
8. Zhao Jiahao. Causes and Treatment of Wireless Electromagnetic Wave Interference[J]. Electronic Technology and Software Engineering, **21**, 1, 2017
9. GB6364—86 Aeronautical Radio navigation Station Electromagnetic Environment Requirements[S]. Beijing. Standard Press of China, 1986
10. GB15707-1995 High-voltage AC Overhead Transmission Line Radio Interference Limit [S]. Beijing. Standard Press of China, 1995

E3S Web of Conferences 64, 05004 (2018)
ICPRE 2018
https://doi.org/10.1051/e3sconf/20186405004

Analysis on the Influence of Transmission Lines span on Passive Interference in Shortwave

Lu Ying[1], *Zhao* Zhibin , *Zhang* Jian gong[2] and *Gan* Zheyuan[2,1]

[1]College of Electrical and Electronical, North China Electric Power University, Beijing 102206, China
[2]China Electric Power Research Institute, Wuhan, China

Abstract. The passive interference of transmission lines to nearby radio stations may affect the effective reception and transmission of radio station signals. Therefore, the accurate calculation of the electromagnetic scattering of transmission lines under the condition of external electromagnetic waves is the basis for determining the reasonable avoidance spacing of the two. For passive stations operating in short-wave frequencies, passive interference is mainly generated by the tower, and span is one of the most significant factors affecting passive interference. This paper uses the method of moments to carry out the passive interference calculations under normal circumstances, expounds the method of calculating the electromagnetic field of transmission line at the same time. And elaborates the method for calculating the electromagnetic field of the transmission line, obtains the space electric field intensity of the transmission line at the same working frequency and space location of the plane wave. Applying the approximate formula to calculate the formula for the span and critical distance between the observation point and the transmission line.

1 Introduction

In recent years, with the rapid development of science and technology, massive construction of infrastructure such as transmission lines and radio stations has led to a close distance between the two[1].In this case, even if the transmission line is not energized, its huge metal structure will affect the reception or transmission of signals from nearby radio stations[2-3]. This effect caused only by the scattering of the metal conductors of the transmission line is called passive interference. How to determine the change of the avoidance distance between the two due to the different transmission span is the key to the consideration of power line construction.

In response to this problem, the United States[4-6],Canada[7],Japan[8,9] and other countries have carried out related research on the influence of power lines on signals in different frequency bands such as MF, VHF/UHV, and m waves since 1960s. The research object involves radio wide broadcast[7], radar navigation[6] and other fields[8]. Since the construction of UHV transmission lines in China, China Electric Power Research Institute Wuhan Branch, North China Electric Power University, and the China Three Gorges University have conducted a systematic and in-depth research on the electric field strength characteristics of UHV transmission lines, and proposed the avoidance standard for 1000kV AC transmission lines.

In recent years, with the gradual improvement of the voltage level of transmission lines in China and the continuous construction of UHV AC and DC transmission lines, the impact Of UHV transmission lines on passive interference of radio stations has become increasingly serious.[10-11] At present, the electromagnetic radiation problem of transmission lines to radio facilities has become one of the most concerned priorities of various departments. According to the correlation and correlation analysis of the electric field strength curves of the short-wavelength transmission lines based on various single macroscopic structures, it can effectively explain that the influence of the range on the passive interference resonance phenomenon in the short-wave band is most significant, that is, from the perspective of macrostructure. the span is the decisive influence factor[8] of the passive interference of the transmission line in the short-wave band. Therefore, in the short-wave frequency range, the study of the influence of the span on passive interference is the key to study the electromagnetic radiation generated by the transmission line in the short-wave frequency.

2 Modeling

2.1. Principle analysis

The passive interference generated by the transmission line is mainly caused by the electromagnetic radiation of the iron tower, which is related to the frequency, polarization mode and relative position of the affected signal. Therefore, the frequency of the electromagnetic wave is set to a certain frequency point, and the

 https://doi.org/10.1051/e3sconf/20186405004

polarization mode is vertical polarization. According to the previous article, the influence of the span on the passive interference resonance phenomenon in the short-wave band is most significant..

The effects of electromagnetic interference are:

$$S = 20\log\frac{E_Y}{E_N}$$

E_Y indicates that the transmission line and the impact of the tower affect the spatial electric field strength of the observation point; E_N indicates that the transmission line and the tower affect the spatial electric field strength of the previous observation point.

2.2 Tower type selection and parameters

Since there is no relevant standard reference, the typical tower type of the horn tower in the AC transmission line is used in calculation. The height of the tower is 98.5m, the structure and parameters of the AC transmission line tower are shown in Figure 1.

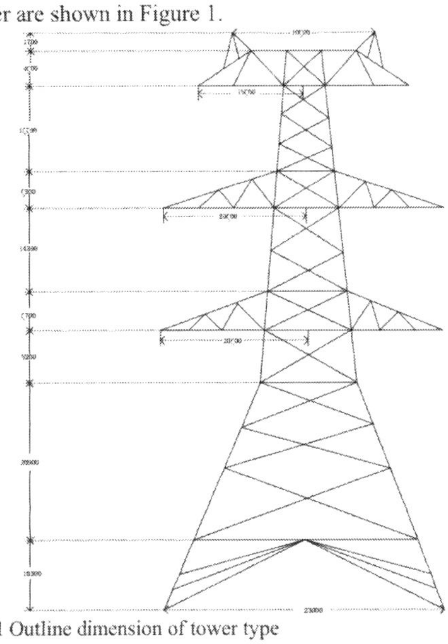

Fig.1 Outline dimension of tower type

3 Comparison of calculation results of spatial electric strength

3.1. principle basis

In the paper of "Research on Passive Interference of UHV AC Transmission Lines to adjacent Stations", the mutual coupling effect between towers is discussed[8]. Since the mutual interference between the towers cannot ignored, here two towers are taken as an example. The electromagnetic interference coupling model as shown in Fig.2 is established. The two towers are respectively regarded as a transmitting antenna and a receiving antenna, and the output power of the transmitting antenna port is P_t ,the input power of the transmitting antenna port is P_r

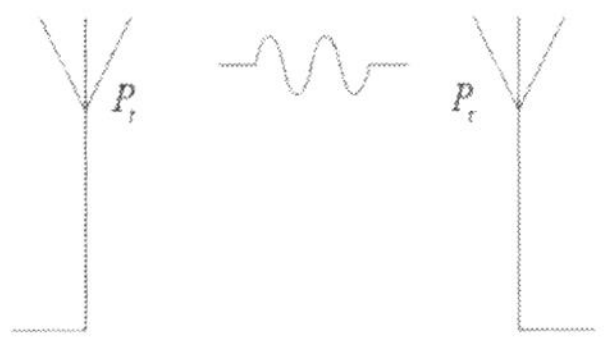

Fig.2 Radiative interference coupling model between two antennas

Define the coupling degree c between the far-field antenna as:

$$c = \frac{P_r}{P_t} \tag{1}$$

Fig.3 is a schematic diagram showing the relative positions of the off-angles between the transmitting and receiving antennas. G_r and G_t :the gain of the receiving and transmitting antenna; θ_1 and θ_2 :the direction in which the radiation direction of the two antenna deviate from each other by the main beam of the opposite party; φ_1 and φ_2 :the angle between the main radiation direction of the two antennas and the horizontal plane; λ :the working wavelength; S :a polarization matching coefficient between the transmitting antenna and the receiving antenna; L :the shortest distance between the two antennas.

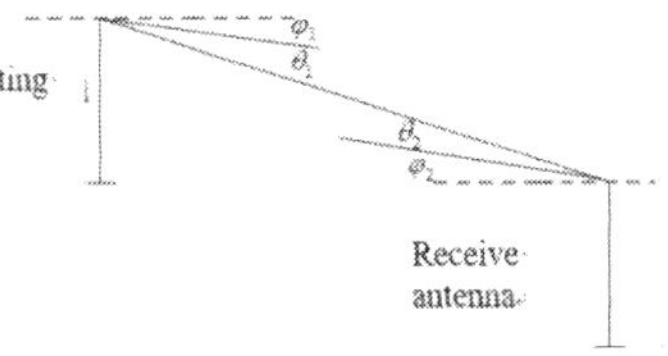

Fig.3 Mutual departure angle between transmitting and receiving antennas

So there are:

$$P_r = \frac{\cos\theta_1 \cos\theta_2 P_t G_t G_r \lambda^2 S}{(4\pi L)^2} \tag{2}$$

Finished up:

$$c = \frac{\cos\theta_1 \cos\theta_2 G_t G_r \lambda^2 S}{(4\pi L)^2} \tag{3}$$

Idealize all the parameters and organize them :

$$c = (\frac{\lambda}{4\pi d})^2 \tag{4}$$

Equation (3)can be derived into the array, only need to d rewritten into d_{ij} . It can be regarded as the interference coupling coefficient of the i-th transmit antenna to the j-th receive antenna, which has the following from:

$$C=\begin{cases}(\frac{\lambda}{4\pi d_{11}})^2(\frac{\lambda}{4\pi d_{12}})^2\cdots\cdots(\frac{\lambda}{4\pi d_{1n}})^2\\(\frac{\lambda}{4\pi d_{21}})^2(\frac{\lambda}{4\pi d_{22}})^2\cdots\cdots(\frac{\lambda}{4\pi d_{2n}})^2\\(\frac{\lambda}{4\pi d_{n1}})^2(\frac{\lambda}{4\pi d_{n2}})^2\cdots\cdots(\frac{\lambda}{4\pi d_{nn}})^2\end{cases} \quad (5)$$

The physical meaning of the elements on the diagonal is the degree of autotransformation of a single antenna element itself, and its own coupling to itself is 1, so matrix C can be rewritten into

$$C=\begin{cases}1\quad(\frac{\lambda}{4\pi d_{12}})^2\cdots\cdots\cdots(\frac{\lambda}{4\pi d_{1n}})^2\\(\frac{\lambda}{4\pi d_{21}})^2(\frac{\lambda}{4\pi d_{22}})^2\cdots\cdots(\frac{\lambda}{4\pi d_{2n}})^2\\(\frac{\lambda}{4\pi d_{n1}})^2(\frac{\lambda}{4\pi d_{n2}})^2\cdots\cdots\cdots 1\end{cases} \quad (6)$$

Taking into account the coupling interference coefficient between the arrays, the electric field strength of the i-th tower is multiplied by the mutual coupling interference coefficient with respect to the electromagnetic radiation field strength of the 0th tower to the far point. Vertically polarized plane wave incident angle is α. E_i is the Field strength for the i-th tower at the required far-field point;r_0, r_i the distance between the 0th and i-th towers to the required far field point。

$$E_i=|E_0|\bullet\frac{r_0}{r_i}e^{j(\frac{2\pi(r_i-r_0)}{\lambda}-\frac{2\pi i*d\sin\alpha}{\lambda})}$$
$$=|E_0|\bullet\frac{r_0}{r_i}e^{j\frac{2\pi(r_i-r_0-i*d\sin\alpha)}{\lambda}} \quad (7)$$

$$E_{all}=E_i\bullet C \quad (8)$$

The method of using the electromagnetic field strength of a single tower as a t=reference and multiplying the array influence coefficient to the obtain the far-field characteristics is accurate and effective from the perspective of simulation calculation. This method can greatly improve the calculation efficiency.

2.1. Electric field strength calculation results

In order to study the influence of transmission line span in passive interference, this paper compared the electric field strength generated by different transmission lines with different spans.as the same time, in order to ensure the universality of the results, this section sets different frequencies. Keep other factors constant.

2.1.1 Analysis of result at a frequency of 5MHz

The calculated electric field strength of the transmission line in the short-wavelength band is plotted as a curve with the horizontal distance between the observation point and the transmission line(referred to as" horizontal distance")and the range, as shown in Figure 4;figure 5(a)shows the Electric field intensity changes with horizontal distance; figure 5(b) Electric field intensity changes with the span. Figure8 is the Critical distance between observation point and transmission line changes with the span(critical distance: a horizontal distance that ensures a monotonous decreasing trend of the electric field strength)

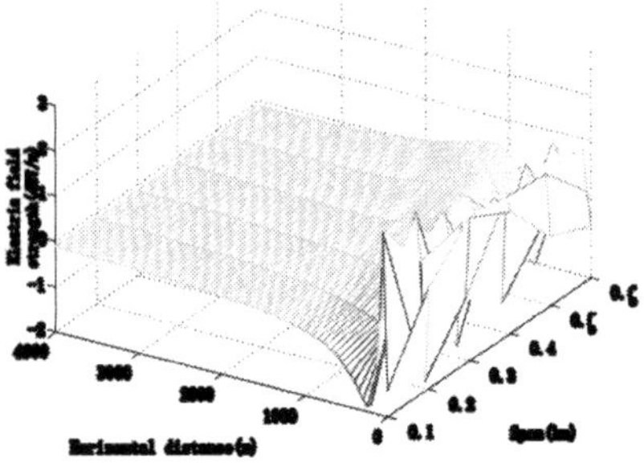

Fig.4 Electric field intensity changes with horizontal distance and the span

As can be seen from the above figure, as the span and horizontal distance increasing, the electric field strength level decreases. Indicating that the horizontal distance and the pan are important factors affecting the electric field strength.

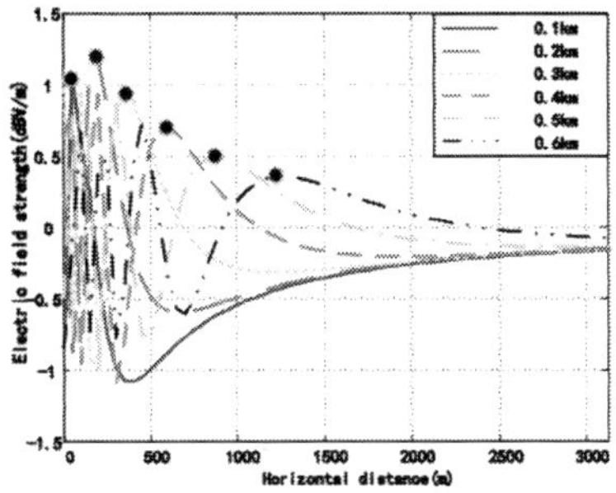

Fig.5(a) Electric field intensity changes with horizontal distance

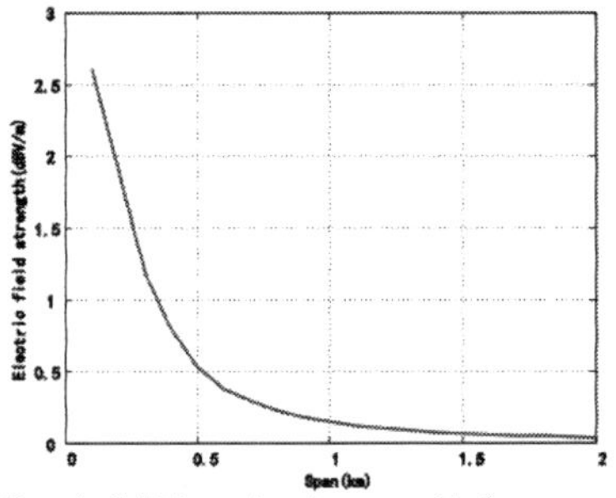

Fig.5(b) Electric field intensity changes with the span

In order to obtain the relationship between the critical distance and the gear distance more accurately, this section studies the towers of difference heights.

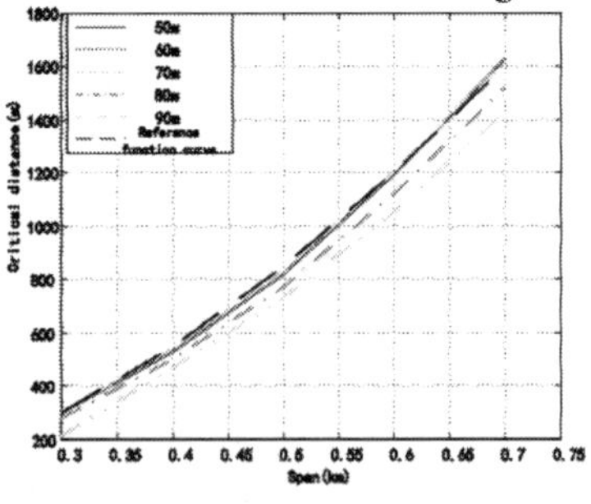

Fig.6 Critical distance between observation point and transmission line changes with the span

As shown in the figure, within the allowable range of error (5%), the curve of the critical distance as a function of the span can be approximated by a function curve. The formula is as follows:

$$L = \frac{d^2}{400} + \frac{3d}{4} - 150(d > 0km) \quad (9)$$

L : ensuring the horizontal distance of the electric field strength in a monotonically decreasing trend, that is, the critical distance; d :span。

It can be seen from the curve change of figure 7(a)and figure 7(b)that the critical distance is $L < 1000m$,or $d \geq 0.5km$,Electric field strength $E < 0.5dBV/m$ ($SE < 0.5$),it can be considered that the influence of the electric field strength generated by the transmission line can be neglected. Therefore, the above horizontal distance and the span can be used as the basis for calculating the avoidance distance of the short-wave radio station and the transmission line.

2.1.2 Calculation results at different frequencies

In order to avoid the above results, this section have studied the results at different frequencies. Figure 7 is the electric field intensity level varies with the frequency.

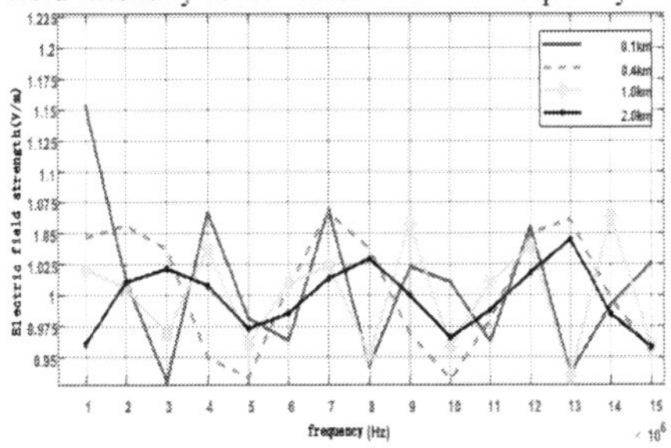

Fig.7 The electric field intensity level varies with the frequency

In order to ensures the universality of the research results, the frequency of the plane wave selected is 12MHz. Figure 8(a) is the electric field intensity changes with horizontal distance; Figure 8(b) is the electric field intensity level varies with the span; and Figure 9 is Critical distance changes with the span

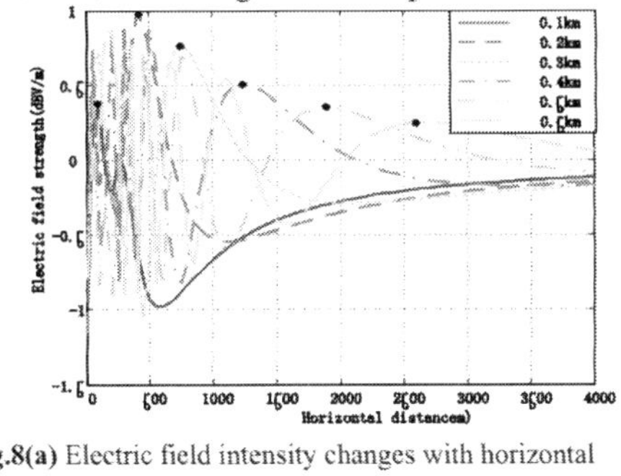

Fig.8(a) Electric field intensity changes with horizontal distance

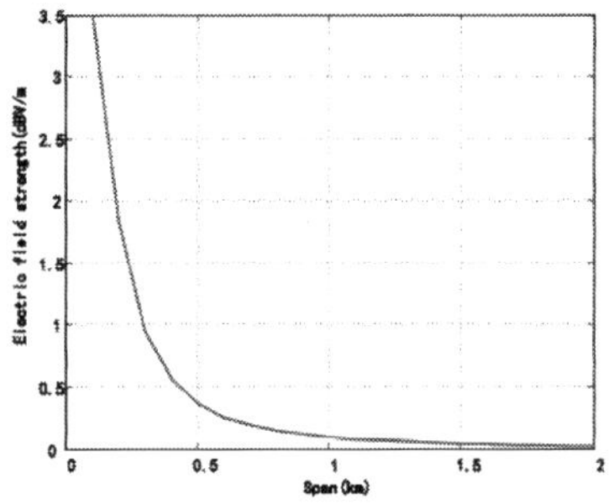

Fig.8(b) The electric field intensity level varies with the span

It can be seen from the curve change of Fig. 9 that as the span increases, the electric field strength decreases. And the results of this study are similar to the simulation results when the plane wave is 5 MHz

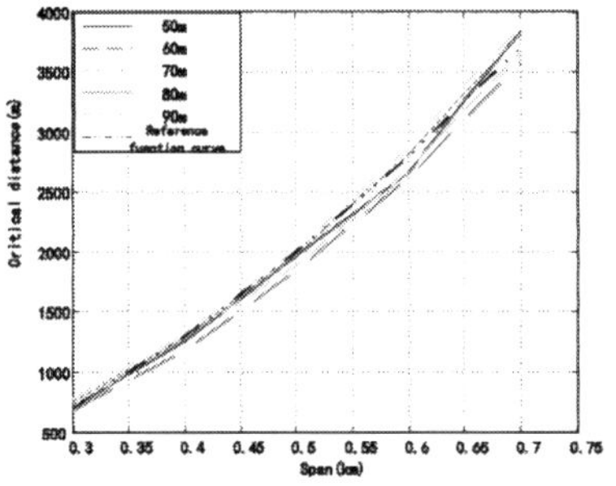

Fig.9 Critical distance changes with the span

It can be seen from the curve change of Fig.11 that when the frequency of the plane wave is 12MHz, the Critical distance changes with the span can be expressed by the following formula.

$$L = 2.4 * (\frac{d^2}{400} + \frac{3d}{4} - 150)(d > 0km) \quad (10)$$

In this case, if the range of the span is 0.3 km~0.6 km, only the critical distance of 1320m is required to ensure the communication quality of the radio station.

This section also simulates the curve of the critical distance with the span when the frequency is 7MHz and 25MHz.

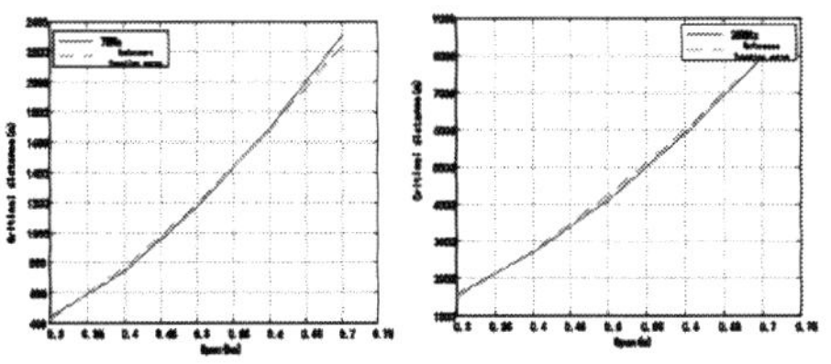

Fig.10 Maximum distance changes with the span with difference frequencies

The calculated fit curve is

$$L = 1.4 * (\frac{d^2}{400} + \frac{3d}{4} - 150)(d > 0km) \quad (11)$$

$$L = 5 * (\frac{d^2}{400} + \frac{3d}{4} - 150)(d > 0km) \quad (12)$$

So, the relationship between the critical distance and the span can be expressed by the formula (13), and this formula is universal.

$$L = \frac{Frequency}{5MHz} * (\frac{d^2}{400} + \frac{3d}{4} - 150)(d > 0km) \quad (13)$$

3 conclusion

The influence of transmission line span on passive interference is different through analysis in this paper. Derive the formula for calculating the distance between the span and the observation point and the transmission line, relationship between the tower height and the span can be summarized follows:

(1)This paper fits the relationship between the transmission line span and the critical distance, as shown in the following equation.

$$L = \frac{\text{Frequency}}{5MHz} * (\frac{d^2}{400} + \frac{3d}{4} - 150)(d > 0km) \quad (14)$$

This formula is verified by comparing the relationship between the actual span and the critical distance: In the short wave band, under the same tower height condition, if the incident frequency and the span of the plane wave are known, the above formula can be used to derive the critical distance at this time, which can predict the guard spacing more accurately when the range is different.

(2)In practical engineering, the distance between the span and the critical distance can be used to study the avoidance distance of transmission lines and radio stations. It greatly reduces the time required for modeling and improves the accuracy of simulation calculations.

Acknowledgement

This work was supported by the Project of SGCC (GY71-16-011)

References

1. Yin hui, Zhang xiaoming, Wang yantao, et a. Research on 3D visualization of electromagnetic interference for Ultra-high Voltage transmission lines. High Voltage Engineering.**40**,12 (2014)
2. Tang bo , Yang jiawei Huang li, et al. Guard spacing of the air surveillance radar station to Ultrahigh Voltage transmission line.**40**.4(2018)
3. Tang bo, Sun rui, Zhao xiaoming, et al. Passive interference solution for UHV transmission line in very high frequency band . Electric Power,**49**,10(2016)
4. Hill P C.Measurement of reradiation from lattice masts at V.H,F.[J].Proc Inst Elec Eng,111,12(1964)
5. Belroses J S,Lavrench W,Dunn J G,et al.The effects of reradiation from high rise buildings and transmission lines upon the radiation pattern of MF broadcasing antenna arrays[C]//Proceedings of AGARD/EPP Meeting .Spatind,Norway:[s.n.],1979.
6. Kamal Sarabandi, Moonsooo Park.A radar cross-section model for power lines at millmeter-wave frequency[J].IEEE Transactions on Antenna and Propagation,**51**,9(2003)
7. Truemen C W, Kubina S J. Scatting from power lines with the skywire insulated from the towers[J].IEEE Transactions on Broadcasting,**40**,2(1994)
8. Kinase. Theoretical analysis of reradiation characteristics of radio wave caused by transmission lines[J] .NHK Technology Research , **20**,3(1968)
9. Takeshita K,Takeshita S,Hashimoto H.Scattering characteristics of VHF/UHF television broadcasting waves by overhead oower transmission conductors[J]..IEEE Transactions on Electromagnetic Compatibility,**21**,1(1979)
10. Chen bin , Tang bo , Cao hongying , et al . Association Analysis of Passive Interference Resonance Impact Factors for Transmission Lines at Short-Wave Frequency,36,6(2015
11. Gan zheyuna, Wang yanhai, Zhang jiangong, et al, Suppression of HF passive interference in transmission lines by using magnetic tube . High Voltage Engineering, **43**,5(2017)

Analysis on the Influence of the Height of Tower on Passive Interference in shortwave

Lu Ying[1], *Zhao* Zhibin[1], *Zhang* Jian gong[2] and *Gan* Zheyuan[2]

[1]College of Electrical and Electronical, North China Electric Power University, Beijing 102206, China
[2]China Electric Power Research Institute, Wuhan, China

Abstract. The passive interference of transmission lines to nearby radio stations may affect the effective reception and transmission of radio station signals. Therefore, the accurate calculation of the electromagnetic scattering of transmission lines under the condition of external electromagnetic waves is the basis for determining the reasonable avoidance spacing of the two. For passive stations operating in short-wave frequencies, passive interference is mainly generated by the tower. This paper uses the method of moments to perform passive interference calculations under normal circumstances, And elaborates the method for calculating the electromagnetic field of the transmission line, obtains the space electric field intensity of the transmission line at the same working frequency and space location of the plane wave. Uses the approximate formula to inductive the formula for calculating height of tower and the protective distance.

1 Introduction

In recent years, with the rapid development of science and technology, massive construction of infrastructure such as transmission lines and radio stations has led to a close distance between the two[1].In this case, even if the transmission line is not energized, its huge metal structure will affect the reception or transmission of signals from nearby radio stations. This effect caused only by the scattering of the metal conductors of the transmission line is called passive interference[3]. How to determine the change of the avoidance distance between the two due to the different transmission span is the key to the consideration of power line construction.

In response to this problem, the United States[4-6],Canada[7],Japan[8,9] and other countries have carried out related research on the influence of power lines on signals in different frequency bands such as MF, VHF/UHV, and m waves since 1960s. The research object involves radio broadcasting[7], radar navigation[6] and other fields[8]. Since the construction of UHV transmission lines in China, China Electric Power Research Institute Wuhan Branch, North China Electric Power University and China Three Gorges University have conducted a systematic and in-depth research on the electric field strength characteristics of UHV transmission lines, and proposed the avoidance standard for 1000 kV AC transmission lines.

In recent years, with the gradual improvement of the voltage level of transmission lines in China and the continuous construction of UHV AC and DC transmission lines, the impact of UHV transmission lines on passive interference of radio stations has become increasingly serious[10-11]. At present, the electromagnetic radiation problem of transmission lines to radio facilities has become one of the most concerned priorities of various departments. The passive interference generated by the transmission line is mainly divided into two aspects: electromagnetic scattering generated by the iron tower and electromagnetic scattering generated by the overhead conductor. According to the correlation and correlation analysis of the electric field strength curves of the short-wavelength transmission lines based on various single macroscopic structures, it can effectively explain that the influence of the range on the passive interference resonance phenomenon in the short-wave band is most significant, that is, from the perspective of macrostructure. the tower is the decisive influence factor[8] of the passive interference of the transmission line in the short-wave band. Therefore, in the short-wave frequency range, the study of the influence of the height of the tower on passive interference is the key to study the electromagnetic radiation generated by the transmission line in the short-wave frequency.

2 Modelling

2.1. Principle analysis

The passive interference generated by the transmission line is mainly caused by the electromagnetic radiation of the iron tower, which is related to the frequency, polarization mode and relative position of the affected signal. Therefore, the frequency of the electromagnetic wave is set to a certain frequency point, and the polarization mode is vertical polarization. According to the correlation degree and correlation analysis of passive interference curves of the short-wavelength transmission lines based on various single macroscopic structures, studying on the passive interference of tower height to radio stations by changing the height of the transmission line. The effects of electromagnetic interference are:

$$S = 20\log\frac{E_Y}{E_N}$$

E_Y indicates that the transmission line and the impact of the tower affect the spatial electric field strength of the observation point; E_N indicates that the transmission line and the tower affect the spatial electric field strength of the previous observation point.

2.2 Tower type selection and parameters

Since there is no relevant standard reference, the typical tower type of the horn tower in the AC transmission line is used in calculation. The height of the tower is 98.5m, the structure and parameters of the AC transmission line tower are shown in Figure 1.

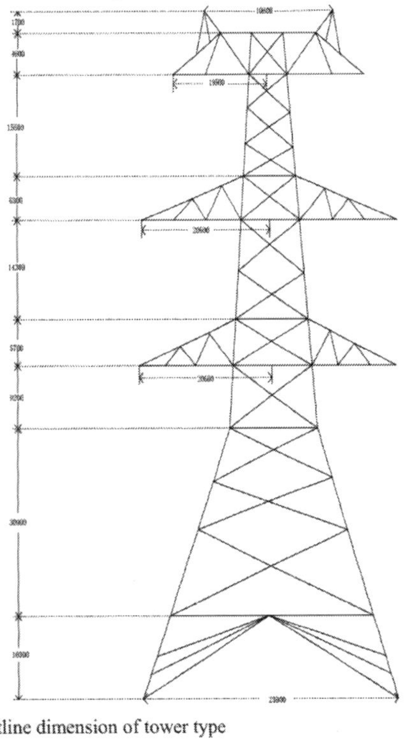

Fig.1 Outline dimension of tower type

3 Comparison of calculation results of electric field strength

3.1. principle basis

In the paper of "Research on Passive Interference of UHV AC Transmission Lines to Neighboring Stations", the mutual coupling effect between towers is discussed[8]. Since the mutual interference between the towers cannot ignored, here two towers are taken as an example. The electromagnetic interference coupling model as shown in Fig.4 is established. The two towers are respectively regarded as a transmitting antenna and a receiving antenna, and the output power of the transmitting antenna port is P_t ,the input power of the transmitting antenna port is P_r .

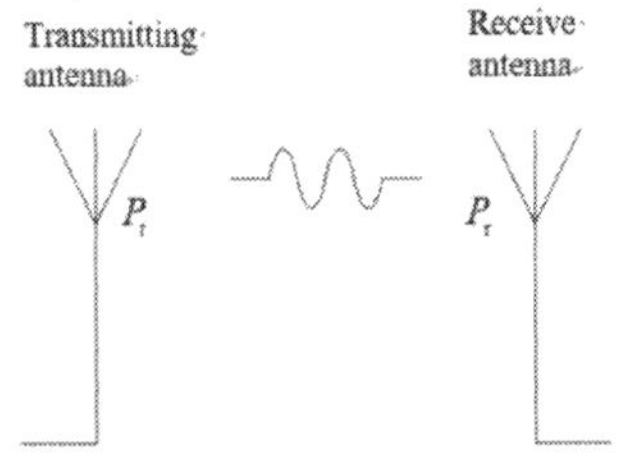

Fig.2 Radiative interference coupling model between two antennas

Define the coupling degree c between the far-field antenna as:

$$c = \frac{P_r}{P_t} \tag{1}$$

Fig.3 is a schematic diagram showing the relative positions of the off-angles between the transmitting and receiving antennas. G_r and G_t :the gain of the receiving and transmitting antenna; θ_1 and θ_2 :the direction in which the radiation direction of the two antenna deviate from each other by the main beam of the opposite party; φ_1 and φ_2 :the angle between the main radiation direction of the two antennas and the horizontal plane; λ :the working wavelength; S :a polarization matching coefficient between the transmitting antenna and the receiving antenna; L :the shortest distance between the two antennas.

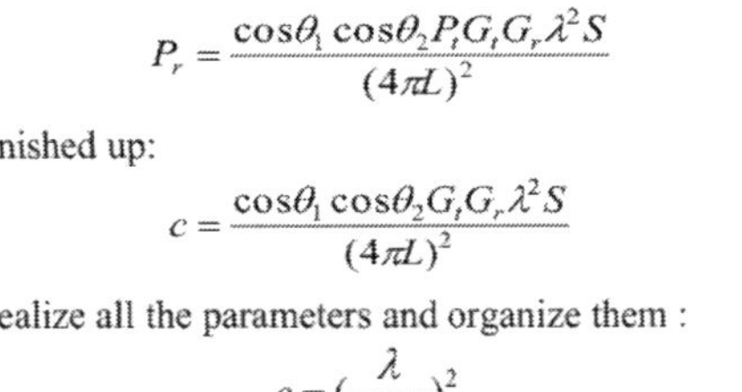

Fig.3 Mutual departure angle between transmitting and receiving antennas

So there are:

$$P_r = \frac{\cos\theta_1\cos\theta_2 P_t G_t G_r \lambda^2 S}{(4\pi L)^2} \tag{2}$$

Finished up:

$$c = \frac{\cos\theta_1\cos\theta_2 G_t G_r \lambda^2 S}{(4\pi L)^2} \tag{3}$$

Idealize all the parameters and organize them :

$$c = (\frac{\lambda}{4\pi d})^2 \tag{4}$$

Equation (3) can be derived into the array, just need to d rewritten into d_{ij} . It can be regarded as the interference coupling coefficient of the i-th transmit antenna to the j-th receive antenna, which has the following from:

$$C=\begin{cases}(\frac{\lambda}{4\pi d_{11}})^2(\frac{\lambda}{4\pi d_{12}})^2\cdots\cdots(\frac{\lambda}{4\pi d_{1n}})^2\\(\frac{\lambda}{4\pi d_{21}})^2(\frac{\lambda}{4\pi d_{22}})^2\cdots\cdots(\frac{\lambda}{4\pi d_{2n}})^2\\(\frac{\lambda}{4\pi d_{n1}})^2(\frac{\lambda}{4\pi d_{n2}})^2\cdots\cdots(\frac{\lambda}{4\pi d_{nn}})^2\end{cases}\quad(5)$$

The physical meaning of the elements on the diagonal is the degree of autotransformation of a single antenna element itself, and its own coupling to itself is 1, so matrix C can be rewritten into

$$C=\begin{cases}1\quad(\frac{\lambda}{4\pi d_{12}})^2\cdots\cdots\cdots(\frac{\lambda}{4\pi d_{1n}})^2\\(\frac{\lambda}{4\pi d_{21}})^2(\frac{\lambda}{4\pi d_{22}})^2\cdots\cdots(\frac{\lambda}{4\pi d_{2n}})^2\\(\frac{\lambda}{4\pi d_{n1}})^2(\frac{\lambda}{4\pi d_{n2}})^2\cdots\cdots\cdots 1\end{cases}\quad(6)$$

Taking into account the coupling interference coefficient between the arrays, the electric field strength of the i-th tower is multiplied by the mutual coupling interference coefficient with respect to the electromagnetic radiation field strength of the 0th tower to the far point. Vertically polarized plane wave incident angle is α. E_i is the Field strength for the i-th tower at the required far-field point;r_0, r_i the distance between the 0th and i-th towers to the required far field point。

$$E_i=|E_0|\bullet\frac{r_0}{r_i}e^{j(\frac{2\pi(r_i-r_0)}{\lambda}-\frac{2\pi i d\sin\alpha}{\lambda})}=|E_0|\bullet\frac{r_0}{r_i}e^{j\frac{2\pi(r_i-r_0-i d\sin\alpha)}{\lambda}}\quad(7)$$

$$E_{all}=E_i\bullet C\quad(8)$$

The method of using the electromagnetic field strength of a single tower as a reference and multiplying the array influence coefficient to the obtain the far-field characteristics is accurate and effective from the perspective of simulation calculation. This method can greatly improve the calculation efficiency.

3.2. Relationship between the protection distance and the height of the tower

In order to ensure that the transmission line does not affect the communication quality of the radio station, this section analyzed the minimum distance between the two (ie protection distance, at this time $SE\le 0.5$). Figure 4 shows the variation of electric field strength with horizontal distance when the tower is at different heights. Point A is the maximum protection distance when the transmission line is at a certain span ($SE\le 0.5$), Figure 5 is the protective distance changes with the to tower.

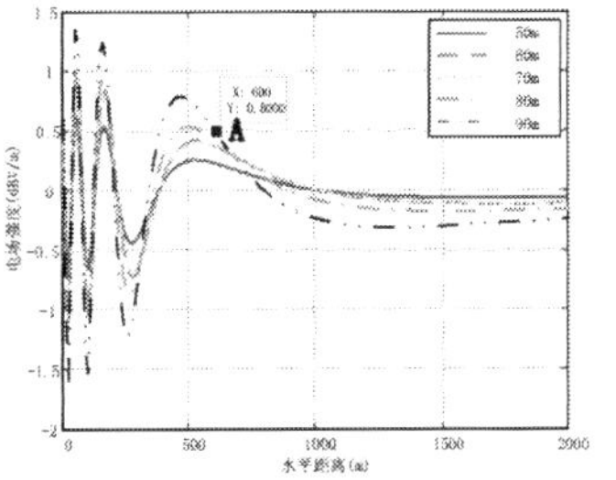

Fig.4 The electric field intensity level changes with horizontal distance

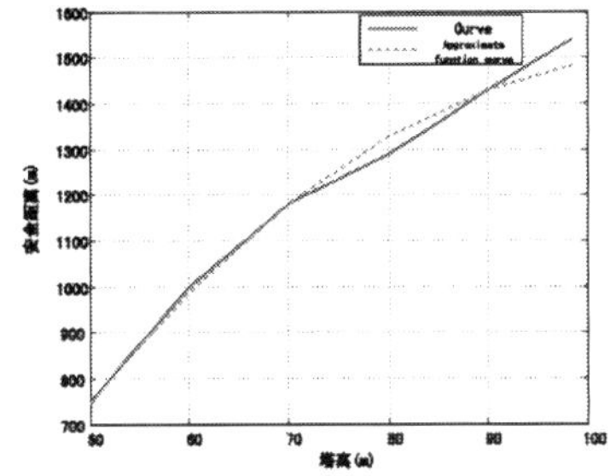

Fig.5 The protective distance changes with the span

In order to ensure that the protection distance is not affected by the height of the tower, this section uses the maximum protection distance. It can be seen from figure 4 that the higher the tower, the greater the protection distance required for the transmission line and the radio station. It can be seen from figure 5 that the relationship between the protection distance and the height of the tower in the error (error≤5%) range is as follows:

$$S=-\frac{9}{40}h^2+\frac{97}{2}h-\frac{2225}{2}\quad(h>0m)\quad(15)$$

S: protection distance between transmission lines and radio stations(ie $E\le 0.5dBV/m$); h :tower height。

In the actual project, if the incident frequency band of the plane wave is the short-wave frequency band, and the remaining factors are unchanged, the above formula can be used to derive the guard spacing at different tower heights. And regardless of the incident frequency of the plane wave, the protection distance curve will show a similar change law.

4 Conclusion

In this paper, the electric field strength calculation is carried out by applying the moment method under normal conditions. At the same time, the method of calculating the electromagnetic field of the transmission line is expounded. The electric field strength of the transmission line with different heights at the same working frequency and space at the same plane is calculated. The influence of tower heights on passive interference is different. Applying the approximation formula to summarizing the relationship between the tower height and protective distance.

(1)The height of the tower is an important factor affecting the protection distance (ie $SE\le 0.5$) of the transmission line and the radio station, but weaker than the span. Under the same span, the distance between the transmission line and the radio station (ie $SE \leq 0.5$) depends on the height of the tower. The higher the height of the tower, the greater the required protection distance. The tower height and the protection distance satisfy the following fitting curve formula

$$S = -\frac{9}{40}h^2 + \frac{97}{2}h - \frac{2225}{2} \quad (h > 0m)$$

(2)In practical engineering, according to the tower height and the protection distance formula, the avoidance distance of the transmission line and the radio station can be more accurately judged. It greatly reduces the time required for modeling and improves the accuracy of simulation calculations.

Acknowledgement

This work was supported by the Project of SGCC (GY71-16-011)

References

1. Yin hui, Zhang xiaoming, Wang yantao, et al.Research on 3D visualization of electromagnetic interference for Ultra-high Voltage transmission lines. High Voltage Engineering.40,12 (2014)
2. Tang bo , Yang jiawei Huang li, et al. Guard spacing of the air surveillance radar station to Ultrahigh Voltage transmission line.40.4(2018)
3. Tang bo, Sun rui, Zhao xiaoming, et al. Passive interference solution for UHV transmission line in very high frequency ban.Electric Power,49,10(2016)
4. Hill P C.Measurement of reradiation from lattice masts at V.H,F.[J].Proc Inst Elec Eng,111,12(1964)
5. Belroses J S,Lavrench W,Dunn J G,et al.The effects of reradiation from high rise buildings and transmission lines upon the radiation pattern of MF broadcasing antenna arrays[C]//Proceedings of AGARD/EPP Meeting .Spatind,Norway:[s.n.],1979.
6. Kamal Sarabandi,Moonsooo Park.A radar cross-section model for power lines at millmeter-wave frequency[J].IEEE Transactions on Antenna and Propagation,51,9(2003)
7. Truemen C W, Kubina S J. Scatting from power lines with the skywire insulated from the towers[J].IEEE Transactions on Broadcasting,**40**,2(1994)
8. Kinase. Theoretical analysis of reradiation characteristics of radio wave caused by transmission lines[J] .NHK Technology Research , **20**,3(1968)
9. Takeshita K,Takeshita S,Hashimoto H.Scattering characteristics of VHF/UHF television broadcasting waves by overhead oower transmission conductors[J]..IEEE Transactions on Electromagnetic Compatibility,21,1(1979)
10. Chen bin, Tang bo, Cao hongying, et al. Association Analysis of Passive Interference Resonance Impact Factors for Transmission Lines at Short-Wave Frequency,36,6(2015
11. Gan zheyuna, Wang yanhai, Zhang jiangong, et al, Suppression of HF passive interference in transmission lines by using magnetic tube . High Voltage Engineering, 43,5(2017)

E3S Web of Conferences 64, 05006 (2018) https://doi.org/10.1051/e3sconf/20186405006
ICPRE 2018

Influence of Transmission Line Pitch on Passive Interference

Lu Ying[1], *Zhang* Jian gong[2], *Gan* Zheyuan[2], *Zhao* Zhibin[1]

[1]College of Electrical and Electronical, North China Electric University, Beijing 102206, China
[2]Wuhan Branch of China Electric Power Research Institute, State Key Laboratory for Environmental Protection of Power Grid, Wuhan 430074, China

Abstract. Passive interference of transmission lines to neighbouring radio stations may affect the effective reception and transmission of radio station signals. Therefore, it is the basis to accurately calculate the electromagnetic scattering of transmission lines including overhead wires and towers under the circumstance of electromagnetic waves from outside. Wherein for radio stations, the overhead conductor is the main factor that affects the radar detection because of the working mode of the radar and other factors. In this paper, in the general case of the application of moment method to carry out passive interference calculations, the overhead lines are assumed to be all the actual wire. The space electric field strength and RCS of the different pitch at different operating frequencies and spaces are obtained. The influence of pitch on passive interference of overhead conductors was analyzed. Thus providing a basis for modeling to save energy and reduce the effects of passive interference provides the modeling basis.

1 Introduction

In recent years, with the rapid development of science and technology, massive construction of infrastructure such as transmission lines and radio stations has led to a close distance between the above two[1].In this case, even if the transmission line is not live working, its huge metal structure will affect the reception or transmission of signals from nearby radio stations, which leads to the increasing electromagnetic radiation from transmission lines to radio stations[2-3]. Therefore the problem that must be considered in power line construction is how to determine the reasonable distance between transmission line and radio station or reduce the passive interference of them.

To solve this problem, the United States[4-6], Canada[7], Japan[8,9] and other countries have carried out related research on the influence of power lines on signals in different frequency bands such as MF, VHF/UHV[10-11], and m waves since 1960s. Which involves radio wide broadcast[7], radar navigation[6] and other fields[8].It is an important practice for the electric power industry to implement this strategic decision to develop ultra-high voltage power transmission, optimize the allocation of energy resources and improve the comprehensive social benefits[12].Based on the mechanism of electromagnetic radiation and energy saving, the passive interference of transmission line is generated by the scattering of transmission line.

2 Passive interference calculation model for transmission lines

2.1. Modelling algorithm for overhead conductors

Since the calculation of passive interference between transmission lines and adjacent radio stations is an open field electromagnetic field calculation problem, most research literatures use Moment Method (MOM) or the rapid multipole method (FMM) based on MOM. In this section, the actual wire is considered as a twisted wire structure, and the modelling method of the stranded wire is given.

2.2 modelling method of actual strand

The actual wire is a twisted wire structure as shown in Figure.1

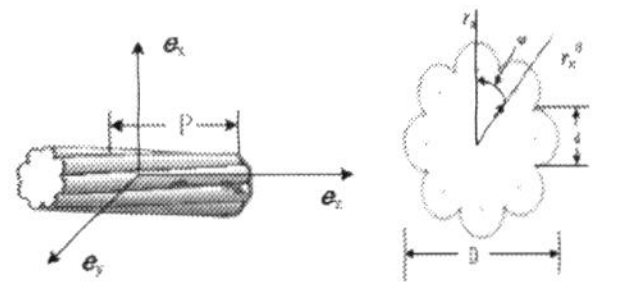

Fig.1 Geometry of a power line and its important parameters

The scattering electric field generated by such a strand also requires the use of a triangular face unit to split the surface. Therefore, the key to the modelling of the actual strand structure is the determination of the coordinates of the points on the surface of the strand. Any one of the sub-wires in the transmission line is composed of a number of steel strands (placed in the center) and a plurality of aluminum hinges (placed on the periphery).

In order to establish a sub-conductor model, the D is overall diameter of the sub-wire, d is the overall diameter of sub-wires, P is Peripheral aluminum strand diameter, φ is an angle parameter slightly larger than 90° between $\boldsymbol{r}_n$ and $\boldsymbol{r}_n^0$, $\boldsymbol{r}_n$ and $\boldsymbol{r}_n^0$ is the point and origin on the n cycle of the sub-wire surface. According to the above parameters, a point on the surface of the wire can be expressed by the following formula [13]:

$$\boldsymbol{r} = \boldsymbol{r}_n^{\,0} + \boldsymbol{r}_n \qquad n=1,2,\ldots,N \qquad (1)$$

Where :

$$\boldsymbol{r}_n^{\,0} = (\frac{D-d}{2})\cos\frac{2\pi n}{N}\boldsymbol{e}_x + (\frac{D-d}{2})\sin\frac{2\pi n}{N}\boldsymbol{e}_y \qquad (2)$$

$$\boldsymbol{r}_n = (\frac{d}{2}\cos\frac{2\pi n}{N}\cos\varphi - \frac{d\sqrt{(\pi D)^2+P^2}}{2P}\times\sin\frac{2\pi n}{N}\sin\varphi)\boldsymbol{e}_x + (\frac{d}{2}\sin\frac{2\pi n}{N}\cos\varphi + \frac{d\sqrt{(\pi D)^2+P^2}}{2P}\times\cos\frac{2\pi n}{N}\sin\varphi)\boldsymbol{e}_y \qquad (3)$$

Table 1. Geometrical Specifications of Four Actual Power Lines

NO	Al/Copper		Steel		D(cm)	P(cm)
	#of St.	d(cm)	#of St.	d(cm)		
1	7	0.400	-	-	1.20	14.60
2	19	0.446	-	-	2.22	23.5
3	54	0.337	7	0.337	3.01	35.56
4	45	0.446	7	0.301	3.52	40.65

As mentioned above, the three-dimensional parameter representation of any point of the sub-wires can be expressed as

$$R = [r_x\cos\frac{2\pi z}{P} - r_y\sin\frac{2\pi z}{P}, r_x\sin\frac{2\pi z}{P} + r_y\cos\frac{2\pi z}{P}, z] \qquad (4)$$

r_x and r_y are the components of $\boldsymbol{r}$, According to the above formula, the coordinates of any point on the outer surface of the power transmission line shown in Fig. 3 can be obtained for use in the division.

The twisted line model given in Fig. 3 is divided into triangular facets, and the electric field strength of each bin in space is the same as that of equation (5).

The calculation formula of the scattered electric field of the surface current of the line surface at any point in space is as shown in equation (5).

$$\boldsymbol{E}_S(\boldsymbol{r}) = j\omega\mu_0 \bullet \int_S g(\boldsymbol{r},\boldsymbol{r}')\boldsymbol{J}_s(\boldsymbol{r}')dS' - \frac{1}{j\omega\varepsilon_0}\nabla\int_S g(\boldsymbol{r},\boldsymbol{r}')\nabla'\boldsymbol{J}_s(\boldsymbol{r}')dS' \qquad (5)$$

$g(\boldsymbol{r},\boldsymbol{r}')$ is the Green's function, $g(\boldsymbol{r},\boldsymbol{r}') = \dfrac{e^{jk|\boldsymbol{r}-\boldsymbol{r}'|}}{4\pi|\boldsymbol{r}-\boldsymbol{r}'|}$;

k is the Wave number, $k = \omega\sqrt{\mu_0\varepsilon_0}$.

In the calculation, in order to simulate the impact of overhead lines on radar operation, it is assumed that the electromagnetic wave direction (k) emitted by the radar is perpendicular to the transmission line. Electric field strength (E) parallel to overhead transmission lines.

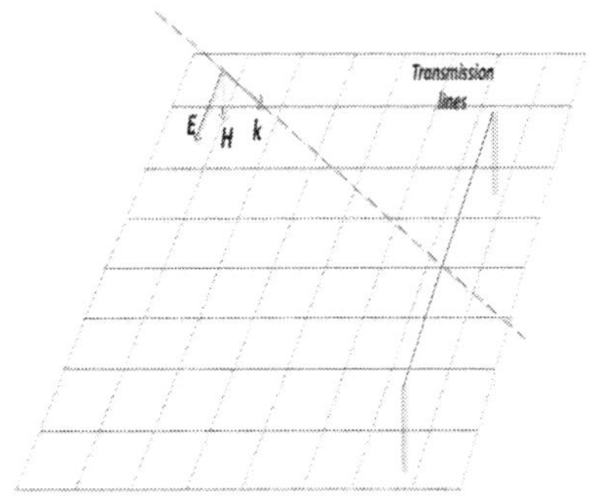

(a) Schematic diagram of the whole model

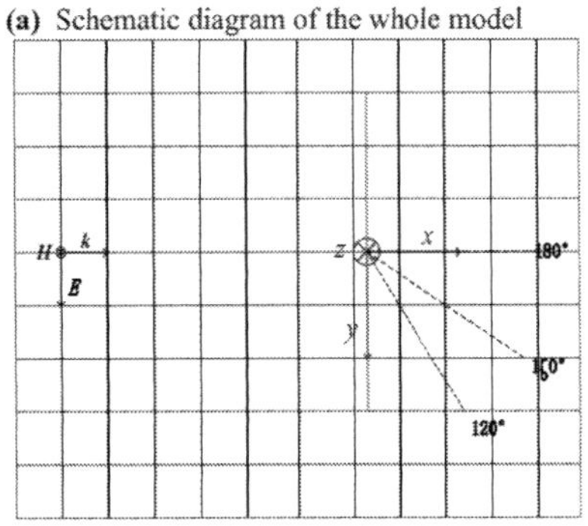

(b) The top view of the model

Fig.2 Passive interference computing model of overhead conductors

3 Comparison of calculation results of spatial electric strength

3.1. Comparison of calculation results of spatial electric strength

In order to compare different calculation models, the calculation results of the spatial electric field in different working frequencies of the radar and the distance from the transmission line, the maximum value of the electric field strength on the plane of different distances of the overhead line is taken to draw the curve shown in Fig. 3 in this section. In order to express the transmission line pitch more clearly, the pitch is replaced by the number of overhead wire windings. During the calculation, the horizontal distance between the observation point and the transmission line is 0.4km, 0.6km, 1.0km and 2.0km, and the spatial electric field strength from 1MHz to 30MHz is calculated.

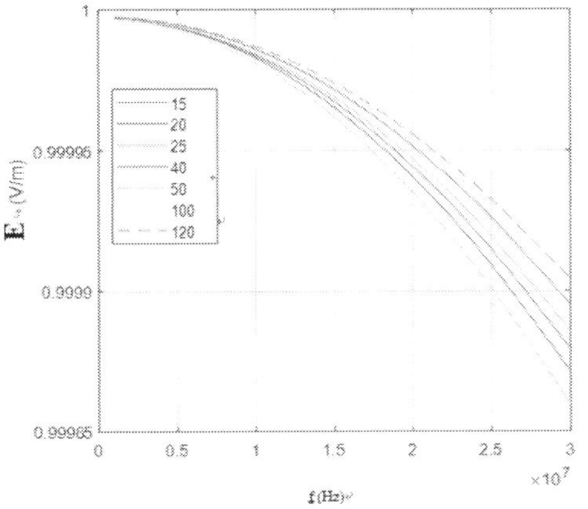

(a) The horizontal distance is 0.4km

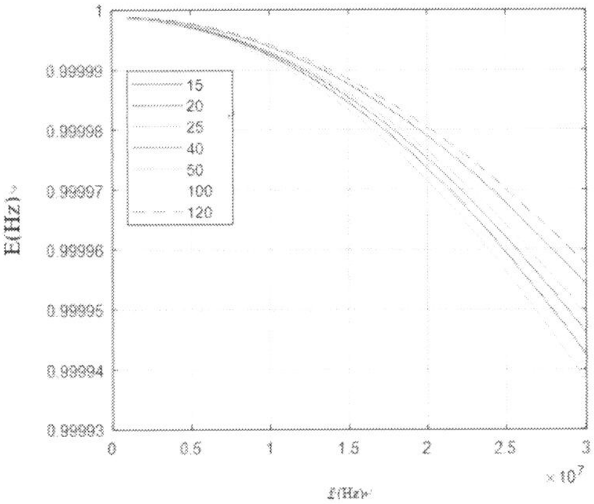

(b) The horizontal distance is 0.6km

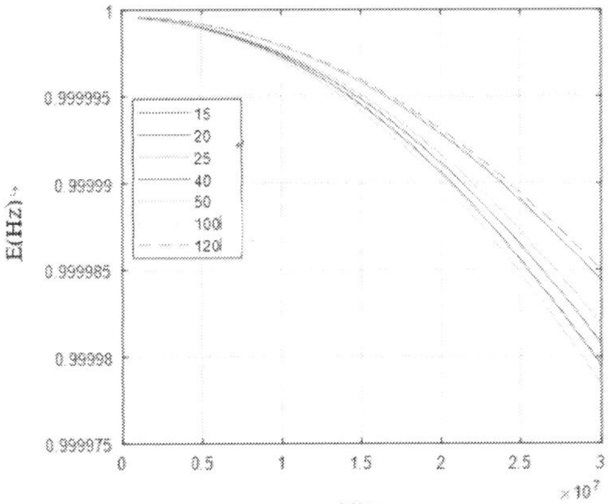

(c) The horizontal distance is 1.0km

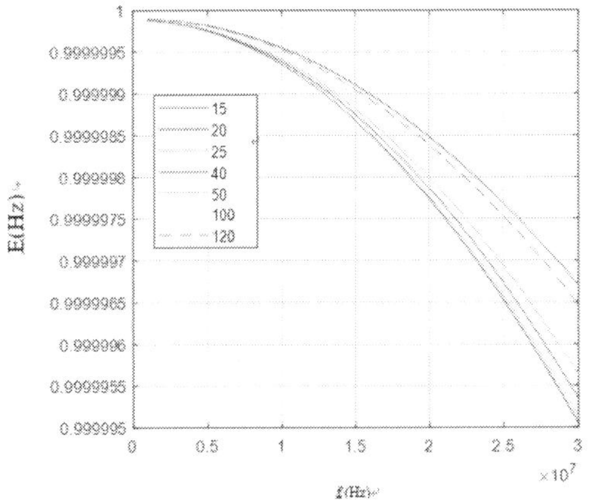

(d) The horizontal distance is2.0km

Fig. 3 Calculation results of space electric field strength

The electric field strength generated by the actual transmission line is:abs($\boldsymbol{E}_{\mathrm{Fig}}$-1) so, It is seen from the figure 3(a), (b), and (c), when the horizontal distance between the observation point and the transmission line is 0.4km, 0.6km, and 1.0km, respectively, the electric field strength of the transmission line with 50 turns is the highest The electric field with 120 turns has the smallest. electric field strength; when the distance between the observation point and the transmission line is 2.0km, the electric field strength of the transmission line with 20 turns is the highest. the electric field with a lap of 40 and 120 produces the smallest electric field strength. In general, regardless of the trend of the pitch of the wire, with the frequency increasing, the electric field strength generated by the transmission line increases.

3.2. Analysis of Relative Deviation of Overhead Conductors with Different Pitches

In order to study the influence of the pitch of the transmission line on the passive interference, the electric field strengths of the transmission lines with different pitches are compared. Due to the limitations of the computer, the plane wave frequency range is set to the short wave range, and the parameters of the transmission line remain unchanged. The calculated electric field strength of the transmission line in the short-wavelength band is plotted as a function of the horizontal distance between the observation point and the transmission line (referred to as "horizontal distance") and the frequency. as shown in Figure 4, this section selects three sets of transmission lines with a difference of 20 pitches. Where the length of the overhead conductor is kept the same, Compare1: the relative deviation of the electric field strength of 25 pitches and 5 pitches of the transmission line; compare2: the relative deviation of the electric field strength of 40 pitches and 20 pitches of the transmission line; compare3: the relative deviation of the electric field strength of 120 pitches and 100 pitches of the transmission line .

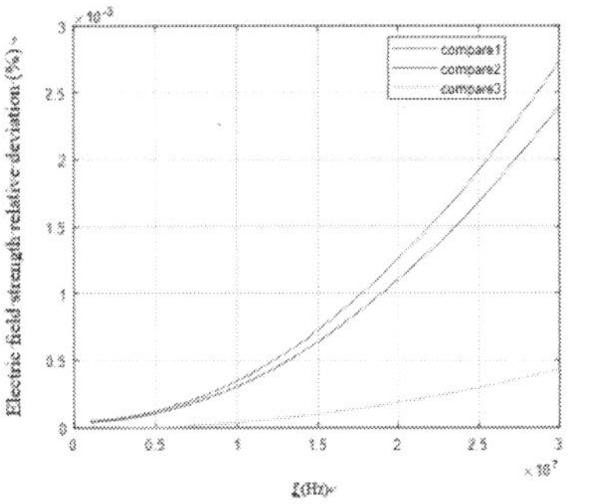

(a) The horizontal distance is0.4km

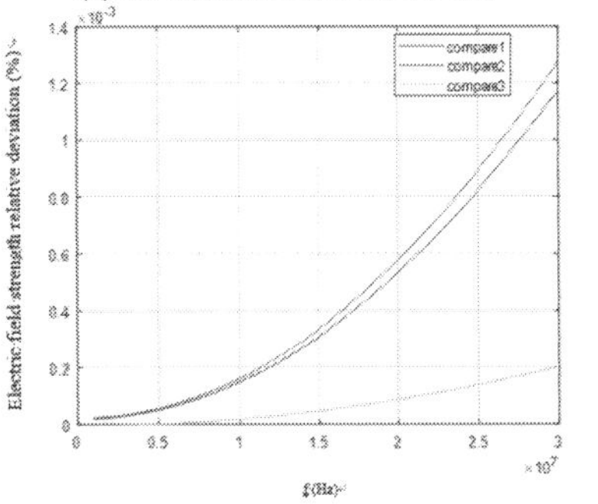

(b) The horizontal distance is0.6km

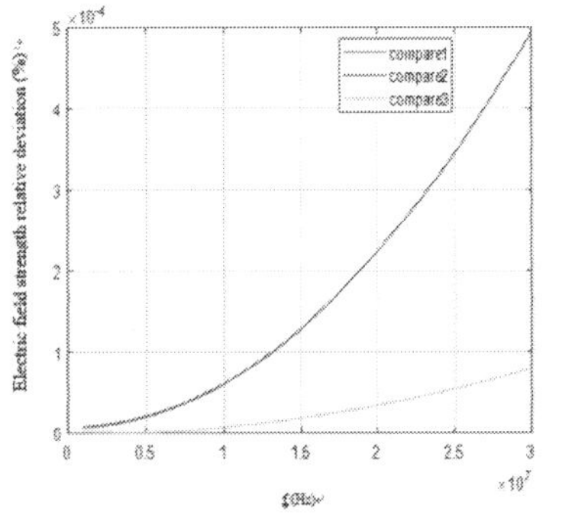

(c) The horizontal distance is1.0km

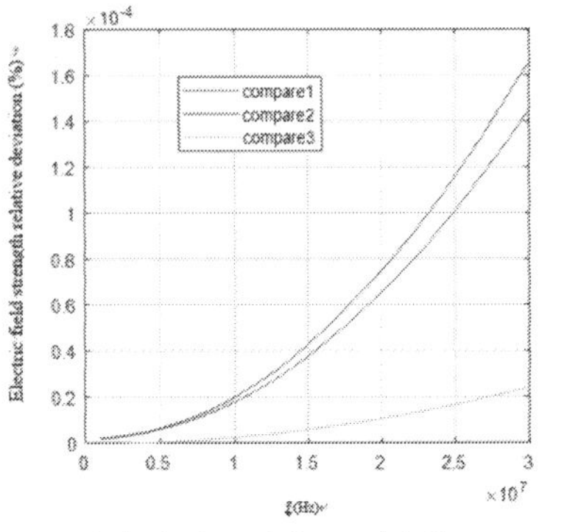

(d) The horizontal distance is2.0km

Fig. 4 Schematic diagram of relative deviation of electric field intensity

It can be seen from Fig. 4 that in the range of 1 km distance, the relative deviation of the electric field strength of the 25 pitches and the five pitches of the transmission line is larger than the relative deviation of the electric field strength of the 40 pitches and 20 pitches of the transmission line; In the range of distance greater than 1km, the former is smaller than the latter; in addition, as the distance changes, the relative deviation of the electric field strength of the 120 pitch and 100 pitch of the transmission line is the smallest among the three groups. In general, as the horizontal distance increases, the relative deviation of the electric field strength generated by the transmission lines of different pitches is gradually decreasing. At the same time, as the frequency increases, the relative deviation of the electric field strength generated by the transmission lines of different pitches is gradually increasing.

4 Conclusion

In this paper, through simulation, the influence of transmission line pitch on passive interference in short-wavelength band is calculated, and the following conclusions are drawn, and the following conclusions are drawn.

1. Regardless of the trend of the pitch of the wire, as the distance between the observation point and the transmission line increases, the electric field strength generated by the transmission line decreases. At the same time, as the frequency increases, the electric field strength generated by the transmission line increases.
2. As the horizontal distance increases, the relative deviation of the electric field strength generated by the transmission lines of different pitches is gradually decreasing. At the same time, as the frequency increases, the relative deviation of the electric field strength generated by the transmission lines of different pitches gradually increases.
3. In the actual project, in order to reduce the influence of the passive interference of transmission line, within the distance of 1 km between the observation point and the transmission line, it is better to choose a 50m overhead line with 40 pitches. Meanwhile, the goal of saving resources is achieved.

References

1. Yin hui, Zhang xiaoming, Wang yantao, et al. Research on 3D visualization of electromagnetic interference for Ultra-high Voltage transmission lines. High Voltage Engineering.**40**,12 (2014)
2. Tang bo , Yang jiawei Huang li, et al. Guard spacing of the air surveillance radar station to Ultrahigh Voltage transmission line.**40**.4(2018)
3. Tang bo, Sun rui, Zhao xiaoming, et al. Three-dimensional Surface Computation model of the Reradiation interference From UHV angle-steel Tower[J]. Proceedings of the CSEE, **31**, 4, 2011.
4. Wu xiong, Wan baoquan, Zhang xiaowu, et al. Study on the influence and protection of the radio station between the two circuits of the 1000kV high voltage exchange[R] .Wuhan, Hubei : Wuhan institute of high pressure, 2008.

5. Hill P C.Measurement of reradiation from lattice masts at V.H,F.[J].Proc Inst Elec Eng,**111**,121964.
6. Belroses J S,Lavrench W,Dunn J G,et al.The effects of reradiation from high rise buildings and transmission lines upon the radiation pattern of MF broadcasing antenna arrays[C]//Proceedings of AGARD/EPP Meeting .Spatind,Norway:[s.n.],1979.
7. Kamal Sarabandi,Moonsooo Park.A radar cross-section model for power lines at millmeter-wave frequency[J].IEEE Transactions on Antenna and Propagation,**51**,9,2003.
8. Truemen C W, Kubina S J. Scatting from power lines with the skywire insulated from the towers[J].IEEE Transactions on Broadcasting,**40**,2,1994.
9. Li cheng, Wang qinjian, Liu youming. Research on passive interference from high-voltage transmission line affecting on medium wave navigation station[J].Modern electronics technique,**39**,7, 2016
10. Tang bo, Sun rui, Zhao xiaoming, et al. Passive interference solution for UHV transmission line in very high frequency band. Electric Power, **49**, 10, 2016.
11. TANG Bo, WEN Yuanfang, ZHANG Xiaowu et al . Key problems of solving radiation interference protecting Distance Between Power transmission Line and Radio Station at MF and SF[J]. Proceedings of the CSEE, **31** ,19, 2011.
12. Sarabandi K, Park M. A Radar Cross-Section Model for Power Lines at Millimeter-Wave Frequencies [J]. IEEE Transactions on Antennas & Propagation, **51**, 9, 2003.
13. Kamal Sarabandi, A Radar Cross-Section Model for Power Lines at Millimeter-Wave Frequencies[J], IEEE Transactions on antennas and Propagation , 2003(9):2353-2360.

Efficiency Rate Optimization of Osmotic Energy Power Plants

L. Stibbe , *A.* Giovanetti, *S.* Khan, and *S.* Fontaine

ECE Paris School of Engineering, Paris, France

Abstract. In a context of world energetic mix diversification and greenhouse effect reduction, attention is focusing on renewable and carbon free sources. Amongst them marine sources are attractive ones by their huge potential today globally evaluated at world scale as high as 120.000TWh/ year. They include osmotic energy one, which has been up-to-now left aside because of technological constraints and prohibitive exploitation cost. Reduction of their impact implies two possible actions : to operate the plants at a maximum efficiency rate, and to improve membrane technology in order to significantly lower exploitation costs. These two actions can be developed independently and it is proposed in the present study to concentrate on the first point to maximize theoretical osmotic plant efficiency rate. The output is a set of relations between system parameters defining the "optimized" line along which the system ought to be operated. This first study opens for a parallel study the technical feasibility of such optimally designed membranes to be installed in the plants.

1 Introduction

The very large amount of energy to produce for human kind development and its exponential growth worldwide is leading countries to investigate alternative renewable sources aside classical fossil ones, in order to also reduce the carbon dependence of present energy mix with its high greenhouse environmental impact. On top of solar, wind, geothermal and river exploitation, interest is now shifting toward new sources from the oceans, in relation with the considerable potential it implies. So is osmotic energy which takes advantage of the pressure difference between waters of different salinity [1], and is now under development. This energy is transformed into electrical ones by the process of Pressure Retarded Osmosis (PRO) [2].

This technique uses semi-permeable membranes to separate the different solutions with their different concentrations. (for instance fresh water and salted water, a solution and another more concentrated one). The membrane lets the solution enter the next compartment and increase its pressure. Water can then be depressurized and be used to move a hydro-turbine generating electricity.

On the base of ocean distribution on Earth's surface, the production potential for osmotic energy is today estimated to 180 TWh for Europe. On a world scale about 1700 TWh could be yearly produced if extended to all river mouths [3], which could theoretically cover a tenth of electricity world demand. To make it worth, it is necessary as for other alternative sources, to make it competitive by lowering both construction constraints and operating costs. In the present case of osmotic energy plants, the fundamental element is the semi-permeable barrier quality which creates the pressure differential between the compartments. Evidently, the higher is the concentration, the larger is the free energy extracted, and the more power can be generated. Along this line, osmotic power looks a promising alternative renewable energy source provided the conversion factor from pressure difference is large enough [3]. However, this technical element ought also to be operated in the best conditions, in the same way as in the past steam engines were gaining from improvement of steel, but also from optimization calculations defining best efficiency rate, which were leading to Carnot cycle. Here it is intended in a similar way to find the best operating line for any osmotic power plant, defined by the highest power output, it can formally deliver. This is based on analysis of a system of equations representing osmotic plant dynamics, from which power optimization is developed and explicit expression of best operating points is given in terms of system parameters. So without calculation, it is now possible to determine a priori if an osmotic power plant is competitive with its given set of nominal parameters, or what should be their modification to reach competitiveness level.

2 Osmotic Power System Production

Let a schematic osmotic power system with J_V , J_S (ms^{-1}) the volume and solute fluxes across the membrane, the solute reflection coefficient, A $(mPa^{-1}s^{-1})$ the membrane fluid permeability, and $E_m = \exp(-[(1-\sigma)J_V L_\omega^{-1}])$ with the solute permeability, see Figure 1.

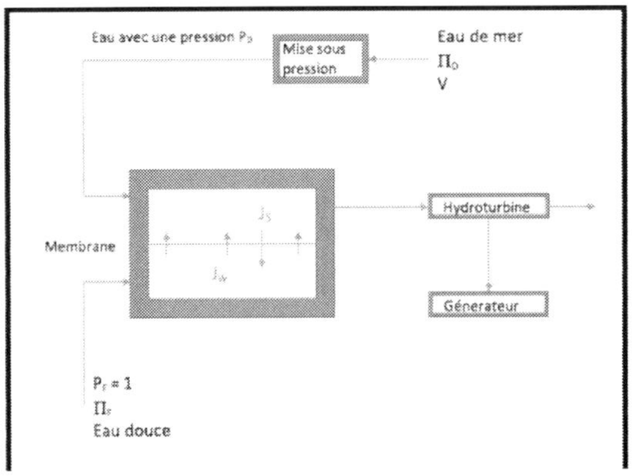

Fig. 1. Osmotic Plant Balance, J_s the Salt Flux across the Membrane

Flux equations for this system [4] can be integrated to get fluxes J_v and J_s flowing through the membrane with pressure retarded osmosis

$$J_V = A(\sigma\Delta\pi - \Delta P)\;; J_S = -(1-E_m)^{-1}(J_V C_d E_m - C_i) \quad (2)$$

Similarly, in the support one gets

$$J_S = -(1-E_s)^{-1}(J_V C_i E_s - C_f) \quad (3)$$

with C_d, C_i and C_f the solute concentration in the draw solution, at the barrier-layer/support interface, and in the feed solution respectively, and $E_s = \exp\left(-\left[J_V \delta D_e^{-1}\right]\right)$ with δ the support thickness. Equality of fluxes in the barrier layer and the support gives from (1,2,3) the non-dimensional equation

$$\propto \{(1-\sigma)(1-E^{1-\sigma})^{-1} + E^{\rho}(1-E^{\rho})^{-1}\} = (1-r)(1-E^{\rho})^{-1} - \sigma \quad (4)$$

With $\propto = 1 - \frac{C_i}{C_d}$, σ the solute reflection coefficient, $r = \frac{C_f}{C_d}$, $\rho = \frac{\delta\omega}{D_e L}$, δ the support thickness, L the barrier layer thickness, D_e the effective diffusion coefficient of the solute, $E = \exp(-X)$, $X = F(\sigma \propto - < \Delta P >)$, $F = ARTvC_d L/\omega$, the salt stoichiometric coefficient and finally $< \Delta P > = (P_d - P_f)/RTvC_d$. For a given dimensionless hydrostatic-pressure difference$< \Delta P >$, (4) is a transcendental equation in $\propto$ (dimensionless concentration difference across the barrier layer) which determines the operating conditions of the osmotic plant for a given set of system parameters.
From these expressions the power produced per surface unit of installed membrane is given by $W = J_m \Delta P$ with W in Wm^{-2}, or in non-dimensional form

$$< W > = \frac{W}{A(RTvC_d)^2} = < \Delta P > (\sigma\alpha - < \Delta P >) \quad (5)$$

$< W >$ is maximum when $< \Delta P > = \sigma\alpha/2$ and is then equal to $< W > = (\sigma\alpha)^2/4$ However it should be verified that this maximum is reachable as and $< P >$ are also linked by (4) which imposes a constraint on system coefficients relating "physical" (concentration performance) to "technical" F (barrier quality). Different limits can be evaluated for possible power outputs from the system [5], but here the general optimum will be directly obtained from (4).

3 Power Plant Optimization

So far, the problem is to determine $< \Delta P >$ in terms of all system parameters from (4). To proceed, it is observed that (4) can be rewritten as

$$\alpha = (1-Y)[(1-r) - \sigma(1-Y^{\Psi})]/[(1-\sigma)(1-Y^{\Psi}) + Y^{\Psi}(1-Y)] \quad (6)$$

$$Y = exp\,[-(1-\sigma)F(\sigma\alpha - <\Delta P>)]$$

with $\Psi = \rho/(1-\sigma)$. Taking now Y ($\in[0,1]$ from its definition) as a dummy variable, (6,7) provides a parametric representation $<\Delta P> =<\Delta P>(Y)$, $\alpha = \alpha(Y)$, of the curve $<\Delta P> =<\Delta P>(\alpha)$ in terms of Y for all other system parameters fixed. If there exists an intersection between the curve $<\Delta P> =<\Delta P>(\alpha)$ and the maximum power output condition $< \Delta P > = \sigma\alpha/2$ from (5) (a straight line in the $\{<\Delta P>, \alpha\}$-plane) then evidently this point will meet all the requirements for being an optimum operating point. From (6) one gets the limits

$$\alpha(0) = 1 - r/(1-\sigma),\ \alpha(1) = (1-r)/(\rho+1) \quad (7)$$

and

$$d\alpha/dY(1) = - \quad /(1+ \quad)(1-\sigma) < 0 \quad (8)$$

$$d\alpha/dY(Y \to 0) = -(1-r)/(1-\sigma) + r \qquad Y^{\Psi-1}/(1-\sigma)^3 \quad (9)$$

In the same way $<\Delta P>$ (0) = $-\infty$ and $<\Delta P>$ (1) = $\sigma\alpha < \alpha$ and both curves can be placed in $\{<\Delta P>, \alpha\}$-plane, see Figure 2

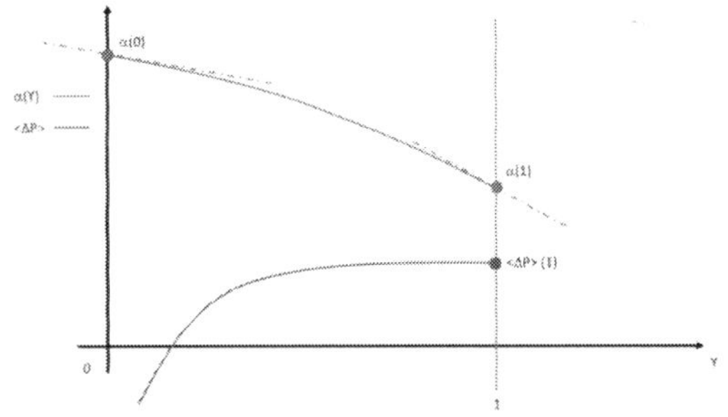

Fig. 2. Schematic Plots of $<\Delta P>$ and α vs Y for $\Psi > 1$ and $\alpha(0) > \alpha(1)$

The curve $<\Delta P> = <\Delta P>(\alpha)$ can be drawn from the two curves, see Figure 3. Next step is to get the intersection point I of this curve with the optimum power output line $< \Delta P > = \sigma\alpha/2$ from (5), which always exists in considered parametric case for all physically acceptable value of the parameters satisfying the existence constraint $\alpha > 0$. Because $<\Delta P> = <\Delta P>(\alpha)$

starts from $\alpha(1) > \sigma\alpha(1)$ and drops asymptotically to $-\infty$ for $\alpha = \alpha(0)$, condition for existence of at least one intersection I is that $<\Delta P> = <\Delta P>(\alpha)$ cuts the abscissa line for positive α. This is always the case here because from (6) $\alpha = 0$ when $Y^{\Psi} = (r + \sigma - 1)/\sigma > 0$ by definition of Y which contradicts the existence condition at 0 $\alpha(0) > 0$, so there does not exist such 0's.

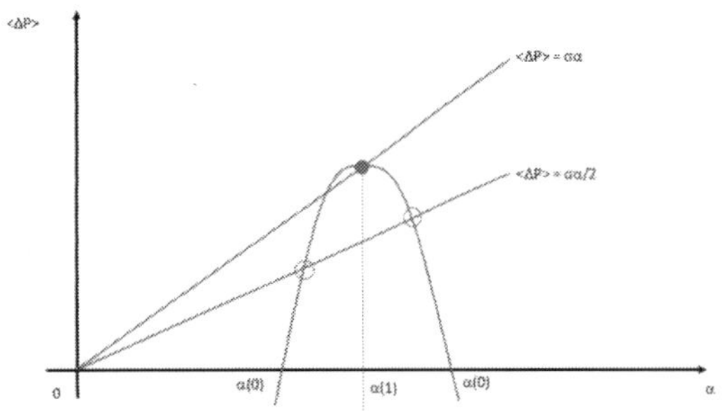

Fig. 3. Sketch of $<\Delta P>$ vs α for $\Psi > 1$: right branch $\alpha(0) > \alpha(1)$ left branch $\alpha(0) < \alpha(1)$. Optimum operating points are marked by circles

A largest power output is obtained when $\alpha(0)$ takes the largest value as shown in Figure 3 (right circle). Being identified, numerical calculation is possible and design of parameters organized to satisfy this constraint. The advantage of this condition is that it is completely explicit in terms of system parameters and is obtained without any calculation of system trajectory dynamics. If $W_{\lim}$ if the lower limit for osmotic plant profitability, one gets from (5) the viability condition

$$W_{lim} = A(RTvC_d)^2 \Delta P^2 \qquad (10)$$

where pressure is evaluated at the optimum point. For typical ΔP = 1.2 MPa (for water) and usual system parameters, the limit value $A \cong$ 1.4mPa ^{1}s ^{1} is still outside the reach of today membranes [6], and extension of simple barriers has been suggested [7].

4 Conclusion

Osmotic pressure systems are recognized as potentially valuable alternative renewable energy sources due in particular to their large distribution over all Earth's surface, when operated in a Pressure Retarded Osmosis mode. To define the parameter domain corresponding to profitable exploitation, the study of their dynamics has been undertaken and it is shown that, aside the improvement of important technical parameters such as semi-permeable membranes, there exists a set of conditions for which the power output is maximized at system level notwithstanding the specific value of its parameters. Their great interest is that they can in full generality be determined explicitly from formulae expressed in terms of system parameters, simplifying considerably the design of the osmotic pressure plant. The net result is that the profitability limit of 5W/m2 is more easily reached when evaluated at identified functioning points corresponding to the optimum by reducing in proportion the technical constraint.

Acknowledgments

The authors are very much indebted to ECE Paris School of Engineering for having provided the necessary setup for the development of the project, and to Pr. M. Cotsaftis for guidance and help in the preparation of the manuscript.

References

1. Loeb, 1975 ; Mishra, 2013 ; Kho, 2010 ; The European Commission ; 2004 ; Kumar et al., 2011, Helfer et al., 2013 ; Skilhagen et al., 2012 ; Skilhagen and Aaberg, 2012
2. Kim and Elimelech, 2013; Helfer et al., 2014; Wang et al., 2012 Lemckert, C., Anissimov, Y.G. (2014). Osmotic Power with Pressure Retarded Osmosis: Theory, Performance, and Trends, *J. Membranes Science*, 453, 337-358
3. Dinger et al., 2012; Braein et al., 2010; Straub et al., 2016 Dinger, F., Troendle, T., Platt, U. (2012) . Osmotic Power Plants, *3rd Osmosis Membrane Summit*, Statkraft, Barcelona: Spain.
4. Spiegler, K.S., Kedem, O. (1966): Thermodynamics of Hyper-filtration (Reverse Osmosis): Criteria for Efficient Membranes, Desalination, And 1,311-326. Spiegler, K.S., Kedem, Banchik, L.D., Sharqawy, M.H., Lienhard, J.H. (2014). Limits of Power Production due to Finite Membrane Area in Pressure Retarded Osmosis, J. Membr. Sci., 468, 81-89.
5. Kim, Y.C., Elimelech, M. (2013). Potential of Osmotic Power Generation by Pressure Retarded Osmosis Using Seawater as Feed Solution: Analysis and Experiments, *J Membrane Sci, 429*, 330-337
6. Cath, T.Y., Elimelech, M., McCutcheon, J. R., Mcginnis, R. L., Achilli, A., Anastasio, D., Brady, A. R., Childress, A. E., Farr, I. V., Hancock, N. T., Lampi, J., Nghiem, L. D., Xie, M. & Yip, N. Yin. (2013). Standard methodology for evaluating membrane performance in osmotically driven membrane processes. *Desalination, 312 (N/A)*, 31-38.
7. Hickenbottom et al., 2016 ; Chen et al., 2016 ; Chou et al., 2013 ; She et al., 2016 ; Banchik et al., 2014; Yip et al

Quantifying the Influences on Probabilistic Wind Power Forecasts

Jens Schreiber , and *Bernhard* Sick

Intelligent Embedded Systems Group, University of Kassel, Kassel, Germany

Abstract. In recent years, probabilistic forecasts techniques were proposed in research as well as in applications to integrate volatile renewable energy resources into the electrical grid. These techniques allow decision makers to take the uncertainty of the prediction into account and, therefore, to devise optimal decisions, e.g., related to costs and risks in the electrical grid. However, it was yet not studied how the input, such as numerical weather predictions, affects the model output of forecasting models in detail. Therefore, we examine the potential influences with techniques from the field of sensitivity analysis on three different black-box models to obtain insights into differences and similarities of these probabilistic models. The analysis shows a considerable number of potential influences in those models depending on, e.g., the predicted probability and the type of model. These effects motivate the need to take various influences into account when models are tested, analyzed, or compared. Nevertheless, results of the sensitivity analysis will allow us to select a model with advantages in the practical application.

1 Introduction

In past years, renewable energy resources have become a fundamental part of the electrical power supply in many countries. In Germany, e.g., renewable resources contribute up to 29 percent to the energy mix [1]. A conventional approach to integrate those volatile resources into current grids is to use *numerical weather prediction (NWP)* data and historical power data as training data for deterministic machine learning models that generate day-ahead forecasts. Here, we call these models deterministic because their output is a point estimate of the future power production.

Recently, probabilistic forecasts can be seen in research as well as in applications [2]. Probabilistic methods forecast distribution of wind power rather than delivering a point forecast of the expected power. *Probabilistic forecasts (PF)* have, in comparison to deterministic approaches, the advantage that power plant operators, grid operators, and energy traders can examine the risk of their day-ahead forecasts. These probabilistic forecasts help decision makers to devise optimal decisions considering the uncertainty of the prediction, mainly caused by the NWP [3].

However, even though many studies investigate influences and interactions of input features on the output of deterministic forecasts, e.g., [4]–[9], there has been little interest in conducting studies that are designed to evaluate influences in probabilistic forecasts. This article examines influences of features on the output of three different probabilistic wind power forecasting models with a focus on NWP data and technical features (e.g., maximum diameter and maximum power generation of a wind turbine) as input (which is typical for day-ahead forecasts). Therefore, we investigate these potential influences with methods from the field of sensitivity analysis regarding the three models as black-boxes.

This analysis allows us to point out a considerable number of potential influences in probabilistic forecasts. Influences are, e.g., the forecasting model, the predicted quantile, as well as selected input features. That is, the causes of forecasts errors might be various. Therefore, we need to take several influences into account when models are tested, analyzed, and compared or simultaneously used within an application. In the end, results of the analysis lead us to a gray-box model, where we have partial knowledge about the internal behavior of the forecasting model. This knowledge will allow us to select a model with advantages in the practical application.

The remainder of this article is structured as follows. First, we give details on related work in Section 2. We continue with explanations on probabilistic forecasts and sensitivity analysis in Section 3. Then the experiment is presented in Section 4. Section 5 summarizes and concludes our work.

2 Related work

Determining influences in forecasting models are an essential research topic in renewable resources. On the one hand, findings can help to improve the forecasting accuracy. On the other hand, insights help decision makers to understand their forecasting model. By understanding the relationship between input and output, it is, e.g., possible to determine potential causes of a forecasting error.

The most common methodology to analyze the influences on the output of the deterministic forecasts model is sensitivity analysis. Sensitivity analysis is, e.g., used in deterministic forecasts for investment calculation [10]. In [4], the authors use sensitivity analysis to proof the feasibility to integrate wind power generators into the Brazilian electricity market. Another application of sensitivity analysis shows the profitability and vulnerability of renewable energy resources [7]. In [5], sensitivity analysis is used to examine influences of features for the placement of floating offshore wind farms for future investments. It is also possible to develop a framework to examine and evaluate features that allow the assessment of potential locations for new wind or solar parks as shown in [6]. The authors of [8] use sensitivity analysis and swarm optimization to obtain the best placement of wind turbines in wind parks [8]. In [9], sensitivity analysis is used to show that weather data has a higher impact than technical features in wind power time series modeling. However, to our knowledge there is no article that attempts to apply sensitivity analysis to the analysis of models used for probabilistic (wind power) forecasting.

3 Methodology

In this section, we give a brief introduction into probabilistic forecasts and sensitivity analysis.

3.1 Probabilistic wind power forecasts

In this section, we focus on the background of probabilistic models, necessary to apply sensitivity analysis later. Probabilistic methods forecast distribution of wind power rather than delivering a point forecast of the expected power. By predicting, e.g., a *cumulative distribution function (CDF)*, we obtain additional information about the uncertainty of our model. Uncertainties arise, e.g., due to the influence of the uncertain *NWP* based model inputs.

Typically, to predict those distributions one either uses parametric or non-parametric methods. Non-parametric methods have the benefit that they do not require assumptions about the mathematical form of the distribution of wind power. Three non-parametric models will be used in this article to forecast the wind power distribution. In particular, the techniques *monotone quantile regression neural network (MQRNN), support vector regression (SVR)* and, *gradient boosting regression tree (GBRT)* are used. For details, please refer to [11]–[14].

Each of the three approaches is considered as a black-box method to forecast an *empirical cumulative distribution function (ECDF)*. In contrast to a CDF, the ECDF is co posed of several quantiles and it is linearly interpolated between the respective quantile estimates.

3.2 Sensitivity analysis

Sensitivity analysis (SA) typically decomposes the uncertainty of a single (deterministic) output to different sources of uncertainty in the input [10]. However, we are interested in applying SA to the predicted ECDF from three black-box models and not to a deterministic output (i.e., point estimate). Therefore, for each predicted quantile SA needs to be applied. Each quantile can be regarded as a single predicted output of the ECDF. This utilization allows us to evaluate each quantile individually without adapting the method itself. In this sense, by applying SA to each quantile, we examine how individual quantiles are influenced differently by the input. Further on, by comparing the results for different quantiles, we can study how these influences are related. For better understanding, we will limit the following explanations to a single output.

One standard SA method to examine influences and interactions of inputs is the *variance based decomposition (VBD)* [10]. VBD uses Monte Carlo simulation to decompose the variance of the output $V(Y)$ w.r.t. the input. VBD, therefore, distinguishes between a *first-* and a *total-order sensitivity index* given by the following formulas:

$$S_i = \frac{V[E(Y|X_i)]}{V(Y)}, \tag{1}$$

$$S_{T_i} = \frac{V[E(Y|X_{\sim i})]}{V(Y)}, \tag{2}$$

where S_i is the first-order sensitivity index and S_{T_i} the total-order sensitivity index. S_i can be interpreted as the degree of influence of feature i on the output. The first-order sensitivity index is the relation between the variance, $V[E(Y|X_i)]$, in the output that is explained by feature i alone and the variance in the output.

S_{T_i} is the first-order index of feature i plus all higher-order interactions of this feature, e.g., the effect of wind speed alone plus the effect together with air pressure. More precise, in the field of SA, the interaction is defined as the effect that cannot be explained by a single effect alone [10]. $X_{\sim i}$ indicates all possible combination of other features with feature i. For further details refer to [10], [15].

Note that a notable difference between S_i and S_{T_i} indicates strong interactions between features that affect the output together.

4 Experimental results

This section investigates the influences of three probabilistic wind power forecasting models, their input features, and the predicted quantiles with sensitivity analysis in three use cases. The SA of one use case along with one forecasting model is called *scenario* for convenience. First, we give our definition of the three use cases. Second, we explain the manual feature engineering and the preselection of features. This set of selected features allows us to get proper evaluations results of all three black-box models as detailed in the next section. The evaluation results are critical to assure that the forecasts of the wind power distribution are reasonable. Finally, we apply SA and get valid results that are summarized and discussed in the final sections.

4.1 Definition of use cases

To examine the relation between input and output in probabilistic wind power generation, we evaluate three different use cases. The evaluation of the different use cases helps to distinguish different influences of the terrain and the relation as modeled by the respective PF model. Similar to other studies of wind power forecast models, see, e.g. [3], we define the following use cases:

- Wind parks located in *non-complex terrain (NCT)* between 200 and 1500 meters above sea level (e.g., farmland).
- Wind parks located in *complex terrain (CT)* between 200 and 1500 meters above sea level (e.g., forest).
- *Offshore (OS)* wind parks located on the ocean.

Thee definitions allow us to evaluate 11 NCT, eight CT and four OS wind parks located in Germany from our data.

4.2 Feature engineering and pre-selection of input data

This section details how we manually engineered features and use feature selection to devise an optimal set of features. This set of features allows us to have reasonable forecast results in each scenario (summarized in the next section). All wind parks of the use cases have a resolution of one hour from 2015-01-01 to 2016-12-31. Initially, the data has the following properties:

- NWP features: *Air pressure (AP), humidity (H), wind direction zonal (WDZ100m) at 100m, wind direction meridional (WDM100m)* at 100m, wind speed (WS10m) at 10m, and *wind speed (WS100m)* at 100m above ground.
- Meta features: Maximum power generation, maximum diameter, maximum hub height, and elevation of the wind park.
- Normalization: Min-max normalization is applied to all NWP features initially. Respectively, derived features are normalized as well. The generated power is normalized with the maximum power generation.
- From the NWP features we derive *variability (V)* features. Variability is defined with $V = \overline{|F(t+k) - F(t)|}$. This derived feature indicates the mean amount of change of a feature F within a time horizon k [16], e.g. of the *previous (P)* hour. For k we used *hour (HR), day (D), week (W), month (M),* and *year (Y)*. Note that, e.g., the variability of wind speed at 100 meters in the previous hour is abbreviated with *VWS100mPHR*.
- We add features for the day, the week, the month of the year, and *hours since last model run (HSMR)* of the NWP to cover, e.g., trends and seasonal effects.

Afterward, the data is split into 80% training (January 2015 to July 2016) and 20% test data-set (August 2016 to December 2016). This data setup is used to pre-select features for each use case on the training data.

We use a combination of *sequential forward selection (SFS)* and two filters (Minimum Redundancy Maximum Relevance and Fisher Score [17]). SFS selects the ten most prominent features with GBRT and so-called *continuous ranked probability score (CRPS)* as selection criterion [18]–[20]. For the sake of this article, one can imagine the CRPS as the mean absolute error from deterministic forecasts. In parallel, we require that results of the SFS to be in the top ten ranked features from one of the filters.

This selection strategy improves the CRPS by a minimum of 1.6 percent and a maximum of 12 percent compared to filter or wrapper feature selection methods. However, the detailed evaluation is out of scope for this article.

4.3 Probabilistic models analyzed by sensitivity analysis

The configuration of the models can be summarized as follows:

- The quantiles 0.1, 0.2, 0.3, 0.4, 0.5, 0.6, 0.7, 0.8, and 0.9 are used and analyzed in all scenarios.
- Input features for the scenario depends on the selected feature for the respective use case.
- All three models are trained and optimized with standard parameters on the train and validation data sets.

For the SA detailed evaluation of the error score is not relevant. However, it may be relevant to know that they achieved reasonable results. GBRT achieved the smallest average CRPS (0.061) and standard deviation (0.019). MQRNN has the second smallest mean (0.07) and standard deviation of the mean (0.025). SVR has the largest standard deviation (0.027) and average CRPS (0.073).

4.4 Sensitivity analysis

The following section summarizes results of the SA. In each use case, the same features are used; those features are selected beforehand by the feature selection strategy as described above to take the specifics of the terrain into account. The models trained on this data (for each scenario) are evaluated here with SA.

For all scenarios, the VBD sensitivity analysis is applied to each quantile separately. The Monte Carlo simulation uses 10.000 sample points. Within a use case, the same sample points are used for all models and quantiles.

Figure 1 shows three examples of the *average quantile first-order index*. By averaging the first-order index (of the same quantile) for all wind parks in a scenario, we obtain the average influence of features for a quantile in this scenario. The three machine learning examples are representative concerning the relation of quantiles. It shows, e.g., how an increase of S_i for WS100m for different quantiles is related to a decrease of S_i for WS10m (see, e.g., Figure 1a).

Table 1 shows the average S_i and S_{T_i} across all quantiles and wind parks within a scenario. In contrast to the average of Figure 1, the first-order or total-order is additionally averaged for all quantiles. This allows obtaining a single value that is representative of the

https://doi.org/10.1051/e3sconf/20186406002

scenario. The difference between these average S_i and S_{T_i}, is used to examine the amount of interaction of features given by an increase between the first and total-order index. As mentioned before, in the field of SA interactions are defined as the effect on the output which cannot be explained by a single feature. The higher the increase from the average S_i to the S_{T_i}, the larger is the amount of interaction of the feature with other features that affect the output together.

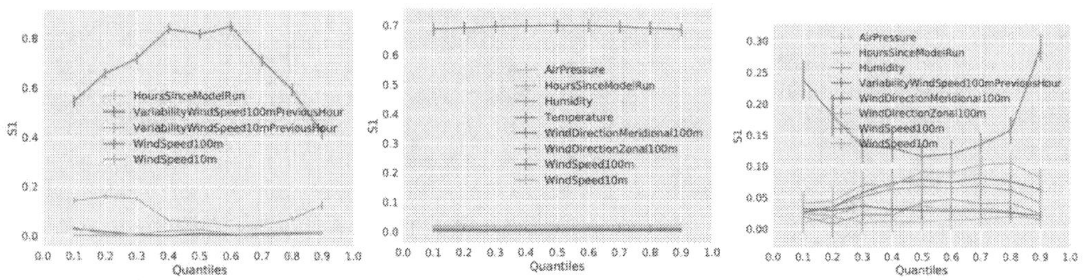

Fig. 1. The figure shows three representative examples of the individual influences of features on the output of the respective model. For each forecasted quantile Eq. **1** is applied in the following three scenarios, from left to right: (a) GBRT offshore scenario, (b) MCQRNN forest, and (c) SVR flatland.

	GBRT						MQRNN						SVR					
	OS		NCT		CT		OS		NCT		CT		OS		NCT		CT	
1	-	-	0.02	0.04	0.02	0.04	-	-	0.02	0.23	0.02	0.14	-	-	0.01	0.23	0.03	0.26
2	0.00	0.01	0.00	0.01	0.00	0.00	0.00	0.08	0.00	0.07	0.00	0.05	0.03	0.31	0.03	0.30	0.02	0.28
3	-	-	0.01	0.03	0.00	0.01	-	-	0.01	0.14	0.00	0.10	-	-	0.03	0.30	0.02	0.25
4	-	-	-	-	0.01	0.03	-	-	-	-	0.01	0.13	-	-	-	-	0.02	0.24
5	0.01	0.03	0.02	0.03	-	-	0.04	0.41	0.01	0.21	-	-	0.02	0.22	0.03	0.22	-	-
6	0.01	0.03	-	-	-	-	0.03	0.38	-	-	-	-	0.04	0.25	-	-	-	-
7	-	-	0.00	0.01	0.00	0.01	-	-	0.01	0.12	0.00	0.08	-	-	0.06	0.33	0.05	0.29
8	-	-	0.00	0.01	0.01	0.03	-	-	0.01	0.13	0.01	0.08	-	-	0.05	0.35	0.05	0.31
9	0.68	0.82	0.82	0.89	0.86	0.91	0.32	0.68	0.49	0.73	0.69	0.84	0.23	0.53	0.17	0.47	0.18	0.49
10	0.09	0.19	0.04	0.07	0.02	0.03	0.07	0.41	0.06	0.27	0.02	0.16	0.10	0.39	0.08	0.35	0.10	0.37
	S_i	S_{T_i}	S_i	S_{T_i}	S_i	S_{T_i}	S_i	S_{T_i}	S_i	S_{T_i}	S_i	S_{T_i}	S_i	S_{T_i}	S_i	S_{T_i}	S_i	S_{T_i}

Table 1. Results of the averaged first- and total-order sensitivity index S_i and S_T for the selected features for each scenario are shown. For each scenario, the sensitivity index is color-coded with green being the highest and red the smallest value. The relevant features are enumerated on the left in the following order AP (1), HSMR (2), H (3), T (4), VWS100mPHR (5), VWS10mPHR (6), WDM100m (7), WDZ100m (8), WS100m (9), and WS10m (10). The dashed line indicates that the specific feature is not selected in the scenario.

Observations related to features:

- As expected, WS100m is the essential feature in all evaluated scenarios. Similar, in most cases WS10m has the second largest total-order index.
- If selected, AP and VWS100mPHR have about the third largest total-order sensitivity index for all scenarios.

Observations related to PF models:

- The average total-order sensitivity index of

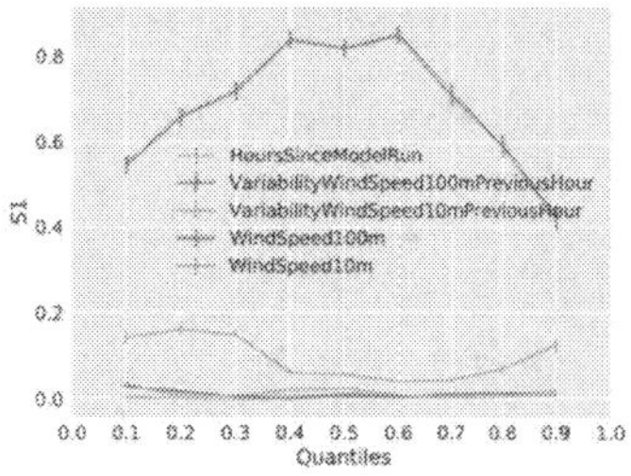

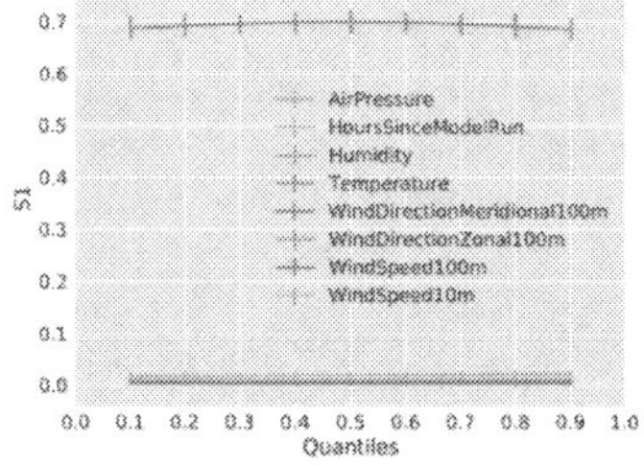

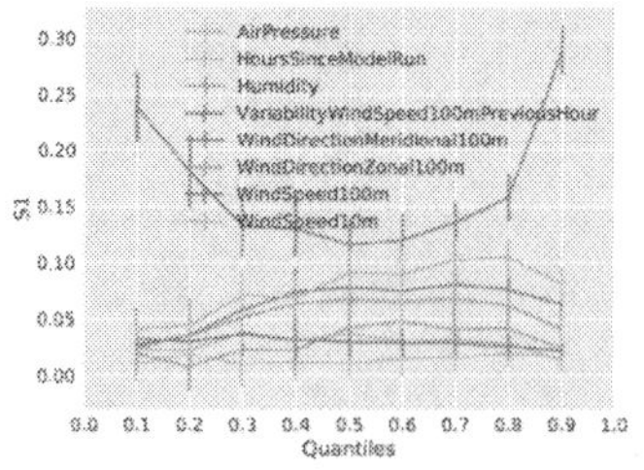

https://doi.org/10.1051/e3sconf/20186406002

Fig. 1. The figure shows three representative examples of the individual influences of features on the output of the respective model. For each forecasted quantile Eq. **1** is applied in the following three scenarios, from left to right: (a) GBRT offshore scenario, (b) MCQRNN forest, and (c) SVR flatland.

	GBRT						MQRNN						SVR					
	OS		NCT		CT		OS		NCT		CT		OS		NCT		CT	
1	-	-	0.02	0.04	0.02	0.04	-	-	0.02	0.23	0.02	0.14	-	-	0.01	0.23	0.03	0.26
2	0.00	0.01	0.00	0.01	0.00	0.00	0.00	0.08	0.00	0.07	0.00	0.05	0.03	0.31	0.03	0.30	0.02	0.28
3	-	-	0.01	0.03	0.00	0.01	-	-	0.01	0.14	0.00	0.10	-	-	0.03	0.30	0.02	0.25
4	-	-	-	-	0.01	0.03	-	-	-	-	0.01	0.13	-	-	-	-	0.02	0.24
5	0.01	0.03	0.02	0.03	-	-	0.04	0.41	0.01	0.21	-	-	0.02	0.22	0.03	0.22	-	-
6	0.01	0.03	-	-	-	-	0.03	0.38	-	-	-	-	0.04	0.25	-	-	-	-
7	-	-	0.00	0.01	0.00	0.01	-	-	0.01	0.12	0.00	0.08	-	-	0.06	0.33	0.05	0.29
8	-	-	0.00	0.01	0.01	0.03	-	-	0.01	0.13	0.01	0.08	-	-	0.05	0.35	0.05	0.31
9	0.68	0.82	0.82	0.89	0.86	0.91	0.32	0.68	0.49	0.73	0.69	0.84	0.23	0.53	0.17	0.47	0.18	0.49
10	0.09	0.19	0.04	0.07	0.02	0.03	0.07	0.41	0.06	0.27	0.02	0.16	0.10	0.39	0.08	0.35	0.10	0.37
	S_i	S_{T_i}	S_i	S_{T_i}	S_i	S_{T_i}	S_i	S_{T_i}	S_i	S_{T_i}	S_i	S_{T_i}	S_i	S_{T_i}	S_i	S_{T_i}	S_i	S_{T_i}

Table 1. Results of the averaged first- and total-order sensitivity index S_i and S_T for the selected features for each scenario are shown. For each scenario, the sensitivity index is color-coded with green being the highest and red the smallest value. The relevant features are enumerated on the left in the following order AP (1), HSMR (2), H (3), T (4), VWS100mPHR (5), VWS10mPHR (6), WDM100m (7), WDZ100m (8), WS100m (9), and WS10m (10). The dashed line indicates that the specific feature is not selected in the scenario.

- for GBRT is small for most features except WS100m and WS10m. In contrast, the values for SVR and MQRNN are more spread among all features. The results for GBRT are potentially caused by the ensemble of weak predictors. The combination of those predictors allows the GBRT to achieve the best forecast results without substantial interaction with other features compared to SVR and MQRNN. For SVR one can assume that the worst evaluation result is related to the extensive amount interactions between all features. These interactions potentially cause too much variability in the output and yield to the worst evaluations results. MQRNN is somewhat in the middle of the evaluation score and the interactions of features. Potentially, the internal data transformation by the MQRNN allows the model to capture interactions of features and keep the variance in the output to a minimum at the same time. Respectively, this data transformation allows MQRNN to achieve better results than the SVR.
- SA shows that influences of features are about the same for identical types of PF methods across different use cases. This observation indicates that the internal structures of the black-box models are more affected by the underlying model than by the data.

Observations related to quantiles:

- For similar PF model, the relation of the first and total-order index for different quantiles are about the same for all use cases. E.g., the initial decrease between the 0.1 and 0.4 quantile in Figure 1c is similar in all SVR scenarios. This observation further motivates that the relationship between input and output is largely depending on the underlying PF model and not the specifics of the input data.
- SA shows that for SVR and GBRT the amount of influence is dependent on the quantile of the predicted ECDF.

4.5 Discussion

Interestingly, SA shows that influences and interactions of features mostly depend on the underlying model. This result partly surprises, because in existing studies of the uncertainty, the forecast error is seen as being largely dependent on the terrain, see, e.g. [3]. Therefore, studies of the uncertainty, as in [21], would probably benefit substantially from SA to understand further causes of the error. On the other hand, it might also be related use case specific selection of the input features.

The individual values (for each PF model) of the sensitivity indexes for different quantiles suggest that it is beneficial to select individual models for different purposes. GBRT, e.g., could be used in a non-complex terrain, to derive a simple model only depending on WS10m and WS100m.

Finally, it seems beneficial to select the model where the influences and the relationship of influences in different quantiles fit our needs. MQRNN, e.g., could be used for probabilistic simulations of the electrical grid for load flow calculations of future energy systems. MQRNN would limit the number of potential influences for different quantiles to provide the simplest possible model for the simulation.

5 Conclusion and future work

In this article, we proposed a simple method to apply SA to ECDF, predicted by PF models, by applying SA to each quantile estimate individually.

By applying the SA to three PF methods and for data of 28 wind parks within three use cases we moved from a black-box to a gray-box probabilistic forecasting model.

We show that influences on quantile estimates and the relationship of those influences for different quantiles depends on the underlying PF model. Further on, we show that influences are more similar for equal PF models in different use cases than they are for different PF models applied to the same use case. The similarity is either related to the pre-selection of input features (specific to the use case) or the internal behavior between input and output modeled by the PF technique. The observed similarity could, e.g., analyzed further by using the same input features for all use cases in the future.

In our future work, we also aim to investigate the differences between model types further. Primarily, we are interested in analysis distinct types of so-called multi-task (MQRNN) approaches such as hard and soft parameter sharing.

Acknowledgment

This work was supported within the project Prophesy (0324104A) funded by BMWi (Deutsches Bundesministerium für Wirtschaft und Energie / German Federal Ministry for Economic Affairs and Energy).

References

1. K. Rohrig *et al.*, "Windenergie Report Deutschland 2016," Kassel, 2017.
2. T. Hong, P. Pinson, S. Fan, H. Zareipour, A. Troccoli, and R. J. Hyndman, "Probabilistic energy forecasting: Global Energy Forecasting Competition 2014 and beyond," *Int. J. Forecast.*, vol. 32, no. 3, pp. 896–913, 2016.
3. P. Pinson, "Estimation of the uncertainty in wind power forecasting," Ecole nationale superieure des mines, Paris, 2006.
4. G. Aquila, P. Rotela Junior, E. de Oliveira Pamplona, and A. R. de Queiroz, "Wind power feasibility analysis under uncertainty in the Brazilian electricity market," *Energy Econ.*, vol. 65, pp. 127–136, 2017.
5. L. Castro-Santos and V. Diaz-Casas, "Sensitivity analysis of floating offshore wind farms," *Energy Convers. Manag.*, vol. 101, pp. 271–277, 2015.
6. A. Bossavy, R. Girard, and G. Kariniotakis, "Sensitivity analysis in the technical potential assessment of onshore wind and ground solar photovoltaic power resources at regional scale," *Applied Energy*, vol. 182, pp. 145–153, 2016.
7. N. E. Mohammad Rozali, S. R. Wan Alwi, and Z. A. Manan, "Sensitivity analysis of hybrid power systems using Power Pinch Analysis considering Feed-in Tariff," *Energy*, vol. 116, pp. 1260–1268, 2016.
8. M. Song, K. Chen, X. Zhang, and J. Wang, "Optimization of wind turbine micro-siting for reducing the sensitivity of power generation to wind direction," *Renewable Energy*, vol. 85, pp. 57–65, 2016.
9. F. Monforti and I. Gonzalez-Aparicio, "Comparing the impact of uncertainties on technical and meteorological parameters in wind power time series modelling in the European Union," *Applied Energy*, vol. 206, pp. 439–450, 2017.
10. D. G. Andrea Saltelli, Marco Ratto, Terry Andres,Francesca Campolongo, Jessica Cariboni and Michaela Saisana and Stefano Tarantola, *Global Sensitivity Analysis. The Primer.* Wiley, 2008.
11. C. Crambes, A. Gannoun, and Y. Henchiri, "Weak consistency of the Support Vector Machine Quantile Regression approach when covariates are functions," *Stat. Probab. Lett.*, vol. 81, no. 12, pp. 1847–1858, 2011.
12. C. Crambes, A. Gannoun, and Y. Henchiri, "Support vector machine quantile regression approach for functional data: Simulation and application studies," *J. Multivar. Anal.*, vol. 121, pp. 50–68, 2013.
13. B. Kriegler and R. Berk, "Boosting the Quantile Distribution: A Cost-Sensitive Statistical Learning Procedure," Los Angeles, 2007.
14. A. Cannon, "Non-crossing nonlinear regression quantiles by monotone composite quantile regression neural network, with application to rainfall extremes," 2017.
15. A. Saltelli, P. Annoni, I. Azzini, F. Campolongo, M. Ratto, and S. Tarantola, "Variance based sensitivity analysis of model output. Design and estimator for the total sensitivity index," *Comput. Phys. Commun.*, vol. 181, no. 2, pp. 259–270, 2010.
16. J. Dobschinski, "How good is my forecast? Comparability of wind power forecast erros," in *Workshop on Large Scale Integration of Wind Power into Power Systems as well as on Transmission Networks for Offshore Wind Farms*, 2014, vol. 13, pp. 1–5.
17. J. Li *et al.*, "Feature Selection: A Data Perspective," p. 45, Jan. 2016.
18. A. Gensler, S. Vogt, and B. Sick, "Metaverication of Uncertainty Representations and Assessment Techniques for Power Forecasting Algorithms including Ensembles," *unpublished.*
19. J. Xie and T. Hong, "Variable Selection Methods for Probabilistic Load Forecasting: Empirical Evidence from Seven States of the United States," *IEEE Trans. Smart Grid*, pp. 1–1, 2017.
20. C. Hübler, C. G. Gebhardt, and R. Rolfes, "Hierarchical four-step global sensitivity analysis of offshore wind turbines based on aeroelastic time domain simulations," *Renewable Energy*, vol. 111, pp. 878–891, 2017.
21. S. Kohler, A.-C. Agricola, and H. Seidl, "dena-Netzstudie II," Berlin, 2010.

Methodology for Calculating the Potential Maritime Wind Park Area by Taking into Account the Needs of Shipping Sector

Astrida Rijkure

Scientific Institute of Economics and Management, University of Latvia, Riga, Latvia

Abstract. Renewable energy sources (wind energy, solar energy, hydroelectricity, ocean energy, geothermal energy, biomass and biofuels) are alternatives to fossil fuel that help to reduce greenhouse gas emissions, diversify energy supplies and reduce dependency on markets of unsustainable and volatile fossil fuels, particularly oil and gas. Wind energy is one of the renewable energy sources and is considered to be self-renewable as it is the result of the Sun's activity. Using wind energy is a rapidly developing industry today, and more and more wind turbines are installed worldwide every year, land-based wind turbines being more widespread than offshore ones. In Latvia, spread of land-based wind parks is hampered by unsettled land ownership rights, while the deployment of wind parks in the sea is a new field for all Baltic States. The neighbouring countries Estonia and Lithuania have developed their own projects for offshore wind parks, therefore the topicality of the development of wind farms in the territorial waters of Latvia has also increased. Experts have proposed best options and their locations. When assessing possibilities for development of wind parks and their capacity, the following economic factors were evaluated: construction and connection costs, potential operational costs and energy prices. The aim of this study is to develop the methodology for calculating the area of a potential wind park by considering the safety distance to shipping routes and height of the wind turbines, as well as for calculating the potential capacity of a wind park.

1 Introduction

Wind energy is electric power that is generated by means of wind turbines. [1] Wind energy is produced, as wind encounters an obstacle on its way and transfers a portion of its kinetic power to such obstacle. This principle is used in wind turbines, or wind generators, where wind moves turbine blades, generating mechanical power, which is then transformed into electricity by means of generators. [2] More and more attention must be paid to methods of generating alternative power. The use of alternative energy allows reducing the costs of heating, it is friendlier for the environment (non-renewable resources that cause environmental pollution from their use, like coal, oil or natural gas, are not used) and increases the autonomy from suppliers of traditional energy resources.

Wind energy is considered to be an energy that is capable of renewing itself, since it is the consequence of the Sun's activity. The use of wind power is an industry that is developing very rapidly. More and more wind turbines are installed worldwide every year. A total of approximately 460 GW of wind power is generated worldwide every day, which amounts to about 4% of the total power generation and consumption worldwide. Wind turbines are frequently installed on land or above water bodies like oceans or seas and, usually, the turbines installed on land are capable of generating power for the cheapest price, as the investment pays off in less than 10 years since installation thereof. While turbines installed in the sea are more expensive and the care for them, as well as their maintenance is more expensive, their power generation properties are also better, because wind blows on the surface of water bodies almost all year round.

Wind turbines are usually considered to be the least predictable method of generating power, since the output of power by one turbine can change even within the limits of 10% over a period of an hour, which can cause different problems. Usually wind turbine parks are developed in a wide area and consist of multiple individual turbines, which makes the total amount of generated wind power more predictable since changes in weather conditions can be predicted a day in advance, thus the total electricity consumption in the region is increased or reduced and the electricity is transferred via the grid to other areas. Wind power can be successfully linked with hydroelectric plants, which can control the generated electric power.

Despite the drawbacks and advantages of wind installations, their use benefits the surrounding environment. Wind energy is a limitless resource that does not produce hazardous waste. Installation of such stationary structures in the sea requires respect for the shipping industry, which is already operating in the seas, meaning that safe distances to wind turbines must be determined to ensure that in the event of wind energy structures falling into water, they would not endanger nearby ships.

2 Safe distance of shipping routes from stationary sea structures

Marine Spatial Planning (MSP) will become an increasingly important issue for the shipping sector over the next few years. Maritime professionals need to engage with other users of waterways space, from both a sea and shore perspective, and to take part in international, regional, national and local MSP debates, to ensure that the needs of the shipping sector are taken into full consideration and that the sector understands the needs of other marine users and resources.

MSP has the potential to address the impacts of all activities in a specific place, so that marine ecosystems can be productive, resilient to change, and accommodate appropriate, responsible economic activities. Several countries are developing MSP approaches to address fragmented management schemes that do not adequately tackle the complex interactions of the myriad activities that occur simultaneously on and in waterways.

Growth in the world economy is expected to result in an increase in ship traffic in certain areas, all in decreasingly available sea space. In addition, there will be challenges for such waters from industries such as oil an gas, offshore renewable energy, commercial fishing, recreational craft, aggregate dredging, mining, fish farms and government imposed restricted areas. [6]

Several provisions, which were adopted at a time when technology was not advanced to the contemporary level, must be observed in order to determine the safe distance to stationary constructions, which among others includes wind turbines.

The following regulations and guidelines have been developed on the international level:

- The General Provisions on Ships' Routeing (GPSR) adopted by International Maritime Organisation in 1974 [7]
- The United Nations Convention on the Law of the Sea (UNCLOS) adopted by the United Nations in 1982 [9]
- International Regulations for Preventing Collisions at Sea (COLREG) adopted 1972 [2]

✓ GPSR 6.4 Course alteration along a route should be as few as possible and should be avoided in the approaches to convergence areas and route junctions or where crossing traffic may be expected to be heavy.

The captain must keep a safe distance from stationary structures. Meanwhile, stationary structures may not be installed in the sea in a manner that forces ships to change their course to keep a safe distance.

✓ GPSR 6.8 Traffic separation schemes shall be designed so as to enable ships using them to fully comply at all times with the International Regulations for Preventing Collisions at Sea (COLREG), adopted in 1972.

Safe distance to the constructions must be determined in the manner that enables the ship to operate in compliance with COLREG at all times – including cases, where the shipping occurs on the delimiting lines.

✓ GPSR 6.10 Traffic lanes should be designed to make optimum use of available depths of water and the safe navigable areas, taking into account the maximum depth of water attainable along the entire length of the route. The width of lanes should take account of the traffic density, the general usage of the area and the sea-room available.

✓ UNCLOS Section 6, article 60. All ships must respect these safety zones and shall comply with generally accepted international standards regarding navigation in the vicinity of artificial islands, installations, structures and safety zones.

The safety zone of 500 m described in Section 6 is intended for the protection of the structure and not safe manoeuvring in accordance with COLREGs.

✓ COLREGS 8 Action taken to avoid collision with another vessel shall be such as to result in passing at a safe distance. The effectiveness of the action shall be carefully checked until the other vessel is finally past and clear.

There are no regulations pertaining to anchoring sites. However, sufficient manoeuvring area must be ensured for safe anchoring: when the anchor is sliding, while getting closer to anchoring sites. The safety research of coast platforms demonstrates that the area that is required for a ship to start the engines and manoeuvre, when the anchor is dragged, is 1.7 sea miles from the safety zone around stationary structures. It has been detected that such distance is sufficient, as ships that us this particular area of water for anchoring are approaching [10].

Standard turning circles for vessels are six times the ship's length. This is a particularly good assumption for vessels on ocean or deep sea passage, which will not have the same manoeuvrability as when engines and systems are prepared for port approach. Requirements for stopping in an emergency must be considered, for example in case of a steering gear failure. The crash stop distance for a large tanker may be up to 3km.

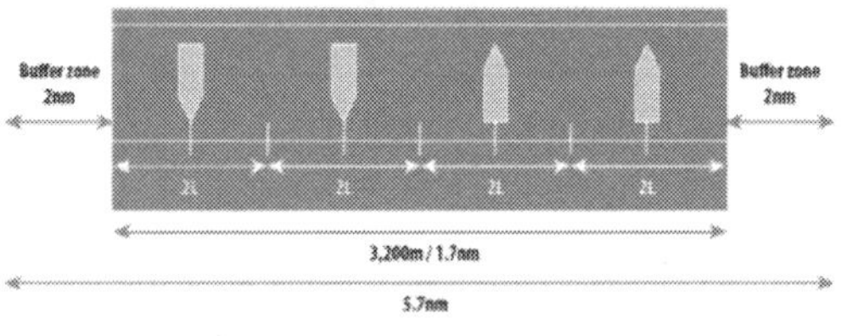

Fig. 1. An example of recommended minimum shipping lane width between two adjacent windfarms considering vessels of 400m in len [6].

It is important to keep a safe distance between shipping corridors and sea wind parks in order to coordinate the interests of shipping and energy sectors. Currently no international document that regulates mandatory and safe distances is in effect. As sea wind parks are developing worldwide, both developers of wind parks, as well as organisations involved in the shipping sector, have to develop a document on an international level, which would regulate safe distances between stationary structures in the sea and a shipping route. The document could be used as a guideline when selecting the potential wind park sites and regulations for ships sailing in the seas close to wind parks.

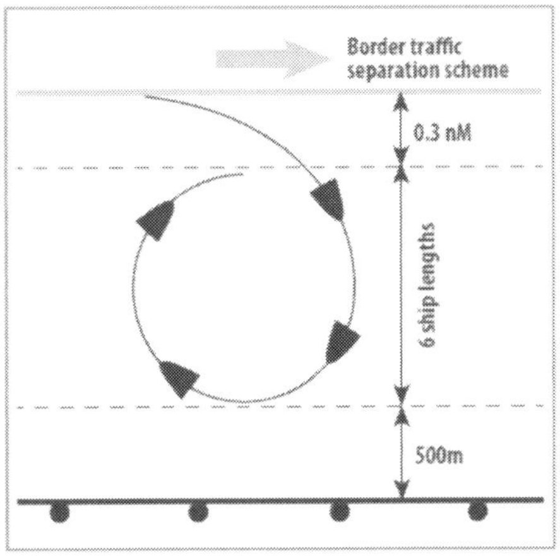

Fig. 2. The width of the shipping route to turn the ship in the opposite direction (SEE COLREGS 8, P13 [2;6]).

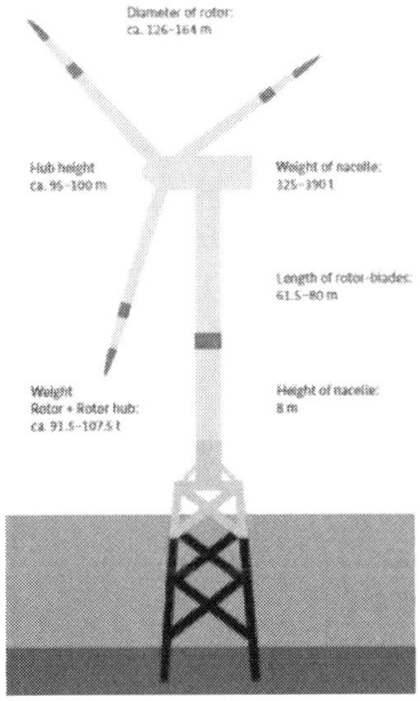

Fig. 3. Standard wind turbine size in Baltic sea region [8].

In the future, greater demands for ships to navigate closer to navigational hazards while ensuring high levels of safety may require new services and technology, in which case serious consideration will need to be given to issues of authority and liability. Evolving navigation technology may provide greater reliability and accuracy of automated electronic position fixing systems. Cheaper communication with greater bandwidth may lead to better provision of critical information and decision support tools for the navigator. Increased traffic density in increasingly constricted water space may require isolation zones for different ocean users such as commercial shipping, fishing and leisure craft [5].

The main components of a wind turbine are the foundation of the tower, the nacelle and the rotor. Although the construction of the turbines is the same on land and at sea, some components differ depending on the varying conditions.

The tower is the longest part of a wind turbine. It supports the nacelle and the rotor. The nacelle houses the components required for the operation of a plant, such as the main bearings, gearboxes, generators, the drive shaft and the control and security systems. The hub and three rotor blades are part of the rotor. The former connects the rotor blades to the rest of the machine and transfers the energy that is generated. The weight of a wind power plant often exceeds 1,000 tonnes.

These days, many operators are planning plants with a capacity of five to seven megawatts. Due to the higher investment costs on the high seas, these systems are installed with a stronger capacity, to generate more energy from the wind. Several manufacturers are working to increase power capacity to further raise energy production between ten and fifteen megawatts per plant. As the capacity increases, the diameter of the rotor becomes larger as well. Rotors used in German projects currently span between 107 and 126 metres, while the latest plants boast rotor diameters of 150 metres.

In order to determine the area required for the falling of one wind turbine during an accident, it is important to consider the height of the wind turbine and calculate the complete perimeter, or the area of the circle that must be allocated for the possible falling of a single wind turbine into the water.

In order to perform such calculations, the area of a circle or one wind turbine safety area, which is calculated in accordance with the formula:

$$S = \pi R^2 \qquad (1)$$

where:

R is the radius of the circle, in this case R= the height of the wind turbine and $\pi \approx 3.14$

Assuming that a safe distance of shipping from stationary structures is the height of the wind turbine R + 500m. The direction of the falling of the wind turbine during an accident cannot be predicted, therefore a circular line around the turbine with the radius equal to the height of the turbine must be drawn.

If the safe perimeter for sailing around one wind turbine, as well as the area of the potential wind park in m^2 is determined, the calculations can be done regarding the quantity of wind turbines that can be installed at safe distances.

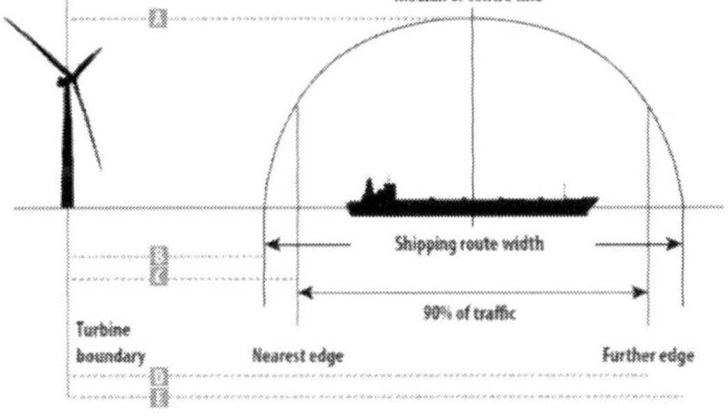

Fig. 4. Wind park safe distance from the shipping route [6].

A - turbine boundary to the shipping route median or centre line;

B - turbine boundary to nearest shipping route edge;

C - turbine boundary to nearest shipping 90% traffic level;

D - turbine boundary to further shipping 90% traffic level;

E - turbine boundary to further shipping route edge.

The number of wind turbines in the wind park can be calculated in accordance with the following formula:

$$W_T = \frac{S_W}{\pi(R+k \geq 500\ m)^2} \quad (2)$$

where:

W_T - the number of wind turbines in the wind park,

R – the height of a wind turbine,

S_W – the area of the potential wind park m^2,

k – the safe distance of shipping from stationary structures in the sea,

$\pi \approx 3.14$

2. the formula has been derived by the author.

The performance of such calculations, while performing research and determining the area that may be allocated to the potential wind park in the sea, enable the planners of a sea area to calculate how many wind turbines can be placed in the planned wind park at safe distance from shipping routes.

As the world presses for greater use of the world's waterways, within a framework of sustainability and economic growth, it is inevitable that established commercial shipping operations will be challenged. Society will need to manage the demands of the multitude of stakeholders all wishing/demanding to use inland waters, coastal and ocean pace. Within the embryonic process of MSP, the many stakeholders involved won't always understand the needs and operational requirements of other stakeholders and the impact changes will have upon them and the society that they support.

This case study demonstrates that our understanding of how maritime activities affect marine ecosystem functioning is not yet sufficient for MSP to resolve conflicts. It also highlights potential interactions between the MSFD and other EU policies such as the EU Renewable Energy Sources Directive, the Common Fisheries Policy, Blue Growth, and the proposed MSP directive. In contrast to the first case study, OWF is in direct competition with other maritime activities for space and, while many environmental effects are on-site, there are a number of potential effects at much larger spatial, including transboundary, and temporal scales.

3 Conclusions

1. The captain of the ship must keep a safe distance from stationary structures in the sea. Meanwhile, stationary structures may not be installed in the sea in a manner that forces ships to change their course to keep such a safe distance. It is important to keep a safe distance between shipping corridors and sea wind parks in order to coordinate the interests of the shipping and energy sectors. The regulatory enactments that are currently effective have been drawn up for the needs of 1972 - 1982, when the shipping intensity and the proportion of stationary structures considerably differed from current figures.

2. Currently there is no international document that regulates the mandatory and safe distance of shipping routes from stationary structures in the sea. As sea wind parks are developing worldwide, both developers of wind parks, as well as organisations involved in the shipping sector, have to develop a document on an international level which would regulate the safe distance between stationary structures in the sea and shipping routes.

3. In order to determine the area required for the falling of one wind turbine during an accident, it is important to consider the height of the wind turbine and calculate the complete perimeter, or the area of the circle that must be allocated for the possible falling of a single wind turbine into the water. The author proposes to use the formula for the determining the area of a circle for such calculations. Assuming that a safe distance of shipping from stationary structures is the height of the wind turbine R + 500m. The direction of the falling of the wind turbine during an accident cannot be predicted, therefore a circular line around the turbine with the radius equal to the height of the turbine must be drawn.

4. The author has deducted her own formula for the planners of sea areas that enables them to calculate how many wind turbines can be placed in the planned wind park at a safe distance from shipping routes, while performing research and determining the area that may be allocated to potential wind parks in the sea.

Acknowledgment

ERDF project No. 1.1.1.2/VIAA/1/16/061 "Long-term impact of blue economy on the increase of the competitiveness of ports in the Baltic Sea region"

References

1. Baltic Sea Region Energy Co-operation analysis "Conditions for deployment of wind power in the Baltic Sea Region"/ [online] [26.02.2018]. Available at: http://basrec.net/wp-content/uploads/2013/09/BASREC-wind-2_strategic-outline_120424.pdf
2. COLREGS - International Regulations for Preventing Collisions at Sea, [online] [10.03.2018]. Available at: http://www.jag.navy.mil/distrib/instructions/COLREG-1972.pdf
3. Dahl K, Mohn C., Maritime Spatial Planning supported by systematic site selection: Applying Marxan for offshore wind power in the western Baltic Sea [online] [15.03.2018] Available at: http://journals.plos.org/plosone/article?id=10.1371/journal.pone.0194362
4. Dirba J., Levins N., Pugacevs V., Wind Energy, Riga: Riga Tehnical University, 2006., p.310.

5. Gilbert A.J., Alexander K., Marine spatial planning and Good Environmental Status: a perspective on spatial and temporal dimensions, 2015, Journal Ecology and Society Vol.20., No.1.
6. Patraiko D. The Shipping Industry and Marine Spatial Planning, 2013 The Nautical Instituts pp.1-14
7. Ships' routeing, Regulation V/10, 1974 SOLAS Convention, [online] [04.03.2018]. Available at: http://www.imo.org/en/OurWork/Safety/Navigation/Documents/Regulation%20V-10And11-ShipsRouteingAndShipReportingSystems.pdf
8. The energy transition – a great piece of work Offshore wind energy - An overview of activities in Germany, [online] [10.06.2018]. Available at: https://www.erneuerbare-energien.de/EE/Redaktion/DE/Downloads/offshore-wind-energy.pdf%3F__blob%3DpublicationFile%26v%3D2
9. United Nations Convention on the Law of the Sea, [online] [21.02.2018]. Available at: http://www.un.org/depts/los/convention_agreements/texts/unclos/unclos_e.pdf
10. Vattenfall's Lillgrund Wind Power Plant Technical Description / [online] [12.03.2018]. Available at: https://corporate.vattenfall.se/globalassets/sverige/om-vattenfall/om-oss/var-verksamhet/vindkraft/lillgrund/technical_description_lillgrund_11336934.pdf

Prospects of Trilateral Flash Cycle (TFC) for Power Generation from Low Grade Heat Sources

Md Arbab Iqbal , *Sohel* Rana, *Mahdi* Ahmadi, *Thomas* Close, *Abhijit* Date, *Aliakbar* Akbarzadeh

School of Mechanical and Automotive Engineering, RMIT University, Melbourne, Australia

Abstract. Despite the current energy crisis, a large amount of low grade heat (below 100°C) is being wasted for the lack of cost effective energy conversion technology. In the case of the conventional Organic Rankine Cycle (ORC) based geothermal power stations, only about 20% of available heat can be utilised due to a technological limitation as there is a phase change in the working fluid involved during the addition of heat which decreases utilisation effectiveness of the system. Therefore, in this paper, a trilateral flash cycle (TFC) based system has been studied to find out its prospect for utilizing more power from the same heat resources as the ORC. The TFC is a thermodynamic cycle that heats the working fluid as a saturated liquid from which it starts its expansion stage. The flash expansion is achieved by feeding the saturated high-pressured liquid working fluid through a convergent-divergent nozzle at which point it undergoes a flash expansion in the low-pressure environment of the generator housing. The momentum of the working fluid is extracted via a Pelton wheel and the cycle is completed with working fluid condensation and pressurisation. The analytical comparative study between the ORC and TFC based system shows that the TFC has about 50% more power generation capability and almost zero contribution on global warming.

1 Introduction

Not only the depletion of conventional non-renewable energy sources i.e. fossil fuel, coal etc. but also the growing concern about the environmental changes forcing the researchers to search for alternative sustainable sources of energy as well as develop system for waste energy recovery. Though modern cutting-edge technology improved the efficiency of power generation system, there is still limitation of recovery of waste heat, specially the low-grade heat having temperature lower than 100oC. For the recovery and conversion of low grade heat energy, the Organic Rankine Cycle (ORC) plays a major role for the simple, compact and low-cost system components with small sizing and the properties of organic fluids that can exploit low and variable temperature heat sources [1-6]. However, because of a phase change in the ORC, not only the highest temperature that the working fluid can reach is quite far from the source temperature [7] but also the lowest temperature that the heat source can reach, remains much higher than the cycle lowest temperature. These facts lead to wastage of a big portion of energy.

In the literature several different solutions have been reported to have a better match between the heat transfer curves of the heat source and the heat recovery system. The Kalina cycle (KCS) was introduced in the 1980s which converts thermal energy to mechanical power using mixture of water and ammonia as working fluid. This cycle was introduced to have a better match between the heat transfer curves by using a non-isothermal evaporation thereby reducing the average temperature difference. The layout of the Kalina cycle, however, is much more complex because of additional separators and heat exchangers which associate with additional cost as well.

Bombarda et al. [8] compared the performance of a Kalina cycle with an ORC. Despite the small advantage in terms of net mechanical power, the high working pressure of water-ammonia mixture demonstrated that this cycle was not applicable in the considered temperature range because of the high equipment costs. Pradeep Varma and Srinivas [9] found the Kalina cycle is a very low efficiency system for heat recovery from low temperature heat sources. They recommended the Organic Flash cycle (OFC) over the ORC and KCS for power generation. Baccioli et al. [6] worked with the OFC and Organic Flash Regenerative Cycle (OFRC) and observed improvement of efficiency with the OFRC as it requires a small heat exchanger and one less throttling process than that of OFC which is a source of major losses.Ho et al. [10, 11] worked on the Organic Flash Cycle (OFC) for power generation from low grade heat and recommended a modified 2-phase OFC which demonstrated 20% more power output, though they observed exergy destruction by the throttling valve for flash evaporation.

In 1993 Smith [12] introduced the Trilateral Flash Cycle (TFC) which is reported to be 14-85% more efficient for low grade heat recovery than the ORC [13, 14].

Figure 1 represents a typical comparison between the ORC and TLC and it is clear from the figure that in the TLC, heat is transferred in a single phase working fluid which is more efficient than multiphase heat transfer like in the ORC.

Basically, the TLC has the same components as the Rankine-cycle engines but unlike the Rankine cycle, it does not evaporate the working fluid during the heating phase; instead expands it from the saturated liquid condition, as a two-phase mixture. The choice of a suitable working fluid for a given application, the feasible operating conditions/ parameters and the system configuration design, are the most essential features of any thermal system design [15-17].

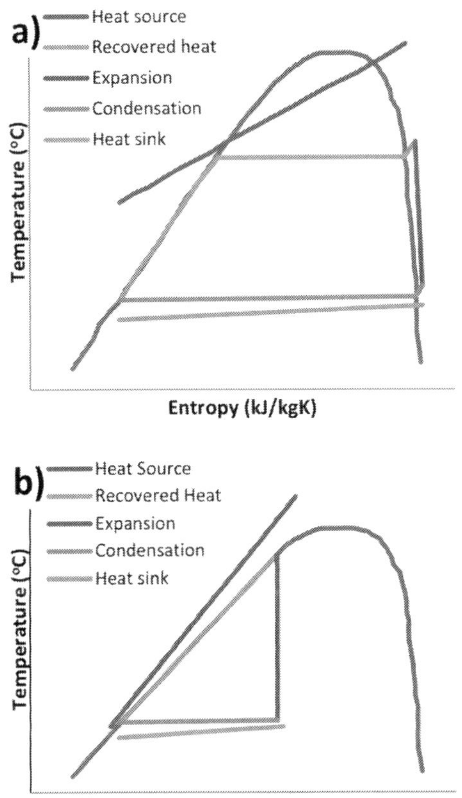

Fig. 1. a) T-s diagram for ORC, b) T-s diagram for TLC.

With the application of a suitable high molecular working fluid, the TFC power plants could recover and convert heat energy efficiently from renewable thermal sources as well as from the by product (waste heat) of numerous non-renewable sources into mechanical or electrical power. The high molecular mass allows efficient exploitation of non-isothermal single-phase heat sources for heat recovery-to-power generation, which allows the net output work of the TLC in a wide range of power capacities [18, 19].

The working fluid adopted for the study is iso-pentane because of its good thermo-physical properties (e.g. relative high critical temperature and pressure), low-cost, and a boiling point slightly above room temperature. It displays a strong positive slope on the T–s diagram and its saturated liquid expansion tends to dry out at temperatures slightly exceeding 453 K [20]. More so, iso-pentane is a dry fluid, whose thermo-physical properties are well-suited for low-grade heat recovery to-power generation and widely used around the world.

In this paper, we would like to study on the technological feasibility of the Trilateral Flash Cycle (TFC) for low grade heat to power generation where we can eliminate the limitations of the ORC and improve overall cycle power output. Here we considered the Birdsville Organic Rankine Cycle (ORC) geothermal power station as our case of study. Figure 2 represents the schematic layout of Birdsville geothermal power station where we can see a larger amount of heat is being wasted as the plant leaving bore water to lagoon at 80°C which carries about 80% of available geothermal energy. For this study, we will simulate the Birdsville ORC system as well as a TFC system for comparative performance analysis and discuss on the prospects of the TFC over the ORC.

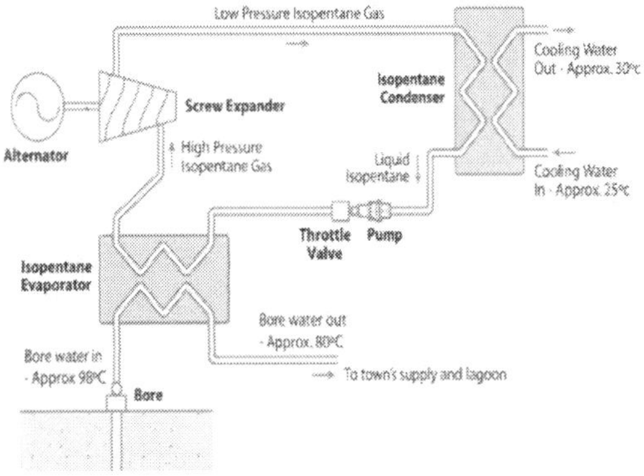

Fig. 2. ORC geothermal power station, Birdsville, Queensland, Australia.

2 Theoretical modelling

2.1 ORC

ORC is the widely used power cycle for waste heat or low-grade heat to power generation. Figure 1.a shows a typical T-s diagram for ORC which complies with pumping, heating, expansion (work out) and heat rejection (condensation).

The heating load of an ORC system can be measured by,

$$Q_{in} = \dot{m}\,(h_3 - h_2) \tag{1}$$

Where, Q_{in} and $\dot{m}$ are heat-in the system and mass flow rate of the working fluid and h_2 and h_3 are enthalpy of working fluid after pump and before expander respectively. If the enthalpy of the working fluid after expander is h_4, the work done by the expander W_{out} can be expressed by,

$$W_{out} = \dot{m}(h_3 - h_4) \quad (2)$$

The cooling load, Q_{out}, and pumping power, P_{in}, can be measured by,

$$Q_{out} = \dot{m}\,(h_4 - h_1) \quad (3)$$

$$P_{in} = \Delta P \times \dot{V} \quad (4)$$

Where, h_1 is the enthalpy of the working fluid at pump inlet, ΔP is the pressure difference between pump inlet and outlet and the $\dot{V}$ is the volumetic flow rate of the working fluid.

2.2 TFC

Figure 1.b represents the T-s diagram of a typical TFC system which is nothing but a modified ORC system where the phase change of the working fluid in the heat exchanger has been skipped. However, in TFC, the expansion is two phase flash expansion rather than single phase expansion of ORC.

In the proposed system, a convergent-divergent (CD) nozzle will aid the flash expansion in the form of high velocity jet of mixture of liquid and vapour. The isentropic quality of jet is the governing operator of the isentropic efficiency of the nozzle. To predict the performance of such kind of TFC system, let us consider, the hot water of T_h temperature from the heat source with a flow rate of m_{hw} is passing through the heat exchanger having effectiveness of $\mathcal{E}_h$ where the iso-pentane is to be heated up. If P_l is considered as the initial condenser pressure and T_{pl} is the corresponding saturated liquid iso-pentane temperature in the condenser, the hot iso-pentane temperature, T_{ph} can be defined as,

$$T_{ph} = \mathcal{E}_h(T_h - T_{pl}) + T_{pl} \quad (5)$$

Maximum heating power can be defined as,

$$H_{max} = m_{hw}C_{pw}(T_h - T_{ph})\mathcal{E}_h \quad (6)$$

Here, C_{pw} is the specific heat of liquid hot water.
The maximum iso-pentane flow rate can be defined as,

$$m_p = \frac{H_{max}}{(T_{ph}-T_{pl})C_{pp}} \quad (7)$$

The isentropic quality at the nozzle exit can be defines as,

$$x_4 = \frac{S_{f,3}-S_{f,4}}{S_{fg,4}} \quad (8)$$

Here, $S_{f,3}$, $S_{f,4}$ and $S_{fg,4}$ are the entropy of the fluid at the nozzle inlet, the entropy of the saturated liquid at nozzle exit conditions and the entropy difference of saturated liquid and vapour at nozzle exit, respectively.
The isentropic enthalpy at the nozzle exit can be defines as,

$$h_4 = h_{f,4} + x_4 h_{fg,4} \quad (9)$$

Here, $h_{f,4}$ and $h_{fg,4}$ are the enthalpy of the fluid at the nozzle exit and the enthalpy difference of saturated liquid and vapour at nozzle exit, respectively.

Now, the isentropic nozzle exit velocity (neglecting inlet velocity and change in potential energy) can be given by,

$$u_4 = \sqrt{2(h_3 - h_4)} \quad (10)$$

The force developed by the nozzle can be given by,

$$F_n = m_p u_4 \quad (11)$$

The work done or power developed by the nozzle $W_{out,N}$ can be expressed by,

$$W_{out,N} = 0.5 \times \dot{m} \times u_4{}^2 \quad (12)$$

Or,

$$W_{out,N} = \dot{m}(h_3 - h_4) \quad (13)$$

3 Proposed TFC system

The TFC implements a heat to power conversion via heat transfer to a pressurised working liquid of low boiling point which is then injected at an optimum temperature (saturation temperature) through a nozzle into a turbine fed generator thus outputting usable electrical energy. The used liquid then returns to a reservoir before it is condensed and forced through the same process again via a high-pressure pump.

4 Prospect analysis

Here we will analyse the prospect of the TFC over the ORC on the basis of net power generation, wastage of heat energy, different component size and environmental impact. From figure 3, the combine T-s diagram of the ORC and TFC, we can see that the TFC has more power output capability than that of the ORC for the same heat source and heat sink though the TFC has a lower thermodynamic efficiency. This phenomenon can be described by the term utilization of energy (UOE) which is the product of co-efficient of utilization (COU) and the efficiency of the thermodynamic working cycle. COU is the ratio of maximum possible available energy that can be recovered by the cycle to the maximum available recoverable energy that is being wasted. If maximum heat source temperature is T_{max} and heat sink temperature is T_{min} and the temperature of the heat carrier leaving the system is T_{out}, the COU can be expressed by the following equation,

$$COU_{system} = \frac{\Delta T_{system}}{\Delta T_{max}} \quad (14)$$

Where,

$$\Delta T_{system} = T_{max} - T_{out}$$

and

$$\Delta T_{max} = T_{max} - T_{min}$$

And if the system thermodynamic is efficiency is η_{system}, the UOE can be expressed by the following equation,

$$UOE_{system} = COU_{system} \times \eta_{system} \quad (15)$$

Figure 4 represents the variation of UOE of the ORC and TFC based system for different heat source and heat sink temperature. For the ORC system, a sharp drop of UOE observed for heat source temperature 92°C to 90°C as there were some changes in constrains to maintain a defined minimum pressure difference (1.7 bar) across the expender. It is also clear from figure 4 that as the heat source temperature decreases, the UOE of both systems falls down and below 90°C, the ORC system is not economically viable.

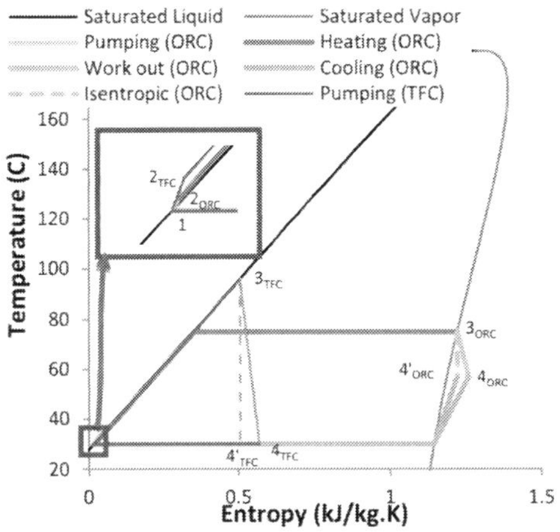

Fig. 3. T-s diagram for TFC and ORC.

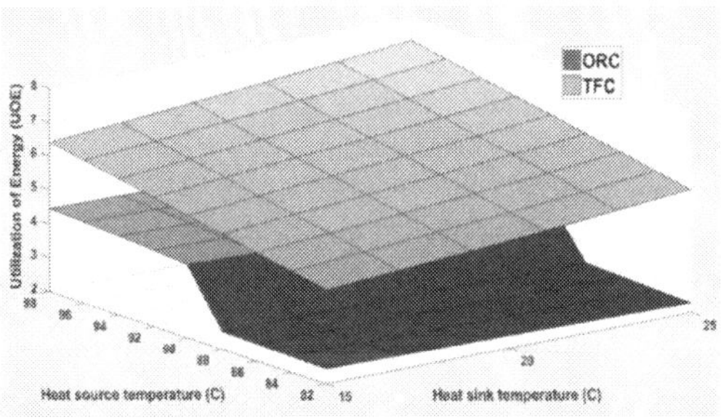

Fig. 4. UOE for ORC and TFC for different heat source and heat sink.

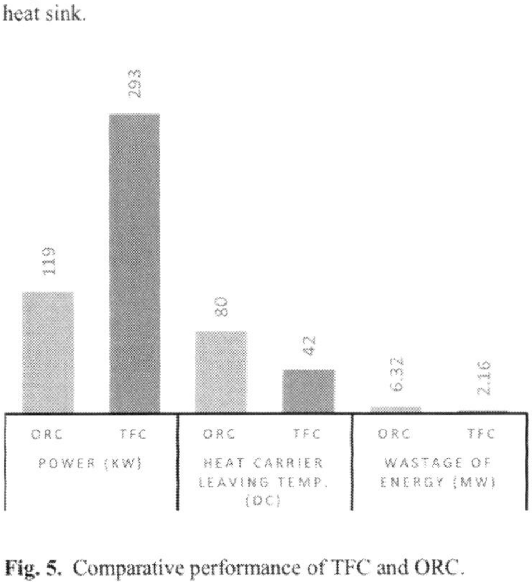

Fig. 5. Comparative performance of TFC and ORC.

Figure 5 shows that in using the TFC instead of the ORC, 174 kW more net power can be produced from the existing Birdsville geothermal resources. Moreover, the TFC system can facilitate to cool down the system leaving hot water temperature close to ambient temperature and reduces overall wastage of energy.

Figure 6 represents the comparison of different system component sizing for the ORC and TFC. It is obvious that the TFC system requires a larger heat exchanger and condenser as it deals with larger amounts of heat than that of the ORC, however, it also requires more pumping power which basically reflects more initial investment which can be compensated by producing additional power. The power developed by the proposed TFC system is directly proportional to the nozzle isentropic efficiency which is presented in figure 7. Though it not practical to achieve 100% efficiency, about 80% efficiency in the nozzle can develop twice as much power than the ORC based system and achieving such a high level of efficiency in the nozzle is hard but not impossible. However, a 65% efficient nozzle which can easily be achieved can produce about 50% more power than the ORC.

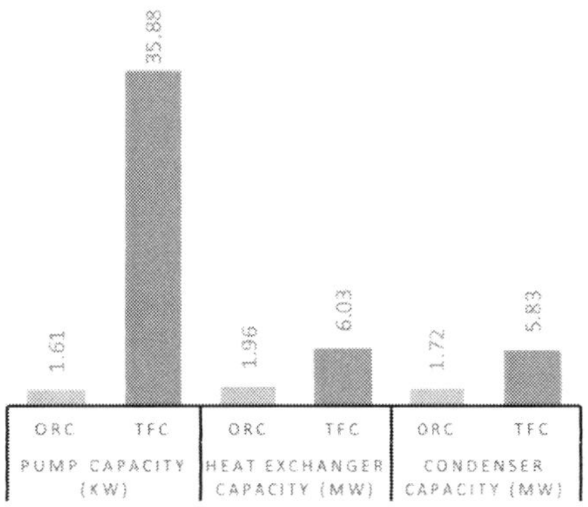

Fig. 6. Comparative component sizing for TFC and ORC.

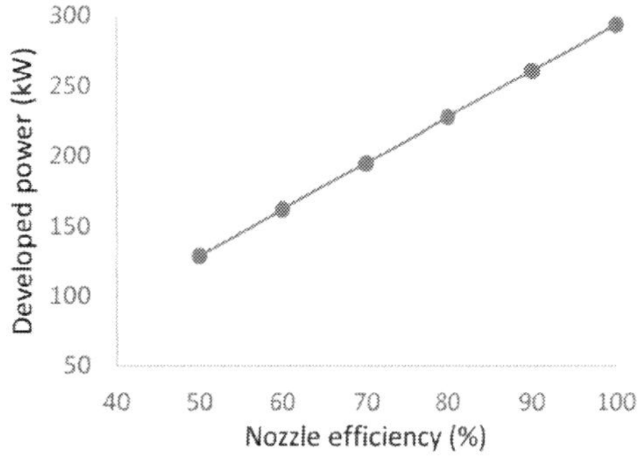

Fig. 7. Variation of developed power with nozzle efficiency.

5 Conclusion

In context of the current energy crisis, wasting energy especially low-grade heat (<100°C) is the major concern and the TFC system can be a promising technology for these waste heat recoveries and power generation. The analytical analysis of this paper shows that replacement of the ORC based current Birdsville geothermal power generation station by a TFC based system can produce more power. The results of this study can be summarised as follows,

- About 50% more power can be produced by the TFC based system than the ORC based system
- for the same heat source and heat sink conditions.
- The TFC system can utilise up to 70% of available thermal power.
- The TFC system can produce a significant amount of power from very low temperature
- heat sources (below 80oC) where with the ORC it is not feasible.
- The TFC system allows the heat carrier to leave the system very close to ambient temperature that contribute almost zero to global warming.

References

1. Liu, B.-T., K.-H. Chien, and C.-C. Wang, *Effect of working fluids on organic Rankine cycle for waste heat recovery.* Energy, 2004. **29**(8): p. 1207-1217.
2. Jing, L., P. Gang, and J. Jie, *Optimization of low temperature solar thermal electric generation with Organic Rankine Cycle in different areas.* Applied Energy, 2010. **87**(11): p. 3355-3365.
3. Pei, G., J. Li, and J. Ji, *Analysis of low temperature solar thermal electric generation using regenerative Organic Rankine Cycle.* Applied Thermal Engineering, 2010. **30**(8): p. 998-1004.
4. Roy, J.P., M.K. Mishra, and A. Misra, *Parametric optimization and performance analysis of a waste heat recovery system using Organic Rankine Cycle.* Energy, 2010. **35**(12): p. 5049-5062.
5. Lakew, A.A. and O. Bolland, *Working fluids for low-temperature heat source.* Applied Thermal Engineering, 2010. **30**(10): p. 1262-1268.
6. Baccioli, A., M. Antonelli, and U. Desideri, *Technical and economic analysis of organic flash regenerative cycles (OFRCs) for low temperature waste heat recovery.* Applied Energy, 2017. **199**: p. 69-87.
7. Nguyen, T.Q., J.D. Slawnwhite, and K.G. Boulama, *Power generation from residual industrial heat.* Energy Conversion and Management, 2010. **51**(11): p. 2220-2229.
8. Bombarda, P., C.M. Invernizzi, and C. Pietra, *Heat recovery from Diesel engines: A thermodynamic comparison between Kalina and ORC cycles.* Applied Thermal Engineering, 2010. **30**(2): p. 212-219.
9. Varma, G.V.P. and T. Srinivas, *Power generation from low temperature heat recovery.* Renewable and Sustainable Energy Reviews, 2017. **75**: p. 402-414.
10. Ho, T., S.S. Mao, and R. Greif, *Comparison of the Organic Flash Cycle (OFC) to other advanced vapor cycles for intermediate and high temperature waste heat reclamation and solar thermal energy.* Energy, 2012. **42**(1): p. 213-223.
11. Ho, T., S.S. Mao, and R. Greif, *Increased power production through enhancements to the Organic Flash Cycle (OFC).* Energy, 2012. **45**(1): p. 686-695.
12. Smith, I.K., *Development of the trilateral flash cycle system Part 1: fundamental consideration.* Proceedings of the Institution of Mechanical Engineers, Part A: Journal of Power and Energy, 1993. **207**(3): p. 179-194.
13. Smith, I., N. Stosic, and C. Aldis, *Trilateral flash cycle system a high efficiency power plant for liquid resources.* Proceedings World Geothermal Congress 1995, 1995: p. 2109.
14. Fischer, J., *Comparison of trilateral cycles and organic Rankine cycles.* Energy, 2011. **36**(10): p. 6208-6219.
15. Dincer, I. and H. Al-Muslim, *Thermodynamic analysis of reheat cycle steam power plants.* Fuel and Energy Abstracts, 2002. **43**(4): p. 264.
16. Chen, H., et al., *A supercritical Rankine cycle using zeotropic mixture working fluids for the conversion of low-grade heat into power.* Energy, 2011. **36**(1): p. 549-555.
17. Sahin, A.Z., et al., *Special Issue: Thermodynamic Optimization, Exergy Analysis, and Constructal Design.* Arabian Journal for Science and Engineering, 2013. **38**(2): p. 219-219.
18. Invernizzi, C., P. Iora, and P. Silva, *Bottoming micro-Rankine cycles for micro-gas turbines.* Applied Thermal Engineering, 2007. **27**(1): p. 100-110.
19. Baral, S. and K.C. Kim, *Thermodynamic Modeling of the Solar Organic Rankine Cycle with Selected Organic Working Fluids for Cogeneration.* Distributed Generation & Alternative Energy Journal, 2014. **29**(3): p. 7-34.
20. Smith, I.K. and R.P.M. da Silva, *Development Of The Trilateral Flash Cycle System Part 2: Increasing Power Output With Working Fluid Mixtures.* Proceedings of the Institution of Mechanical Engineers, Part A: Journal of Power and Energy 1990-1996 1994. **208**: p. 135-144.

https://doi.org/10.1051/e3sconf/20186406005

Power generation from low grade waste heat using thermoelectric generator

Sohel Rana[1], *Arbab* Iqbal[1], *Abhijit* Date[1], *Aliakbar* Akbarzadeh[1]

[1]Energy Conservation and Renewable Energy, School of Mechanical and Automotive Engineering, RMIT University, Bundoora East Campus-3083, Australia

Abstract. Thermoelectric technology is thought to be a great solution in near future for producing electrical power and recovering low grade waste heat to cut the cost of power generation because of its consistency and eco-friendly affability. Though commercial accessibility of TEG is available currently but heat to electricity conversion efficiency is still low and cost of the module is reasonably high. It's essential to use the modules competently which is strongly depends on suitable heat exchanger design and selection of proper operating conditions. In this work, TEG module has been selected from the commercially available modules with efficiency of 1.91% for the targeted low-grade waste heat temperature of T_h=90°C and T_c=15°C which validated by experiment. Mathematical model has been proposed to simulate TEG based power generation system; the model can predict maximum net power, choose optimum operating conditions and dimensions of heat exchanger. Lab scale design with channel length 1 m, width 0.08 m and gap size 9 mm which is suitable for 50 TEG module (4 mm x 4 mm) have been simulated using proposed mathematical model. For above temperature range, predicted optimum net power was 76.45 W with optimum flow rate 0.94 L/s (56.4 L/min). This lab scale setup will be used for experimental validation of the proposed mathematical model. The obtained results from experiments and simulation are closely matched.

1 Introduction

In recent years, there are clear attempts to recoup the waste heat and produced electrical power using different source of waste energy such as cars, power plant, telecommunication, industries, automobiles, cooling systems, solar, geothermal, solar ponds and so on. These attempts have been led towards lessening the generation cost of electricity and diminishing the air contamination worldwide. Though, petroleum is still principle hotspot for producing electrical power. I major parts of this consumed energy is wasted as heat [1, 2].

Based on temperature, waste heat can be classified into low grade (below 230°C), medium (between 230°C to 630°C) and high temperature (above 630°C) and a typical example of waste heat and its work potential has shown in figure 1 [3-6].

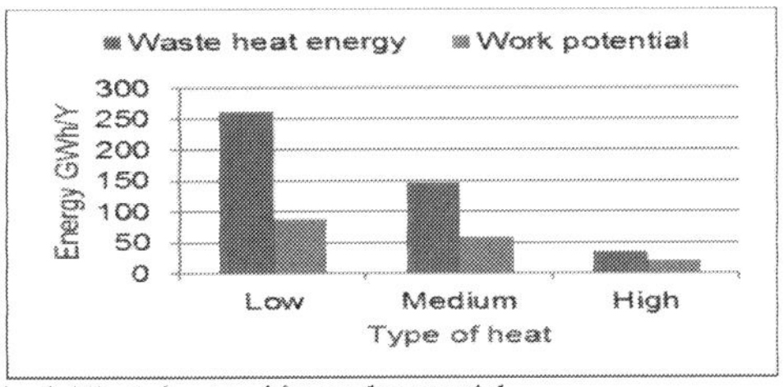

Fig. 1. Waste heat and its work potential

There are several technologies can be found to generate electricity from low grade waste heat such as organic Rankine cycle, thermoelectric, water desalination and piezoelectric. Thermoelectric technology is supposed to be an alternative way to recover low grade waste heat; which can convert heat to electrical power and vice versa. Though thermoelectric has low conversion efficiency and costly; it has no moving parts and does not require maintenance which attracting to use this technology.

Low grade waste heat recovery using thermoelectric has been studied theoretically as well as experimentally from last few decades. From literature, some important related findings have been studied and summary has been listed in Table 1.

Table 1: Summary from related literature

Author(s) & year	Type of work	Hot side & cold side temperature	TEG module number	Achievements
Rowe et al (1997)[7]	Exp	Hot water=90°C, Cold water=14°C	36	Max. power =95W
Crane et al (2004)[8]	Exp	Hot water=90°C, Cold air=25°C	6	Max. power=0.54 W
Niu et al (2009)[9]	Exp	Hot fluid=90°C, Cold fluid=25°C	56	Max power =146.5W
Gou et al (2010)[10]	Exp	Hot water=80°C, Cold air=24°C	10	Max. power =0.85W

 https://doi.org/10.1051/e3sconf/20186406005

Kim et al (2011) [11]	Exp	Hot water =95°C, Cold air=45°C	72	Max. power = 75W
Gou et al (2013)[12]	Exp	Hot water=90°C, Cold air=28.5°C	18	Max. power =6.5W
B. Orr et al (2016)[13]	Exp	Hot exhaust air =218°C, Cold air=31°C	8	Max. power =38W
Remeli et al (2016)[4]	Exp	Hot air =160°C, Cold air=65°C	1	Max. power =1.72W

All these studies were experimental and some of them were able to generate significant amount of electrical power. So, above literature only concentrated on testing commercially available TEG performances; none of current literature make known to an optimization technique to maximize the net electrical power by considering losses due to thermal resistance and parasitic power losses. In this works, a system has considered which consisted thermoelectric modules sandwiched between hot water and cold-water channel. TEG can produced electricity from the temperature difference between hot and cold water. This power generation is increased with the higher temperature difference and higher mass flow rate, but there is a penalty for higher mass flow rate called parasitic loss. This parasitic loss also depends on the size of the heat exchanger. So, the amount of net power/ useful power from this system depends on size of channel, mass flow rate, efficiency of TEG modules, number of modules and parasitic power. For chosen TEG modules, efficiency is fixed and number of TEG has considered also fixed for this study. So, maximum net power depends on proper selection of length, width and gap height of water channel and depends on suitable selection of mass flow rate. The novelty of this study is proposed mathematical model to predict gross power, parasitic power and net power where mass flow rate, gap height, length and width have taken as variables. Number of TEG and its property (such as dimensions, efficiency and thermal resistance) have considered as fixed. As the number of TEG and its dimensions fixed, channel length and width are interrelated [6]. Proposed model will help to predict maximum net power, optimized channel dimensions and suitable flow rate for defined conditions. Predicted results from proposed model will be validated by experiments. This model will be useful for system optimisation.

2 Selection of thermoelectric generator

Based on market research, we have carefully selected five different high performance thermoelectric modules presented in figure 2. These five thermoelectric modules are Seebeck based (generator) have been chosen based on the performance data sheet given by manufacturer and price. A number of experiments have been done using TEG testing setup shown in figure 3. Experiment has been run for similar operating conditions with hot side temperature T_h= 90°C and cold side temperature T_c=15°C and recorded generated power from selected modules. Since module size and price are not identical, we have been selected two properties (kW/m^2, $/Watt) for an uniform comparison. Power generation per square meter (kW/m^2) will give an idea about the power generation capability TEG module and $/W is required to get an idea about power production cost of per unit electricity. These two properties can be calculated from experimentally measured data and module price. Dimensions of five modules, maximum operating temperature, calculated properties for targeted temperature has been shown in Table 2.

Fig. 2. Selected thermoelectric module

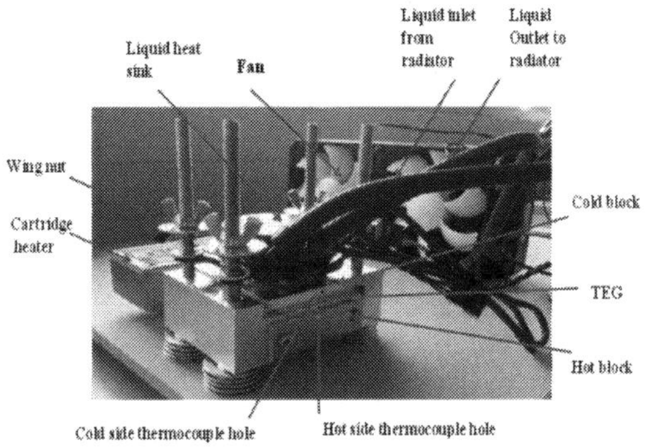

Fig. 3: TEG testing setup

Table 2: Experimental performance of selected TEG

Module	L x W (mm)	h (mm)	T_h(°C)	T_c(°C)	kW/m^2	$/W	Max (ΔT)(°C)
A	30	3.7	90	15	0.73	37	140
B	30	3.8	90	15	0.66	29	140
C	40	3.2	90	15	1.08	12	90
D	55	3.8	90	15	0.63	40	170
E	62	6.3	90	15	0.16	18	270

Another important parameter is efficiency of TEG under the specified operating condition; where TEG efficiency can be defined as the conversion ability of heat to electrical power. Another experiment has been undertaken using TEG testing rig to measure efficiency of selected thermoelectric module. The obtained efficiency for five modules has plotted in figure 3. By analysing table 2 and figure 4, we can conclude that

https://doi.org/10.1051/e3sconf/20186406005

module C is utmost suitable for stated temperature range as higher Power generation per square meter (kW/m^2) and efficiency; and lower power production cost of per unit electricity ($/W)

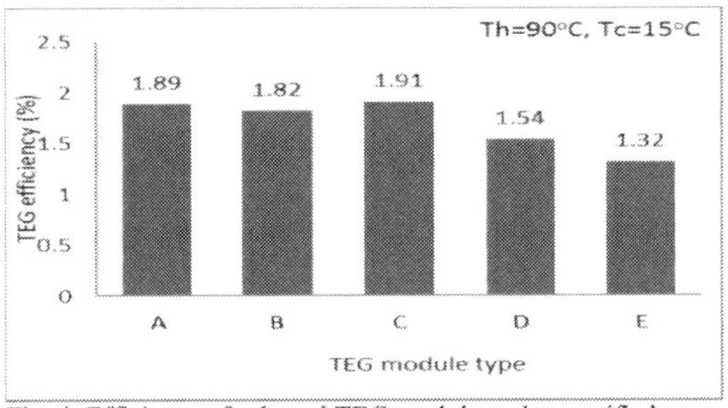

Fig. 4: Efficiency of selected TEG module under specified operating temperature

3 Mathematical modelling

A conceptual model has been shown in figure 5 for power generation from low grade waste heat using TEG. The model contains hot water channel and cold-water channel with counter flow arrangement; thermoelectric module is sandwiched between two channels. Thermal paste has been used between water channel and thermoelectric module to avoid the air gap and to accelerate heat transfer rate. The system is surrounded by thermal insulation to reduce the heat loss.

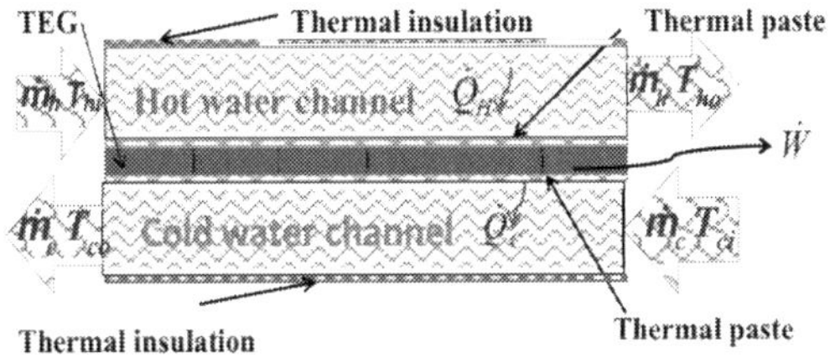

Fig.5: Conceptual design for power generation

Energy balance equation for hot and cold fluids are shown in equation (1) and (2) respectively.

$$\dot{Q}_H = \dot{m}_h c_{ph}(T_{hi} - T_{ho}) \quad (1)$$

$$\dot{Q}_C = \dot{m}_c c_{pc}(T_{co} - T_{ci}) \quad (2)$$

The energy balance equation can be written in following form

$$\dot{Q}_H(d,\dot{m},w) = U(d,\dot{m},w)A(d,w)(T_{hi} - T_{co}) \quad (3)$$

$$\dot{Q}_C(d,\dot{m},w) = U(d,\dot{m},w)A(d,w)(T_{ho} - T_{ci}) \quad (4)$$

U is the overall heat transfer coefficient and A is Surface area. Here heat transfer rate has represented as function of all variables such as channel thickness (d), flow rate ($\dot{m}$) and length/width (w). More explanation can be found in next section on this paper.

Consider amount of power generation by TEG is negligible compared to total of heat transfer in hot or cold-water channel. By equating equation 3 and 4

$$T_{hi} - T_{co} = T_{ho} - T_{ci} \quad (5)$$

On the other hand, from equation (1) and (4)

$$\dot{m}_h c_{ph}(T_{hi} - T_{ho}) = U(d,\dot{m},w)A(d,w)(T_{ho} - T_{ci}) \quad (6)$$

Outlet temperature of hot water channel can be calculated from equation 6 and denoted by equation 7 as function of variables.

$$T_{ho}(d,\dot{m},w) = \frac{\dot{m}_h c_{ph} T_{hi} + U(d,\dot{m},w)A(d,w)T_{ci}}{\dot{m}_h c_{ph} + U(d,\dot{m},w)A(d,w)} \quad (7)$$

Consider hot side flow rate and cold side flow rate is equal ($\dot{m}_h = \dot{m}_c = \dot{m}$) and specific heat is also equal ($c_{ph} = c_{pc} = c_p$). Equation (7) can be written as

$$T_{ho}(d,\dot{m},w) = \frac{\dot{m} c_p T_{hi} + U(d,\dot{m},w)A(d,w)T_{ci}}{\dot{m} c_p + U(d,\dot{m},w)A(d,w)} \quad (8)$$

So, outlet temperature of cold channel can be found from equation (5)

$$T_{co}(d,\dot{m},w) = T_{hi} - T_{ho}(d,\dot{m},w) + T_{ci} \quad (9)$$

Total heat transfer rate through TEG can be found from equations 1 and 8. For simplicity, hot water and cold-water flow rate have considered equal. So, total heat transfer rate in both channels would be identical if heat to electricity conversion has considered as negligible. Total heat transfer rate can be estimated from following equation 10.

$$\dot{Q}_H = \dot{Q}_C = Q(d,\dot{m},w) = \dot{m} c_p \frac{U(d,\dot{m},w)A(d,w)(T_{hi} - T_{ci})}{\dot{m} c_p + U(d,\dot{m},w)A(d,w)} \quad (10)$$

Q amount heat transfer through TEG and a portion of this heat converted to electricity by TEG which depends on the efficiency of thermoelectric module. Power Generated by Thermoelectric generator (TEG) is

$$P_{TEG}(d,\dot{m},w) = Q(d,\dot{m},w)\eta_{TEG} \quad (11)$$

η_{TEG}= Heat to electrical power conversion efficiency of TEG.

3.1 Calculation of overall heat transfer coefficient (U)

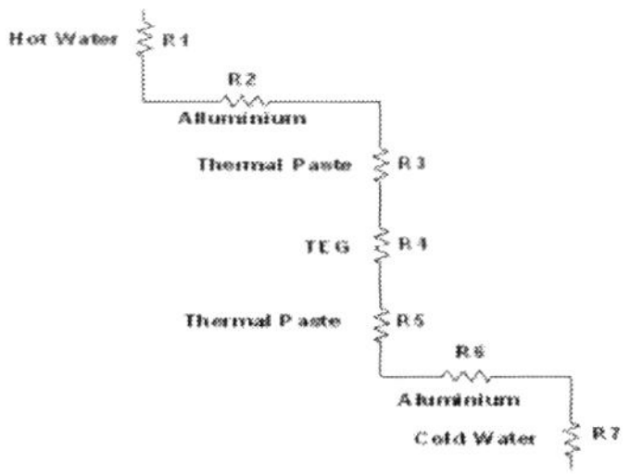

Fig.6: Thermal resistance for one unit

According to figure 5, the thermal resistance for one unit has been plotted in figure 5. Now, total thermal resistance is algebraic sum of all resistances which can be found from following equation 12. So, overall heat transfer coefficient can be defined by equation 13.

$$R_t(d,\dot{m},w) = R_1(d,\dot{m},w) + R_2 + R_3 + R_4 + R_5 + R_6 + R_7(d,\dot{m},w) \quad (12)$$

$$U(d,\dot{m},w) = \frac{1}{R_t(d,\dot{m},w)} \quad (13)$$

Convection resistance of hot water and cold water can be calculated by using following equation;

Equation to calculate Reynolds's number

$$\mathrm{Re}(d,\dot{m},w) = \frac{\rho v(d,\dot{m},w) D(d,w)}{\mu}$$

Reynolds's number is identical for both hot water and cold-water channel as flow rate and dimensions considered as same.

Friction factor;

$$F(d,\dot{m},w) = (0.790 * \log(\mathrm{Re}(d,\dot{m},w)) - 1.64)\text{^}(-2)$$

If 3000 <Re<5x10^5

$$F(d,\dot{m},w) = 64 / \mathrm{Re}(d,\dot{m},w)$$

If Re <3000

Nusselt number can be found by below equations

$$Nu(d,\dot{m},w) = \frac{\frac{F(d,\dot{m},w)}{8}(\mathrm{Re}(d,\dot{m},w) - 1000) * P_r}{1 + 12.7 * (F(d,\dot{m},w)/8)^{0.5}(P_r^{2/3} - 1)}$$

If 3000 <Re<5x10^5

$$Nu = 3.66$$

If Re <3000

So, heat transfer coefficient for hot water and cold water can be calculated by using below equation

$$h(d,\dot{m},w) = \frac{N_u(d,\dot{m},w) K_w}{D}$$

Now convection resistance of hot water and cold water is

$$R_1(d,\dot{m},w) = R_7(d,\dot{m},w) = \frac{1}{h(d,\dot{m},w)}$$

Conductive resistance of aluminium in both hot and cold side can be calculated by following equation

Where, a_t= thickness of aluminium and K_m= thermal conductivity of aluminium.

$$R_2 = R_6 = \frac{a_t}{K_m}$$

3.2 Parasitic Power calculation

In this study, pump will be required to maintain the flow rate in hot and cold-water channel. A significant amount of power is losses due to friction which depends on rate of water flow and the dimensions of channel. Since flow rate and dimensions of both channel considered as same, the parasitic power losses also will be identical; which can be estimated by following equation

$$p_h(d,\dot{m},w) = p_c(d,\dot{m},w) = \frac{\dot{m}}{\rho} \Delta P(d,\dot{m},w) * \eta_p \quad (14)$$

Where ΔP= pressure difference and η_p =Pump efficiency

Total parasitic power is sum of required power for hot and cold-water pumping power

$$P_L(d,\dot{m},w) = p_h(d,\dot{m},w) + p_c(d,\dot{m},w) \quad (15)$$

3.3 Effective net power

Effective net power can be calculated by subtracted parasitic power from total gross power. Net power can be optimized by maximize gross power and reduced parasitic power; both are depends on flow rate and dimensions of channel. So, proper heat exchanger/channel design and selection of appropriate flow rate is imperative to optimize ne power.

$$P_{Net}(d,\dot{m},w) = P_{TEG}(d,\dot{m},w) - P_L(d,\dot{m},w) \quad (16)$$

4 Model validation by experiments

An experimental setup has been prepared to validate the mathematical model. The dimensions of used hot and cold-water channel were identical with length=1m, width=0.08 m and gap height=9 mm which shown in figure 7.

Fig. 7: Size of lab scale hot and cold-water channel

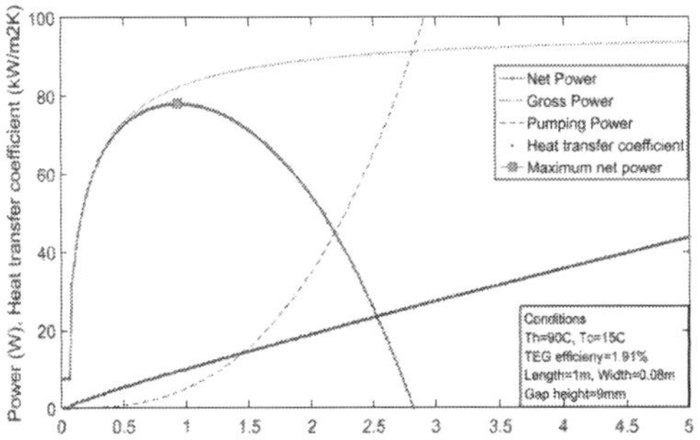

Fig. 8: Power and heat transfer coefficient

Figure 8 shows the estimated performance of TEG based power generation system using the proposed mathematical model. Gross power from TEG upsurges with the increment of flow rate as heat transfer coefficient increases. Net power also increases with the rise of flow rate up to 0.94 L/s. Interestingly, if we increase the beyond this point, gross power rises but net power started to diminish which turn into zero for the flow rate 2.8 L/s. This happen due to pumping power tremendously increase beyond the flow rate 0.94 L/s, pumping power is equal to gross power for flow rate 2.8 L/s which turn into zero gross power. If flow rate increased beyond this point, pumping power is higher than gross power and net power goes negative.

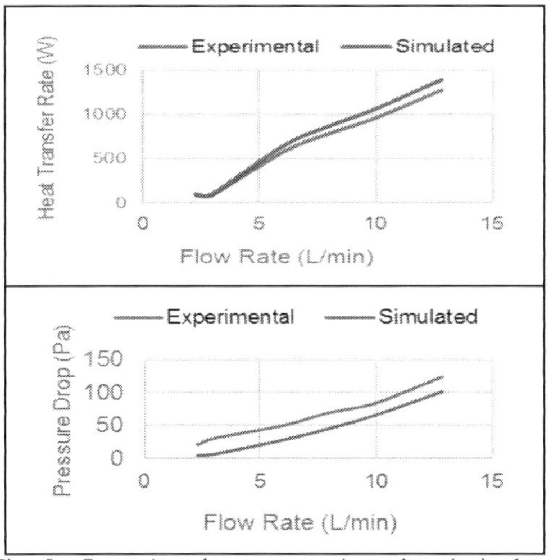

Fig. 9: Comparison between experimental and simulated results

We have run few initial experiments using the channel shown in figure 7. Though thermoelectric modules were not installed yet, we were able to run some experiments to find heat transfer rate from hot stream to cold stream and pressure drop in both channel. Experimental and simulation results are compared in figure 9. The water flow was varied between 1 L/min to 13 L/min with hot inlet temperature of 52.5 °C and cold inlet was 21.5 °C. The obtained heat transfer rate from experiments was 5-10% less than simulated results, which happened due to poor thermal insulation. In the case of pressure drop in channel, experimental value was 20% higher compared to simulated results due to friction losses for hose connection. The surface inside the channel is not fully smooth which increase the friction loss hence increased the friction loss.

5 Conclusion

In this work, five TEG modules have been selected from wide range of commercially available thermoelectric generator. An experiment has been undertaken to select the most suitable thermoelectric module for focused operating conditions. A theoretical model has been proposed to estimate the generated power from thermoelectric based power generation system. This model can calculate the gross power from TEG, parasitic power and net power for all possible combinations and able to predict the optimum net power for specified temperature range and conditions. The optimum operating conditions such as flow rate, heat exchanger dimensions can be found from the proposed model. An experiment has been undertaken to verify the simulated results. Simulated and experimental results are closely matched which can give confident about proposed mathematical model for further use.

References

1. Ebrahimi, K., G.F. Jones, and A.S. Fleischer, *A review of data center cooling technology, operating conditions and the corresponding low-grade waste heat recovery opportunities*. Renewable and Sustainable Energy Reviews, 2014. **31**: p. 622-638.
2. Orr, B., et al., *A review of car waste heat recovery systems utilising thermoelectric generators and heat pipes*. Applied Thermal Engineering, 2016. **101**: p. 490-495.
3. Remeli, M.F., et al., *Experimental investigation of combined heat recovery and power generation using a heat pipe assisted thermoelectric generator system*. Energy Conversion and Management, 2016. **111**: p. 147-157.
4. Matsuda, K., *Low heat power generation system*. Applied Thermal Engineering, 2014. **70**(2): p. 1056-1061.
5. Rana, S., et al., *Modelling and Optimization of Low-temperature Waste Heat Thermoelectric Generator System*. Energy Procedia, 2017. **110**: p. 196-201.
6. Rowe, M.D., et al. *Thermoelectric recovery of waste heat-case studies*. in *Energy Conversion Engineering Conference, 1997. IECEC-97., Proceedings of the 32nd Intersociety*. 1997. IEEE.
7. Crane, D.T. and G.S. Jackson, *Optimization of cross flow heat exchangers for thermoelectric waste heat recovery*. Energy Conversion and Management, 2004. **45**(9): p. 1565-1582.
8. Niu, X., J. Yu, and S. Wang, *Experimental study on low-temperature waste heat thermoelectric generator*. Journal of Power Sources, 2009. **188**(2): p. 621-626.
9. Gou, X., H. Xiao, and S. Yang, *Modeling, experimental study and optimization on low-temperature waste heat thermoelectric generator system*. Applied energy, 2010. **87**(10): p. 3131-3136.
10. Kim, S., et al., *A thermoelectric generator using engine coolant for light-duty internal combustion engine-powered vehicles*. Journal of electronic materials, 2011. **40**(5): p. 812.
11. Gou, X., et al., *A dynamic model for thermoelectric generator applied in waste heat recovery*. Energy, 2013. **52**: p. 201-209.
12. Orr, B.G. and A. Akbarzadeh, *Experimental testing of a car exhaust heat recovery system utilising TEGs and heat pipes*. SAE-A Vehicle Technology Engineer-Journal, 2016. **2**(1).

 https://doi.org/10.1051/e3sconf/20186406006

Parameter Identification and Effect of Partial Shading on a Photovoltaic System

Amin Hajizadeh[1] and *Jishnu* Warrier Anil Kumar[1]

[1]Department of Energy Engineering, Aalborg University, Esbjerg 6700, Denmark

Abstract: Partial shading cause significant losses to the performance of a photovoltaic (PV) system. So, it is imperative to study effects of partial shading; for that a two-diode model of the experimental setup made. Upon verifying the model with the experimental parameters, a MATLAB/Simulink model is made based on this model. Various shading patterns, the effect of bypass diodes; the effect of overlapping bypass diode is studied on this MATLAB/Simulink model. It is found out that the reduction in power loss is depended on the location of the shaded cell but not the area of the shaded cell. Also, the over-lapping bypass diode configuration has a slight improvement in the PV performance compared to the non-overlapping bypass diode configuration. An experimental test is also conducted by applying different shading pattern and they proves the results are compatible with the simulated results.

1 Introduction

Most of the present photovoltaic system's efficiency can be increased by the proper optimization. But these optimizations can vary from system to systems since optimization parameters vary according to the geographical conditions. The simplest approach is the one diode model [1]. But in a single diode method, the effect of recombination is neglected which happens in the depletion region [2]. In a two-diode model this effect is considered with the help of the second diode. Iterative programming method is used to calculate the values of R_s and a in [3] and in [4] fuzzy logic is used to resolve these values. But in many works the identity factor a_1 and a_2 is considered are *1* and *2* respectively which is not correct during every case [5]. Partial shading is the scenario in which one or more solar cell is shaded from the solar irradiation compared to the other cells in a module. Each of the cells in a module is connected in series, the power output from the module will be depended on the level of current passing through the weakest cell. This is one of the characteristics of series the connection. So even a small shading can adversely affect the performance of the system [6]. As a result, the bypass diode controls the voltage drop across the cell [7]. A module without a bypass diode will have greater power loss under shaded the condition when it is compared to a module which has a bypass diode [8].

2. Two Diode Model

One of the assumption which is considered in the one-diode model is that there is no recombination loss at the depletion region. But in a two-diode model, this effect is also considered. Even though the two-diode model is more complicated, it is more accurate method to model a PV system. The circuit diagram of a two-diode model can be seen in Fig.1. In this model diode one, D_1 represents the diffusion process while the second diode D_2 represents the effect of recombination in the cell. The series resistance R_s represents the initial losses caused by the current flow and parallel resistance R_p represents the modelling of the reverse saturation current. The total current I generated by the system is given by the Eqn.3.

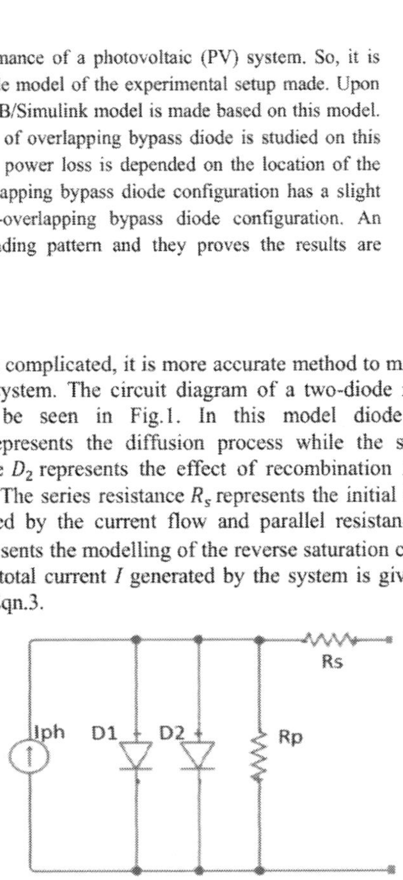

Fig. 1. Two- Diode Model

$$B = I_{01}exp\left(\frac{V+R_sI}{V_{t1}a_1} - 1\right) \quad (1)$$

$$C = I_{02}exp\left(\frac{V+R_sI}{V_{t2}a_2} - 1\right) \quad (2)$$

$$I = I_{ph} - B - C - \frac{V+IR_s}{R_p} \quad (3)$$

The terms B and C are used for the convenience. Here I_{01} represents the reverse saturation current through diode one and I_{02} represents are reverse saturation current through the diode two. V_{T1} and $V_{T2} = N_skT/q$ are the thermal voltage of the module with N is the number of cells, q is the electron charge and k is the

Boltzmann's constant of diode one and two. Also a_1 represents the ideality factor of diode one and a_2 represents the ideality factor of diode two. The value of light generated current I_{ph} with respect to the temperature and irradiation is given by the Eqn.4 which was described in [9].

$$I_{ph} = [I_{sc} + K_i(T - T_n)]\frac{G}{G_n} \quad (4)$$

Here K_i represents the short circuit current coefficient, T_n represents the temperature at standard testing condition, G represents the irradiance and G_n represents the irradiance at standard testing condition. An assumption is made here as the diode current from diode one I_{01} is equal to the diode current from diode two I_{02}.

$$I_{01} = I_{02} = I_0 \quad (5)$$

So, the equation for saturation current for two-diode model is described in [10].

$$I_0 = \frac{I_{sc}+K_i(T-T_n)}{D-1} \quad (6)$$

$$D = exp\left[\frac{V_{oc}+K_v(T-T_n)}{\frac{a_1+a_2}{p}V_T}\right] \quad (7)$$

The value for ideality factor is assumed as $a_1 = 1$ and $a_2 = 2$ The ideality factor $a_1$represents the diffusion of current components and ideality factor a_2 represents the recombination of the current components.

2.1 Calculation of R_s and R_p values

Finding the value of R_p is a stepwise procedure. The procedure that is used here is like that of [11]. As a first step of finding the value of R_p, the value of R_s is assumed as zero in Eqn.9. By this we get the value of R_p and this value of R_p is substituted in Eqn.8. The result is cross checked with the P_{max} value in the data sheet. If the values are different, then it is necessary that the same procedure should be repeated. Each time a very small increment is given for the value of R_s. This procedure is continued until the value of P_{max} reaches the P_{max} value described in the PV module data sheet provided by the manufacturer.

$$P_{max} = V_{max}I \quad (8)$$

$$R_p = \frac{V_{mp}+I_{mp}R_s}{I_{pv}-I_0[E+F-\frac{P_m}{V_m}]} \quad (9)$$

$$E = exp\frac{V_{mp}+I_{mp}R_s}{V_T} \quad (10)$$

$$F = exp\frac{V_{mp}+I_{mp}R_s}{(p-1)V_T+2} \quad (11)$$

The terms E and F are used only for the convenience. Without a substitution the equation for R_p is too long to be represented.

3. Experiment Validation

To verify the two-diode model, an experimental test is conducted, and an electronic load is used for the experiment. The PV module from which the measurements are made can be seen in Fig.2. Since the location is prone for the temperature changes throughout the year, the experiment is conducted under two different temperatures. An electronic load and MATLAB/Simulink software is used for the experiment. Upon completing the experiment, simulation is conducted in MATLAB/Simulink software under similar temperature and irradiation values as of the experiment. The Table.1 shows the comparison between the experimental values and the simulation values for two different days. In Fig.3 a QQ-plot of has been made for the residual value between the experimental results and the simulated results. The data points almost follow a straight line, hence it is safe to arrive at a conclusion that this is a normal distribution and the model represents the real-world system in a good way.

Fig.2. Real PV Module

4. Shading Effect

The effect of partial shading is very significant for a photovoltaic system and even a small level of shading can affect the performance adversely [6]. Using the parameters from the two-diode model, a MATLAB/Simulink model is made and the effect of partial shading and effect of number of bypass diodes is analysed. The effect of shading under a series and parallel connections are explained individually. In all the above-mentioned cases the bypass diode configuration is assumed as a non-overlapping configuration.

4.1 Shading in a series connection

At first a single module with *60* cells which are connected in series is considered. These cells are sub-divided into three strings, i.e. each *20* cells will have one bypass diode which is connected in parallel to them. In the Case A three cells from a string is shaded and the two other strings are unshaded. The simulations are conducted at a temperature of 25^oC the unshaded irradiation is set to $1000W/m^2$ while the irradiation for shaded cells are $500W/m^2$. The Fig.4 illustrates the shading pattern which is applied for studying.

Table.1. Comparison of Simulated and experimental data for two different days

	Two-Diode	Experimental
Day 1		
H	627	627
T	17	17
I	17.90	17.70
V	41.6	41.2
I	17.041	17.184
V	32.8	32.5
R_S	0.006	-
R_p	32	-
P_{max}	558.581	554.178
Day 2		
H	358	358
T	4	4
I	10.217	10.217
V	41	41.2
I	9.5819	9.7165
V	36.4	36.4
R_S	0.006	-
R_p	32	-
P_{max}	348.7453	353.6821

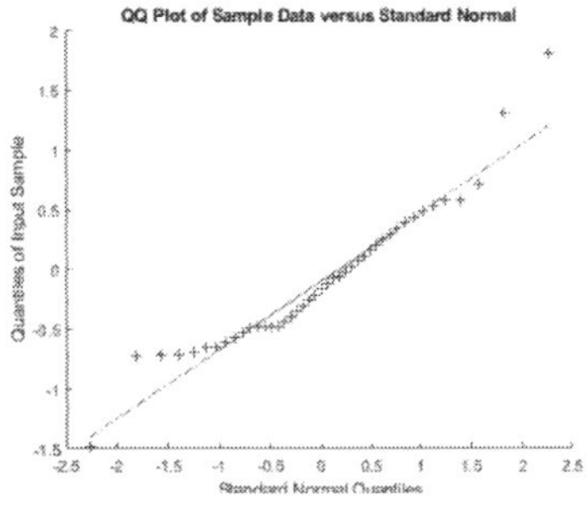

Fig.3. QQ-plot

The Fig.5 and Fig.6 illustrates the PV and IV curve for both Case A and Case B simulations. The blue curve represents the unshaded module. From the IV curve of Case A it looks as if the entire string has been shaded but it must be noted that only three cells has been shaded. But since the bypass diodes are placed the two unshaded strings operate. From the PV curve the power has reduced by *40%*. In Case B, one cell from each of the strings are shared while the rest of the cells are unshaded. In this case bypass diodes have no effect on the strings as all three of them are shaded and due to this the value of current is quite low compared to the previous case.

4.2 Shading in a parallel connection

In this case three modules that are connected each other in a parallel connection is considered. This consideration is made because the experimental setup also has the very same configuration, so that it is possible to have a comparison between them. The shading pattern applied here is very similar to that of the shading pattern applied in a series connection. The results of this test can be seen in the Fig.7 and Fig.8.

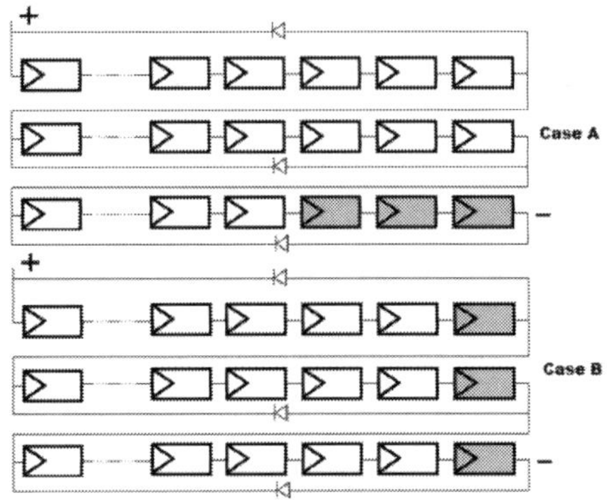

Fig. 4. Shading pattern

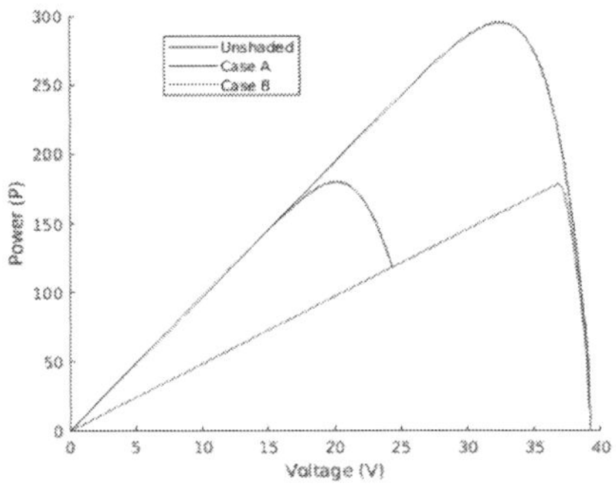

Fig. 5. PV curve when 3 cells from a module is shaded

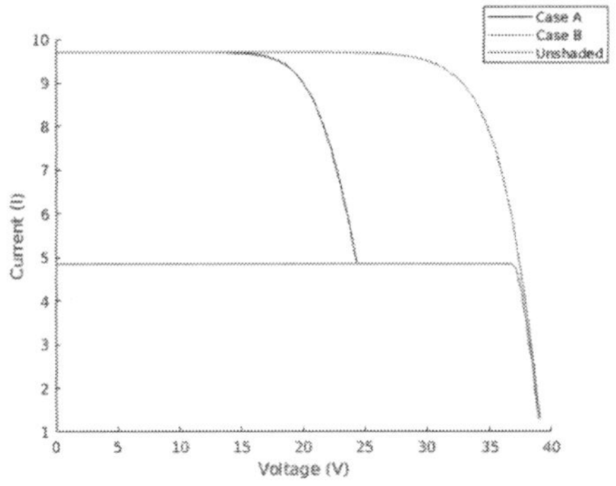

Fig. 6. IV curve when 3 cells from a module is shaded

4.3 Number of bypass diodes and its effects on parallel connection

The effect of having bypass diode is studied by considering four cases. A module without a bypass diode will have greater power loss under shaded condition when it is compared to a module which has bypass diode [8]. In Case-A were an entire module is connected to one bypass diode in parallel. In Case-B the module is sub-divided into two strings and two bypass diodes are present here. Case-C consist of a module divided into three strings, meaning three bypass diodes

https://doi.org/10.1051/e3sconf/20186406006

are placed. Finally, in Case-D the module is divided into six strings and each string is connected to a bypass diode in parallel. In each of the cases three modules are connected in parallel. This study is done for understand how the number of bypass diodes would affect the performance of the system. The Fig.9 and Fig.10 represents the PV and IV curve when different number of bypass diodes were considered in a parallel connection of modules. It is clear from both the figures that as the number of bypass diodes were increased the performance of both IV and PV curves improved considerable. In each of the cases the number of cell that are partially shaded is six.

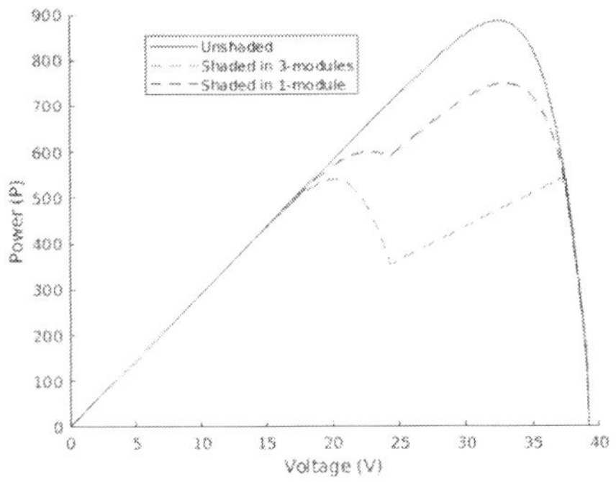

Fig. 7. PV curve when 3 cells shaded in 3 module

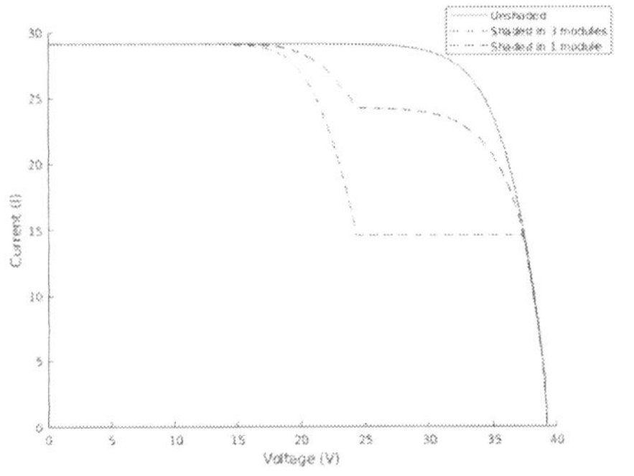

Fig. 8. IV curve when 3 cells shaded in 3 module

From the Table.2 it can be clearly seen that both the maximum power and fill factor values has increased as more bypass diodes are added into a module. It is evident that for the first three cases the values don't change much but in the fourth case there is a significant increase in both the fill factor value and in the maximum power value.

4.4 Shading on an overlapping bypass diode

In an over-lapping bypass diode configuration, some cells inside the module will have an over-lapping bypass diode connection. That means those cells are connected to two bypass diodes. A module of *60* cells are considered with bypass diode BD-1 is connected in parallel between cell *1-29*, BD-2 is connected between the cell *20-50* and BD-3 is connected between cell *39-60*. So according to this configuration the cell between *20-50* are connected to two bypass diodes and this configuration can be seen in Fig.11. It is important to note that under an overlapping configuration it is possible to have three kinds of paths for the current to flow when an overlapping cell is shaded [12].

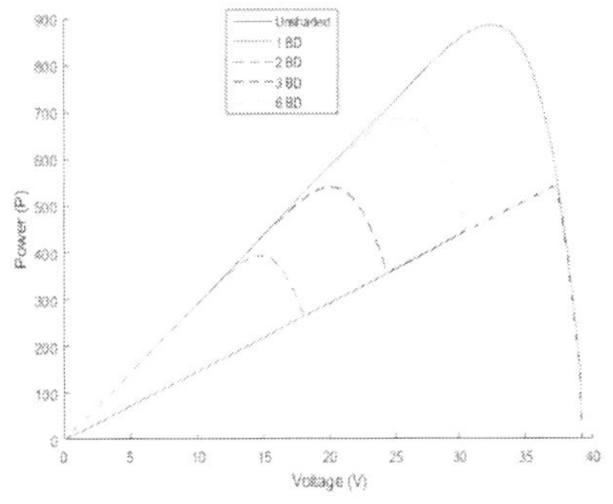

Fig. 9. PV curves of the bypass diode test

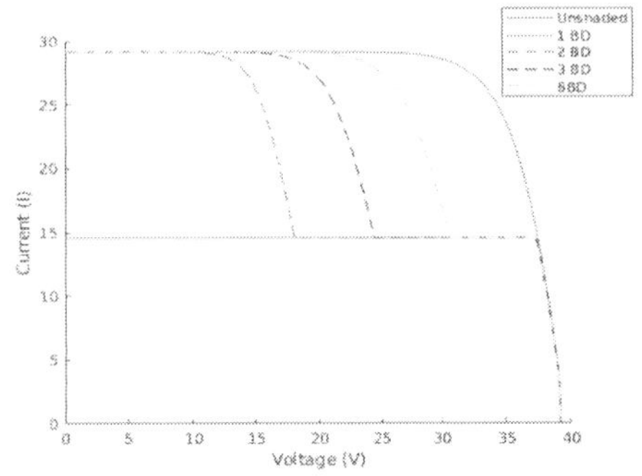

Fig. 10. IV curves of the bypass diode test

Table. 2. Comparing the effect of bypass diodes

	Case A	Case B	Case C	Case D	Case D
Imax	14.54	14.55	14.69	25.14	28.1
Vmax	37	37.17	37.4	26.47	32
Pmax	540.9	540.92	545.2	681.3	880
FF	0.4864	0.488	0.496	0.601	0.755

In Fig.12 the IV curve of an overlapping diode configuration and a non-overlapping diode configuration. In this case the shading is done in such a way that two overlapping cells and one non-overlapping cell is shaded. From the Fig.12 the short-circuit current I_{oc} for overlapping is very high than that of the short-circuit current for an unshaded module. This happens because at a low voltage both the BD-1 and BD-2 are conducting at the same time. So, because of this the current value adds up. But as the voltage value increases the BD-1 turns off, as a result the current value drops down. According to [13] this kind of effect only happens in a small-scale system; this effect is not applicable to the large scale photovoltaic systems.

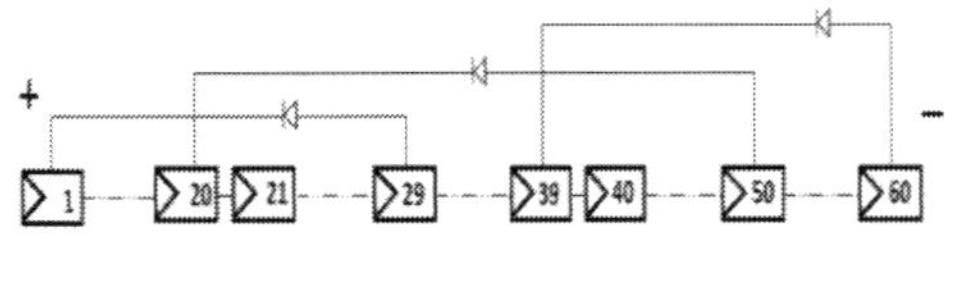

Fig. 11. Shading pattern for overlapping configuration

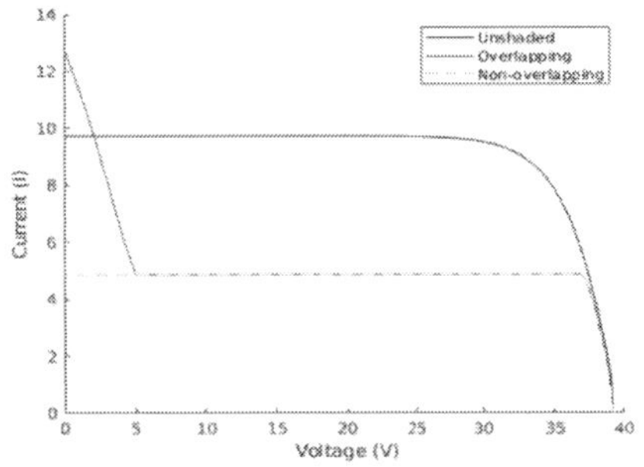

Fig. 12. IV curve under an overlapping bypass diode

4.5 Experimental Validation of Shading Effect

The experiment is conducted on a three modules system which are connected in parallel to each other and this setup can be seen in Fig.2 One cell from each module is shaded. Since there are three modules so three cells are shaded as a whole. The cells are shaded using four sheets of papers, this is done to make sure that the cells are not shaded completely. The complete shading is not done due to the fact that in most of the cases the cells are not shaded completely, instead they would be shaded partially. The IV can be seen in Fig.13 and the results are in good terms with the simulated results. While the test was conducted the ambient atmospheric temperature was 16^0C and the irradiation value was at $681W/m^2$.

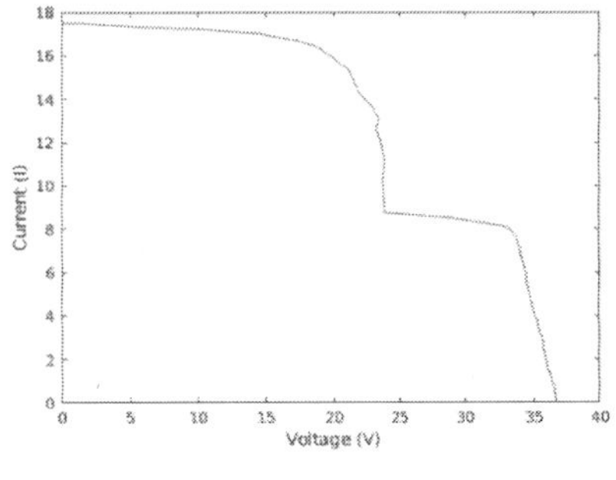

Fig. 13. IV curve of experimental test

5. Conclusion

The two-diode model based on the real-world system is made and validated by comparing it to the experimental values. The simulations for partial shading is conducted based on the two-diode model. From these simulations it is understood that the position of the shading has more importance compared to the area of shaded cell. It was observed that the power loss is different by changing the location of the shading effect while the total surface area remains the same. Adding more bypass diodes reduces the effect of shading on a system. Finally, an over-lapping bypass diode configuration has a slight advantage over a non-overlapping bypass diode configuration.

Reference

[1] E. E. Yousef Mahmoud, "Accuracy Improvement of the PV Model," in *IEEE TRANSACTIONS ON SUSTAINABLE ENERGY, VOL 3*, July 2015.

[2] H. S. Rauschenbachr, Solar Cell Array Design Handbook, New: Van Nostrand Reinhold, 1980.

[3] J. Perlin, From Space To Earth(The Story Of Solar Electricity), London: Harvard University Press, 1999.

[4] J. R. G. a. E. R. F. Marcelo Gradella Villalva, "Comprehensive Approach to Modeling and Simulation," in *IEEE TRANSACTIONS ON POWER ELECTRONICS VOL. 24, NO.5*, May 2009.

[5] J. G. E. F. M.G. Villalva, "Comprehensive approach to modeling and simulation of photovoltaic arrays," in *IEEE Trans. power electron. 24 (5) 1198–1208*, 2009.

[6] J. F. M.C Alonso Barc, "Experimental study of mismatch and shading effects in the IV characteristic of a photovoltaic module," in *Solar Energy Materials & Solar Cells 90 329340*, 2009.

[7] G. M, Page-485,Renewable and Efficient Electric Power Systems, A JOHN WILEY & SONS.

[8] A. S.-G. C. C. a. J. C. E. Díaz-Dorado, "Influence of the shadows in photovoltaic systems with different configurations of bypass diodes.," in *International Symposium on Power Electronics, Electrical Drives, Automation and Motion*, 2010.

[9] R. T. P. R. Dexso Sera, "PV panel model based on datasheet values," in *IEEE Int. Symp. Ind.Electron. (ISIE)*, 2007.

[10] A. C. d. W. a. W. G. v. S. Boudewijn B. Pannebakker, Photovoltaics in the shade: one bypass diode per solar, PROGRESS IN PHOTOVOLTAICS: RESEARCH AND APPLICATIONS Prog. Photovolt: Res. Appl., 2017.

[11] Y. R. F. M. Bellia H, "A detailed modeling of photovoltaic module using MATLAB," in *Astronomy and Geophysics, 3, 5361*, 2014.

[12] R. T. P. R. D. Sera, "PV panel model based on datasheet values," in *Proceedings of the IEEE International Symposium on Industrial Electronics (ISIE) 7, pp. 2392–2396*, 2007.

[13] J. G. E. F. M.G. Villalva, "Comprehensive approach to modeling and simulation of photovoltaic arrays," in *IEEE Trans. power electron. 24 (5) 1198–1208*, 2009.

Offline Photovoltaic Maximum Power Point Tracking

Fyali Jibji-Bukar[1], and *Olimpo* Anaya-Lara[1]

[1]University of Strathclyde, Department of Electrical and Electronics Engineering, 16 Richmond Street, G1 1XQ, Glasgow, United Kingdom

Abstract. As more renewable energy sources are connected to the electrical grid, it has become important that these sources participate in providing system support. It has become needful for grid-connected solar photovoltaics to participate in support functions like frequency support. However, photovoltaic systems need to implement a maximum power tracking algorithm to operate at maximum power and a method for de-loading photovoltaic systems is necessary for participation in frequency support. Some conventional maximum power tracking techniques are implemented in real time and will not adjust their output fast enough to provide system support while other may respond fast but are not very efficient in tracking the maximum power point of a photovoltaic system. This paper presents an offline method to estimate the maximum power voltage and current based on the characteristics of the photovoltaics module available in the datasheet and using the estimated values to operate the photovoltaics at maximum power. The performance of this technique is compared to the conventional technique. This paper also describes how the photovoltaic system can be de-loaded.

1 Introduction

Power in electrical systems is increasingly supplied from diverse generation sources. As sources such as photovoltaics gain a larger share of power generation and replace conventional generators, the inertia of power systems will be reduced. The could lead to a high rate of change of frequency when there is a difference in generation and demand [1]. This problem is exacerbated in island systems where there are very few generators and high renewable energy penetration because they have reduced inertia and variability in power generation which could lead to more frequency deviation events [2]. As a result of this, it has become important for different kinds of generation sources to participate in frequency support if more renewable generation is to be added to electrical grids.

Wind turbines have been demonstrated to contribute to frequency support by providing inertia support and providing primary frequency support similar to the support obtainable from conventional power plants in [3] and [4]. Energy storage systems have also been shown to provide frequency support. In [5], flywheel energy storage system is shown to provide frequency support. Battery storage systems can also be used to provide frequency support as shown in [6]. This is because they are fast-acting and can increase active power supply in the time-scale of inertia response.

Like battery storage systems, PV systems can support grid frequency with the appropriate control and operation method. However, PV systems require a maximum power point tracking method. Common methods used for operating PV at maximum power include perturb and observe, incremental conductance and fractional open-circuit voltage [7]. These methods are not suitable for systems required to provide frequency support because they either require significant computation which makes them slower or are fast but lead to significant power losses.

Various methods for estimating PV power have been proposed in [8] and [9]. [8] uses artificial neural network but this will require historical information on the performance of the PV system and the method used in [9] will require some real-time calculation to estimate maximum power. This paper proposes a method to operate the PV systems at maximum power by obtaining the maximum power voltage from the PV current-voltage (I-V) curve for the entire operating range of the PV module using the characteristics of the PV module from the PV datasheet.

Section 2 describes the effect of temperature and irradiance on the open-circuit voltage (V_{OC}) and short-circuit (I_{SC}) and how V_{OC} and I_{SC} can be estimated for any combination of temperature and irradiance. Section 3 describes how the maximum power voltage and current can be calculated from the I-V curve and how the PV system can be operated at the desired de-loading level. Section 4 presents the operation of the proposed method and compares the result with operation using incremental conductance.

2 Estimating V_{OC} and I_{SC}

2.1 Effect of Changing Irradiance

The I_{SC} current varies directly with the irradiance up to very high levels of irradiance [10]. This is because the generated photocurrent is a direct result of the amount of available sunlight. This implies that for any given temperature, the I_{SC} can be readily determined. The I_{SC} current at standard testing condition (STC) is available in the datasheet. The V_{OC} is also affected - to a lesser extent -by changing irradiance. The V_{OC} changes logarithmically to the irradiance. Figure 1 shows the I-V at different irradiances and constant temperature.

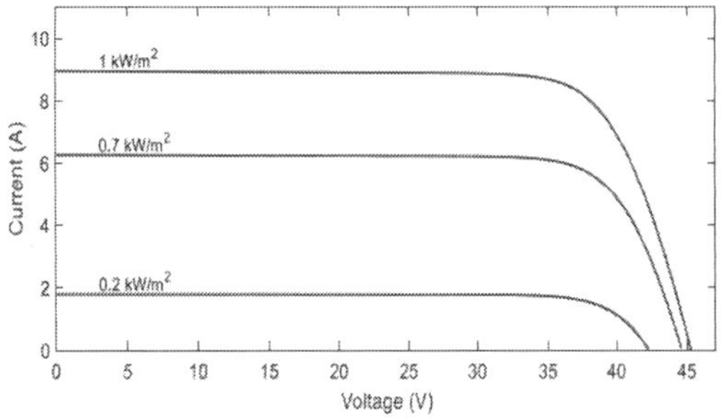

Fig. 1.Effect of Changing Irradiance on V_{OC} and I_{SC}

2.2 Effect of changing temperature

A change in temperature results in a change in both the V_{OC} and I_{SC}. As temperature increases, the I_{SC} increases while the V_{OC} reduces. This results in an overall loss of efficiency with increasing temperature. This is because the increase in temperature leads to an increase in the bandgap of the semiconductor which leads to an increase in the I_{SC} [11]. V_{OC} reduces with an increase in temperature because of the increase in the dark current density [11].

The effect of temperature on V_{OC} and I_{SC} is defined by the temperature coefficient of the PV module and is provided by the manufacturer in the module datasheet. Figure 2 shows the effect of changing temperature on the I-V curve of the PV module.

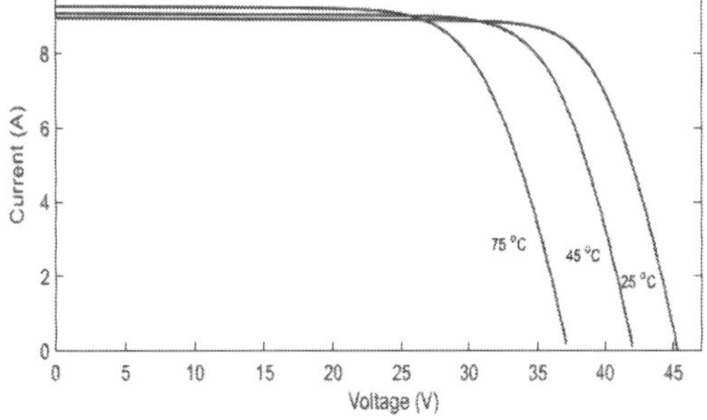

Fig. 2.Effect of Changing Temperature on V_{OC} and Isc

2.3 Effect of series and shunt resistances

The series and shunt resistances of the PV module/cell make up the parasitic resistances. They reduce the efficiency of solar cells. The series resistance is as a result of the resistance of the electrical contact of the cell [12]. The series resistance does not affect the V_{OC} except at high values. The effect of the series resistance can be discounted in estimating V_{OC}.

The shunt or parallel resistance is as a result of leakage current from the side of the cell which leads to loss of power and should be as high as possible to reduce losses [12]. The shunt resistance has no significant effect in measuring V_{OC} and I_{SC}. The parasitic resistances result in a shift in the maximum power point and must be considered in obtaining the I-V curve.

Figure 3 below shows the I-V curve of a Trina Solar TSM310PD14 cell with the series and shunt (parasitic) resistances and without the series and shunt resistances. The V_{OC} and I_{SC} changed very little but there is a significant shift in the maximum power point. The maximum power voltage dropped from 0.5536V without the parasitic resistances to 0.5033V with the parasitic resistances while the maximum power current dropped from 8.4533A to 8.3256A. This presence of the parasitic resistances resulted in a maximum power loss of 0.4895W which is about 10% of the power without the parasitic resistances. The shunt and series resistances are not given in the module datasheet but can be calculated using the method described in [13].

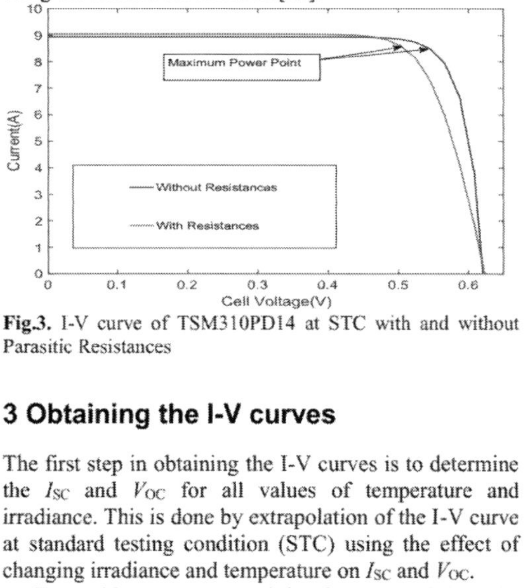

Fig.3. I-V curve of TSM310PD14 at STC with and without Parasitic Resistances

3 Obtaining the I-V curves

The first step in obtaining the I-V curves is to determine the I_{SC} and V_{OC} for all values of temperature and irradiance. This is done by extrapolation of the I-V curve at standard testing condition (STC) using the effect of changing irradiance and temperature on I_{SC} and V_{OC}.

Two tables are generated for the V_{OC} and I_{SC}. The range of irradiance considered is from 0-1700w/m^2 in intervals of 50w/m^2 while the range of temperature considered is from -40°C to 85°C in intervals of one. The PV module used is the Trina Solar TSM-310PD14.

3.1 Calculating V_{OC}

The Effect of temperature on V_{OC} is given by the temperature coefficient which is given in the module datasheet. At a given the irradiance, V_{OC} is given by equation 1.

$$V_{OC}(t) = V_{OC}(STC) + \alpha(t\text{-}25) \qquad (1)$$

Where α is the temperature coefficient of the open-circuit voltage and t is temperature. To determine the V_{OC} for any irradiance (I) at a given temperature, equation 2 is used.

$$V_{OC} = mk_B T \ln\left(\frac{I_{SC}}{J_0} + 1\right) \quad (2)$$

Where m is the diode ideality factor, k_B is the Boltzmann constant, J_0 is the diode saturation current and T is the temperature in Kelvin. To obtain the V_{OC} voltage for any combination of irradiance and temperature, equation 1 and 2 are combined resulting in equation 3. Equation 3 gives the V_{OC} for any temperature and irradiance combination.

$$V_{OC}(I,T) = mkBT \ln\left(\frac{I_{SC}(I,T)}{J_0} + 1\right) + \alpha(T - 25) \quad (3)$$

3.2 Calculating I_{SC}

The I_{SC} should be calculated first because it is needed in equation 3 to calculate the V_{OC}. I_{SC} varies linearly with the irradiance for any given temperature. The I_{SC} for a given irradiance at any temperature is given in equation 4.

$$I_{SC}(t) = I_{SC}(STC) + \alpha(t\text{-}25) \quad (4)$$

Where α is the temperature coefficient of the short-circuit current. To obtain the I_{SC} for any combination of temperature and irradiance, the I_{SC} at STC have to first be adjusted for the irradiance and then adjusted for the difference in temperature from 25°C. Equation 5 can be used to calculate the short-circuit current for any temperature and irradiance.

$$I_{SC}(I,T) = \frac{I}{1000} \times I_{SC}(STC) + \alpha(T - 25) \quad (5)$$

3.3 Calculating maximum power voltage/current

To determine the maximum power point at any irradiance and temperature, the I-V and P-V curves need to be obtained. PV cell can be modelled as a current source connected in parallel to a diode and the shunt resistance (R_{SH}) and in series to the series resistance (R_S). The five parameter model of a PV cell is given in figure 4.

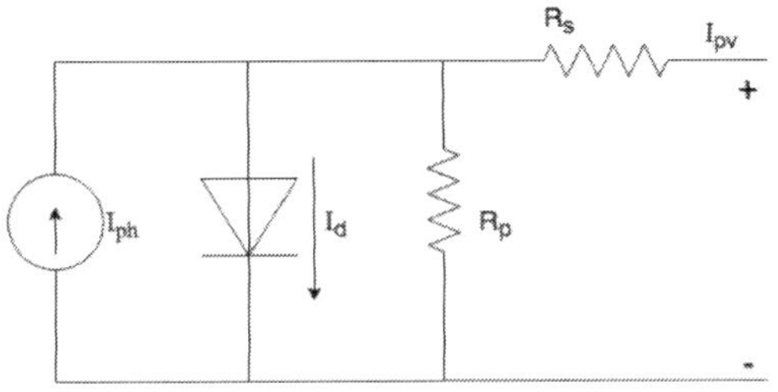

Fig. 4. Five parameter model of a solar cell

The I-V curve can be obtained by solving for the current in the terminal of the module. The current from PV module is given by equation 6.

$$i = I_{SC} - J_0\left(e^{q(V+JR_S)/kT} - 1\right) - \frac{V - JR_S}{R_{SH}} \quad (6)$$

Where i is the current across the terminals of the PV, R_{SH} is the shunt/parallel resistance, R_S is the series resistance, V is the voltage and q is the electric charge.

The I-V curve can be obtained by solving the equation 6. The voltage should be taken in small steps from 0 to V_{OC} and the current computed to obtain the I-V curve. The step should be as small as possible as the maximum power point is usually around the end of the curve. This will increase the accuracy of the curve. However, equation 6 is an implicit equation and have to be solved numerically. For every voltage from 0-V_{OC}, the value of current which satisfies the equation is determined numerically. This was done using the fzero function in Matlab. Figure 5 shows the I-V curve at different points using the described method.

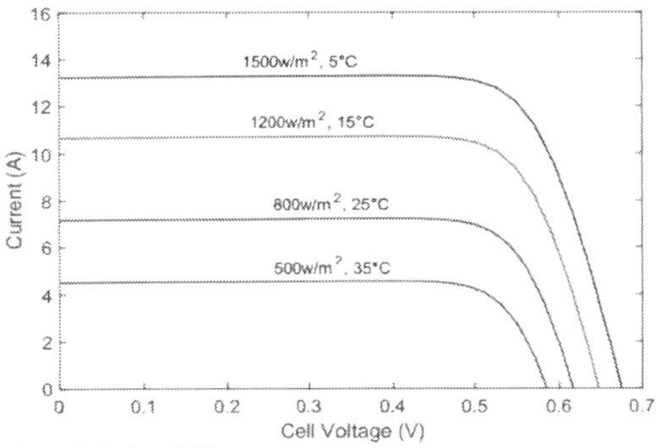

Fig. 5. Calculated IV curves

The maximum power point can be determined by plotting the graph of voltage against power. It is the highest point in the PV graph. Table 1 shows the maximum power voltage of one module of Trina Solar TSM 310 PD14 at different temperature and irradiance using the described process.

Table 1: Calculated maximum power voltages

Irradiance	-20°C	25°C	45°C	65°C
500	42.12	36.4	33.72	31.13
1000	42.04	36.8	33.4	30.3
1500	41.41	36.2	33.15	30.61
1700	41.63	35.4	32.25	29.78

4 Operating PV system at maximum power

To test the proposed method, a PV system was modelled in Simulink. The maximum power points at different operating conditions are calculated and stored in a lookup table. Because of the number of possible combinations of temperature and irradiance, it is

 https://doi.org/10.1051/e3sconf/20186406007

important that only a manageable number of power points are calculated for. For any irradiance and temperature value not stored in the lookup table, the closest approximation should be used.

For any combination of temperature and irradiance, the lookup table will give the maximum power voltage as the reference voltage. The reference voltage will then be used to obtain the duty cycle which will be the input of the DC-DC converter. The operation of the system is shown in figure 6 below.

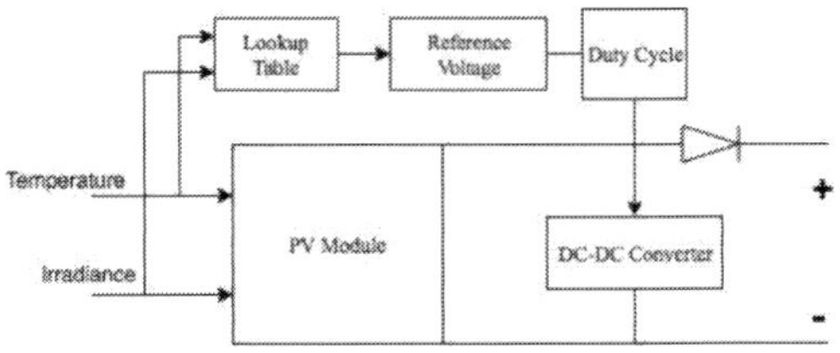

Fig. 6. Offline MPPT Implementation

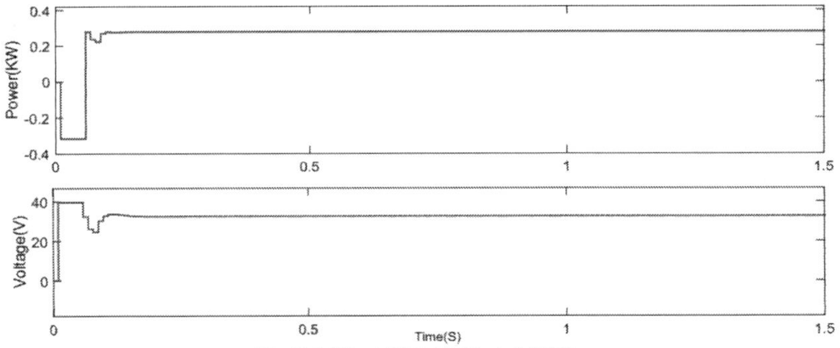

Fig. 7. Offline MPPT (1000w/m^2 50°C)

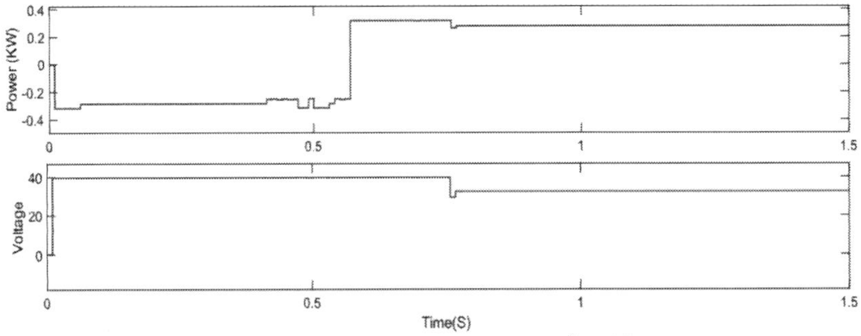

Fig.8. Incremental conductance (1000w/m^2, 50°C)

The performance of the offline method is compared to the performance of the system using incremental conductance. In figure 7, the PV system is using the offline maximum power point tracking while in figure 8, the PV system is using incremental conductance maximum power point tracking technique. The time taken to reach maximum power by the PV system using the proposed offline method is about 100ms while the time taken to reach maximum power using incremental conductance is about 700ms.

The system can be de-loaded by storing a percentage of the maximum power voltage in the lookup table as the reference voltage. This implies that if the system is required to operate with a 10% reserve the maximum power voltage will be multiplied by 0.9 before being stored in the look-up table. This is because the voltage is approximately directly proportional to the power up to the maximum power point. The system can also be de-loaded by operating it higher than the maximum power voltage but it is more difficult to get it to operate at the desired percentage of maximum power because the power drops steeply after the maximum power point.

This method depends on the behaviour of the PV module and as a result, the accuracy of the system is affected by degradation of the PV module. This can be factored into the calculation of the maximum power voltage.

5 Conclusion

This paper describes and tests a fast and accurate method for operating PV systems at maximum power or at maximum power with a predetermined reserve. The V_{OC} and I_{SC} are first determined using the temperature coefficients of the PV module and the effects of changing irradiance on the PV. Then the maximum power is determined by obtaining the I-V curve. The points of the I-V curve are obtained by numerically solving the diode equation. This maximum power voltage is stored in a lookup table and is the reference voltage for a given irradiance and temperature. The implementation of the offline method shows that the system reaches maximum power quickly and is faster than when incremental conductance is used. This main advantage of this method is the speed at which the reference voltage is determined. The method presented in this work will find application in the use of converter connected sources such as PV and grid-scale batteries in providing frequency support and thus will enable more converter connected sources in electrical grids.

References

1. N. Masood, N. Modi, *'Low Inertia Power Systems: Frequency Response Challenges and a Possible Solution'*, Australasian Universities Power Engineering Conference (AUPEC), 25-28 September, (2016).
2. Y.K Wu, *'Frequency Stability for an Island Power System: Developing an Intelligent Preventive-Corrective Control Mechanism for an Offshore Location'*, IEEE Industry Application Magazine, Vol. **23**, Issue 2, pp. 74-87. March-April, (2017).
3. J. Morren, J. Pierik, S.W.H. de Haan, *'Inertia Response of Variable Speed Wind Turbine'*, Electrical Power Systems Research **76**. Pp. 980-987, (2006).
4. L. Wu, *'Provision of Power System Frequency Response in the Context of High Wind Penetration'*, PhD Thesis, (2014). Retrieved from: http://oleg.lib.strath.ac.uk/R/?func=dbin-jump-full&object_id=24376
5. S. S Sami, *'Virtual Energy Storage for Frequency and Voltage Control'*, Ph.D Thesis, July (2017).
6. S. S Sami, M. Cheng, J. Wu, *'Modelling and Control of multi-type, Grid-scale Energy Storage for Power System Frequency Response'*, IEEE 8th International Power Electronics and Motion Conference (IPEMC-ECCE Asia), (2016).
7. M. A. Eltawil, Z. Zhao, *'MPPT Techniques for Photovoltaic Applications'*, Renewable and Sustainable Energy Reviews. Vol **25.** Pp. 793-813. September (2013).
8. L. M. Elobaid, A. K. Abdesalam, E. E. Zakzouk, *'Artificial Neural Network Based Maximum Power Point Tracking Technique for PV Systems'*, IECON 2012- 38th Annual Conference on IEEE Industrial Electronics Society, October 25-28, (2012).
9. A. Hoke, E. Muljadi, *'Real-time Photovoltaic Plant Maximum Power Point Estimation for use in Grid Frequency Stabilization'*, Controlling and Modelling for Power Electronics (COMPEI), IEEE 16th workshop 12-15 July (2015).
10. J. Nelson, *'The Physics of Solar Cells'*, London, Imperial College Press, (2003). Pg. 172-173
11. J. Nelson, *'The Physics of Solar Cells'*, London, Imperial College Press, (2003). Pg. 173
12. J. Nelson, *'The Physics of Solar Cells'*, London, Imperial College Press, (2003). Pg. 14-15
13. M. G. Villalva, J. R. Gazoli, *'Comprehensive Approach to Modelling and Simulation of Photovoltaic Arrays'*, IEEE Transactions on Power Electronics, Vol **24**, Issue 5. Pp 1198-1208, (2009).

Monitoring and Analysis of Power Quality in Photovoltaic Power Generation System

JI Cui[1], *HUA* Siming[1], *ZOU* Bingbing[1], *ZHANG* Hua[1], *DIAO* Chang[1], ZHOU Guofa[2], MIN Bo[2]

[1]State Grid Shanghai Fengxian Electric Power Company, Shanghai Fengxian 201499, China.
[2]Shanghai Puhai Qiushi Electric Power High Technology Co.,Ltd., Shanghai 200090, China.

Abstract. Solar photovoltaic (PV) has been developed rapidly due to its clean and green renewable characteristics. The connection of photovoltaic power generation to the traditional grid system is bound to bring power quality problems. Based on above, this paper introduces the power quality testing method of photovoltaic grid-connected power grid in detail. And then takes the Shanghai Qingpu Nanrong distributed photovoltaic power generation for example, which is the largest single roof photovoltaic power station in Shanghai by far, to use the method to test its power quality. The power quality index include harmonic, voltage imbalance, and frequency deviation. The test and analysis provide a reference for the monitoring and analysis of power quality of photovoltaic grid-connected power grid.

1 Introduction

At present, China is still based on thermal power generation. Although the proportion of nuclear power, wind power and solar power is improved, the capacity of thermal power installed capacity still accounts for most of the total installed capacity of electric power, and the proportion of thermal power generation accounts for more than 80% of the total power generation [1]. When generating electricity, a large amount of coal needs to be burned, which not only demands for large coal resources, but also causes destruction to the environment. Despite of the decreasing of the coal when thermal power generating electricity [1], the environmental pressure brought by it is still not small. Under the pressure of energy depletion and environmental pollution, people have to turn their attention to clean and renewable new energy sources [2]. Solar photovoltaic power generation is inexhaustible and clean and environmentally friendly, so it has attracted wide attention [3-7].Under the influence of sustainable development and national energy strategy layout, more and more new distributed energy sources have been applied to modern power grid, and even many countries have the tendency to replace the conventional energy with distributed energy. Solar energy is the most abundant resource on the earth. The annual radiation from the sun to the earth's surface reaches to 5.4*1024J, equivalent to 1.8*1014t standard coal. If only 0.1% of the energy is converted to electricity by 5% conversion rate, the annual generating capacity can reach to 5600TWh, equivalent to 40 times the current world energy consumption. If we can make rational use of solar energy, we will greatly alleviate the resource pressure facing the world at present [3].

The principle of solar photovoltaic power generation is the photoelectric effect. Photoelectric effect refers to the semiconductor material in the solar energy under the action of light, its internal electronic directional movement and form an electric current. The formation of a large scale photovoltaic array is to encapsulate a single photovoltaic cell and then connect a wire to a circuit according to the need to meet the requirements of a larger DC power output.

Photovoltaic power generation uses clean energy solar energy to generate electricity, which does not consume fuel and has no pollution to the environment. Its installation is not restricted by the region and can be flexible to set up its installation capacity as needed. At the same time, photovoltaic power plant operation is safe and reliable, and later maintenance is relatively simple [3].

However, after solar PV grid connected, the fluctuation of PV will lead to the power quality problems, such as voltage fluctuation, voltage flicker, frequency fluctuation and so on. At the same time, the photovoltaic inverter components will lead to the power quality problems such as harmonic and DC component [4,8]. Therefore, the grid connected photovoltaic has to consider the power quality problems brought by it.

Based on the above, this paper will introduce the method of measuring the power quality of photovoltaic grid connected power, and test the power quality of the power station with the largest single roof photovoltaic power station in Shanghai, Shanghai City, Qingpu District, and carry out the power quality test of the power station, and carry out the harmonic, voltage imbalance, frequency deviation and other electrical energy for the test results. The quality index analysis provides a reference for monitoring and analyzing the power quality of grid connected PV system.

2 Power quality testing process

2.1 test tool

This test adopts the PW3198 power quality analyzer of Japan electric motor Co., Ltd.

2.2 test environment

Because of the sensitivity of photovoltaic components and photovoltaic inverters to environmental conditions, the environmental conditions of the test have a great influence on the test results. So the test should have appropriate environmental conditions. Under normal circumstances, the recommended test environment temperature is -10~50 degrees C, the ambient humidity is not more than 90%.

2.3 Test location

Test location: PV power station network terminal and point of common coupling (PCC). As shown in Figure 1.

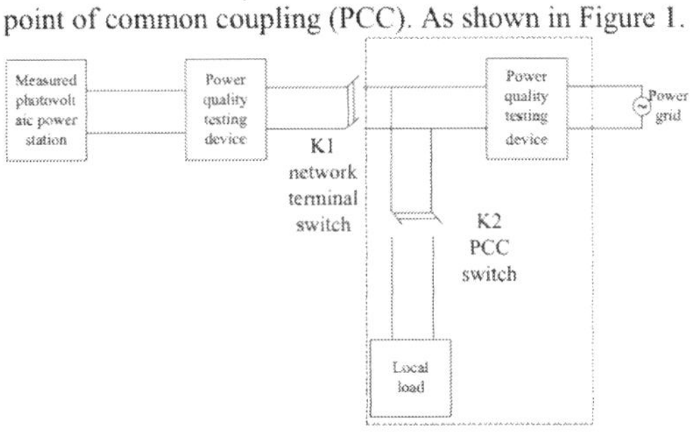

Fig.1 Power quality test schematic diagram

2.4 test procedure

1) The power quality test points should be located in PV power stations and outlets and public connection points.

2) Check the capacity of the photovoltaic power station actually put into the power grid.

3) The electric energy quality test device is used to measure the parameters of the energy quality index of the test point. Under the normal operation of the system, the continuous measurement is at least one day (with a complete irradiation period).

4) Read the test data of power quality test equipment and analyze it to determine whether to meet the relevant requirements.

2.5 Theoretical calculation of power quality

1) Harmonic

The main theoretical calculation formulas for harmonics are shown as follows [9].

$$HRU_h = \frac{U_h}{U_1} \times 100\% \tag{1}$$

Harmonic voltage content of hth:

$$U_H = \sqrt{\sum_{h=2}^{\infty} (U_h)^2} \tag{2}$$

Total voltage harmonic distortion rate:

$$THD_u = \frac{U_H}{U_1} \times 100\% \tag{3}$$

2) Voltage imbalance

The severity of voltage imbalance is often quantified by (negative sequence) voltage unbalance. It is defined as the ratio of negative sequence voltage U_{-1} to positive sequence voltage U_{+1}. It is generally expressed as a percentage [5]:

$$K_d = \frac{U_{-1}}{U_{+1}} \times 100\% \tag{4}$$

Sometimes, the regional distribution company uses the difference between the maximum and minimum of the steady phase voltage or line voltage to quantify the unbalanced voltage.

3 General situation of photovoltaic power station

The capacity of the PV module is 5983kWp. The annual PV power generation is 3647640kWh, and the maximum user demand is 309080kWh.The grid type is first to meet self-need and the surplus electricity supplies to the grid. According to the regulations and site conditions of *Typical design of distributed photovoltaic power generation access system in State Grid*, this project is divided into 6 connection points ,and the capacity of the PV module are 4 groups of 1247.75kWp and 2 groups of 496kWp respectively. The PV module is partially connected to the low voltage distribution cabinet of the low-voltage side bus of the user distribution station, and some of the power stations are connected to the user distribution station. High voltage power distribution cabinet with high voltage side bus.

The wiring diagram of the photovoltaic power station is shown in Figure 2.

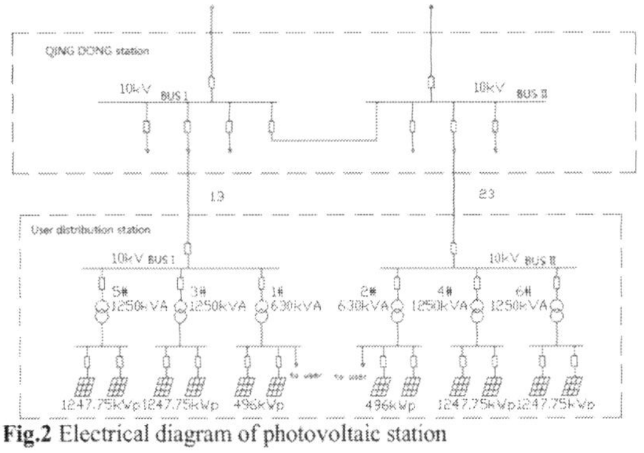

Fig.2 Electrical diagram of photovoltaic station

 https://doi.org/10.1051/e3sconf/20186406008

4 Test and analysis

This article mainly carries on the power quality test to the high voltage side and the low voltage side of #1-#6 transformer in Qingpu NGO distributed photovoltaic power station.

4.1 harmonic analysis

The harmonic content of the photovoltaic power station should be controlled in the range of the allowable value stipulated in the *GB14549-1993 power quality public power grid harmonics*. The total harmonic distortion rate of the 10 kV bus voltage is not more than 4%.The odd harmonic content is not more than 3.2%, and the ratio of the even subharmonics is not more than 1.6%. Otherwise, the corresponding filter device should be equipped to avoid harmonic pollution to the public power grid.The harmonic test result of the high voltage side and low voltage side of each transformer in Qingpu photovoltaic power station are shown in Table 1. It can be seen from table 1 that the voltage values of each harmonic are small and all within the limit specified by the state. Fig. 3 to Fig.6 show the voltage waveforms of the 2# and 4# transformers on the high voltage side and the AC side of the inverter and the corresponding harmonic voltage fraction. As shown in Fig. 3 to Fig.6 that the voltage waveform is smooth, and the harmonic voltage fraction is low, and there is no abnormal condition.

Tab.1 Harmonic result

Transformer name	Rated capacity (kVA)	High voltage side		Low voltage side	
		THD(%)	*HRU*(%)	*THD*(%)	*HRU*(%)
1#	630	2.39	2.23	2.26	1.99
2#	630	1.51	1.42	1.64	1.43
3#	1250	2.50	2.38	2.15	1.89
4#	1250	1.45	1.34	1.84	1.65
5#	1250	2.30	2.10	2.15	1.89
6#	1250	1.47	1.38	1.70	1.54

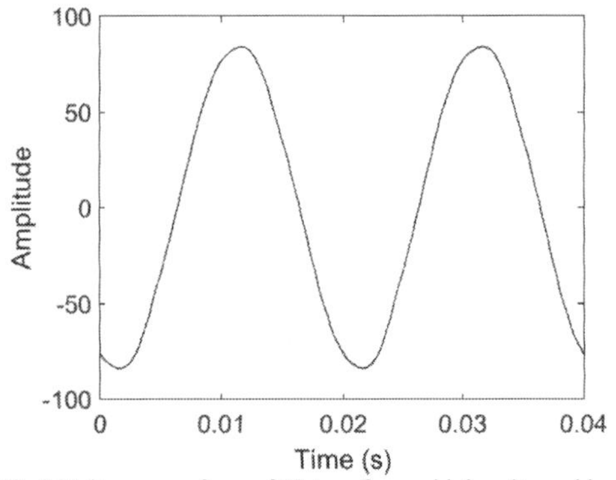

Fig.3 Voltage waveform of #2 transformer high voltage side

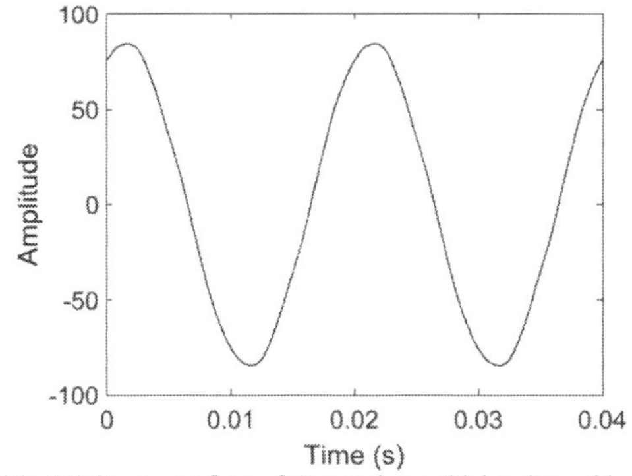

Fig.4 Voltage waveform of #4 transformer high voltage side

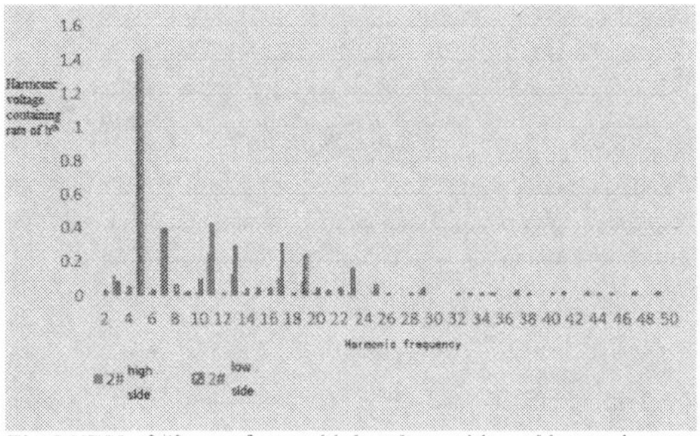

Fig.5 HRU of #2 transformer high voltage side and low voltage side

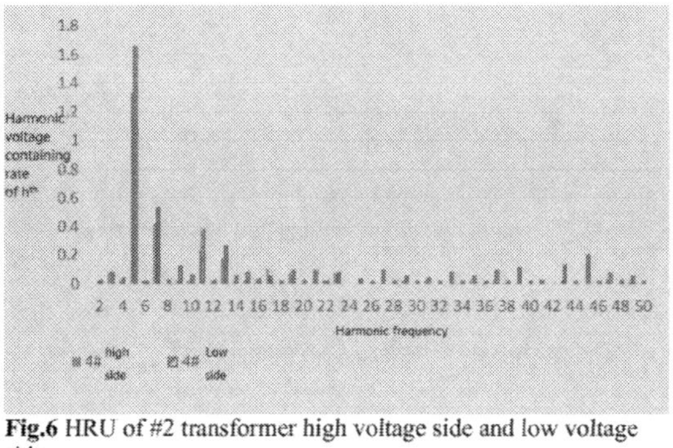

Fig.6 HRU of #2 transformer high voltage side and low voltage side

4.2 Analysis of voltage imbalance

When the PV power station is connected to the power grid, the voltage imbalance of the PCC should not exceed the requirements of *GB/T 15543-2008 power quality three-phase voltage imbalance*.

The negative sequence voltage unbalance test values of high voltage side and low voltage side of each transformer in Qingpu NGO photovoltaic power station are shown in Table 2. As shown in Table 2, the PV power station satisfies the requirement that the negative sequence voltage imbalance of the common connection point is less than 2%.

https://doi.org/10.1051/e3sconf/20186406008

Tab.2 Voltage imbalance result

Transformer name	Rated capacity (kVA)	Voltage imbalance (%)	
		High voltage side	Low voltage side
1#	630	0.2	0.28
2#	630	0.2	0.16
3#	1250	0.21	0.22
4#	1250	0.17	0.11
5#	1250	0.29	0.28
6#	1250	0.14	0.17

4. Frequency deviation

After the PV power station is connected to the power grid, the frequency deviation of the PCC should not exceed the requirement of *GB/T 15943-2008 power quality power system frequency deviation*. The frequency deviation test values of high voltage side and low voltage side of each transformer in Qingpu NGO photovoltaic power station are shown in Table 3. It can be seen from table 3 that the PV power station satisfies the requirement of frequency deviation of the PCC less than ±0.2Hz.

Tab.3 Frequency deviation result

Transformer name	High voltage side frequency (Hz)	High voltage side frequency deviation (Hz)	Low voltage side frequency (Hz)	Low voltage side frequency deviation (Hz)
1#	50.026	0.026	50.002	0.002
2#	49.971	0.029	49.980	0.020
3#	49.964	0.036	49.963	0.037
4#	49.964	0.036	50.009	0.009
5#	49.975	0.025	49.981	0.019
6#	49.976	0.024	49.997	0.003

5 Conclusion

In this paper, by introducing the method of measuring the quality of grid connected photovoltaic power, and combining with the largest photovoltaic power station in Shanghai, the Qingpu District of Shanghai, the power quality test of the power station is carried out, and the test results are analyzed, such as harmonic, voltage imbalance, frequency deviation and so on. The results show that the PV system is running well, and its access has no adverse effect on the power quality of the grid. But in the actual operation process, we should also take more precaution. With the increasing capacity of photovoltaic access, the problem of power quality will become more and more prominent. It is very important to master the method of power quality testing. It is necessary to test the power quality of the grid system regularly, to find problems early and solve problems early, and to provide security for the safe and stable operation of the power grid.

References

1. YANG Yongping, YANG Zhiping, XU Gang, WANG Ningling.Situation and Prospect of Energy Consumption for China's Thermal Power Generation[J].Proceedings of the CSEE,2013,33(23):1-11.
2. DING Ming, WANG Weisheng, WANG Xiuli, SONG Yunting, CHEN Dezhi, SUN Ming.A Review on the Effect of Large-scale PV Generation on Power Systems[J].Proceedings of the CSEE ,2014,34(1):1-14.
3. CHEN Wei,AI Xin,WU Tao,LIU Hui.Influence of grid-connected photovoltaic system on power network[J].Electric Power Automation Equipment,2013,33(2):26-32.
4. SANG Jingjing.Study of Impacts of Grid-connected PV on Distribution System[D]. Taiyuan University of Technology,2010.
5. XU Qun. Effects and Assessment on Power Quality of Distribution network with Distributed Generation[D]. North China Electric Power University,2012.
6. LIU Wei, PENG Dong, BU Guangquan, SU Jian.A Survey on System Problems in Smart Distribution Network With Grid-Connected Photovoltaic Generation[J]. Power System Technology,2009,33(19):1-6.
7. Wang Wenjing, Wang Sicheng .Status and Prospect of Chinese Distributed Photovoltaic Power Generation System[J].Journal of the Chinese academy of sciences,2016,31(2):165-172.
8. Han Fujia, Wang Chun.Influences of the distributed photovoltaic grid-connected generation system on the distribution network power quality based on Matlab[J].Electrical Measurement & Instrumentation,2015,52(15):16-21.
9. GB14549-93, Power quality common grid harmonics[S].State technical supervision bureau of the People's Republic of China,1994.

E3S Web of Conferences **64**, 06009(2018) https://doi.org/10.1051/e3sconf/20186406009
ICPRE 2018

Optimal design of a coproduction system of electricity and hydrogen to manage imbalances resulting from forecast errors in photovoltaic outputs

Ryuji Matsuhashi and *Tsuyoshi* Yoshioka

Department of Electrical Engineering and Information Systems, Graduate School of the University of Tokyo, 113-8656 Tokyo, Japan

Abstract. Renewable power sources are increasing mainly because of economic institutions such as renewable portfolio standard or feed-in tariff program. In Japan, the feed-in tariff program triggered explosive growth of photovoltaic systems because of its high tariff level. Although mass introduction of photovoltaic systems certainly contributes to reduce CO2 emissions, it causes instability issues in power systems. One of the most serious issues is management of imbalances resulting from forecast errors in photovoltaic outputs. On the other hand, power-to-gas technologies are attracting our attention, since these technologies could convert surplus of renewable energy to other energy carriers. In particular, hydrogen is efficiently produced from electricity using electrolysis. We could use hydrogen to manage the imbalances by the system, in which uncertain parts of photovoltaic outputs are used to produce hydrogen. In this paper, we propose a coproduction system of electricity and hydrogen to reduce the imbalances. For this purpose, a novel mathematical model is developed, in which we determine the structure of the coproduction system with a mixed integer linear programming method. Evaluated results indicated that the coproduction system is economical under appropriate capacity of the electrolyzer.

1 Introduction

Renewable power sources are increasing in the world, mainly because of economic institutions such as RPS, renewable portfolio standard or FIT, feed-in tariff program. In Japan, PV, photovoltaic systems have been explosively increasing by the FIT program since 2013.

Increase of renewable power sources certainly contributes to CO_2 reductions to mitigate climate change. However, mass introduction of renewable power sources is causing serious instability issues; PV outputs are uncertain, causing power system instability. In unit commitments to schedule starts and stops of power plants in advance, operators have to forecast PV outputs in future. But this forecast inevitably includes errors due to uncertainty in future weather conditions. [1] As results, operators have to manage the imbalances resulting from the forecast errors, which are rapidly increasing as the explosive growth of PV systems. Thus management of the imbalances is one of the most serious issues in present power systems.

On the other hand, local power retail companies are increasing nowadays in various countries including Japan. These companies are mainly procuring renewable power sources such as PV systems. It is therefore significant for the local power retail companies to reduce the imbalances by themselves. If they can appropriately manage and compensate the imbalances, that also contributes to the stable management of entire power systems. Regarding analyses in this field, Takahashi [2] developed a mathematical model to operate power systems in the uncertainty. But it is not for local power retail companies, but for large electric utilities.

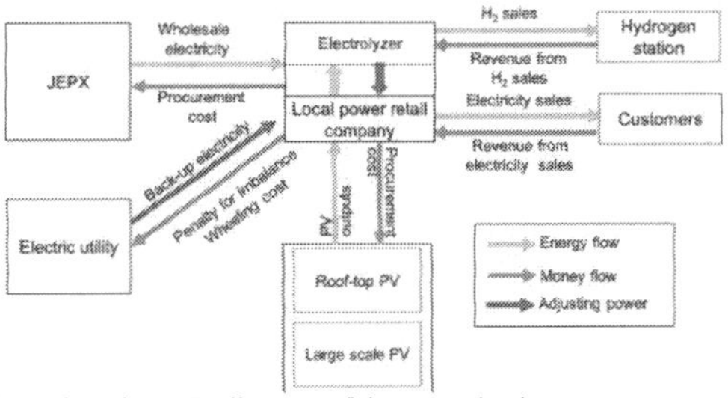

Fig.1 The schematic diagram of the coproduction system

We therefore intend to develop a novel system to manage the imbalances for the local power retail companies. In this respect, power-to-gas technologies are attracting our attention, since these technologies could convert surplus of renewable energy to other energy carriers. In particular, hydrogen is efficiently produced from electricity using electrolysis. We could use hydrogen to manage the imbalances, if uncertain parts of

https://doi.org/10.1051/e3sconf/20186406009

PV outputs are used to produce hydrogen. This paper aims at designing a novel coproduction system, in which certain parts of PV outputs are sold as electricity, and uncertain parts of PV outputs are sold as hydrogen. Figure 1 shows the schematic diagram of the coproduction system.

There are past researches utilizing hydrogen to control fluctuating wind power outputs. For instance, the article [3] analyzes the profitability of a wind power plant in combination with a hydrogen storage using decision tree approach, while the article [4] addresses the problem of high imbalance costs for effective trading of wind power, and presents the operation strategy for the minimization of imbalance costs by the use of hydrogen storage. Although there are past researches utilizing hydrogen to control fluctuating wind power outputs, very few papers exist regarding to hydrogen utilization for managing imbalances in a local power retail company, which is investigated in this article.

2 Methodologies adopted in this article

This chapter deals with an objective function, constraint functions and other assumptions in the mathematical model for simulating power systems of the evaluated local power retail companies. Table 1 lists the variables used in the objective function and the constraint functions.

Table 1. Variables used in the mathematical model.

P_{tot}	Cash flow of the local power retail company
R_{tot}	Cash inflow of the local power retail company
C_{tot}	Cash outflow of the local power retail company
R_L	Cash inflow from low-voltage consumers
R_H	Cash inflow from high-voltage consumers
R_{H2}	Cash inflow from hydrogen sales
C_{PV}	Procurement cost of electricity from PV
C_{BU}	Procurement cost of electricity based on the ordinary back-up contract
C_{WR}	Wheeling cost
C_M	Procurement cost of electricity from Japan Electric Power Exchange
C_{ELE}	Cost of electrolyzer
C_{IB}	Expected cost of penalty for imbalance
$R_{L,base}$	Basic charge by low voltage retail contracts
$R_{H,base}$	Basic charge by high voltage retail contracts
$RP_L(t)$	Amount-based charge by low voltage retail contracts
$RP_H(t)$	Amount-based charge by high voltage retail contracts
$PP_{PV}(t)$	Procurement rate of electricity from PV
$C_{BU,base}$	Basic rate of ordinary back-up contract
$PP_{BU}(t)$	Procurement rate of electricity based on the ordinary back-up contract
$C_{WR,base}$	Basic charge of wheeling
$PT_H(t)$	Amount-based rate of high voltage wheeling
$PT_L(t)$	Amount-based rate of low voltage wheeling
$PP_M(t)$	Wholesale rate of electricity in Japan Electric Power Exchange
$PE_{IB}(t)$	Penalty rate for imbalances
$L_L(t)$	Planned value of low voltage demand
$L_H(t)$	Planned value of high voltage demand
$G_{PV}(t)$	Planned value of procuring PV outputs
$G_{BU}(t)$	Planned value of procuring ordinary back-up electricity
$WR_H(t)$	Planned value of wheeling at high voltage
$WR_L(t)$	Planned value of wheeling at low voltage
$G_M(t)$	Planned value of procuring electricity from JEPX
$IB_{EXP}(t)$	Expected value of Imbalance
$H_2(t)$	Production amount of hydrogen
$P_{H2}(t)$	Selling price of hydrogen

2.1 The objective function and the constraint functions in this analysis

In this section, we explain the objective function and the constraint functions in our numerical model, in which the power system is represented with a mixed integer linear programming method.

In this study, the power system is determined, so that the total profit including revenues and costs takes the maximum as shown in the equation (1).

$$\text{Maximum } P_{tot} = R_{tot} - C_{tot} \tag{1}$$

The revenues come from electricity retail for high-voltage customers, low-voltage customers and hydrogen sales, which are shown in the equation (2), (3) and (4), respectively.

$$R_L = R_{L,base} + \sum_t (L_L(t) \times RP_L(t)) \tag{2}$$
$$R_H = R_{H,base} + \sum_t (L_H(t) \times RP_H(t)) \tag{3}$$
$$R_{H2} = \sum_t (H_2(t) \times P_{H2}(t)) \tag{4}$$

The costs consist of procurement of electricity from the contracted solar power generators, wholesale electricity traded in JEPX, Japan Electric Power Exchange, ordinary back-up electricity from Kyushu Electric Utility, the cost of electrolyzer, the wheeling payment and the penalty cost of imbalance as shown in the equation (5).

$$C_{tot} = C_{PV} + C_M + C_{BU} + C_{WR} + C_{ELE} + C_{IB} \tag{5}$$

The wheeling costs paid to the Kyushu Electric Utility are also included as shown in the equation (6), since the local power retail company uses the power transmission and the distribution lines owned by the utility.

$$C_{WR} = C_{WR,base} + \sum_t (WR_H(t) \times PT_H(t) + WR_L(t) \times PT_L(t)) \tag{6}$$

The equations (7), (8), (9) and (10) express how to estimate the procurement cost of electricity from the contracted solar power generators, the wholesale electricity traded in JEPX, the ordinary back-up electricity, and the penalty cost of imbalance, respectively.

$$C_{PV} = \sum_t (G_{PV}(t) \times PP_{PV}(t)) \tag{7}$$
$$C_M = \sum_t (G_M(t) \times PP_M(t)) \tag{8}$$
$$C_{BU} = C_{BU,base} + \sum_t (G_{BU}(t) \times PP_{BU}(t)) \tag{9}$$
$$C_{IB} = \sum_t (IB_{EXP}(t) \times PE_{IB}(t)) \tag{10}$$

The sum of electricity demand must be equal to the sum of electricity supply in each time slot as shown in the equation (11).

$$L_L(t) + L_H(t) = G_{PV}(t) + G_{BU}(t) + G_M(t) \qquad (11)$$

2.2 Imbalances resulting from forecast errors in PV outputs

In this section, we explain how to deal with errors in forecasting solar power outputs. Takahashi [1] estimated RMSE, root mean square error, in the forecast. According to this article, the probability density function of the forecast errors follows a Laplacian distribution, in which the value of RMSE is approximately 10% in the one-day-ahead forecast.

Therefore, we assumed the Laplacian distribution as in the equation (12) with 10% of RMSE in the forecast errors. Then we numerically divided the Laplacian distribution into the following six sets, S(1), S(2), S(3), S(4), S(5), S(6).

$$\mathrm{P(dx)} = \frac{1}{\sqrt{2}\sigma} \cdot \mathrm{e}^{-\frac{\sqrt{2}|dx|}{\sigma}} \qquad (12)$$
$$\mathrm{S(1)} = \{\mathrm{dx}| -4.1\sigma \le \mathrm{dx} \le -2.1\sigma\} \qquad (13)$$
$$\mathrm{S(2)} = \{\mathrm{dx}| -2.1\sigma \le \mathrm{dx} \le -0.8\sigma\} \qquad (14)$$
$$\mathrm{S(3)} = \{\mathrm{dx}| -0.8\sigma \le \mathrm{dx} \le 0\} \qquad (15)$$
$$\mathrm{S(4)} = \{\mathrm{dx}| \quad 0 \le \mathrm{dx} \le 0.8\sigma\} \qquad (16)$$
$$\mathrm{S(5)} = \{\mathrm{dx}| \quad 0.8\sigma \le \mathrm{dx} \le 2.1\sigma\} \qquad (17)$$
$$\mathrm{S(6)} = \{\mathrm{dx}| \quad 2.1\sigma \le \mathrm{dx} \le 4.1\sigma\} \qquad (18)$$

Probabilities, for which forecast error, dx is included in S(1), S(2), S(3), S(4), S(5) and S(6), are assumed to be 2.5%, 13.6%, 33.9%, 33.9%, 13.6% and 2.5%, respectively.

Figure 2 depicts the forecast errors, in which the numbers, 1, 2, 3, 4, 5 and 6 imply the probabilistic areas of S(1), S(2), S(3), S(4), S(5) and S(6), respectively.

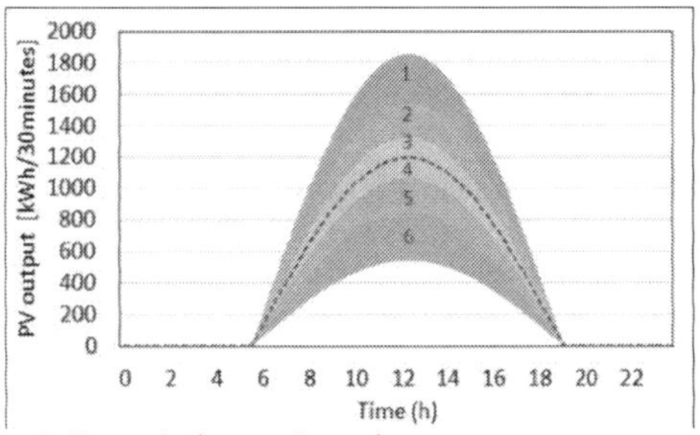

Fig. 2. Errors in forecasting solar power outputs.

Then we determine a baseline for coproduction of electricity and hydrogen as follows. That is, we use PV outputs higher than the baseline to produce hydrogen by electrolysis, and lower than the baseline to sell as electricity. In this article, we set three kinds of baselines, which are the lower limits of S(4), S(5) and S(6). These are defined as baselines 4, 5 and 6, respectively. In chapter 3, we evaluate the coproduction systems with these baselines as case studies.

2.3 Assumptions on electricity demand

Regarding electricity demand, we assumed the high and low voltage demand on the basis of an actual local power retail company. [5]

2.4 Assumptions on prices of electricity procured and sold by the local power retail company

Wholesale prices, by which the local power retail company procure electricity, is assumed based on the price traded in JEPX. [6]

On the other hand, we assumed the retail prices of electricity by the local power company referring to actual data. [5]

3 Evaluated results and discussions

We first evaluated the relationships between capacities of the electrolyzer and annual cash flows. Figure 3 depicts the cases, in which the baselines of electrolysis are placed on 4, 5 and 6, respectively. Evaluated results show that the cash flow is lowest in the baseline 6, when the capacity of electrolyzer is 0. It is natural because, revenue from electricity sales is the least among three cases. As the capacity of the electrolyzer increases, the cash flow in the baseline 6 is drastically improved. Although cash flows in the baselines 4 and 5 are also improved, they are not so sharp as in the baseline 6. As results, the highest cash flow point is acquired by 600kW of electrolyzer in the baseline 6.

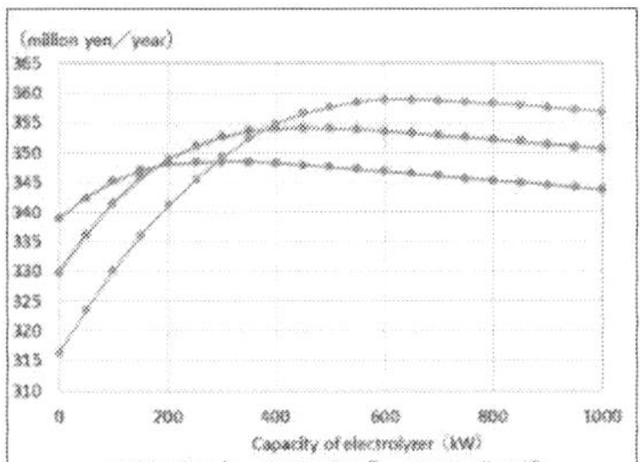

Fig. 3. Relationships between annual cash flows and capacities of electrolyzers.

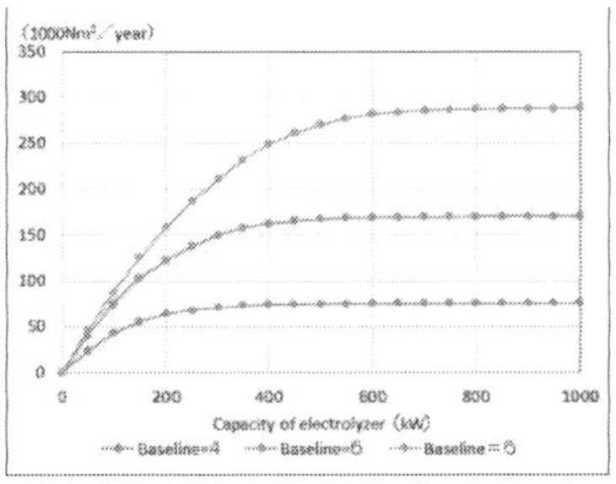

Fig. 4. Relationships between annual revenues from hydrogen sales and capacities of electrolyzers.

Figure 4 depicts relationships between annual revenues from hydrogen sales and capacities of electrolyzers, while relationships between annual penalties and the capacities are shown in figure 5. From figure 4, hydrogen sales are shown to saturate after certain capacities of electrolyzers.

In particular, the hydrogen sale in the baseline 6 most drastically increases and then saturates as the capacity increases. From figure 5, imbalances are shown to steeply decrease as capacities of electrolyzers increase. The imbalance cost in the baseline 6 is the highest when the capacity is zero. But it drastically decreases and becomes the least with the capacity of electrolyzer higher than 700kW.

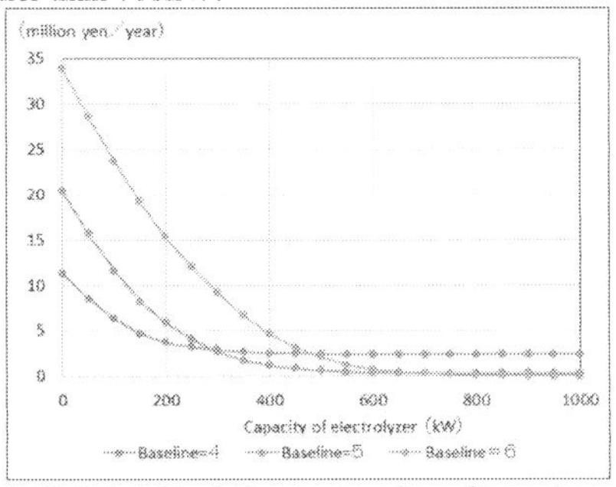

Fig. 5. Relationships between annual penalty costs and capacities of electrolyzers.

Figure 6 shows relationships between factors influencing annual cash flows in the baseline 6 and the capacity of electrolyzer. The hydrogen cost increases and the imbalance cost decreases and then saturate, as the capacity increases. On the other hand, annual cost of the electrolyzer is proportional to the capacity. Taking these trends into consideration, it is natural that the highest cash flow point is acquired in the baseline 6.

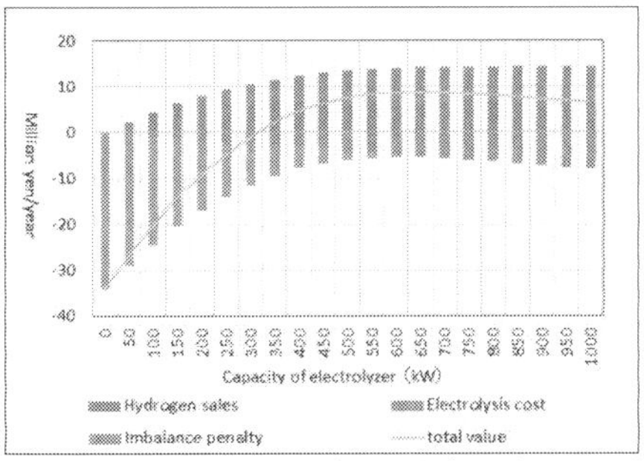

Fig. 6. Relationships between factors influencing annual cash flows in baseline 6 and capacities of electrolyzers.

The above analyses show that the coproduction system of electricity and hydrogen is economical under certain condition.

4 Conclusion

Although increase of renewable power sources certainly contributes to decarbonization of energy systems, explosive growth of them threatens stability of power systems. One of the serious issues is how to manage imbalances resulting from forecast errors in solar power generation.

In this paper, we propose a coproduction system of electricity and hydrogen to manage the imbalances. For this purpose, we developed a mathematical model to design the coproduction system dealing with the imbalances. In this model, the imbalances are assumed to be generated probabilistically according to Laplacian distribution. We determine the structure of the coproduction system to maximize the cash flow of a local power retail company, for which the model is represented with a mixed integer linear programming method.

Evaluated results using the above mathematical model indicated that the coproduction system is economical under certain capacities of the electrolyzer and the appropriate baselines of PV outputs, above which hydrogen is produced using electrolysis.

This study was conducted as a part of 'Evaluation and verification of effects to reduce CO2 emissions utilizing hydrogen in 2018 fiscal year' by Ministry of Environment, Japan.

References

1. M. Takahashi and R. Matsuhashi, "Area-wide Total Wind and Photovoltaic Power Forecasting Using Multiple Regression Technique and Analysis of Forecast Error Characteristics", *Journal of Japan Society of Energy and Resources*, **38**, 5, 2017.
2. M. Takahashi and R. Matsuhashi, "A Cost Reduction Analysis of Introduction of Battery Energy Storage and Controllable Heat Pump Water Heaters by Operation Planning Model of Power Generation System Considering the Uncertainty in Renewable Power Generation", *Transactions on Power and Energy (Journal Sector B)*, Japan Institute of Electrical Engineering, **137**, 12, 2017.
3. Schuster, Martin and Walther, Thomas, Valuation of Combined Wind Power Plant and Hydrogen Storage: A Decision Tree Approach, IEEE Xplore: Proceedings of the 14th International Conference on the European Energy Market (EEM), Dresden 2017.
4. Thakur, T., Goyal, S., Gambhir, J., and Kaur, I. "Optimization of Imbalance Cost for Wind Power Marketability using Hydrogen Storage", 2008 Joint International Conference on Power System Technology and IEEE Power India Conference, 2008.
5. Miyama Smart Energy. [Online]. Available: http://miyama-se.com/electrical/sim-menu/sim-ans-index. [Access date: 2017.11.16]
6. JEPX, *JEPX trading information*, [Online]. Available: http://www.jepx.org/market/index.html. [Access date: 2017.11.17].

Wake and Turbulence Analysis for Wind Turbine Layouts in an Island

Amrender Singh Bachhal[1], *Klaus* Vogstad[2], *Mohan Lal* Kolhe[1], *Abhijit* Chougule[1], *Hans George* Beyer[3]

[1]Faculty of Engineering and Science, University of Agder, PO Box 422, NO 4604, Kristiansand, Norway
[2]Wind Farm Designs, Kristiansand, Norway
[3]Department of Science and Technology, University of the Faroe Islands, Torshavn, Faroe Islands

Abstract. There is a big wind energy potential in supplying the power in an island and most of the islands are off-grid. Due to the limited area in island(s), there is need to find appropriate layout / location for wind turbines suited to the local wind conditions. In this paper, we have considered the wind resources data of an island in Trøndelag region of the Northern Norway, situated on the coastal line. The wind resources data of this island have been analysed for wake losses and turbulence on wind turbines for determining appropriate locations of wind turbines in this island. These analyses are very important for understanding the fatigue and mechanical stress on the wind turbines. In this work, semi empirical wake model has been used for wake losses analysis with wind speed and turbine spacings. The Jensen wake model used for the wake loss analysis due to its high degree of accuracy and the Frandsen model for characterizing the turbulent loading. The variations of the losses in the wind energy production of the down-wind turbine relative to the up-wind turbine and, the down-stream turbulence have been analysed for various turbine distances. The special emphasis has been taken for the case of wind turbine spacing, leading to the turbulence conditions for satisfying the IEC 61400-1 conditions to find the wind turbine layout in this island. The energy production of down-wind turbines has been decreased from 2 to 20% due to the lower wind speeds as they are located behind up-wind turbine, resulting in decreasing the overall energy production of the wind farm. Also, the higher wake losses have contributed to the effective turbulence, which has reduced the overall energy production from the wind farm. In this case study, the required distance for wind turbines have been changed to 6 rotor diameters for increasing the energy gain. From the results, it has been estimated that the marginal change in wake losses by moving the down-stream wind turbine by one rotor diameter distance has been in the range of 0.5 to 1% only and it is insignificant. In the full-length paper, the wake effects with wind speed variations and the wind turbine locations will be reported for reducing the wake losses on the down-stream wind turbine. The Frandsen model has been used for analysing turbulence loading on the down-stream wind turbine as per IEC 61400-1 criteria. In larger wind farms, the high turbulence from the up-stream wind turbines increases the fatigues on the turbines of the wind farm. In this work, we have used the effective turbulence criteria at a certain distance between up-stream and down-stream turbines for minimizing the fatigue load level. The sensitivity analysis on wake and turbulence analysis will be reported in the full-length paper. Results from this work will be useful for finding wind farm layouts in an island for utilizing effectively the wind energy resources and electrification using wind power plants.

1 Introduction

Wind turbines generate power by converting the kinetic energy of wind into electricity[1]. When the wind passes through an operating wind turbine, the wind speed of the downstream turbine decreases due to the wake from upstream turbine. Thus, due to the reduced wind speed, the power generated by the turbine operated downstream in the wake reduces compared to that of the leading turbine. As second effect, the increase in turbulence intensity increases the fatigue loading of the downstream turbine and thus reduces the lifespan of that wind turbine. Both effects are necessary to analyse for prediction of energy gain as well as lifetime of a wind farm configuration [2].

The purpose of this analysis is to determine the magnitude of wake losses when the effective turbulence criteria is satisfied assuming a certain ambient turbulence. We demonstrate the wake loss sensitivity of two turbines that are spaced such that they satisfy the effective turbulence criteria. As per effective turbulence criteria, the value of the effective turbulence σ_{eff} must be lower or equal than the value of the turbulence standard deviation σ_1 that is given by the 90-percentile value of the wind turbulence intensity I_{90} for the given hub height wind speed V_{hub} [3, 4], such that:

$$\sigma_{eff} \leq \sigma_1, \quad (1)$$

where, σ_1 is the standard deviation of the longitudinal component of the wind velocity, and σ_{eff} is effective turbulence intensity.

We consider V_{hub} wind speed at hub height 15 m/s as per IEC standards. The comparison is only necessary in a wind range from 60% of rated wind speed at hub height $V_{hub} < 0.6V_r$. Where, V_r is the rated wind speed averaged over 10 min.

To predict the wind velocity deficit, many studies have been done to compare different engineering wake models to analyse the performance. In a recent case study, Frandsen and Jensen models are explained in detailed, have been compared with Computational Fluid Dynamics (CFD) simulations (see in Methods) [5].

In the following sections, a detailed study has been done on both Jensen and Frandsen analytical models. Both the models are compared to check wake loss sensitivity as well as effective turbulence criteria, respectively. The total power of the downstream wind turbine is compared with upstream turbine to analyse the amount of wake loss effect. Wake loss sensitivity is demonstrated by effective turbulence criteria.

2 Methods

Two models N.O. Jensen and Frandsen [6] are used to calculate the normalised wake loss for downstream wind turbine and effective turbulence intensity I_{eff} .

2.1 N.O. Jensen wake model

Wake loss is estimated by using N.O. Jensen Wake Model. The model uses the linear expansion of the wake radius with the downstream spread distance [7], the model describes the single turbine wake.

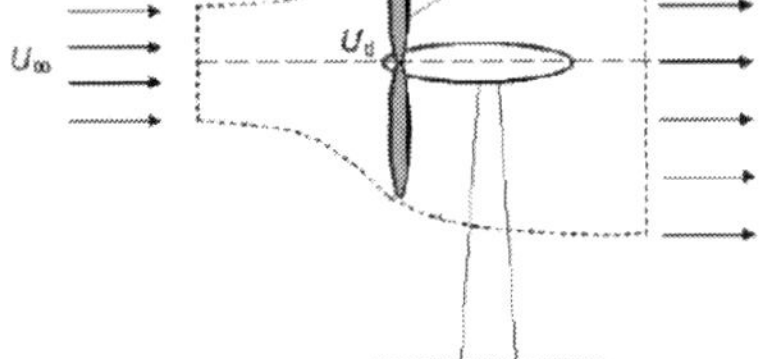

Fig. 1. Schematic of Jensen Wake Model [8].

The velocity deficit in wind wake, can be expressed as [9]:

$$1 - \frac{U_w}{U_\infty} = \frac{1-\sqrt{1-C_t(V_{hub})}}{(1+2k_w x)^2}, \quad (2)$$

where, U_w is the in-wake velocity, U_∞ is the ambient free stream velocity, C_t is the thrust coefficient, k_w the wake decay coefficient [10] (Onshore value of k_w = 0.075 and for offshore value of k_w = 0.04 [11]) *and* x is the turbine spacing, normalized by the rotor diameter.

2.2 Frandsen model

The Frandsen model [12] defines the effective turbulence as a combination of ambient and wake generated turbulence integrated overall directions in a way that accounts for accumulation of fatigue using material properties. The effective turbulence is calculated using the 90th percentile of ambient turbulence as per IEC61400-1 edition-3 2010 amendment [13].

For each wind Turbine Generation position in the calculation, the Frandsen model needs the following inputs[12]:

1. $\hat{\sigma}(\theta, V_{hub})$ and $\hat{\sigma}_\sigma(\theta, V_{hub})$ – Ambient turbulence (mean and standard deviation functions of direction and wind speed)
2. $W(A_i, k_i)$ and $f(\theta_i)$ – Weibull distributions and sector-wise frequencies
3. C_T – Turbine thrust curve and park geometry
4. m – Relevant material fatigue property Wöhler exponent [14].

Input 1 is used to calculate the ambient characteristic turbulence, i.e. the 90th percentile.

Input 2 is used to calculate the directional wind speed distribution conditioned on wind speed.

Input 3 is used to calculate the wake generated contribution to turbulence.

Input 4 is used in the fatigue weighted combination model of single directions to obtain an omnidirectional effective turbulence as a function of wind speed only.

Effective turbulence is calculated as function of wind speed only. This is done by integrating the directional variation of turbulence over all directions for each wind speed bin. However, effective turbulence is not a measurable quantity as it combines the directional contributions with a special weighting that accounts for material fatigue via use of the material parameter, the Wöhler exponent. The estimated wake added turbulence (σ_{wake}) contribution is combined with the 90th percentile of the ambient turbulence at each wind turbine generation. The normal turbulence model is illustrated below for each of the three turbulence classes. To solve for effective turbulence as a function of wind speed at hub height $\sigma_{eff}(V_{hub})$:

$$\sigma_{eff}(V_{hub}) = [\sum_s \sigma_{total}(s, V_{hub})^m . f(s)]^{\frac{1}{m}}, \quad (3)$$

where, s is the number of sectors, (in our case s=1), f is the frequency sector-wise, m is the material parameter Wöhler exponent (in our case m=10), and σ_{total} is the total turbulence.

If we consider only one wind direction and Wöhler coefficient (m=10), then the effective turbulence, σ_{eff} reduces as:

$$\sigma_{eff}(V_{hub}) = \sigma_{tot}(V_{hub}) \quad (4)$$

2.3 Wake added turbulence (σ_{wake})

Wind turbines in their normal operation periods within wind farm changes the ambient turbulence that causes wake from the neighbouring or nearest turbines. We calculate the wake added turbulence (σ_{wake}), for distance with less than 10RD, to verify with $I_{ref} = 0.14$. We verify the effective turbulence variations over distance less than 10RD, such as:

$$\sigma_{wake} = \begin{cases} \dfrac{V_{hub}}{1.5+0.8\times\dfrac{x}{\sqrt{C_T(V_{hub})}}}, & x < 10\text{RD} \\ \sigma_{wake} = 0, & x > 10\text{RD} \end{cases}, \quad (5)$$

where, RD is rotor diameter.

The turbulence criteria given by equation (5) solve for the distance that satisfies the effective turbulence criteria of the turbine class (in our case class B) at the given wind speed (=15 m/s) [12], and we calculate the wake loss at this distance. Then we verify the changes in wake loss if we increase the distance by one rotor diameter.

2.4 Total Turbulence

The total turbulence σ_{tot} is different from effective turbulence σ_{eff} , is calculated in each direction combining of characteristic turbulence (σ_c) and calculated wake added turbulence σ_{wake} [12].

$$\sigma_{tot} = \sqrt{\sigma_c{}^2 + \sigma_{wake}{}^2} \quad (6)$$

The characteristic turbulence(σ_c) is calculated as the 90th percentile of the estimated turbulence. The correction factor 1.28 is estimated from the normal distribution curve of standard deviation, where the value of 90% percentile is sought [15].

$$\sigma_c = (\hat{\sigma} + 1.28\hat{\sigma}_{\sigma}) \quad (7)$$

The measured turbulence($\hat{\sigma}$) and standard deviation of the measured turbulence($\hat{\sigma}_{\sigma}$) estimated from time series data on 10-min interval at wind speed of 15 m/s. Downstream turbine is positioned such that the effective turbulence criteria is satisfied.

2.5 Turbine Classes

The external conditions to be specified for the design of wind turbines, i.e., extreme wind conditions, material properties etc. In general, when designing a wind farm, it is important to choose a certified turbine as per IEC 61400-1 edition 3 [13, 16], which defines the standard wind turbine design classes as:

Table 1. Wind Turbine Classes as per IEC 61400-1 [16]

Wind Turbine Class	I	II	III	S
V_{ref} [m/s]	50.0	42.5	37.5	Values Specified by the designer
A I_{ref} [-]	0.16			
B I_{ref} [-]	0.14			
C I_{ref} [-]	0.12			

The selection of wind turbines must be made in terms of its capacity of withstanding with severe wind conditions, and structural component of wind turbine such as the rotor blades, must comply with certain load capacity, that is basically choosing a suitable turbine model to develop a wind farm.

2.6 Estimation of annual energy production

To calculate the annual energy production, it is important to know the power curve. The power curve is function of wind speed, is obtained from given wind turbine model S-108 3MW is shown in Fig. 3.

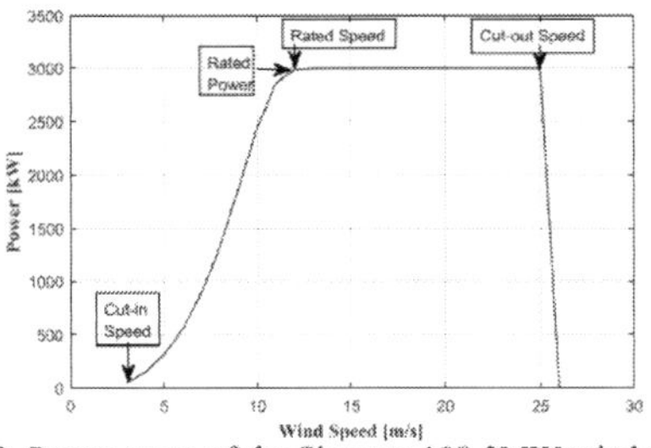

Fig. 3. Power curve of the Siemens 108 3MW wind turbine, for a range of wind speeds. The turbine starts producing power at cut-in wind speed of 3 m/s, but turbine reaches at its maximum power at wind speed of 13 m/s. When turbine reaches at maximum wind speed of 25 m/s, it stops operating.

The main reason for stopping the wind turbine at cut out speed is safety, because components of the turbines are not designed to handle the loads created by wind speed higher than the cut-out speed.

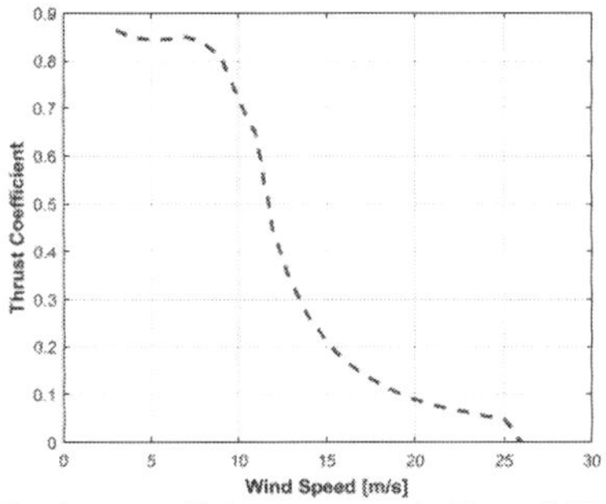

Fig. 4. The thrust coefficient curve, typically available from the manufacturer. Both the wind turbine power curve and thrust coefficient curve for the Siemens 108 turbine are also used for the study of wake loss effects.

Then, the overall annual energy (E) production of each wind turbine can be calculated as:

$$E(V_{hub}) = \int_3^{25} P(V_{hub}) W(V_{hub}, A, k) dV, \quad (8)$$

where, P is the Power curve of turbine as a function of wind speed at hub height, W is the Weibull distribution function, the k is Weibull shape parameter (equal to 2), A is Weibull scale parameter that is scaled at mean wind speed, 8.11 m/s and calculated as:

$$A = \frac{\overline{U}}{\Gamma\left(1+\frac{1}{k}\right)}, \quad (9)$$

where, $\overline{U}$ is the mean wind speed and Γ the gamma function.

The annual energy loss is estimated over the whole range of wind speed from 3-25 m/s. The energy loss is calculated for the downstream turbine due to the wake effects.

$$E_w\left(V_{def}\right) = \int_3^{25} P(V_{hub})W(V_{hub}, A_W, k)dV \quad (10)$$

Jensen wake model uses the assumption that the momentum is conserved inside the wake. Weibull parameter A_W is corrected by using sensitivity of the wake loss (ΔU) for downstream turbine to check the energy loss. Whereas, sensitivity of the wake loss ΔU or the velocity in the fully developed wake is calculated as:

$$\Delta U = \overline{U} \times \frac{1-\sqrt{1-C_t(V_{hub})}}{(1+2kx)^2} \quad (11)$$

The total power loss is estimated by:

$$\Delta E(V) = E(V_{hub}) - E_w\left(V_{def}\right), \quad (12)$$

where, E is the energy from upstream turbine [kWh] and E_w the energy from downstream turbine [kWh].

3 Results and discussion

3.1 Effective Turbulence Criteria

Effective turbulence is a simplified way of performing load calculations, recommended in the IEC 61400-1 standard. It is based on Frandsen's model. The effective turbulence criteria is checked by using normal turbulence model (NTM) as per IEC 61400-1 [13]. The parameter for the effective turbulence σ_{eff} is calculated and compared with value of the turbulence standard deviation σ_1, that is given by the 90 percentiles for the given hub height wind speed, V_{hub}(=15m/s), see Fig. 5. This value of the standard wind turbine class is described as [13]:

$$\sigma_1 = I_{ref}(0.75V_{hub} + 5.6), \quad (13)$$

where, I_{ref} is referenced value of turbulence intensity at certain hub height at 10 min average wind speed of 15 m/s. All values of the V_{hub} must be between the wind speed $0.2V_{ref} - 0.4V_{ref}$, (V_{ref}= 42.5, for class B turbines).

The effective turbulence σ_{eff} is verified with standard deviation turbulence intensity σ_1 over distance less than 10RD, see Fig. 6. Increase in loading are result of wake effects, and that is accounted for using effective turbulence.

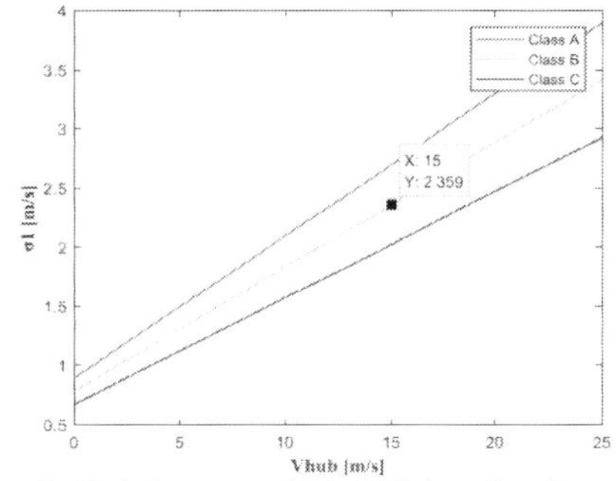

Fig. 5. Turbulence standard deviation for the normal turbulence model (NTM).

The value of turbulence is larger due to the higher wake of upstream turbine from x = 2RD -6RD, where it violates the condition of σ_1, is in order of 4.5 to 2, see Fig. 6. The effective turbulence criteria σ_{eff} is satisfied when turbine spacing at x = 8.40RD.

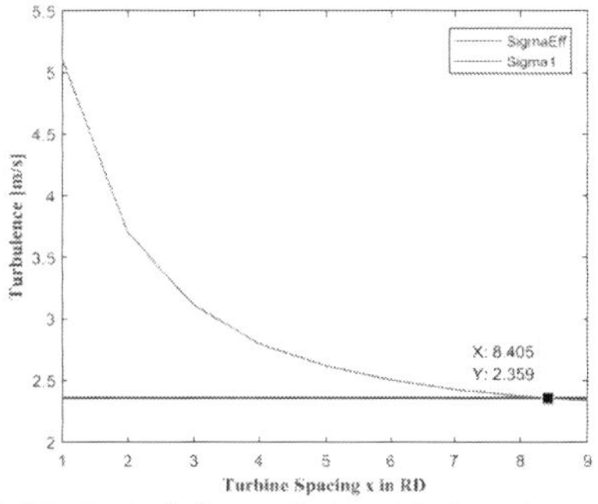

Fig. 6 Effective turbulence criteria satisfy the optimum turbine distance at x = 8.405RD

The normalized distance x, where criteria for effective turbulence is satisfied at $\sigma_{eff} = \sigma_1 = 2.359$. High effective turbulence from upstream turbines causes excessive fatigue on the blade of downstream turbine. Effective turbulence must satisfy the condition at certain distance from upstream turbine. We can check also that wake loss is almost insignificant at the optimum distance satisfied by effective turbulence criteria.

Table 2. Wake Loss at satisfied turbulence criteria

E(V) [kWh]	E($V_{deficit}$) [kWh]	ΔE[%]	Distance
1417	1368	3.38	8.405RD

Wake Loss with respect to distance is demonstrated where the total power production from freestream or upstream turbine is 1417 kWh, while power production from downstream turbine is decreased to 49 kWh, due to the wake loss effects at normalized distance x = 5RD.

Table 3. Wake Loss with Respect to Distance

E(V) [kWh]	E($V_{deficit}$) [kWh]	ΔE [%]	Distance [RD]
1417	1368	7.26	5
1417	1337	5.6	6
1417	1353	4.5	7
1417	1365	3.65	8
1417	1373	3	9
1417	1380	2.6	10

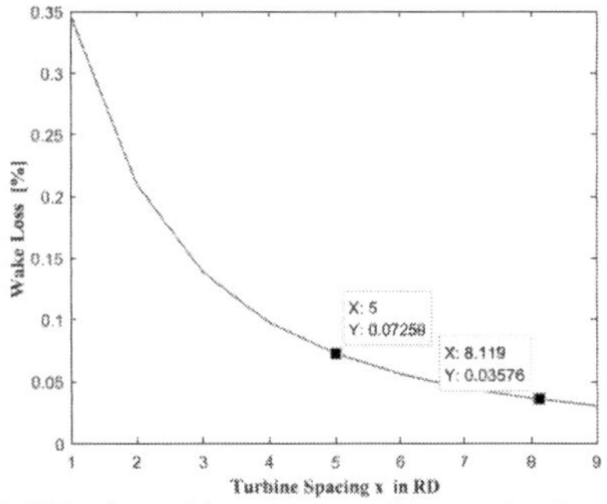

Fig.7. Wake loss with respect to turbine spacing in total, the energy loss at the distance of 5RD, is 7.26 %.The marginal change in wake losses is 0.5 to 1 % at 5RD. The losses are not significant.

In complex terrain, the "standard" wake decay coefficient $k_w = 0.075$ is lower, because of more turbulent mixing, which results in a faster decay. WindPro [16] has included the relationship of wake decay coefficient as a linear relationship of ambient turbulence σ and wake decay coefficient, k_w= 0.37 at 8.11 m/s [12].

$$k_w = 0.47.\sigma(V_{hub}, x)/V_{hub} + 0.04. \quad (14)$$

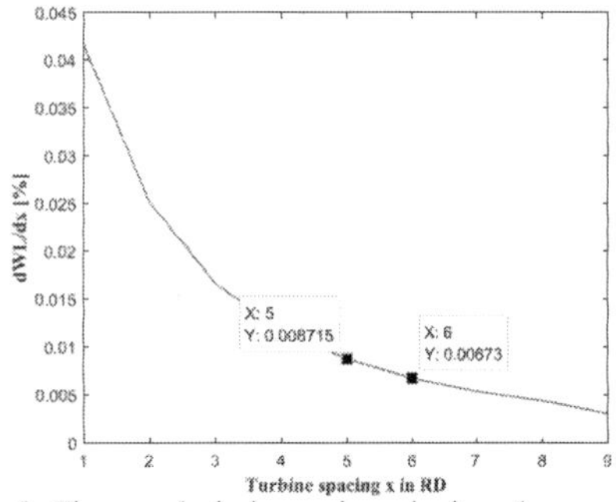

Fig. 8. The marginal change in wake loss, by moving the turbine one RD further downstream is:dWL/dx, where, WL is the wake loss function or a given mean wind speed (in our case 8.11 m/s). the marginal change in wake losses by accounting for wake losses at rotor distances around 5 RD, is in the order of less than 1 %.

Wake losses as function of rotor distances are in the order of 5-15%. The results assume Weibull distributions, and the SWT-108 power curve. If we consider an optimized layout without accounting for wake effects, moving the turbine downstream will gain some more energy (in the order of 2% from reduced wake losses), but will also lose some energy due to lower ambient wind conditions, and it is quite probable that this would reduce the potential gain from including wake effects down to 1% or less.

The other consideration is that positioning the turbine downstream from an optimization that does not consider the wake loss will be a less windy spot. The marginal change in energy is the sum of the reduced wake loss from moving the turbine downstream, plus the reduced energy production from a less windy location. We would assume this effect by reducing the effect of wake losses by 50%.

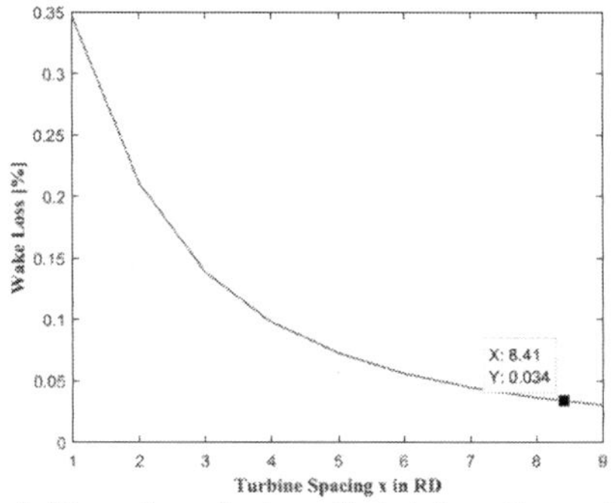

Fig. 9. Shows the wake loss effect at the optimum distance satisfied by effective turbulence criteria.

4 Conclusion

This paper demonstrates an analysis tools for windfarms in view of both the energy gain and the restriction of the fatigue loads within acceptable limits. As example, for a two turbine along-wind case the dependence of both wake losses and turbulence conditions to turbine spacing is analysed. As a result, it can be stated that for the case analysed the turbine spacing that results in turbulence conditions follow IEC 61400-1 minor wake losses. However, the required distance is elevated with 6 rotors diameters.

Wind turbines are sensitive to the wake losses at distances when they are operating in a large wind farm. N.O. Jensen wake model is used to check the wake effects from the upstream turbines. From the experimental results, we estimated that the marginal change in wake loss by moving downstream turbine by one rotor diameter distance is in order from 0.5 to 1% only. This marginal change is insignificant. On the other hand, if wake effects are not considered to optimize the wind farm layout, result can be less windy place for turbine, that will reduce the wake losses by half margin.

By using Frandsen model, we analysed the increased loads on downstream turbine during their normal operations as per IEC 61400-1 criteria (WFDs uses the same approach). In large wind farm, the high turbulence from upstream turbine increases the fatigue damage levels. We satisfy the effective turbulence criteria at a certain distance between upstream and downstream turbines to minimize the fatigue load level.

To sum up, it can be estimated from the analytical results that wake losses are not so important in load compliant optimizations. However, we must satisfy the effective turbulence criteria at certain distance to avoid fatigue for lifetime damages

In more general applications, this tool can be used to optimize the wind farm layout that provide fast and accurate wind turbine suitability assessment, maximizing the annual energy production including load constraints as per IEC 61400-1 standard.

References

1. K. E. Diamond and E. J. Crivella, "Wind turbine wakes, wake effect impacts, and wind leases: Using solar access laws as the model for capitalizing on wind rights during the evolution of wind policy standards," Duke Envtl. L. & Pol'y F., vol. 22, p. 195, 2011.
2. A. J. Brand, J. Peinke, and J. Mann, "Turbulence and wind turbines," Journal of Physics: Conference Series, vol. 318, no. 7, p. 072005, 2011.
3. P.H. Madsen, "Introduction to the IEC 61400-1 standard," Risø National Laboratory, Technical University of Denmark, 2008.
4. T.S. Leu, J.M. Yo, Y.T. Tsai, J.J. Miau, T.C. Wang, and C.C. Tseng, "Assessment of IEC 61400-1 Normal Turbulence Model for Wind Conditions in Taiwan West Coast Areas," International Journal of Modern Physics: Conference Series, vol. 34, p. 1460382, 2014.
5. S. J. Andersen, J. N. Sørensen, S. Ivanell, and R. F. Mikkelsen, "Comparison of engineering wake models with CFD simulations," Journal of physics: Conference series, vol. 524, no. 1, 2014.
6. P. Y. Zhang, "Topics in wind farm layout optimization: Analytical wake models, noise propagation, and energy production," University of Toronto, 2013.
7. M. Nielsen, H. E. Jørgensen, and S. T. Frandsen, "Wind and wake models for IEC 61400-1 site assessment,"European Wind Energy Conference and Exhibition, 2009.
8. M. Bilgili, A. Yasar, A. Ilhan, & A. Sahin, "Aerodynamic Characteristics of a Horizontal Axis Wind Turbine in Belen-Hatay, Turkey," International Journal of Natural and Engineering Sciences, vol. 9, no. 1, pp. 54-58, 2015.
9. J. Kollwitz, "Defining the Wake Decay Constant as a Function of Turbulence Intensity to Model Wake Losses in Onshore Wind Farms," ed. 2016.
10. M. Nielsen. Site assessment with WAT. Available: http://www.wasp.dk/wat#site-assessment
11. T. Sørensen, M. L. Thøgersen, P. Nielsen, and N. Jernesvej, "Adapting and calibration of existing wake models to meet the conditions inside offshore wind farms," EMD International A/S. Aalborg, 2008.
12. WindPro 2.9 - Site compliance module,. https://www.emd.dk/windpro/windpromodules/load-modules/, 2013.
13. International Eelectrotechnical Commission, "IEC 61400-1: Wind turbines part 1: Design requirements," International Electrotechnical Commission, 2005.
14. N. J. Tarp-Johansen, "Examples of fatigue lifetime and reliability evaluation of larger wind turbine components," 2003.
15. M. A. Saleh, A. A. Ani, and F. A. Hadi, "Ambient Turbulence Intensity Calculation for Al-Nasiriyah Province in Iraq." Iraqi Journal of Science, Vol 55, No.2A, pp:561-571, 2014.
16. E. I. A/S. windPRO Modules. Available: http://www.emd.dk/windpro/windpro-modules/

Influence of the DC offset on the DFT-based frequency estimation for noised multifrequency signals in PV systems with a DSP processor

Dariusz Kania[1], *Józef* Borkowski[1]

[1]Wrocław University of Science and Technology, Chair of Electronic and Photonic Metrology, B. Prusa 53/55, 50-314, Poland

Abstract. Digital signal processing is present in many areas of industry and science. One of them is analyzing multifrequency signals, e.g. in photovoltaic systems. This paper focuses on the frequency estimation of pure signals and signals distorted by AWGN noise in the presence of a DC voltage offset. The used IpDFT estimation method is based on the FFT procedure, I class Rife-Vincent time windows and three points of the spectrum taken to calculations. Measurement time was limited only up to two cycles of a tested signal and the method is very accurate even below one cycle. Obtained results show that additional DC component negatively affect the accuracy. The paper can be very useful because it shows properties of the method in real measurement conditions for various values of parameters.

1 Introduction

Signal processing is a key part of many industrial and scientific applications. One of them is analyzing a multifrequency signal which consists of many sinusoidal components. Such signals occur as mechanical vibrations, telecommunication signals (e.g. passive intermodulation issue), radar signals, the grid signal, etc. [1 – 4]. Often, determining parameters is crucial for the whole measurement or power system. Nowadays, renewable energy systems are very popular and the estimation of grid frequency is often necessary to produce energy. Moreover, the faster and more accurate the method the better performance of the system. Using solar energy in photovoltaic systems is one of the most popular way to produce so called green energy [5]. The basic configuration here includes photovoltaic panels, the DC/DC converter, the DC/AC inverter and the control unit with a DSP (Digital Signal Processor) processor (Fig. 1). The value of frequency is obtained and used e.g. to drive the inverter, in the FLL (Frequency Locked Loop) algorithm, etc.

Several factors affect the energy quality in the grid and one of them is the DC voltage offset which emerges in some cases and its value depends on the application [6 – 10]. Main factors of this situation in the grid are e.g. offsets in voltage sensors, A/D converters, switching devices like power converters and lack of transformers to separate power systems. Usually, it is not a desired component in the signal because it distorts the waveform and causes e.g. saturation of transformers, generation of even harmonic, heating power receiving devices, etc. In the grid signal it is usually up to 0.1% of the amplitude values [6].

Nowadays, there can be found many various signal processing techniques to estimate the frequency. It is worth to note such methods like: subspace methods, Prony's methods, parametric modeling methods or DFT-based methods [11 – 13]. Methods based on the DFT spectrum are very fast but with not the highest accuracy. Some ways to improve it are: linear interpolation of DFT LIDFT [14] or spectrum interpolation methods (IpDFT) where several points of the spectrum are taken into calculations [15]. If these points have their weights methods are called MWIDFT (Multipoint Weighted Interpolation DFT). Other than DFT-based methods are usually very accurate but much slower which can cause problems in real time applications [16].

This paper presents estimation results of a newly developed IpDFT method that is applied for the grid signal frequency estimation in photovoltaic systems. Due to some international standards the maximum reaction time of the system is 2.5 cycle [17] (50 ms for 50 Hz signal). In this paper, the estimation time is limited up to 2 cycles to spare time for other algorithms implemented in the DSP processor. In this method, the Fast Fourier procedure to obtain the spectrum is used and also the 3-point interpolation and I class time windows are the base here. The tested signal was pure or with additional AWGN (Additive White Gaussian Noise) noise sinusoid with the DC voltage offset. Obtained results show properties of the method in real measurement conditions and for a short measurement time.

Section 2 gives some information about the estimation method and the signal processing in the photovoltaic system. Section 3 contains simulation results and conclusions are presented in Section 4.

2 Estimation method

The presented method was developed to estimate values of frequency in the model of the multifrequency signal:

$$x(t)=\sum_{i=1}^{M} A_i \sin(2\pi f_i t+\varphi_i) \quad (1)$$

where: A_i – the amplitude of the i-th component, f_i – the frequency of the i-th component, φ_i – the phase of the i-th component, M – the number of sinusoidal components.

Time windows of H-order used for the method are I class Rife-Vincent windows which belong to the cosine windows family defined as [18 – 20]:

$$w_n=\sum_{h=0}^{H-1}(-1)^h a_h \cos\left(\frac{2\pi nh}{N}\right), \quad n=0, ..., N-1 \quad (2)$$

The spectrum for these windows can be approximated for $H > 1$, $N \gg 1$ and $N \gg \lambda$ as [20]:

$$W(\lambda)=\frac{D(\lambda)}{P(\lambda)} \quad (3)$$

where

$$D(\lambda)=\frac{N(2H-2)!}{\pi 2^{2H-2}}\sin(\pi\lambda)e^{-j\pi\lambda} \quad (4)$$

$$P(\lambda)=\lambda\prod_{h=1}^{H-1}(h^2-\lambda^2)=\frac{\prod_{h=0}^{H}(h^2-\lambda^2)}{(-\lambda)(H-\lambda)(H+\lambda)} \quad (5)$$

and $\lambda = fNT$ is a normalized frequency respect to the measurement time NT, where N is the number of samples in a measurement window and T is the sampling period. This value means also the number of signal cycles in the time window taken into calculations – *CiR* (*Cycle in Range*). According to IEC 61727 norm the length of the measurement window is up to 2 cycles in this paper [17].

The signal processing of the grid signal contains a few steps (Fig. 1). Firstly, the analog signal has to be sampled in the time window using an A/D converter. Later, samples are taken for the FFT procedure to obtain the signal spectrum. After that, an equation to obtain the value of frequency is used:

$$\hat{\lambda}_1=\mathrm{Re}\left\{\sqrt{\frac{\Pi_1}{\Pi_2}}\right\} \quad (6)$$

where

$$\Pi_1=\begin{vmatrix}(2H-1)H & (2H-1) & X_{k-1}-X_k\\ -k^2-H^2 & 2k & X_k\\ (2H-1)H & -(2H-1) & X_{k+1}-X_k\end{vmatrix} \quad (7)$$

$$\Pi_2=\begin{vmatrix}1 & -(2H-1) & X_{k-1}\\ 1 & 0 & X_k\\ 1 & (2H-1) & X_{k+1}\end{vmatrix}=(2H-1)\begin{vmatrix}1 & X_{k-1}-X_k\\ -1 & X_{k+1}-X_k\end{vmatrix} \quad (8)$$

where $X(\lambda)$ are spectrum points for $\lambda=k-1$, k, $k+1$ (values X_{k-1}, X_k, X_{k+1}) around the main lobe.

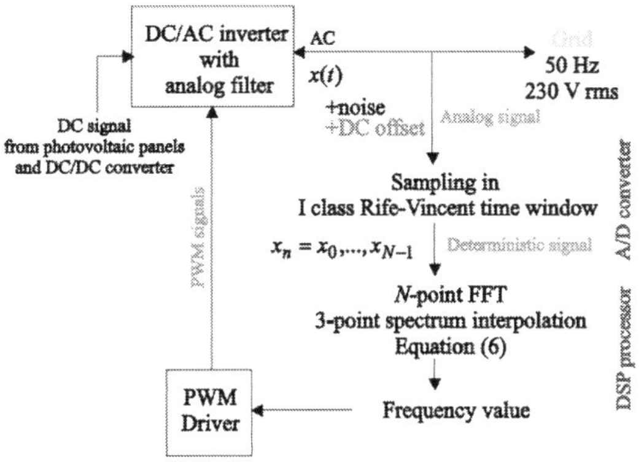

Fig. 1. Grid signal processing to obtain the frequency value in the control of the photovoltaic system.

3 Simulation research

Presented in this paper simulation results were obtained in the MATLAB environment. Estimation errors were determined as maximum errors of the entire range of φ_1 (0 – 2π in steps of 0.01 rad) for each *CiR* value. In the first part a pure grid signal was tested (without disturbances) with additional DC voltage offset (marked as DC in figures) up to 0.1% according to [6]. The second part contains results for the signal with added AWGN noise and also with the DC component.

Systematic errors for the signal without the offset decrease when number of N, H increases (errors are inversely proportional to N^{2H}) (Fig. 2, Fig. 4). Also, in the tested range of $CiR < 2$ the most accurate results are obtained for $k = 1$. After adding the 0.1% dc voltage offset the accuracy is changing because the offset distorts the spectrum and the method uses three initial samples of the spectrum due to very short measurement time. For $H = 2$ the most accurate results are obtained for $k = 3$ which means that among the three point of the spectrum taken into calculations there is not the point from the zero frequency where the offset occurs and that the spectrum distortion has smaller effect on results (in Fig. 2 curves for $H = 2$ are the same for the signal with the offset and without it). When the H increases, the accuracy is worse because of the wider main lobe of the time window in the spectrum domain and hence the spectrum is longer distorted by the offset (while increasing the *CiR*) and the DC has bigger effect on results. In this situation, to increase the accuracy *CiR* should be much longer than 2 (but then the time limit for photovoltaic systems is exceeded) or k should be much bigger than 1 (at least 8 for $H = 7$). Of course, the bigger value of the DC offset the least accurate the results (Fig. 3). Bigger number of samples N in this case does not improve the accuracy (Fig. 4) and it is approximately 10^{-4} Hz/Hz for 512 samples in the measurement window, $CiR = 0.5$, $k = 2$ and DC = 0.02%.

In the second part of simulations the *eMSE* (*empirical Mean Square Error*) error was calculated as the estimation error defined in [21]. Results were compared to the Cramér-Rao (*CRB*) bound (Fig. 5) for various number of H, DC and k as a function of *SNR*

(*Signal to Noise Ratio*). The bigger *SNR* the more accurate the method for the signal without the DC offset or for $k = 3$. However, in the presence of the DC curves for $k = 1$, 2 are on the constant level from some *SNR* value. In this case, the systematic error of the method (for the signal without noise) is bigger than the error caused by the AWGN noise. So for example, for $k = 2$, $H = 7$ and DC = 0.1% increasing the *SNR* value (the noise has lower power) does not improve the accuracy from about 60 dB. Also, the bigger H value the less accurate results.

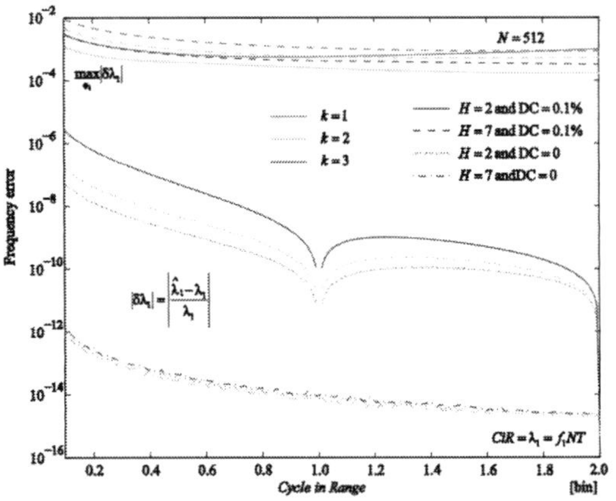

Fig. 2. The frequency estimation for the different number of k, H and DC as a function of *CiR*.

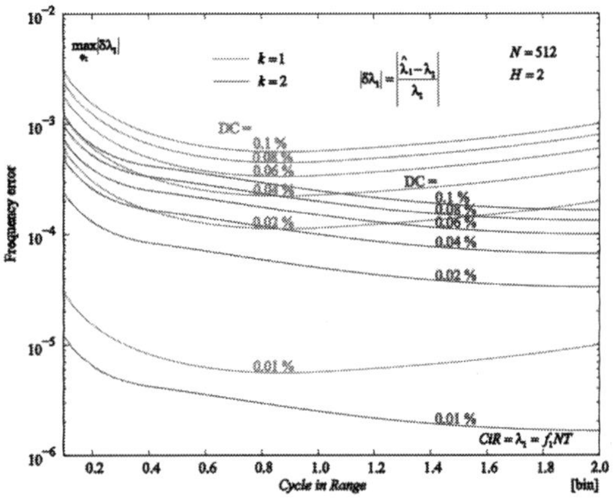

Fig. 3. The frequency estimation for the different number of k for the non-zero DC component.

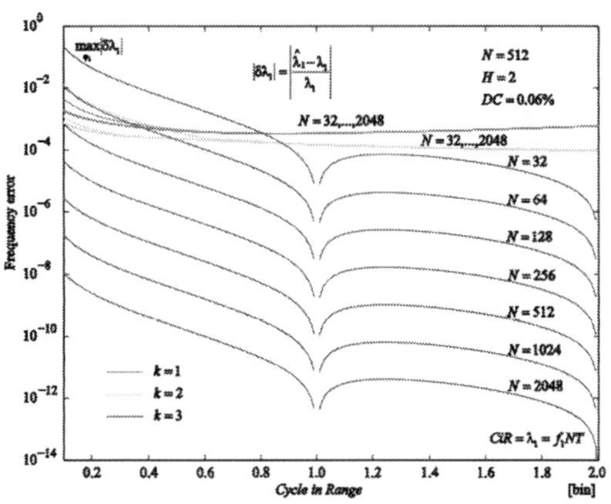

Fig. 4. The frequency estimation for the different number of k and samples N in the measurement window.

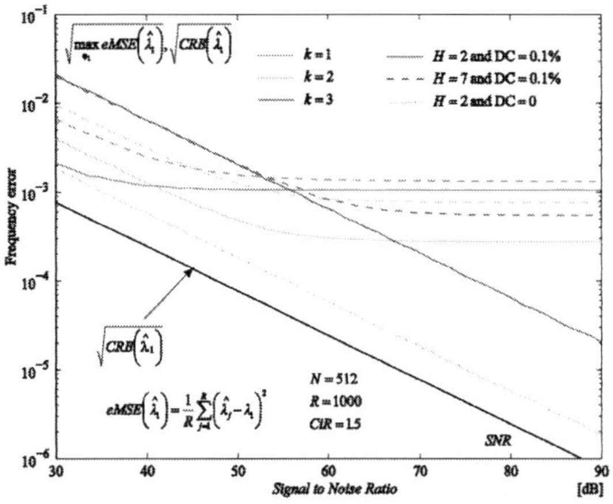

Fig. 5. Statistical properties of the method in the presence of the DC voltage offset.

3 Conclusions

This paper presents simulation results of a DFT-based estimation method of multifrequency signals in various measurement conditions with additional DC voltage offsets for the short measurement time up to 2 cycles.

The method without the DC offset is very accurate and very fast (even below one period of the tested signal) in various measurement conditions. However, the additional DC component in the signal causes bigger estimation errors and it is necessary to extend the measurement time or increase the index k in the (6) to improve the accuracy. Also, a very good option is filtering the signal before the estimation process to reject the DC component from the spectrum.

Obtained results show the accuracy in various measurement conditions in the presence of the DC offset in the signal which usually takes place in practice. Further research will be focused on the elimination of this component using digital filters with a short group delay which is important to fulfill maximum time conditions in photovoltaic systems.

Acknowledgement

This work was supported by Wrocław University of Science and Technology, the project *Młoda Kadra* (ZSI: 0402/0128/17) and partially by the National Science Centre, Poland under decision 2015/19/B/ST7/00497.

References

1. C. Correia, J.P. Veran, G. Herriot, *Advanced vibration suppression algorithms in adaptive optics systems*, Journal of the Optical Society American, **29**, 185-194 (2010)
2. R.E. Delgado, O. Ozdamar, S. Rahman, C.N. Lopez, *Adaptive noise cancellation in a multimicrophone system for distortion product optoacoustic emission acquisition*, IEEE Trans. on Biomedical Eng., **47**, 1154-1164 (2000)
3. D. Hoppe, M. Sadakata, P. Desain, *Developement of real-time visual feedback assistance in singing training: a review*, Journal of Comp. Assisted Learning, **22**, 308-316 (2006)
4. D. Kania, J. Borkowski, *The Rejection of Vibrations in Adaptive Optics Systems Using a DFT-Based Estimation Method*, SPIE Photonics Europe Conference, Brussels (2016)
5. N.A. Rahim, J. Selvaraj, K.H. Solangi, *Energy policy to promote photovoltaic generation*, Renewable and Sustainable Energy Reviews, **25**, 44-58 (2013)
6. IEEE Recommended Practice for Monitoring Electric Power Quality. IEEE Std 1159-2009.
7. S.N. Vukosavic, L.S. Petric, *High-Precision Sensing of DC Bias in AC Grids*, IEEE Trans. on Power Delivery, **30**, 1179-1186 (2015)
8. F. Wu, D. Sun, L. Zhang, J. Duan, *Influence of Plugging DC Offset Estimation Integrator in Single-Phase EPLL and Alternative Scheme to Eliminate Effect*, IEEE Trans. on Industrial Electronics, **62**, 4823-4831 (2015)
9. F. Wu, L. Zhang, J. Duan, *Effect of adding DC-offset estimation integrators in there-phase enhanced phase-locked loop on dynamic performance and alternative scheme*, IET Power Electronics, **8**, 391-400 (2015)
10. L. Zheng, H. Geng, G. Yang, *Fast and Robust Phase Estimation Algorithm for Heavily Distorted Grid Conditions*, IEEE Transactions on Industrial Electronics, **63**, 6845-6854 (2016)
11. A. Ahmad, F.S. Schlindwein, G.A. Ng, *Comparison of computation time for estimation of dominant frequency of atrial electrograms: Fast fourier transform, blackman tukey, autoregressive and multiple signal classification*, Journal Biomedical Science and Engineering, **3**, 843-847 (2010)
12. H.J. Koglin, Z. Leonowicz, T. Lobos, *High-Resolution Spectrum Estimation Methods for Signal Analysis in Power Systems*, IEEE Transactions on Instrumentation and Measurement, **55**, 219-225 (2006)
13. J. Zygarlicki, J. Mroczka, *Variable-frequency Prony method in theanalysis of electrical power quality*, Metrology and Measurement Systems, **19**, 39-48 (2002)
14. J. Borkowski, J. Mroczka, *LIDFT method with classic data windows and zero padding in multifrequency signal analysis*, Measurement, **43**, 1595-1602 (2010)
15. J. Borkowski, D. Kania, J. Mroczka, *Interpolated DFT-Based Fast and Accurate Frequency Estimation for the Control of Power*, IEEE Trans. on Industrial Electronics, 61, 7026-7034 (2014)
16. D. Kania, J. Borkowski, *Estimation Methods of Multifrequency Signals with Noise and Harmonics for PV Systems with a DSP Processor*, International Conference on Telecommunications and Signal Processing, Barcelona (2017)
17. Characteristics of the utility interface for photovoltaic (pv) systems. IEC 61727-2002.
18. A.H. Nuttall, *Some Windows with Very Good Sidelobe Behavior*, IEEE Transactions on Acoustics, Speech, and Signal Processing, **29**, 84-91 (1981)
19. D.C. Rife, G.A. Vincent, *Use of the Discrete Fourier Transform in the Measurement of Frequencies and Levels of Tones*, Bell System Technical Journal, **49**, 197-228 (1970)
20. D. Belega, D. Dallet, *Multifrequency signal analysis by Interpolated DFT method with maximum sidelobe decay windows*, Measurement, **42**, 420-426 (2009)
21. P. Handel, *Properties of the IEEE-STD-1057 four-parameter sine wave fit algorithm*, IEEE Transactions on Instrumentation and Measurement, **49**, 1189-1193 (2000)

Electrical Heating Emissions on the Island of Ireland

William D. Kerr, *David* M. Laverty, and *Robert* J. Best

School of Electronics, Electrical Engineering and Computer Science, Energy, Power and Intelligent Control (EPIC), Queen's University Belfast, Belfast, Northern Ireland

Abstract. This paper shows the effect on household greenhouse gas emissions if standalone or supplementary electric heating was to replace conventional heating methods, based on the present day electrical grid. While having the capability to improve future grid effectiveness and dynamic stability through the potential incorporation of demand side management (DSM). The All-Ireland system has been used in this paper as an example of a network which has been experiencing a significant increase in renewable generation. To show the potential of the electric heating methods the characteristics of existing domestic heating systems will be discussed, in terms of their heat output against their exhaust emissions (gCO2e/kWh). This will then be compared to that of the grid CO2 Intensity, showing the frequency and duration of the possible emission savings involved when using electricity as a main or supplementary heating source.

1 Introduction

As the EU has introduced strict emissions polices, and the Paris Agreement has been implemented, both the Republic of Ireland (ROI) and Northern Ireland (NI) have set targets of 40% RES-E (Renewable Energy Sources – Electricity) by 2020 [1], [2]. This is part of the strategy to obtain 16% of all energy consumed from renewable sources. To achieve this RES-E figure both Irish jurisdictions aim to provide 37% of their electric energy from wind generation [3].

Operating with such a high infeed of wind generation presents challenges with regards to system stability, due to low inertia. Presently, the system operator limits 'system non-synchronous penetration' (SNSP), of which wind is a leading contributor.

The SNSP problem is amplified on the island of Ireland as the island currently only has two low capacity DC interconnectors with Great Britain (GB). Meaning unlike countries such as Denmark which can achieve extremely high SNSP levels due to their heavily interconnected system [4] and ability to export large volumes of energy, while maintaining a high level of synchronous generation online [5]. The All-Ireland grid currently operates with an SNSP of up to 65%, at this point the SNSP is curtailed due to stability concerns [5].

Plans are underway to increase the SNSP limit up to 75%. A 5% increase is proposed to be trialled in the first quarter of each year [5] to identify the technical challenges that must be addressed. This transition of the system (generation and demand) is necessary, as without this the 2020 RES-E targets would be economically impossible to reach due to the amount of wind dispatch-down. This paper considers that the dynamic stability problem faced by the island, managed by the SNSP limit, is not that of oversupply of wind generation, but underutilisation of the potential RES-E resource.

The use of installed electric heating to provide support for the grid, in addition to its role in replacing conventional household heating methods, may help with this SNSP problem. With the use of a communication network in the future, electric heaters could be used for demand side management (DSM) and help push the island towards the 75% target.

Presently in Ireland, residential space and water heating accounts for approximately 19% of the total energy consumption of the country [6], [7]. Of this heating there is a modest amount of installed domestic electric heating on the island, with the latest census data showing the Rep. of Ireland having 13.5% electric central heating and Northern Ireland having 5% electric central heating [8], [9].

This paper looks at the effect of implementing electric heating on household emissions in the immediate term. A fuller picture of the CO_2e intensity per kWh of the All-Ireland grid, in terms of the Well-to-Socket (WTS) emissions has been shown, covering all aspects associated with electrical generation, giving a more realistic emissions figure of the electricity used within the household. The emissions figures in this paper have also been broken down into the different tariff categories showing that contrary to some popular opinions, the off-peak emissions are actually higher than that of on-peak.

2 Potential Emission Benefits of Electrification

 https://doi.org/10.1051/e3sconf/20186407001

On the island of Ireland, the lack of mains gas infrastructure and a sparse and rural population density means a high percentage of households use oil heating. ROI has approximately 40% oil central heating households and NI has around 74% oil central heating households (this NI figure considers households with multiple heating source, one of which is oil) [8], [9]. Due to the high use of this carbon intense fuel, electric heating may provide a viable alternative to conventional central heating systems as a means to reduce emissions in the domestic energy market. To determine the benefits, the authors have performed a comprehensive study of fuel types, combustion appliances and electricity generation trends and time of use. For both electrical and combustion heating, the authors consider well-to-tank emissions as well as the emissions arising from the fuel itself.

The electricity grid data was provided by EirGrid, including data obtained via their 'Smart Grid Dashboard' [10]. The heating system efficiency data was taken from the latest 'Building Research Establishment Domestic Energy Model' (BREDEM) document [11]. The data for both grid and heating systems was used alongside DECC data for calculating an overall emissions picture [12].

2.1 Point of use electricity breakdown

Previous studies which have investigated the effect of household electrification on CO_2 emission have made some crucial misjudgements [13], [14], [15]. In their comparisons they have only considered the emissions due to the burning of the actual fuel. This may be slightly misleading, as the process of electrical generation and distribution has much more energy loss than that of fuels burnt at the point of use, such as gas and oil. This study considers emissions of electrical energy produced on the island. If all aspects are considered from acquirement of fuel right to point of use, the losses in the electricity supply process increases the emissions by approximately 25% above that of the generation alone (based on data from 2014 and 2015, shown in Table 1). Whereas if all aspects are considered for fuels burnt at the point of use for conventional heating methods, gas emissions increase by only 13.4%, with oil increasing by 20.7%. This means that the emission saving shown in these studies, and many others, are not necessarily true in practice, as they have increased emissions in other economic categories and then not considered them in their specific area.

For actual comparison purposes, it is reasonable to use All-Ireland data due to the fact the island is considered to be one entity for SNSP level. From the data retrieved from EirGrid, it was found that the CO_2 intensity did not include 'house load' (power to run generation plant). Due to this, the generation and emission data was used to calculate a more accurate figure. Emissions data was provided in the form of tCO_2/hr, mean only CO_2 intensity could be found. DECC produces data in CO_2e, so considers nitrous oxide (N_2O) and methane (CH_4) as the other contributors to the global warming effect of combustion emissions. To convert this to CO_2e intensity, figures from DECC for the UK were used to scale the EirGrid data [12], this scaling method was used considering the similar climate and largely similar generation fuel mix of the UK.

The calculated figure for CO_2e intensity (gCO_2e/kWh) was for the transmitted generation, meaning the power directly exported on to the network from generation plant. For comparison purposes, the emissions from the well to the point of use needed to be considered, as is the case for fossil fuel heating technologies (shown in Section 2.3). Again, using DECC figures the full breakdown of CO_2e intensity was found and can be observed in Table 1. This table shows the Well-to-Socket (WTS) emissions; this considers not only generation emissions, but the additional emissions attributed to electricity on the All-Ireland grid due to transmission and distribution losses in the network as well as the emissions attributed to the Well-to-Tank (WTT) aspects of the fuel. WTT losses are emissions associated with extraction, refining and transportation of the raw fuel sources to an organisation's site (or asset), prior to their combustion [16].

From Table 1 the average WTS emissions for 2015 can be observed. To help show a more appropriate emissions picture the average daily seasonal trend of CO_2e intensity for 2015 per kWh is shown in Fig. 4. From this it can be seen that during the warmer seasons the CO_2e intensity is considerably higher during the off-peak periods. This is due to the reduced demand, in turn meaning the high-carbon fuels used for base load make up a larger portion of the overall generation, thereby increasing the CO_2e intensity. This supports the outcome of the off-peak energy tariff emissions shown in Table 2.

Table 1. Average WTS Electricity CO_2e Breakdown

Year	EirGrid (gCO_2e/kWh)	Transmission Losses	WTT	Total WTS (gCO_2e/kWh)
2014	537.76	(8.25%) 44.35	(16.98 %) 91.29	673.39
2015	491.89	(8.37%) 41.16	(16.52 %) 81.26	614.31

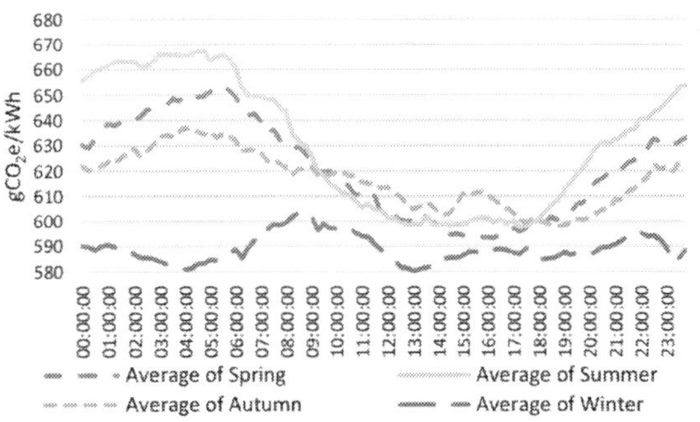

Fig. 1. Average daily seasonal trend of CO_2e intensity per kWh for 2015 (WTS of the All-Ireland Grid).

2.2 Off-peak tariffs and CO_2e trends

There are two domestic off-peak energy tariffs available on the island of Ireland, 'Economy 7' used in NI and

'NightSaver' used in ROI. 'Economy 10' has also been considered in this paper, however it is currently only available in GB. The off-peak times when each tariff is in effect, and the normal occupancy hours for domestic properties, is shown below. Economy 7 and NightSaver tariffs do not adjust for summer time and consequently are one hour behind. Some suppliers adjust Economy 10 for summer time, others do not.

- **Economy 7:**
  - 01:00-08:00 (summer 02:00-09:00)
- **NightSaver:**
  - 23:00-08:00 (summer 00:00-09:00)
- **Economy 10:**
  - 00:00-05:00, 13:00-16:00, 20:00-22:00
- **Normal occupancy hours:** [17]
  - Weekday: 07:00-09:00, 17:00-23:00
  - Weekend: 08:00-10:00, 16:00-00:00

Table 2. All-Ireland generation CO_2e intensity for different electricity tariffs at point of use.

Tariff Type	Emissions (gCO2e/kWh)
Economy 7	629.33
NightSaver	628.78
Economy 10	616.59
Normal Occupancy Hours	611.47

The average annual emissions for each tariff have been calculated and are shown in Table 2. During the off-peak tariff hours, grid emissions are in excess of the average (shown in Table 1), whilst during normal occupancy hours grid emissions are slightly below average. This would seem counterintuitive as wind is stochastic, meaning due to lower off-peak demand wind would likely make up a larger percentage of generation. However, as generation decrease due to lower off-peak demand, the high-carbon fuels used for base load capacity generally reduce by smaller margin than other generation methods. This results in them make up a larger percentage of the overall generation meaning CO_2e intensity actually increases, this can be seen further on in Figure 1. showing the average daily seasonal trends.

The figures from Table 2 indicate the importance of considering the time of use of electricity in determining the effectiveness of electric heating for carbon emissions reduction.

2.3 Heating system CO_2e intensities

Table 3. BREDEM – space heating efficiency.

Space heating system	Winter efficiency, η_{winter}
Typical natural gas/LPG boiler	80%
New natural gas/LPG boiler	90%
Typical oil boiler	80%
New oil boiler	92%
New electric storage heaters	100%

For a comprehensive set of efficiency data for a known boiler or system setup, reference should be made to the latest 'Standard Assessment Procedure' (SAP) document [18] and to the SAP database found at [19], in conjunction with the BREDEM model [11]. Efficiency figures for combustion heating appliances are shown in Table 3.

In SAP and BREDEM documents, winter efficiency is used for the season in which space heating is necessary. In the UK and ROI this is considered to be the whole year, apart for the summer months of June, July and August, whenever usually only water heating is required.

The SAP/BREDEM efficiency figures only consider the heating efficiency of the boilers, as seen in equation 1. To allow for a better comparison against electric heaters the carbon benefits ratio (CBR) was considered, seen in equation 2 [20]. The CBR considers the ratio between the kWh input to kWh output of the boilers, when both primary fuel and running electricity are considered.

- $\sigma = useful\ heat\ delivered\ [kWh]$
- $\omega =\ fuel\ burnt\ [kWh]$
- $\gamma = electricity\ used\ [kWh]$
- $CEF = Carbon\ Emissions\ Factor\ [gCO_2e/kWh]$

$$Heat\ Efficiency\ (\%) = \frac{\sigma \times 100}{\omega} \tag{1}$$

$$CBR\ (\%) = \frac{(\sigma \times CEF_{fuel}) \times 100}{(\omega \times CEF_{fuel}) + (\gamma \times CEF_{electricity})} \tag{2}$$

A value for the amount of electricity used by the boilers was calculated using data from a consumer website for the most popular systems of 2015 [21], yielding an average electricity consumption per kWh of heat output. Boilers will generally use a similar amount of electricity at any output load, meaning this is only accurate when the boilers are running at 100% load. However, the emission increases due to this are minimal as there is such a high ratio between heating output and electricity use.

The average seasonal CO_2e intensity of the fossil fuel heating methods based on CBR can be observed in Table 4, along with their average power outputs and electrical power consumption. The values from Table 3 has been used along with equations 1 and 2 as well as the data collected from the consumer website [21] to calculate these values.

Table 4. 2015 – Average Fossil Fuel Heating CO_2e Intensity Using CBR and Table 3 Figures

Type	Power Output (kW)	Electrical Power Consumption (kWh)	CBR Emissions (gCO2e/kWh)
New Gas	25.5	0.105	234.98
Typical Gas	25.5	0.105	264.04
New Oil	25.0	0.225	328.13
Typical Oil	25.0	0.167	376.10

For the comparisons in this paper basic electric heating and night storage heaters have been assumed to be 100% efficient, meaning their emissions are the same as that of the grid per kWh. However, this is not strictly accurate for night storage heaters. The actual heating efficiency of the storage medium may be 100%, but all of this energy is not stored until the time of demand. Some heat will escape, reducing useful effectiveness. These night storage heater problems are greatly reduced when considering newer models, due to their much higher heat retention capabilities. Some vendors of new models of night storage heater claim that their products use 22% less energy, while being 27% cheaper to run due to better use of off-peak energy [22].

For the heat pump comparisons in Table 6, the coefficients of performance (COP) have been based on an A16M Dimplex A-Class heat pump running at water output temperatures of 35°C and 65°C [23]. Some modern heat pumps have been designed to work in different climates with companies making models specifically for the UK and ROI. This means that their performance and efficiencies are designed to work best at typical winter temperatures of between -2°C to 10°C [24].

3 Benefits of Widespread Uptake

When considering the potential emissions savings of the household, the heating demand of all systems has been based on a typical occupancy of a family of four with active weekday occupancy between 07:00-09:00 and 17:00-23:00, and an active weekend occupancy of 08:00-10:00 and 16:00-24:00, as described in [17]. No space heating demand has been considered during the summer months of June, July and August, as per SAP.

The yearly space heating demand of an average household is considered to be 11,704 kWh (based on the average household floor area in ROI [25] together with the average heating energy per floor area in the UK [26]). Based on this space heating demand and the degree days, the average hourly power use per month has been calculated in Table 5 for occupancy and off-peak tariffs, during both weekdays and weekends. These figures have been calculated using historcial data from MET Éireann [27] together with the average space heating demand. The average degree days each month were calculated using the averaged value of 25 weather stations positioned across the ROI, this was then used to scale the yearly space heating demand down to hourly heating demand for each month.

Table 5. 2015 - Average Hourly Heating Demand per Tariff (kWh).

Tariff	Occupancy Heating		Economy 7		NightSaver		Economy 10	
	WD	*WE*	*WD*	*WE*	*WD*	*WE*	*WD*	*WE*
January	7.12	7.12	8.13	10.17	6.33	7.91	5.69	7.12
February	7.44	7.44	8.50	10.62	6.61	8.26	5.95	7.44
March	6.42	6.42	7.34	9.17	5.71	7.14	5.14	6.42
April	5.05	5.05	5.77	7.22	4.49	5.61	4.04	5.05
May	3.92	3.92	4.49	5.61	3.49	4.36	3.14	3.92
June	0.00	0.00	0.00	0.00	0.00	0.00	0.00	0.00
July	0.00	0.00	0.00	0.00	0.00	0.00	0.00	0.00
August	0.00	0.00	0.00	0.00	0.00	0.00	0.00	0.00
September	2.32	2.32	2.65	3.32	2.06	2.58	1.86	2.32
October	3.43	3.43	3.92	4.89	3.05	3.81	2.74	3.43
November	4.46	4.46	5.09	6.37	3.96	4.95	3.57	4.46
December	4.99	4.99	5.70	7.13	4.44	5.55	3.99	4.99
Heating Hours	8	10	7	7	9	9	10	10

WD = Weekday, WE = Weekend day

For this paper, typical oil heating has been used as the heating source with which to compare the alternative electric heating methods, meaning it produces 100% equivalent emissions observed in Table 6. This is due to the large portion of homes on the island of Ireland which use oil heating, as discussed in Section 2, and the fact that its CO_2e emissions are the worst of the fossil fuel methods discussed in this paper. The methodology used throughout this paper for the different heating method and emissions figures has been used together with Table 5, to provide an annual heating emissions comparison with respect to typical oil heating as shown in Table 6.

From Table 6 it can be observed that the use of basic electric heating as a method of reducing household emissions is currently not very promising, even if off-peak electricity tariffs were considered. The reason for this can be observed in Table 4, looking at the average seasonal emissions for fossil fuels, it can be seen that they are considerably lower than that of the annual grid average (Table 1) or even that of the average hourly seasonal values (Figure. 1). This means that only in very rare occasions is wind generation high enough to reduce WTS emissions sufficiently to offset home heating emissions when compared to fossil fuels.

Table 6. 2015 – Annual Space Heating Emissions Comparison. Emissions of Various Heating Methods with Respect to a Typical Oil Boiler

Tariff	Type	Emissions
Occupancy Heating	Electric Heating	161.6%
	Typical Oil	100.0%
	New Oil	87.5%
	Typical Gas	70.2%
	Heat Pump 65°C	68.1%
	New Gas	62.5%
Economy 7	Electric Heating	164.7%
NightSaver	Electric Heating	164.4%
Economy 10	Electric Heating	162.1%
	Heat Pump 35°C	39.5%

If a conventional oil fired boiler is augmented with an electric boiler, then some emissions savings would be seen. However, with the current electrical generation fuel mix, the electric boiler would only be used for around 10.5 hours in a given year and would only save approximately 0.54 kg of CO_2e emissions per year during occupancy hours, shown in Section 2.2.

From Table 6, heat pumps have significant emissions savings over typical oil across all tariffs. However, if heat pumps were to be used in the traditional way during occupancy hours and countries such as the UK adopted 100% heat pumps, it is believed the electricity demand would increase by 25%, and peak electricity demand by 65% [28] necessitating the need for smart grid / DSM techniques. To overcome this problem concrete underfloor heating together with a heat pump running at 35°C with the Economy 10 tariff [29] offers promise in terms of emissions reduction and reduced grid stress. However, the 35°C figure in Table 6 is improbable as an increased demand would be necessary to run heating in this way as it is not an efficient heating storage technique like that of night time storage heaters.

4 Conclusion

The concept of implementing electric heating to help reduce household emissions has been considered, using governmental and industrial data. It has been demonstrated that the All-Ireland grid's CO_2e intensity has not yet reached the point at which it is comparable to fossil fuel heating when using basic electric heating methods. Potential emission savings were shown for an augmented system using an electric boiler together with a typical oil boiler. However, this method proved to provide very little emissions benefit at the present electrical grid CO_2e intensity. Currently the best option in terms of emissions savings alone is for consumers with an older oil fired boiler to replace this with a new fossil fuel boiler or with a heat pump. It has been shown in this paper that although renewable penetration is high at night, the All-Ireland grid CO_2e intensity is actual greater during the night than during the day. This is due to low system demand at night fulfilled using high-carbon fuels for system base load. However, the latest generation of night storage heaters are currently running below their potential. The inclusion of a communication network to allow them to be used by the utilities for DSM has potential in helping to increase the SNSP limit. This would mean claims about their use of low carbon electricity could be put into practice through optimisation and scheduling. Furthermore, even though the use of night storage heaters may not reduce household heating emissions, it has a huge effect on the grid's overall running efficiency and emissions by helping to level the demand curve throughout the day. The full extent of the effect/benefit will be considered in future work.

Acknowledgments

This work was supported and funded by the Engineering and Physical Sciences Research Council (EPSRC) and the Centre for Advanced Sustainable Energy (CASE) project DINOSAURS.

The authors would like to thank EirGrid for providing data to enable this study as well as the CASE grant DINOSAURS. Mr. Kerr would like to thank the EPSRC for funding his PhD studentship.

References

1. Sustainable Energy Authority of Ireland, "Energy targets FAQ," [Online]. Available: http://www.seai.ie/Energy-Data-Portal/Frequently-Asked-Questions/Energy_Targets_FAQ/. [Accessed 2016 March 15].
2. M. Howley, M. Holland and K. O'Rourke, "Renewable energy in Ireland 2012," Sustainable Energy Authority of Ireland, Dublin, 2012.
3. E. McGarrigle, J. Deane and P. Leahy, "How much wind energy will be curtailed on the 2020 Irish power system?," *Renewable Energy,* vol. 55, pp. 544-553, 2013.
4. A. Neslen, "Wind power generates 140% of Denmark's electricity demand," The Guardian, 10 July 2015. [Online]. Available: https://www.theguardian.com/environment/2015/jul/10/denmark-wind-windfarm-power-exceed-electricity-demand. [Accessed 06 November 2016].
5. EirGrid Group, "DS3 Programme," EirGrid Group, [Online]. Available: http://www.eirgridgroup.com/how-the-grid-works/ds3-programme/. [Accessed 23 October 2017].
6. Sustainable Energy Authority of Ireland, "Residential energy roadmap," [Online]. Available: http://www.seai.ie/Renewables/Residential_Energy_Roadmap.pdf. [Accessed 21 January 2016].
7. M. Howley, M. Holland and D. Dineen, "Energy in Ireland," Sustainable Energy Authority of Ireland, Dublin, 2014.
8. Central Statistics Office, "Domestic building energy ratings quarter 2 2015," [Online]. Available: http://www.cso.ie/en/releasesandpublications/er/dber/domesticbuildingenergyratingsquarter22015/. [Accessed 2016 03 16].
9. Northern Ireland Neighbourhood Information Service, "Central heating: KS404NI (Settlement 2015)," [Online]. Available: http://www.ninis2.nisra.gov.uk/public/PivotGrid.aspx?ds=6665&lh=74&yn=2011&sk=136&sn=Census%202011&yearfilter=. [Accessed 2016 March 16].
10. EirGrid, 06 May 2016. [Online]. Available: http://smartgriddashboard.eirgrid.com/.
11. J. Hart and J. Henderson, "BREDEM 2012 – A technical description of the BRE Domestic Energy Model," BRE, Watford, 2015.

12. Department of Energy & Climate Change (DECC), "Government emission conversion factors for greenhouse gas company reporting," Department for Business, Energy & Industrial Strategy, 01 June 2016. [Online]. Available: https://www.gov.uk/government/collections/government-conversion-factors-for-company-reporting#conversion-factors-2016. [Accessed 21 October 2016].
13. F. Pallonetto, S. Oxizidis, R. Duignan, O. Neu and D. Finn, "Demand Response Optimisation of All-Electric Residential Buildings in a Dynamic Grid Environment: Irish Case," in *13th Conference of International Building Performance Simulation Association*, Chambéry, 2012.
14. J. A. Kelly, M. Fub and J. P. Clinch, "I home heating: The potential for air source heat pump technologies as an alternative to solid and liquid fuels," *Energy Policy*, vol. 98, pp. 431 - 442, 2016.
15. B. Morvaj, R. Evins and J. Carmeliet, "Decarbonizing the electricity grid: The impact on urban energy systems, distribution grids and district heating potential," *Applied Energy*, vol. 1991, pp. 125-140, 2017.
16. C. D. R. W. K. J. Nikolas Hill, "Methodology Paper for Emission Factors Final Report," Department of Energy and Climate Change, London, 2015.
17. N. J. Kelly, P. G. Tuohy and A. D. Hawkes, "Performance assessment of tariff-based air source heat pump load shifting in a UK detached dwelling featuring phase change-enhanced buffering," *Applied Thermal Engineering*, vol. 71, no. 2, p. 809–820, 2014.
18. BRE, "SAP 2012 - The government's standard assessment proceedure for energy rating of dwellings," Watford, 2014.
19. BRE, "Building energy performance assessment support website," [Online]. Available: http://www.ncm-pcdb.org.uk/sap/index.jsp. [Accessed 24 March 2016].
20. G. Orr, T. Lelyveld and S. Burton, "Final Report: In-situ monitoring of efficiencies of condensing boilers and use of secondary heating," GASTEC at CRE Ltd, Cheltenham, 2009.
21. Direct Heating Supplies, "Best boilers to buy 2015," Direct Heating Supplies, 17 December 2014. [Online]. Available: http://www.directheatingsupplies.co.uk/news/best-boilers-buy-2015/. [Accessed 14 January 2016].
22. heatershop, "Product Focus - Dimplex Quantum Storage Heater," heatershop, 02 February 2017. [Online]. Available: https://www.heatershop.co.uk/product-focus-quantum-storage-heater. [Accessed 16 06 2017].
23. Dimplex, "A-Class - head and shoulders above the competition," [Online]. Available: http://www.dimplex.co.uk/a-class/features/highoutputtool.htm. [Accessed 13 December 2016].
24. WDS green energy, "Dimplex A Class Air Source Heat Pump," WDS green energy, [Online]. Available: http://www.wdsgreenenergy.co.uk/dimplex-a-class-air-source-heat-pump/. [Accessed 17 May 2017].
25. A. Teicu, "Average Home Sizes of the World Gathered in a Chart," Freshome, 01 August 2011. [Online]. Available: http://freshome.com/2011/08/01/average-home-sizes-of-the-world-gathered-in-a-chart/. [Accessed 26 July 2017].
26. OVO Energy Ltd, "How much energy do you use to heat your home?," OVO Energy Ltd, [Online]. Available: https://www.ovoenergy.com/guides/energy-guides/how-much-heating-energy-do-you-use.html. [Accessed 26 July 2017].
27. Met Éireann, "Display and Download Historical Data," Met Éireann, [Online]. Available: http://www.met.ie/climate-request/. [Accessed 14 November 2017].
28. T. Fawcett, R. Layberry and N. Eyre, "Electrification of heating: the role of heat pumps," in *BIEE Conference*, Oxford, 2014.
29. P. Dunbabin and C. Wickins, "Detailed analysis from the first phase of the Energy Saving Trust's heat pump field trial," Department of Energy and Climate Change, London, 2012.

Simulation of Annual Functionality of Roof Turbine Ventilator

Giedrė Streckienė, *Violeta* Motuzienė, *Dovydas* Rimdžius, *Vytautas* Martinaitis[1] and *Juozas* Bielskus

Department of Building Energetics, Vilnius Gediminas Technical University, Vilnius, Lithuania

Abstract. Ventilation systems using renewable energy enable to reduce electricity demand. However, their operation directly depends on the stability of the renewable energy source. In this study, the wind driven roof turbine ventilator (RTV) is analysed. As a rule, this equipment is selected based only on the average annual wind speed and there exists a lack of date related to functionality of RTV. The case study presented in the paper seeks to assess functional operation of the RTV within the whole year. Simulations, performed with TRNSYS software, are based on the empirical equation for the ventilation flow rate extracted by the tested turbine ventilator. Results provide the number of RTV operational hours and share (%) of the time, when the RTV operates. Most of the time RTV operates at partial required load, however, there are periods when air flow rates are excessive and this should be considered as storage potential. The presented results could help to determine more accurately functional operation of RTV in the selected room/building and estimate demand for additional ventilation solutions as well energy storage potential.

1 Introduction

Despite that a growing number of countries have put in place policies to improve building energy performance, average energy consumption per person in the global buildings sector still remains practically unchanged since 1990 [1]. Assertive action is needed now across all countries to improve global average energy use per capita by at least 10% by 2025 using energy-efficient and low-carbon building technologies and this is possible without changing comfort levels or requiring households to reduce their purchases of appliances and other electronic equipment [2].

In building's sector main energy consumers are heating and cooling systems. They consume half of the European Union's (EU) energy and much of it is wasted [3]. The EU has set itself a long-term goal of reducing greenhouse gas emissions by 80-95%, when compared to 1990 levels, by 2050. Renewable heating and cooling are vital to decarbonisation [4].

Ventilation is the key issue for providing suitable indoor air quality (IAQ). This can be done either using fans – mechanical ventilation, or without interaction of any mechanical means – natural ventilation. Mechanical ventilation generally uses electricity in order to provide proper ventilation rate. As Chenari et al. [5] have shown in their review paper – up to 90 % of fans electricity can be saved using natural ventilation. Therefore, natural ventilation is now one of the fundamental methods in the energy efficient design of buildings.

The two fundamental principles of natural ventilation are stack effect and wind driven ventilation. Rooftop turbine ventilator (RTV) is one of the wind driven ventilation devices, which is cost effective and environmentally friendly, used to extract airflow from a building [6]. A RTV is a wind-driven air extractor which combines features of wind turbine and extract fan. RTV is not only widely accepted as industrial ventilation, but also has become a common ventilation feature used in other types of buildings including institutional, commercial and residential [7].

There is a significant number of studies related to the experimental measurements or CFD simulations of RTV. Some of them are concentrating on the influence of height [6] or diameter [8], [9] on performance of the ventilator, the others – on different constructions and/or blades [10] or actual applications in buildings and improvement of IAQ [11]. Newer tendencies of research in the field of RTV are related to additional benefits – production of electricity using RTV [12], [13] or exploitation of solar energy to increase the efficiency of the device [14], [15]. Also, some studies show that wind driven ventilators can play an important role in the design of a cost effective and energy efficient ventilation system inside a building as they have low cost, require no maintenance and no electricity to run [6,8,11].

The aim of the study presented in the paper is to define annual ventilation efficiency of the room and potential for energy storage at local climate conditions, using TRNSYS simulation model, based on experimentally defined ventilator's performance curve. Simulations are performed for 6 different European cities with different climate conditions.

2 Methodology

This study analyses functionality of RTV for different local conditions. The turbine ventilator used in this investigation is H100 (Hurricane®) produced by

Edmonds® [16]. It is constructed from corrosion resistant aluminium and consists of a rotating portion (Rotor) with 8 curved blades and a stationary portion – base of cylindrical shape. The dimensions of this RTV are shown in Fig. 1. A rotating RTV can be compared to a vertical axis wind turbine in its rotor configuration and in the sense that it also uses wind energy directly for its rotation [17].

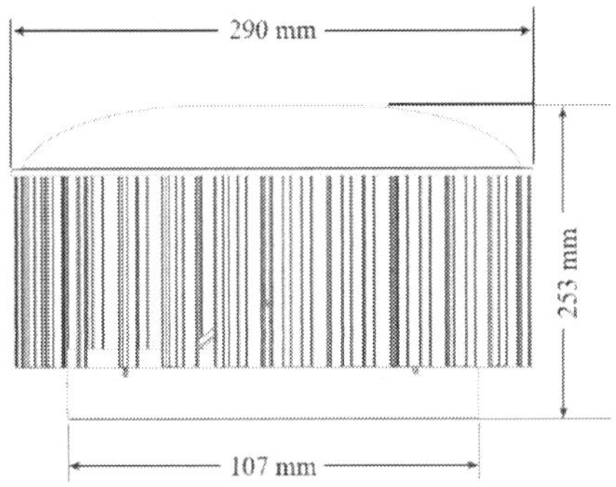

Fig. 1. Schema of analysed RTV.

Seeking to analyse functionality of the certain RTV, following steps must be performed:

1. Collection of the required model input data:
   - Definition of the building type, area, requirements for the IAQ;
   - Wind speed data for the analysed locations (if meteorological station is not at the analysed location, additional evaluation of the wind speed is required). In this analysis wind speed data in height of 10 m is used (from Meteonorm tm2 file);
   - RTV properties (dependency of the extracted air flow rate on wind speed) are required. When exact location of the RTV on the roof is known, additional coefficient to assess impact of the roof shape on the performance of RTV can be introduced. For this CFD simulation would be required. This study, does not take into account influence of the roof construction.
2. Creating the model in TRNSYS software, including all input data and system's/device's evaluation criteria. In this step also simulation time period (1 year) and step (1 min.) are defined.
3. Simulation and analysis of the results. Output data of the simulation at each time step is processed and used further for the RTV's operation efficiency evaluation within the year. If analysis shows that RTV is not able to ensure proper IAQ for the defined period, model input data must be changed, e.g. another RTV can be chosen or number of RTV's increased.

A theoretical warehouse room of 50 m^2 floor area is chosen as a case study. It is assumed that thermal comfort criteria are satisfied and only ventilation rates are analysed which are independent on season. This analysis assumes the minimum ventilation air flow rates in breathing zone based on ASHRAE Standard [18] – 0.3 l/(s m^2) for the warehouses. The outdoor airflow required in the breathing zone of space in a ventilation zone is assumed to be equal to extracted airflow. In accordance with [18], air flow rate is determined to be 15 l/s for the warehouse space as normally it is unoccupied.

In this analysis, stack effect is neglected, just wind caused ventilation effect is evaluated as stack effect depends a lot on building internal temperature, geometry, location and size of the air supply openings. RTV's operation functionality is evaluated based on two criteria: 1) number of RTV operational hours and 2) share (%) of the time, when the RTV operates. For the base case simulations weather data for Vilnius (Lithuania) is used. Additionally, five biggest EU cities – London, Berlin, Madrid, Rome and Paris are simulated for the comparison of the RTV's operation in different locations. Tm2 weather data files (Meteonorm data in TRNSYS software) are employed in simulations.

Possible extracted air flow rates from the warehouse room are modelled by TRNSYS v. 17 software. The software has a modular structure with a main visual interface. The main components of the system are described below and presented in Fig. 2.

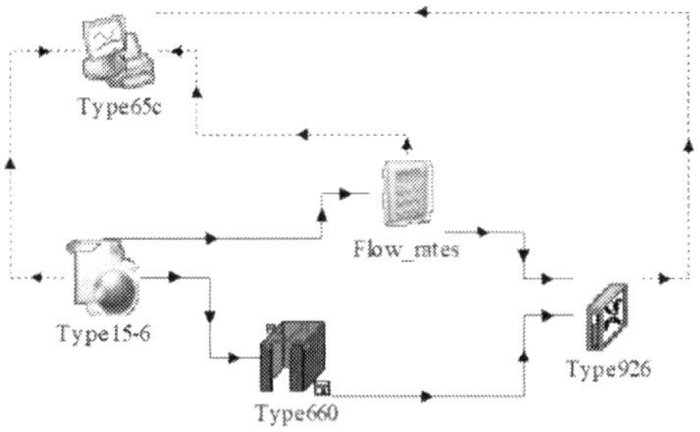

Fig. 2. TRNSYS model.

Type15-6 is weather data processor used to determine local climatic conditions. Type660 is multi-zone building model component. It enables to describe ventilated space. Type926 is a component for variable speed fan. Type65c is graphical plotter with output file enabling to analyse calculation results and "Flow_rates" is type to insert equations.

The main equation used in this study is based on experiments executed in the Laboratory of Buildings Energy and Microclimate Systems in Vilnius Gediminas Technical University for the exact type of RTV. The extracted air flow rate (m^3/h) is simulated in TRNSYS using experimental equation:

$$V_{ex} = 0.0185 \cdot v^2 + 11.819 \cdot v - 6.3174, \quad (1)$$

where v is a wind speed, m/s.

It was defined by experiment, that the minimum wind speed of 2 m/s is required to start operation of the analysed RTV to extract air using only wind energy (stack effect is not included).

3 Results and discussion

https://doi.org/10.1051/e3sconf/20186407002

3.1 Base case study – Vilnius city

The comparison between simulated flow rate extracted by turbine ventilator and total required ventilation flow rate (15 l/s) against wind speed is shown in Fig. 3. It is evident that wind speed of 5.1 m/s enables RTV to extract total required ventilation air flow rate for the studied room. The area above 100 % shows wind energy storage potential. The stored energy could be used for the periods when wind speed is low.

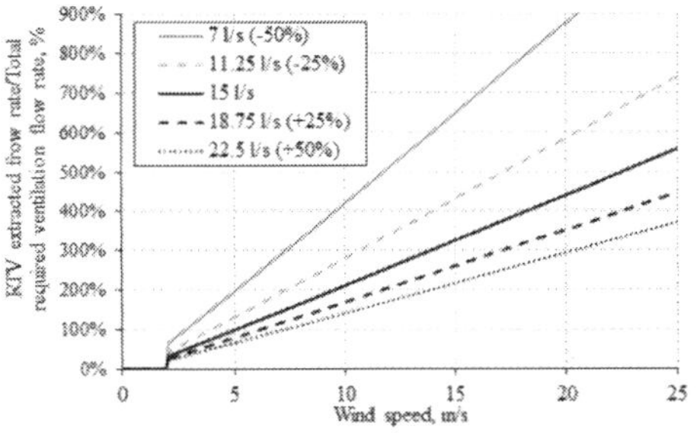

Fig. 3. Comparison between the flow rate extracted by RTV and total required ventilation flow rate vs wind speed.

Fig. 3. also shows sensitivity of the results on the required ventilation demand, e.g. if we reduce area of the room or use more turbine ventilators, a higher share of extracted air flow rate could be achieved. For the floor area of 25 m^2 RTV could extract total ventilation flow rate (7 l/s) at wind speed of 2.9 m/s. However, higher flow rates, e.g. 18.75 l/s or 22.5 l/s (+25% or 50% of the case study) will need wind speed of 6.2 m/s and 7.4 m/s, respectively.

Analysis of annual operation of RTV in a warehouse room of 50 m^2 in Vilnius city shows that RTV can fully cover required ventilation demand ~2218 h/year (25.3% of the time). Also, during this time period RTV could extract more air than needed (has energy storage potential). It is calculated, that this unit does not operate 2325 h/year (~26.5%) at all and other technological solutions are needed when wind speed is less than 2 m/s. Most of the time RTV operates at the partial performance (48.2% of time).

When RTV operates at partial required performance, it can cover from 30 to 100 % of the ventilation demand (Fig. 4), at the wind speeds varying from 2 m/s to 5 m/s. At the low wind speeds (<2 m/s), RTV practically does not operate, therefore low air flow rates are not extracted (0-30% of total ventilation air flow rate).

From Fig. 4, it is obvious that most of the time RTV extracts 30-80% of the total ventilation rate (15 l/s) as the wind speed in Vilnius usually fluctuates between 2 m/s and 4 m/s. To satisfy higher extracted air flow rates, more units for the same floor area must be designed.

Fig. 5 shows detailed analysis of the partial extraction of air flow. Lower columns in Fig. 5 show that when the time of partial operation mode of RTV decreases, the RTV can extract all required air flow rate. If we have a situation, when air flow rate is twice higher (0.6 l/s/m^2), it is seen that most time of the year RTV extracts just 10-50% of the required air flow rate. Meanwhile, when ventilation demand is twice lower, RTV operates at full required performance most of the time.

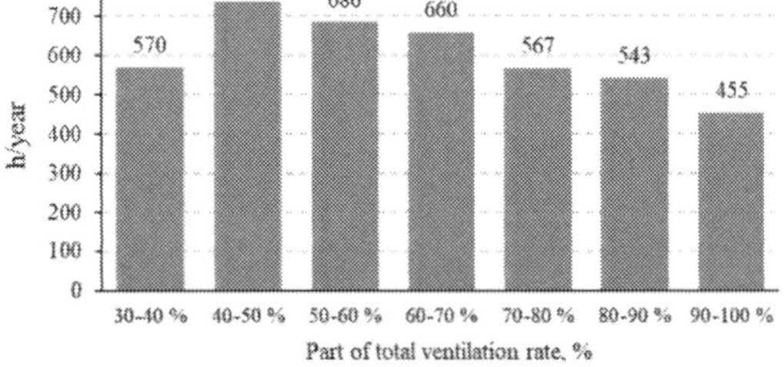

Fig. 4. RTV part load operation duration (h/year).

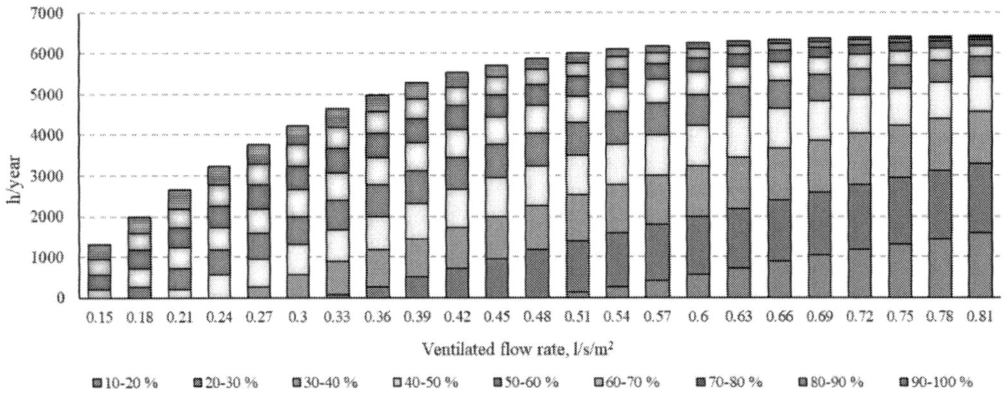

Fig. 5. Dependency of the RTV part load operation on required air flow rates

The dependency of extracted (fully and partially) ventilation air flow rate on required ventilation flow rate (l/s/m^2) is presented in Fig. 6. When wind speed is ≤ 2 m/s, RTV does not extract the air. Meanwhile, fully (100%) satisfied extracted ventilation air flow rate increases almost exponentially when ventilated flow rate decreases in the range from 0.7 l/s/m^2 to 0.1 l/s/m^2 (Fig. 6). At the same time, the partial extracted flow rate by RTV decreases.

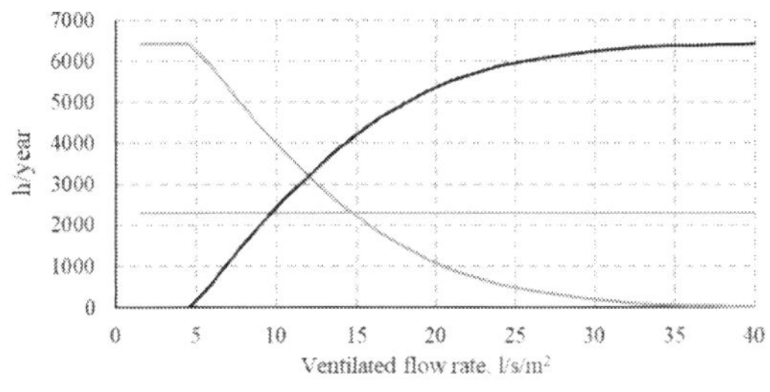

Fig. 6. Dependency of fully and partially ventilation flow rate by RTV on required ventilation flow rate.

Analogous charts as Fig. 5 and Fig. 6 could be used to determine operation duration of RTV for the specified volume of the room.

3.2 Comparison of RTV operation in other EU cities

As the main RTV operation efficiency influencing parameter is wind speed, its duration time (h/year) in different EU capitals is presented in Fig. 7. From this figure it is obvious that Madrid is characterized by low wind speeds, Vilnius and London have similar wind speeds. Paris and Berlin have the highest wind speeds in comparison with other analysed cities.

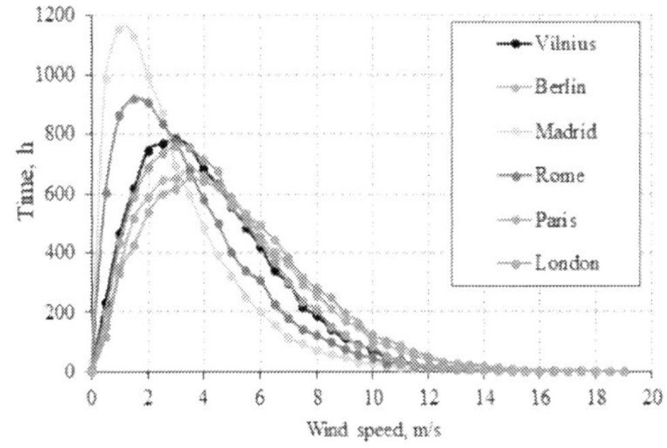

Fig. 7. Analysis of wind speed in different cities.

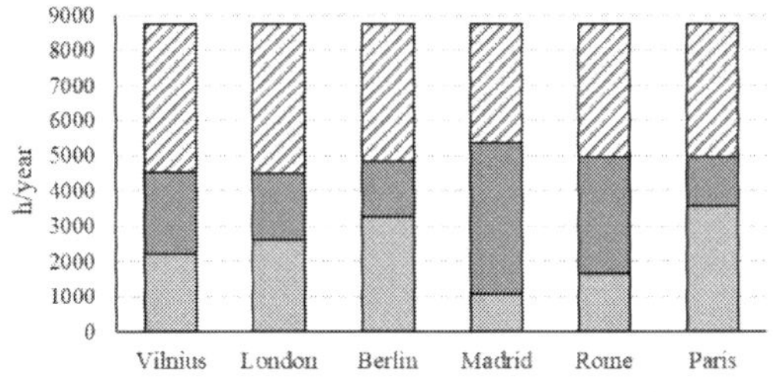

Fig. 8. Comparison of RTV operation in six EU capitals.

Knowing the actual operation of the RTV and wind speed in a certain location enables to evaluate operation duration of the ventilator in different modes (Fig. 8).

Fig. 8 clearly shows – if analysed RTV uses only wind kinematic energy, Madrid has the worst results for RTV operation (48.9% of the time the RTV does not use wind speed energy) and Paris – the best ones (only 15.9% of the time is not satisfied). Respectively, not satisfied time (h/year) by the RTV is 26.5% in Vilnius, 21.6% in London, 18.1% in Berlin and 37.4% in Rome. These values show that the same RTV unit cannot be chosen for the same purposes in different geographical locations. Also, partial operation mode together with full operation mode (100% and more of need is satisfied) should be analysed for energy storage possibilities.

Even though average wind speed in the cities can be similar, annual distributions differ. This causes differences in RTV's annual functionality. For example, about 570-590 h/year the RTV in Vilnius and Rome extracts only the 30-40% of the total ventilation flow rate. Meanwhile, the RTV in Paris operates only 423 h for the same extracted flow rate. For the highest extracted flow rates (90-100%), Paris is leading (518 h/year), respectively, Berlin has 504 h/year, London – 489 h/year, Vilnius – 455 h/year, Rome – 349 h/year and Madrid only 271 h/year. It is simulated that RTV could extract more air than required from 12% (Madrid) to 41% (Paris) time of the year.

The further step of this analysis will be evaluate stack effect of natural ventilation when the RTV does not operate at low wind speed and introduce variable ventilation flow rates.

4 Conclusions

A study of a roof turbine ventilator functional operation using TRNSYS simulation and empirical correlation for air flow rates has been carried out. The influence of wind conditions in six EU's cities on the flow rate extraction has been discussed. Results show that the same RTV unit cannot be used for the same purposes in different locations as there is strong influence of wind speed on extracted air flow rates. Case study of the warehouse room shows that for different locations using just kinetic energy of wind, RTV is able to extract more air than required from 12% to 41% time of the year. As a perspective, storage technologies could be used during this time period. It has also been found that RTV cannot use wind energy from 15.9% (Paris) till 48.9% (Madrid) of the time for the selected cities. Most time of the year RTV operates just at the partial required performance and does not satisfy required ventilation air flow rates. At these periods, auxiliary energy systems (e.g. mechanical ventilators) are required or stored wind energy can be employed.

As a future work, the proposed model can be improved and adapted for the buildings/rooms where there is no demand for permanent ventilation and there exists a potential to utilize stored wind energy.

Acknowledgement

This research is funded by the European Regional Development Fund according to the supported activity 'Research Projects Implemented by World-class Researcher Groups' under Measure No. 01.2.2-LMT-K-718 (project grant No 01.2.2-LMT-K-718-01-0016).

References

1. IEA, *Energy efficiency indicators* (2017).
2. IEA, *Transition to Sustainable Buildings*, Paris (2013).
3. European Comission, A*n EU Strategy on Heating and Cooling* 2016, Brussels (2016).
4. European Comission, *Energy Roadmap*, 2050, Brussels (2011).
5. B. Chenari, J. D. Carrilho, and M. Gameiro, Sustain. Energy Rev., **59**, 1426–1447 (2016).
6. S. J. Lien and N. A. Ahmed, Build. Environ., **45**, 1808–1815 (2010).
7. M. Ismail, A. Malek, and A. Rahman, J. Sust. Development, **5**, 121–131 (2012).
8. G. K. Jadhav, P. M. Ghanegaonkar, and S. Garg, Journal of Building Engineering, **6**, 196–202 (2016).
9. C. Lai, Energy Build., **35**, 188, 927–932 (2003).
10. N. Khan, Y. Su, S. B. Riffat, and C. Biggs, Renew. Energy, **33**, 2441–2447 (2008).
11. J. Lien and N. Ahmed, Procedia Engineering, **49**, 124–134 (2012).
12. C. Torasa and N. Sermsri, Procedia - Social and Behavioral Sciences, **197**, 1690–1696 (2015).
13. R. Shinde, V. Lavhale, A. Nair, S. Pawar, and R. Mahajan, Int. J of Current Engineering and Technology, **6**, 89–93 (2016).
14. P. Valeh-e-sheyda, M. Rahimi, A. Parsamoghadam, and M.M. Masahi, Energy Build., **73**, 115–119 (2014).
15. A. Suresh, B. Jacob, L. F. Sharaf, and S. Gomathy, IOSR Journal of Electrical and Electronical Engineering, 88–91 (2017).
16. Hurricane® Roof mounted ventilators online: last access 08-05-2018. https://www.edmonds.com.au/commercial-industrial-ventilation/auditoriums/hurricane/specifications
17. P. M. Pardalos, S. Rebennach, M. V. F. Pereira, N. A. Iiadis. *Handbook of Wind Power Systems*. Springer. (2013).
18. ASHRAE Standard 62.1-2013. *Ventilation for Acceptable Inddor Air Quality* (2013).

Experimental Evaluation of Turbine Ventilators Performance under Different Test Conditions

Dovydas Rimdžius , *Juozas* Bielskus, *Vytautas* Martinaitis, *Violeta* Motuzienė, and *Giedrė* Streckienė

Department of Building Energetics, Vilnius Gediminas Technical University, Vilnius, Lithuania

Abstract. One of the most popular wind driven ventilation devices is a turbine ventilator. Its main construction components usually are air suction duct and impeller. However, there are many different configurations of such devices – different shape of vanes and ratios between main dimensions. In order to develop turbine ventilators there is demand for deeper fundamental researches. Major part of previous experimental investigations compares turbine ventilators performance. The aim of presented experimental investigation is to understand key parameters affecting wind driven ventilators performance. For this purpose, straight and curved vane devices under 4 different test conditions are tested. All tests are performed in modified wind tunnel based on the following methodology: 1) testing devices under normal operation conditions; 2) testing without impeller; 3) testing while impeller is stopped; 4) testing while air suction duct is sealed. Experimental results have shown that the biggest part of extracted air flow rate is impacted due to ejection of wind and air suction duct interaction process. A slight increase in performance of turbine ventilators compared to open duct column has been confirmed. Experiments also revealed differences between tested turbine ventilators rotational speed trends. Presented empirical equations of experiments could be used for design of turbine ventilators or other researches purposes.

1 Introduction

Nowadays energy consumption of the modern buildings heating is highly reduced, meanwhile air flow required for ventilation persists the same or even gets higher. Therefore, electricity consumed by ventilation fans, constitutes a significant share in energy efficient buildings total energy demand. In order to maintain proper indoor air quality in buildings and ensure high efficiency of ventilation system, new solutions are required. Currently there are attempts being made to use technological solutions employing renewable energy. One of possible solutions is to use wind driven technologies. Recently wind driven rooftop turbine ventilators (RTV) is a spreading technology which has a simple construction and is relatively cheap.

Lai [1] has performed experiments in low-speed wind tunnel, seeking to define efficiency of RTV. He has confirmed that it is an effective technology to ventilate buildings and extracted air flow rate of the RTV depends on its size. However, influence of the size is not as significant, as expected.

The results of Bay *et al.* showed that when buoyancy effects were dominant, the open duct column performed better than roof turbine ventilator [2]. There were also tests of combined column-turboventilator prototype. Performance of the combined device was found to be better than that of standard RTV due to stack effects gained by an increased suction duct height.

Jadhav *et al.* observed that RTV with larger suction duct diameter gives more mass flow rate and less rotational speed compared to small size ventilator [3]. They also created CFD model and determined that CFD mass flow rate results are observed to be within 12 % to 15 % higher than the experimental results.

Rashid *et al.* studied aerodynamic forces on straight vane rotating wind driven ventilator and determined that tested ventilator was more efficient in operation at low wind speed than at high speed [4]. It is also concluded the increase of wind flow force on ventilator construction and flattening of lift force curves towards larger speed. Researchers concluded that partial flow separation might have been occurring on the blades at those speeds.

Investigation of Khan *et al.* concludes that turbine ventilator may not necessarily be the cheapest and most efficient option [5]. The standard vent hats showed similar abilities to the tested turbine ventilator comparable in size. However, researchers claim that turbine ventilator is more weather resistant and most importantly prevents wind-induced down draughts.

Lien *et al.* experimental and numerical analysis results indicate that the efficiency and reliability of a turbine ventilator operation may require boosting by hybrid system by incorporating other power sources such as solar power to overcome the inertia and friction forces acting on the body to initiate ventilator rotation at lower wind speed [6]. Lai additionally installed an inner fan powered with photovoltaic panels, at the bottom of RTV and found increase of ventilation air flow rate at wind speed up to 5 m/s [7]. However, ventilation rate was not increased by installing an inner fan at higher wind flow velocities.

Review of recent researches indicates that most experiments of turbine ventilators are targeted at their performance investigation. However, many experiments are performed under different pressure conditions and it is difficult to make proper comparison of results. Other problem is lack of fundamental researches of phenomenon occurring on these devices. The interaction between characteristic processes of turbine, ventilator and ejector takes place there. Due to complex processes of air flows mixing inside turbine ventilator, it is difficult to understand different flow impact on its performance.

To solve problem related to comparison of results and in order to better understand key parameters of the turbine ventilator's performance, experimental evaluation of two different type RTV's is performed under different test conditions. Presented experimental results could be useful for comparison purposes and for turbine ventilator future development. The methods for better performance comparison between devices of different sizes and calculation of extracted air flow ratios are suggested.

2 Test devices

During experiment 2 different types of turbine ventilators are tested. One of them has straight vanes (on the left in Fig. 1) and another has curved vanes (on the right in Fig. 1). The first one is manufactured by Australian company "Edmonds" and its model is "Hurricane 100". The second one is distributed by Polish company and its model is called "SX-WO130MAXAL/CENTRO".

Fig. 1. Photos of tested turbine ventilators.

Because of different construction of vanes, tested devices are similar size but not identical. RTV with straight vanes has a bigger impeller while with curved vanes a bigger air suction duct hole. Main dimensions of the analysed RTV's are specified in Table 1.

Table 1. Dimensions of tested turbine ventilators.

Dimensions	Straight vane turbine ventilator	Curved vane turbine ventilator
Air suction duct diameter, mm	110	130
Impeller diameter, mm	280	190
Impeller height, mm	92	130

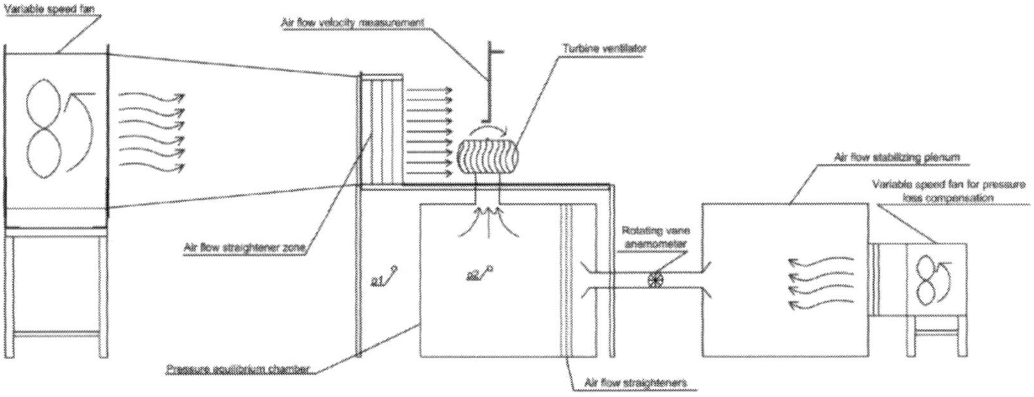

Fig. 2. Schematic of experimental setup and measurement points.

3 Methodology

3.1 Experimental setup

Experimental setup illustrated in Fig. 2 is based on AS/NZS 4740:2000 [8] standard and ideas of experiments published by other authors [3, 5, 9]. The construction base is made of modified open circuit wind tunnel with a variable speed fan. Tested RTV's are placed in the centre of direct wind flow and attached to the chamber of imitated ventilation room (in Fig. 2 it is called pressure equilibrium chamber). This chamber was well insulated with sealant and before experiments pressurized and tested with fumes. There wasn't almost any visual leakage while it was pressurized up to more than 300 Pa. During experiments it is important to ensure uniform wind flow. For this purpose, air flow straightening zone with specific screens is integrated.

Experimental setup has pressure measurement point outside of experimental stand (p_1) and 2 pressure measurement points on side walls of chamber (p_2). For better accuracy last measurement points are attached by piezometric ring principle. For sucked air flow measurement 5 identical rotating vane anemometers "Ahlborn FVAD15S140" of 1.85 cm diameter, which are capable to measure up to 40 m/s air velocity are used.

These rotating vane anemometers are connected between pressure equilibrium chamber and air flow stabilizing plenum. The latter plenum is attached to another variable speed fan and is used for pressure loss compensation.

Another important parameter of experiments is wind flow velocity. In this experimental setup it is measured by Pitot tube. Also, it is necessary to measure rotational speed of turbine ventilator. For this task used optical tachometer "Extech RPM33".

3.2 Test conditions

Prior the measurements being taken, a wind tunnel variable speed fan calibration was performed. The purpose of this was to determine wind flow velocity on specific test rig fan rotational speeds. For equal measuring conditions wind flow velocity is measured without turbine ventilator in apparent central position of it. Measurements were taken with pitot tube in 7 different positions and then averaged. It was determined that such test setup is capable to blow up to 9 m/s wind flow.

In order to determine air flow components of tested devices, experiment was split in 4 different series. Further Table 2 details every type of experimental series and shows which parameters were measured.

The first experimental series was performed without turbine ventilator. Instead equal diameter air suction duct which height is up to turbine ventilator impeller bottom was used. During second experiment ventilator is connected to chamber of imitated ventilation room and it is tested under normal operation conditions. Purpose of the next experiment is to determine turbine ventilator's extracted air flow rate while rotation of impeller is stopped. The aim of the last experiment configuration is to compare RTV's rotational speed performance while its air suction duct is closed.

During operation turbine ventilator creates negative pressure in room imitating chamber. However, such phenomenon in reality is highly dependent on room's tightness therefore it is difficult to compare devices performance results of other installations. To solve this problem all experiments are performed with pressure equilibrium conditions. It is done by equalizing static pressure in chamber and outside of experimental stand (p_1=p_2). For this purpose, the second variable speed fan is used.

Since tested devices are not of equal size, for comparison purpose specific air flow calculation index is used. The equation of this calculation:

$$Q_s = \frac{4 \cdot Q}{\pi \cdot d_i^2} \tag{1}$$

Where Q_s is a specific extracted air flow rate in $m^3/(h\ m^2)$, Q is an absolute extracted air flow rate in m^3/h and d_i is an internal diameter of air suction duct.

Table 2. Configurations of experiment.

Test conditions	Experimental series			
	Without turbine ventilator	**Turbine ventilator operates normally**	**Stopped rotation of turbine ventilator**	**Closed air suction duct**
Control of wind flow velocity	+	+	+	+
Closed air suction duct				+
Attached to pressure equilibrium chamber	+	+	+	
Measured parameters				
Extracted air flow, m^3/h	+	+	+	
Rotational speed of turbine ventilator, RPM		+		+
Wind speed velocity, m/s	+	+	+	+
Static pressure outside of experimental stand, Pa	+	+	+	
Static pressure in pressure equilibrium chamber, Pa	+	+	+	

Another calculation is performed to compare tested turbine ventilators extracted air flow rate ratios to wind flow rate:

$$\eta = \frac{Q}{v_w \cdot A_{imp} \cdot 3600} \tag{2}$$

Where η is turbine ventilator extracted air flow rate ratio to wind flow rate perpendicular to impeller projection, v_w is wind flow velocity in m/s and A_{imp} is area of impeller projection perpendicular to wind flow in m^2.

4 Experimental results

In this chapter results of turbine ventilator air flow components and comparison of 2 tested devices performance are presented. Empirical equations attached to graphs are valid only for studied turbine ventilators.

The first experimental results are presented in Fig. 3 and show extracted air flow rate against wind flow velocity under 3 different test conditions: without turbine ventilator (marked open duct), normal and stopped rotation.

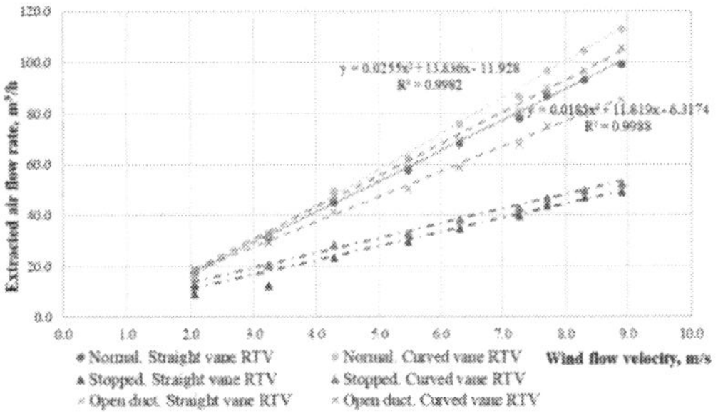

Fig. 3. Extracted air flow rate against wind flow velocity.

During experiments it was observed that rotation of turbine ventilators starts only when wind flow velocity is higher than 2 m/s. At lower velocity, used measurement devices weren't fairly accurate and because of that it was made an assumption that there isn't any air suction.

Analysis of the results shows that in both cases of tested devices the best extracted air flow performance is reached under normal test conditions. However, curved vane turbine ventilator extracts up to 12 % more air flow compared to straight vane RTV at the same wind flow velocity.

It is evident that stopping of turbine ventilators does not completely block air suction because of ejection. Though such operation highly impacts their performance. In this case the impact of stopped ventilator test condition reduces nominal performance more than 50 %. It could be ventilation problem if rotation mechanism is damaged.

This graph illustrates minor difference between open duct and turbine ventilator performances. However, differently from other investigation results [5], turbine ventilators perform better than open duct. Looking more detail to extracted air flow curves it can be seen that rotation of straight vane RTV has more impact for sucked air flow than curved vane RTV. Turbine ventilators can extract up to 18 % more air flow than open duct and curved vane ventilator only up to 7 % on maximum tested wind flow velocity. Although curved vane ventilator impeller is higher, such difference could be mainly due to bigger straight vane ventilator impeller diameter and different aerodynamics.

From experimental data results it can be written these empirical equations for extracted air flow calculation:

$$Q_{st} = 0.0185 \cdot v_w^2 + 11.819 \cdot v_w - 6.3174 \quad (3)$$

$$Q_{cu} = 0.0255 \cdot v_w^2 + 13.836 \cdot v_w - 11.928 \quad (4)$$

Where Q_{st}, Q_{cu} are extracted air flow rates in m^3/h accordingly of straight vane and curved vane RTV's while static pressure is in equilibrium between room and atmosphere; v_w is wind flow velocity in m/s.

Other experimental data results are plotted on rotational speed (RPM) against wind flow velocity graph in Fig. 4. This time there are presented only normal and closed air suction duct of turbine ventilators test conditions results.

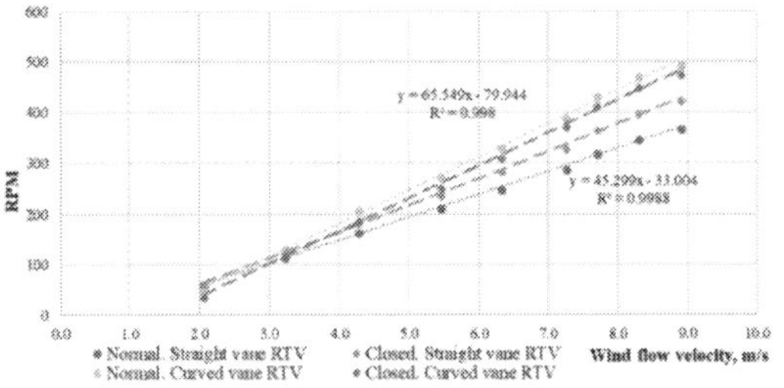

Fig. 4. Rotational speed against wind flow velocity.

Experimental results show that in both test series, curved vane ventilator reaches higher rotation speed. Under normal operation condition it is up to 34 % higher compared to straight vane RTV's RPM.

However, as can be seen from all plotted curves the results are controversial. Analysing results of curved vane ventilator rotational speed, it is observed that under normal test conditions such device has a little bit higher (up to 4 %) RPM compared to case while its air suction duct is closed. On the other hand, rotational speed of straight vane ventilator has the opposite trend of change. While its air suction duct hole is sealed such device starts to rotate faster. During this test series, straight vane ventilator reached up to 15 % higher RPM while its air suction duct was closed.

Supposedly such phenomenon could be explained by difference of such devices aerodynamics. During normal operation of curved vane ventilator its extracted air flow adds lift and drag forces to vanes and by this ensures faster rotation. However, construction of straight vane ventilator impeller creates different effect. This may be because during normal operation part of extracted air flow creates different direction aerodynamic drag forces and that stops rotation of impeller or creates air flow separation.

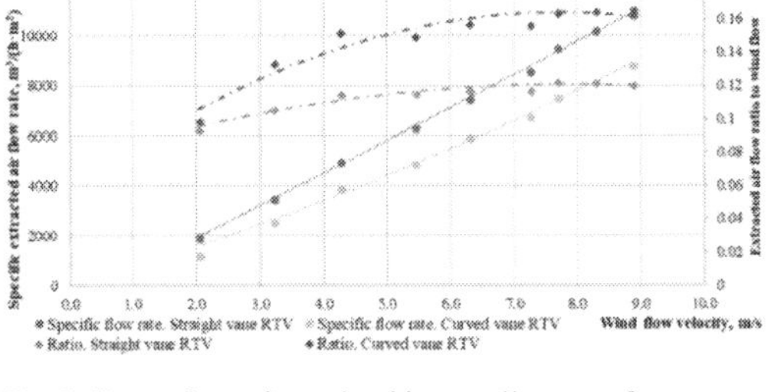

Fig. 5. Comparison of tested turbine ventilators performance.

Previous results (Fig. 3) showed better extracted air flow performance of curved vane ventilator. It should be kept in mind that these devices air suction duct sizes are not equal and as can be seen from figure 3 open ducts experimental results, sizes have significant impact. Inner diameters of air suction ducts are 108 mm and 128 mm, but area differs more than 40 %. Therefore, comparison of their absolute air flow rate isn't objective and calculation of specific extracted air flow is used. The calculations are done by using (1) equation and the results plotted in Fig. 5.

Results of specific extracted air flow rate calculations show opposite trend compared to absolute values. It can be seen that this time straight vane turbine ventilator has better performance (up to 23 %) but the change of trends remains similar.

Another important factor to RTV's performance is impeller size. Based on this factor, additionally there is calculated extracted air flow rate ratio to wind flow rate by (2) equation and the results plotted in the same Fig. 5. However, results of such calculations show better ratio of curved vane RTV. Analysis of curves indicates that performance of devices is similar on low wind flow velocity and difference gets higher up to 6 m/s. On higher wind speed difference between ratios remains almost the same.

5 Conclusions

Experiments of turbine ventilators under different test conditions revealed that the main effect for extracted air flow rate is due to ejection of wind and air suction duct interaction process. This is confirmed by test condition while ventilators are stopped and between differences during devices operate normally.

Controversial to some experimental investigations of such devices, it is determined ventilation flow rate increase while turbine ventilator is used, compared to open duct case. However, the performance increase is minor and reaches up to 18 %. Also, rotation of straight vane turbine ventilator has more impact on extracted air flow rate than curved vane ventilator. It could be mainly due to bigger straight vane ventilator impeller diameter and different aerodynamics. Different test conditions revealed that during periods when there isn't any rotation of turbine ventilator it makes huge impact to extracted air flow rate.

Comparing results of RTV's rotational speed under different test conditions, opposite trends are observed. During conditions while air suction duct is closed, straight vane turbine ventilator reaches higher RPM and curved vane turbine ventilator shows different trend. Supposedly such phenomenon could be explained by difference of such devices aerodynamics and flow separation.

Experimental results show that curved vane RTV reaches up to 12 % higher absolute extracted air flow rate compared to straight vane RTV. Calculation of extracted air flow ratio to wind flow also indicates better results of curved vane RTV. However, recalculation to specific extracted air flow rate shows opposite results. Evaluation of impact due to different open duct area identifies better performance of straight vane RTV.

Acknowledgement

This research is funded by the European Regional Development Fund according to the supported activity 'Research Projects Implemented by World-class Researcher Groups' under Measure No. 01.2.2-LMT-K-718 (project grant No 01.2.2-LMT-K-718-01-0016).

References

1. C. M. Lai, Energy Build., **35**, 9, 927–932 (2003).
2. N. T. Bay and H. Phuong, Int.J. Eng. **6**, 2, 86–95, (2012).
3. G. K. Jadhav, P. M. Ghanegaonkar, and S. Garg, J. Build. Eng., **6**, 196–202 (2016).
4. D. M. H. Rashid and N. A. Ahmed, Wind Eng., **27**, 1, 63–72 (2003).
5. N. Khan, Y. Y. Su, S. B. Riffat, and C. Biggs, Renew. Energy, **33**, 11, 2441–2447 (2008).
6. J. Lien and N. A. Ahmed, *Wind Driven Ventilation for Enhanced Indoor Air Quality*, 2002.
7. C. M. Lai, Energy Build., **38**, 3, 174–180 (2006).
8. AS/NZS 4740:2000, *Australian/New Zealand Standard. Natural ventilators-Classification and performance*, p. 46 (2000).
9. Y. Sik, D. Hun, H. Chung, H. Jeong, and S. Choi, Energy Build., **139**, 232–241 (2017).

Multi-Resonant based Output Voltage Control of Autonomous Distributed Generators

Abdelhakim Saim[1], *Rabah* Mellah[2], and *Azeddine* Houari[3]

[1]Faculty of Electrical Engineering, University of Science and Technology Houari Boumedien (USTHB), BP n°32, El Alia - Bab Ezzouar-16111, Algiers, Algeria
[2]L2CSP Laboratory, Faculty of Electrical and Computing Engineering, University Mouloud Mammeri of Tizi-Ouzou, 15000, Tizi-Ouzou, Algeria
[3]IREENA Laboratory, University of Nantes, 44602, Saint-Nazaire, France

Abstract. The distributed or decentralized generation electricity constitutes the central stone of various recent energy models, such as the intelligent electrical supply networks. The quality of energy in this type of structure depends primarily on the strategy of control adopted, in order to guarantee proper operation according to the international standards. The strategy of control proposed in this paper uses two control loops. An internal control loop aims at deadening the phenomena of resonance while ensuring the required control dynamic for fast disturbance rejection. As for the external control loop, a multi-resonant composed of a stabilizing state feedback that ensure an asymptotic tracking of the voltage reference with a weak rate of harmonic distortion. Finally, experimental results are presented to show the performances and feasibility of the proposed control strategy.

1 Introduction

The autonomous systems of distributed generation appropriate to electricity require an adequate and permanent control of the tension of the bus AC in order to ensure a quasi sinusoidal tension with a weak rate of harmonic distortion (THD), and this, independently of the dynamics of the fed electric charges. Moreover, the autonomous electrical supply networks with low tension (BT) are regarded as fragile structures because of the difficulties related to the maintenance of a bus tension with amplitude and frequency constants. These difficulties are exacerbated by the proliferation of the electric charges with strong not linearity, as by the intermittency which characterizes the majority of the sources of renewable energies. The face values of the amplitude and the frequency of tension are imposed by the requirements related to the electric charges (domestic or industrial) and those of the common electrical supply network. However, the variations can be introduced by the primary control of flows of power in the case of several inter-connected distributed generators.

Several strategies of controllers are proposed in the literature with a common aim to maintain the output voltage of the inverters on satisfactory levels [1, 2]. Among the principal strategies of control which exist in the literature, the resonant controllers (PR) can bring a solution to the problems related to the follow-up of periodic references. Indeed, resonant control provides profits very high to a precise frequency what makes it possible to eliminate, in closed loop, any static error at this frequency. Nevertheless, the selective character of this type of control with regard to the frequency of resonance conduit with the drastic deterioration of the performances of control even during light variations of the frequency of the bus AC [3-4]. Thus, application of the resonant controllers in the control of the autonomous systems of distributed generation of electricity must face the recurring variations of frequency in the micro grid network [6]. Moreover, research in progress, is focused on the design of a multi resonant controller who is more robust with regard to the fluctuation of the frequency all while maintaining the effectiveness of the multi resonant controllers in the follow-up and the rejection of periodic signals.

In this context, the work developed in this paper aims to bring solutions allowing the improvement of the quality of the energy of an autonomous micro grid comprising several distributed units of generation.

Control level relates to the regulation of the output voltage of the generator via the concept of resonant control. Indeed, resonant control positions from its performances among the techniques of control the most adapted for the follow-up and/or the rejection of periodic references, but suffers nevertheless certain gaps dependent primarily on its instability and its sensitivity to the variation of the point of operation (frequency) [7]. This can affect these performances harshly and limit its application in the systems of generation distributed considering the wave of the tension in these systems is the subject constantly of recurring variations in frequency and amplitude [8]. In this direction, a solution is proposed in order to benefit from the concept of resonant control without limiting its application. This solution consists in using a multi-resonant controller on stabilizing feedback states [9], whose profits of control are obtained from the minimization of a quadratic criterion.

Moreover, a bench of experimental test is built to validate the performances of the strategy of control developed using many experimental results before finishing with some conclusion.

2 Description of studied system

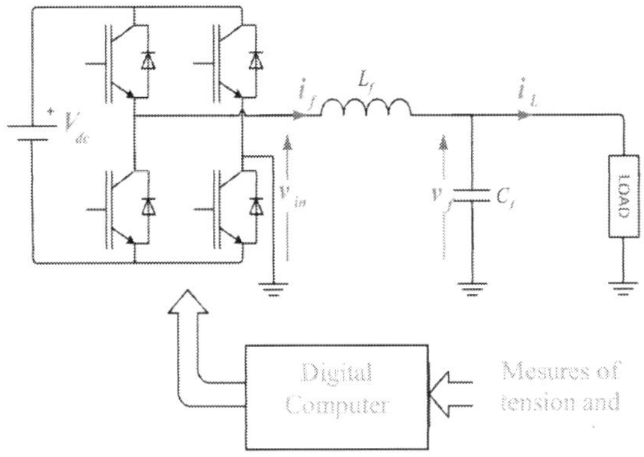

Fig. 1. Power circuit of an autonomous generator distributed.

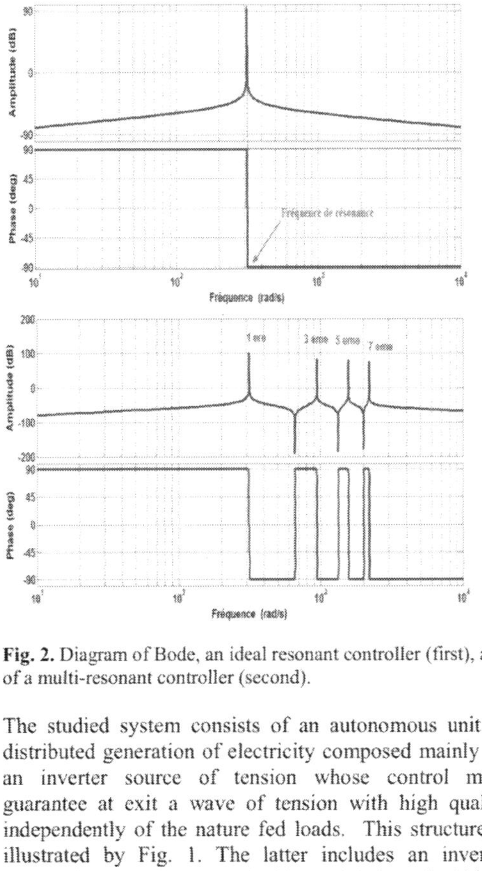

Fig. 2. Diagram of Bode, an ideal resonant controller (first), and of a multi-resonant controller (second).

The studied system consists of an autonomous unit of distributed generation of electricity composed mainly by an inverter source of tension whose control must guarantee at exit a wave of tension with high quality independently of the nature fed loads. This structure is illustrated by Fig. 1. The latter includes an inverter single-phase current controlled in modulation of width of impulse, MLI, a filter LC of exit whose role is to reduce the high frequency harmonic contents due to cutting as well as a bus AC represented by an unknown load [10-11].

The circuit proposed does not consider the use of a transformer considering the use of this last is preferred in three-phase systems, because not only supporting the capacities of insulation of the system and providing a means of transformation but also to provide a way appropriate to the homopolar component of the currents. Thus, the distributed generator proposed is designed to deliver directly on the bus of the loads. The operation of the converter of power is controlled by a strategy of adequate control implemented in a digital computer whose objective is to generate the optimal states of commutations in order to ensure at exit the desired signals.

3 Resonant controllers

The resonant corrector takes then, Like illustrated by Fig. 2, a theoretically infinite value at the frequency determined by the indicated poles. The principle of the resonant corrector rises from the theory of the internal models which stipulates that to facilitate the follow-up of the references and the rejection of the disturbances, it is necessary to integrate the models of these signals in the loop of control [12-16]. The transfer function of this corrector is given by:

$$C_R(s) = \frac{k_R s}{s^2 + \omega_R^2} \tag{1}$$

The transfer function of the ideal resonant corrector (3.1) has purely imaginary poles. With the result that it is able to correct any statistical error of the same frequency than its frequency from resonance, than it is in the follow-up or the rejection of alternating signals. Where ω_R the angular frequency of resonance of corrector and k_R the gain of control.

An assembly of several resonant controllers of various frequencies is possible. What increases the attractively of this type of control in the systems of distributed generation of electricity. Consequently, the control of the fundamental component of the output signals must be accompanied by control by the harmonic components. From where the transfer function of the corrector:

$$C_{Multi-R}(s) = \sum_{h=1,odd}^{n} \frac{k_{R,h} s}{s^2 + (h \times \omega_R)^2} \tag{2}$$

4 Multi resonant control with stabilizing feedback states

The strategy of control proposed is contingent, as Fig. 3 shows it, on a structure in cascade where the control of the tension v_f of output will be done by ensuring the control of the current i_f. Thus, each loop of control will be seen charged to provide the adequate reference to the loop which follows.

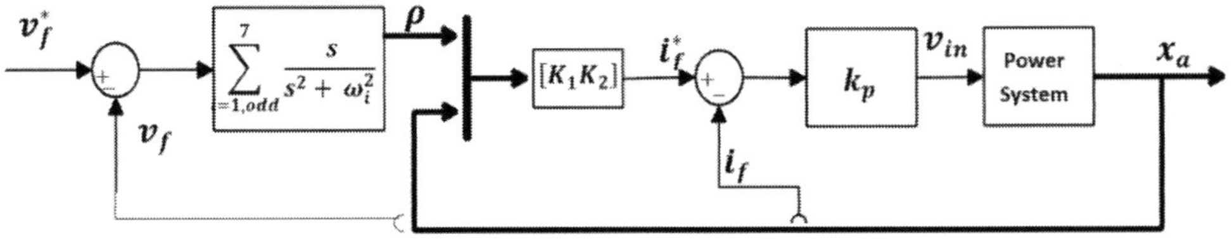

Fig. 3. Multi-resonant Controller with stabilizing return of states.

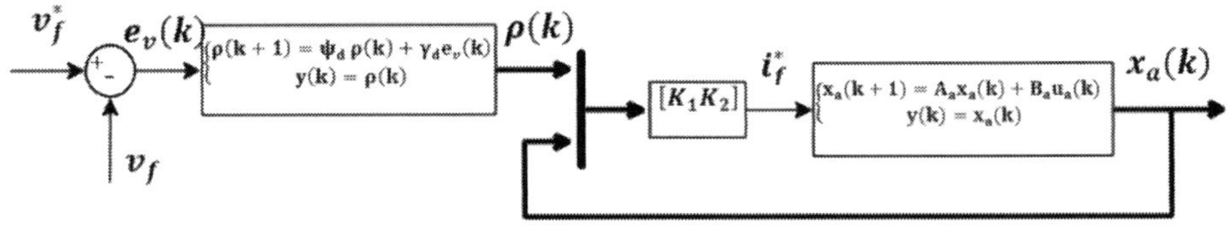

Fig. 4. State representation of the global system.

The internal loop of control aims to deadening the phenomena of resonance while ensuring dynamics necessary for a fast rejection of the disturbances. This loop uses the following law of control:

$$v_{in}(k+1) = k_p\left(i_f^*(k) - i_f(k)\right) \quad (3)$$

Where k_p is the proportional gain of the controllor and i_f^* is the reference of the output current.

The external loop of control considers for it the use of a multi resonant controller with a complete stabilizing state feedback. It aims to ensure an asymptotic follow-up of the reference voltage standard with a reduced rate of harmonic distortion.

This loop employs the following law of control:

$$i_f^* = \boldsymbol{K_1 x_a} + \boldsymbol{K_2}\left[\sum_{i=1,odd}^{7} \frac{S}{S^2+{\omega_i}^2}\right] e_v \quad (4)$$

The error of follow-up is defined by: $e_v = v_f^* - v_f$. The centre frequency of the resonant controllers is $\omega_i = 2\times\pi\times f_0\times i$, where f_0 represent the fundamental frequency of the system, and the argument i refer to fundamental or the order of the harmonic concerned. The choice of the vectors of gains $\boldsymbol{K_1}$ and $\boldsymbol{K_2}$ is done in an optimal way, after the resolution of a quadratic criterion. $\boldsymbol{x_a}$ represent the state vector of the increased system.

4.1 State representation of the increased system

The state representation of the increased system is given by:

$$\begin{cases} x_a(k+1) = A_a x_a(k) + B_a u_a(k) + E_a d(k) \\ y(k) = C_a x_a(k) \end{cases} \quad (5)$$

The state vector $x_a(k) = [v_f(k) \quad i_f(k) \quad v_{in}(k)]^T$
The control signal $u_a(k) = i_f^*(k)$

$$A_a = \begin{bmatrix} a_{11} & a_{12} & b_1 \\ a_{21} & a_{22} & b_2 \\ 0 & -k_p & 0 \end{bmatrix}, B_a = \begin{bmatrix} 0 \\ 0 \\ k_p \end{bmatrix} E_a = \begin{bmatrix} e_1 \\ e_2 \\ 0 \end{bmatrix} \text{and}$$

$$C_a = [C \quad 0]$$

4. 2 State representation of multi resonant controller

The state form of the multi resonant controller used is given as follows:

$$\frac{d\rho}{dt} = \psi\rho + \gamma e_v \quad (6)$$

Where

$$\rho = [\rho_1 \quad \rho_3\rho_5 \quad \rho_7]^T$$

$$\psi = diag[\psi_1 \quad \psi_3\psi_5 \quad \psi_7]$$
$$\gamma = [\gamma_1 \quad \gamma_3\gamma_5 \quad \gamma_7]^T$$

And

$$\psi_i = \begin{bmatrix} 0 & 1 \\ -\omega_i & 0 \end{bmatrix}, \gamma_i = \begin{bmatrix} 0 \\ 1 \end{bmatrix}, \text{for } i = 1,3,5,7.$$

The discrete state representation of the multi resonant controller is given by:

$$\rho(k+1) = \psi_d\rho(k) + \gamma_d e_v(k) \quad (7)$$

Where

$$\psi_d = e^{\psi T_s} \text{et} \gamma_d = \int_0^{T_s} e^{\psi(T_s-\tau)}\gamma d\tau$$

4.3 Total representation of the system.

The optimal strategy of control employed corresponds to a structure of linear quadratic regulation (LQR) of which the resolution requires, in first, the total state representation of the system (Fig. 4) in order to carry out the placement of the eigenvalues of the state matrix via a single choice of the gain vectors $\boldsymbol{K}_1$and $\boldsymbol{K}_2$

Thus, the total state representation of the system can be written as follows:

$$\hat{x}(k+1) = \hat{A}\hat{x}(k) + \hat{B}u_a(k) + \widehat{E_1}d(k) + \widehat{E_2}v_f^*(k) \quad (8)$$

Where $u_a(k) = i_f^*(k)$ represent the control signal, the state vector is define by $\hat{x}(k) = [x_a(k) \quad \rho(k)]$, while, the matrix coefficients are described as follows

$\hat{A} = \begin{bmatrix} A_a & 0 \\ -\gamma_d C_a & \psi_d \end{bmatrix}$, $\hat{B} = \begin{bmatrix} B_a \\ 0 \end{bmatrix}$, $\widehat{E_1} = \begin{bmatrix} E_a \\ 0 \end{bmatrix}$,

$\widehat{E_2} = \begin{bmatrix} 0 \\ \gamma_d \end{bmatrix}$, and $C_a = [C \quad 0]$.

Finally, the control signal is define by:

$$u_a(k) = -\boldsymbol{K}^T\hat{x}(k) = -[\boldsymbol{K}_1 \quad \boldsymbol{K}_2]\begin{bmatrix} x_a(k) \\ \rho(k) \end{bmatrix} \quad (9)$$

The choice of the vector gain $\boldsymbol{K}$ must ensure the stability of the closed loop system while maintaining an asymptotic follow-up of the reference. This vector results from the minimization of the following quadratic criterion:

$$J(u_a(k)) = \sum_{k=1}^{\infty} \hat{x}(k)^T Q\, \hat{x}(k) + u_a(k)^T R u_a(k) \quad (10)$$

The matrix Q is semi definite positive, while the coefficient $R > 0$.

5 Experimental results

As illustrated by Fig.5, a bench of experimental test was built to validate the strategy suggested for control in tension of an autonomous unit of generation distributed intended to feed from the electric charges of various natures. The bench of test consists of a rectifier intended to operate in source of continuous tension, an inverter SEMIKRON with filter LC of output, as well as panoply of electric charges. The two loops associated with both control current and tension is implemented in a platform dSPACE MicroAutoBox and requires only the measurement of the both output current and voltage. The execution time of the whole of the controller is about 20.1 µs, with the result that the strategy suggested does not require an excessive computing time.

The effectiveness of the strategy of control proposed in the continuation of the trajectories and the rejection of disturbance is shown through Fig. 6, and Fig.7. Those figures respectively characterize the response of the system for the linear load, the nonlinear load, and the level of load. The results obtained corroborate perfectly with those anticipated. The linear load is represented by a resistive load of $18\,\Omega$, while the nonlinear load is consisted a rectifier with diodes coupled with a circuit RLC $(R = 123\Omega, L = 500\mu H, C = 1mF)$. The measurement of the harmonic rate of distortion of the output voltage is equal to 0.8% for the linear load, and to 2.7% for the nonlinear load, which enables us to conclude the effectiveness of the control strategy proposed.

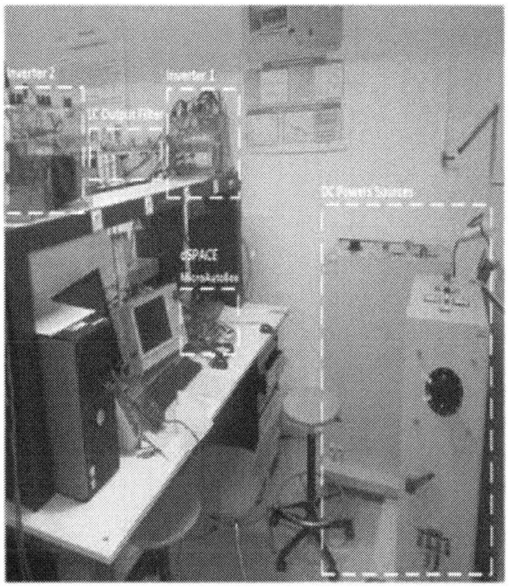

Fig. 5. Bench of experimental test.

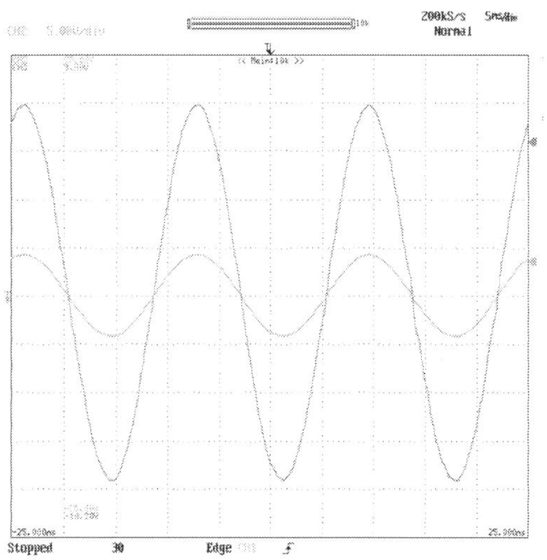

Fig. 6. Experimental results: Response of the system to a linear load, voltage curve [Pink (axis-X: 5ms/div; axis-Y: 40V/div)] and of the charging current [Green: (axis-X: 5ms/div; axis-Y: 10A/div)].

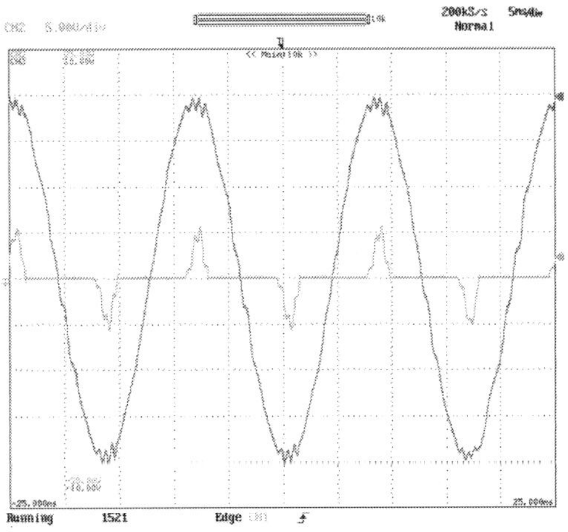

Fig. 7. Experimental results: Response of the system to a nonlinear load, voltage curve [Pink (axis-X: 5ms/div; 40V/div)] and of the load current [Green: (axis-X: 5ms/div; 10A/div)].

6 Conclusion

In this paper, we had to deal with the problematic appropriate to the voltage control of an autonomous distributed generator using a control strategy in nested loops has been proposed. This strategy aims to perform the synthesis in an optimal way, based on a multi-resonant control with stabilizing feedback state. Several experimental results are obtained to confirm and validate the proposed control strategy. In perspective, we plan to use the same concept of resonant control while developing an adaptive control structure in frequency and gain for a micro-array application.

References

1. R. Ortega, E. Figueres, G. Garcerá, C. Trujillo, and D. Velasco, "Control techniques for reduction of the total harmonic distortion in voltage applied to a single-phase inverter with nonlinear loads: Review," *Renewable and Sustainable Energy Reviews,* vol. 16, pp. 1754-1761, 2012.
2. M. R. Miveh, M. F. Rahmat, A. A. Ghadimi, and M. W. Mustafa, "Control techniques for three-phase four-leg voltage source inverters in autonomous microgrids: A review," *Renewable and Sustainable Energy Reviews,* vol. 54, pp. 1592-1610, 2016.
3. A. Hasanzadeh and H. Mokhtari, "A simplified droop method implementation in parallel UPS inverters with proportional-resonant controller," *Iranian Journal of Science and Technology,* vol. 33, p. 163, 2009.
4. A. R. Monter, E. J. Bueno, A. García-Cerrada, F. J. Rodríguez, and F. M. Sánchez, "Detailed analysis of the implementation of frequency-adaptive resonant and repetitive current controllers for grid-connected converters," *Electric Power Systems Research,* vol. 116, pp. 231-242, 2014.
5. N.F Guerrero-Rodriguez and A.B Rey-Boue, "Adaptive-frequency Resonant Harmonic-Compensator structure for a 3-phase grid-connected photovoltaic system," *Energy Conversion and Management,* vol. 87, pp. 328-337, 2014.
6. J. M. Guerrero, L. Hang, and J. Uceda, "Control of distributed uninterruptible power supply systems," *IEEE Transactions on Industrial Electronics,* vol. 55, pp. 2845-2859, 2008.
7. R. S. Rao, K. Ravindra, K. Satish, and S. Narasimham, "Power loss minimization in distribution system using network reconfiguration in the presence of distributed generation," *IEEE transactions on power systems,* vol. 28, pp. 317-325, 2013.
8. A. Saim, R. Mellah, A. Houari, M. Machmoum, and A. Djerioui, "Adaptive resonant based multi-loop control strategy for parallel distributed generation units in standalone microgrid application," *Electric Power Systems Research,* vol. 143, pp. 262-271, 2017.
9. P. Sreekumar and V. Khadkikar, "A New Virtual Harmonic Impedance Scheme for Harmonic Power Sharing in an Islanded Microgrid," IEEE Trans. Power Deliv., vol. 31, no. 3, pp. 936–945, Jun. 2016.
10. A. D. Paquette and D. M. Divan, "Providing Improved Power Quality in Microgrids: Difficulties in Competing with Existing Power-Quality Solutions," IEEE Ind. Appl. Mag., vol. 20, no. 5, pp. 34–43, Sep. 2014.
11. M. Aamir, K. A. Kalwar, and S. Mekhilef, "Review: Uninterruptible Power Supply (UPS) system," *Renewable and Sustainable Energy Reviews,* vol. 58, pp. 1395-1410, 2016.
12. M. Kale, M. Karabacak, W. Kruschel, F. Kilic, and P. Zacharias, "Chattering free robust control of LCL filter based shunt active power filter using adaptive second order sliding mode and resonant controllers," *International Journal of Electrical Power & Energy Systems,* vol. 76, pp. 174-184, 2016.
13. J. Rodriguez, J. Pontt, C. A. Silva, P. Correa, P. Lezana, P. Cortés, and U. Ammann, "Predictive current control of a voltage source inverter," *IEEE Transactions on Industrial Electronics,* vol. 54, pp. 495-503, 2007.
14. M. Rivera, V. Yaramasu, A. Llor, J. Rodriguez, B. Wu, and M. Fadel, "Digital predictive current control of a three-phase four-leg inverter," *IEEE Transactions on Industrial Electronics,* vol. 60, pp. 4903-4912, 2013.
15. P. Cortes, J. Rodriguez, C. Silva, and A. Flores, "Delay compensation in model predictive current control of a three-phase inverter," *IEEE Transactions on Industrial Electronics,* vol. 59, pp. 1323-1325, 2012.
16. M. Pichan, H. Rastegar, and M. Monfared, "Deadbeat control of the stand-alone four-leg inverter considering the effect of the neutral line inductor," *IEEE Transactions on Industrial Electronics,* 2016.

FEM Based Prototype of Moving-Coil Coreless Linear-Generator (MCCLG) for Wave Energy Extraction

Raed Althomali[1] and *Mohammed* Alsumiri[1]

[1]Electrical and Electronics Engineering Department, Yanbu Industrial College, Yanbu, Saudi Arabia.

Abstract. Linear generators are one of the most efficient sea wave's energy extraction setup. In the linear generator setup the energy is taken out by the vertical movement of waves. In most of the energy conversion systems, the energy is extracted by rotating motion by using the rotary generators. This paper is based on the prototype design of moving coil linear generator. The analysis of the prototype is done using the Maxwell finite element software. Finite-Element-Method (FEM) is used to analytically calculate the induced voltage and cogging forces. In this research, the FEM model of a coreless moving coil P.M linear generator has been outlined with an air cored stator. The main aim of the prototype is to convert marine energy directly without conversion from one form of motion to another, in regions the wave height is around 50 cm. A detail machine design formula is derived for the bases of initial prototype design. Prototype is modelled using Maxwell 2D FEM analysis. Hardware setup results were discussed to confirm the design, and to explore the working of the prototype as a generator. The movement of coils is accomplished by buoyancy mechanism of sea waves. The EMF generated is 6 Vp-p according to finite element simulation software. The prototype is able to generate EMF of about 2.5 Vp-p.

1 Introduction

In the recent decades the exhaustion of fossil fuels and rise of concerns regarding the changes in global weather and the increase level of carbon gases in atmosphere, leading cause to focus on renewable energy sources for electricity. The importance of research in the field of renewable energy is increasing day by day. At the present, the need for renewable power generation has increased dramatically. Two of the most focused renewable energy resources are Solar and Wind energies. While the sea wave energy is side lined, although it is actually one of the most concentrated and constant energy sources [1]. It is estimated that the world wide wave energy reserve is about 140-750 TWh/year keeping in mind the existing wave energy technologies that are developed. There are numerous topologies of wave energy converter (WEC) concepts [2], [3]. Previously several prototypes were presented to convert the oceans energy to generate electricity. They can be summarized as the followings [4]

1) Wave energy converters capture the vertical movement of ocean waves and with help of generators transforms the motion into electricity. This system consists of water-column or buoy mechanism to captures the motion of sea wave in some types it uses sub surface buoy systems [5].

2) Tidal Energy systems converts the energy of sea waves and tides under the surface and transform them into electrical energy. These systems are also known as underwater turbines, wither horizontal or vertical. These rotate with sea wave these are like under water wind mills [6].

Fig. 1. Wave Energy Converters [5].

3) Ocean temperature difference energy technology is based on the phenomena that there is density difference between the warmer water of sea and colder water. The systems generate electricity through the temperature differential between surface water and deep water. This system is a type thermal generator.

Fig. 2. Tidal/Current Systems [6].

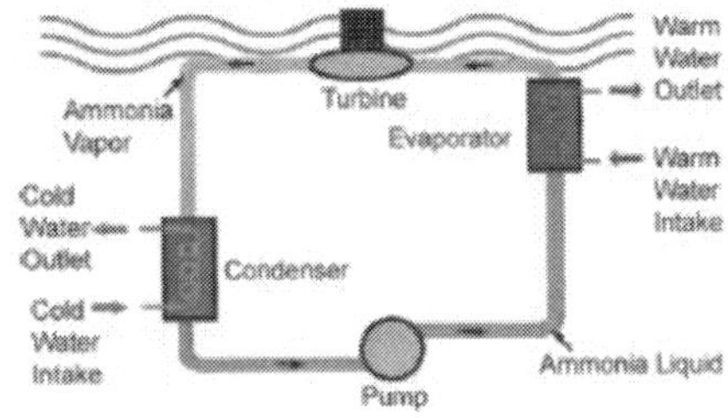

Fig. 3. Ocean Thermal Energy System [7].

4) Marine biomass system uses under water algae which are producing fuels in the shape of gases or liquids. These marine bio fuels are used for electrical energy generation. The earth surface is covered by 70% oceans. The ocean surface can be used for solar farms as an alternative for on land farms.

There is large number of devices used to convert the wave energy, but many of these devices are in immature or at the stage needs more improvement to able to compete with existing energy conversion devices. There are few devices been tested at large scale and deployed in the oceans. The LIMPET Land Installed Marine Power Energy Transmitter oscillating water column (OWC), is installed at Islay, Scotland, in 2000 a connected to grid able to produce 250KW. In September 2008, other commercial wave energy conversion setup was installed and operating in Northern Portugal [9].

Wave energy conversion system has advantages over the other sources. These advantages are highlighted as below:

1) The energy density of water waves is higher among renewable energy sources [11], [12]. It is well known that sea-waves are results of movement of wind. The wind flow due to change in temperature which is produces by solar energy. This energy strength is of typically 0.1 to 0.3 kW/m2 along the horizon is able to produce the winds flow power up-to the strength of 2 to 3 kW/m2.

2) The bad effects on the environmental by using wave energy are minimized. It has been noted that the devices which is offshore have the lowest negative effects.

3) In moderate climate, there is a correlation between the due to seasons and the requirement of electrical energy. This is very clear from the report that sea wave conversion can operate and generate electricity up to 90 % of the time. In the comparison of the wind and solar power devices which is up-to 20 to 30% [12].

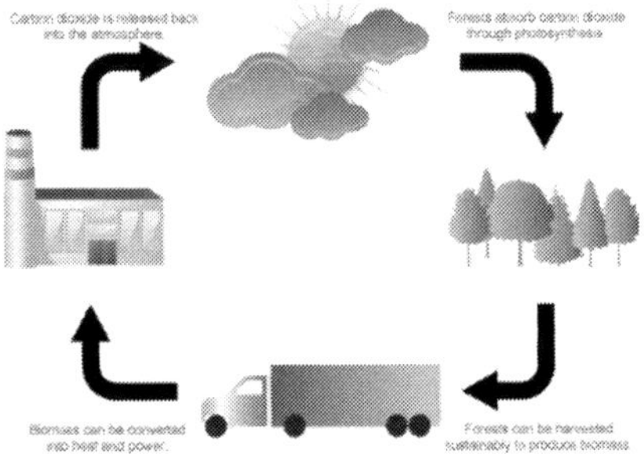

Fig. 4. Marine Biomass [8].

In this research, the performance and FEM model of a prototype coreless moving coil permanent magnet linear generator machine has been outlined with a nonmagnetic stator. The main aim of the prototype is to convert marine energy directly without conversion from one form of motion to another, in regions the wave height is around 50 cm. Detail machine design formula is derived for the bases of initial prototype design. Prototype is modeled using Maxwell 2D FEM analysis. Hardware setup results were discussed to confirm the design, and to explore the working of the prototype as a generator. The movement of coils is performed by buoyancy mechanism of the sea waves. Linear Generators are used to convert the linear motion directly into electromotive force (EMF). In conversion from linear to rotational tremendous amount of energy is lost and hence the conversion efficiency of rotary is less as compared to linear generator. In this project the permanent magnet arranged as stationary and coil is moved through the magnetic field which generates EMF. The EMF generated is 6 Vp-p according to finite element simulation software. Project concludes with some suggestions for future developments

2 Linear generator

Permanent Magnet Linear generator comprises of a magnets and coils. The portion which is subjected to the movement is called as translator. The portion that is not moving is called as stator. These machines can be categorized based on the construction, magnetic field arrangement, and magnets arrangements. In this paper the linear generators are categorized according to its construction. There are three categories of linear generators i.e. Moving-Coil-Linear-Generators (MCLM),

Moving-Iron-Linear-Generators (MILG) and Moving-Magnet-Linear-Generators (MMLG). MCLG are the machines in which the coil is moving with respect to magnets. MILG are the machines in which the flux lines path is changed by changing the reluctance of the path. The change in reluctance of the magnetic path is provided by moving the low reluctance magnetic iron. MMLG are the machines in which the magnets are moving while the coils are stationary. In case of alternator the translator is reciprocated with the help of mechanical force. Permanent Magnet (PM) flux lines cuts the stator coil and voltage is generated in stator. The current is injected into the stator coils of linear motor the field is generated by the current. The field due PM and due to stator coil interact with one another and exert force on translator. Drawbacks in this system is that it has low power factor and the magnet is expose to thermal and impact energy which causes demagnetization of the magnet [13].

3 Types of linear generator

3.1 Permanent magnet linear generator (MMLG)

These machines are very desirable for most of the application now days having benefits over the other types of linear MMLG have advantages such as high power density, and strength of field. MMLG category there are lot of variation and can be design by numerous methods. To understand this configuration a simplified MMLG is presented in Fig. 5.

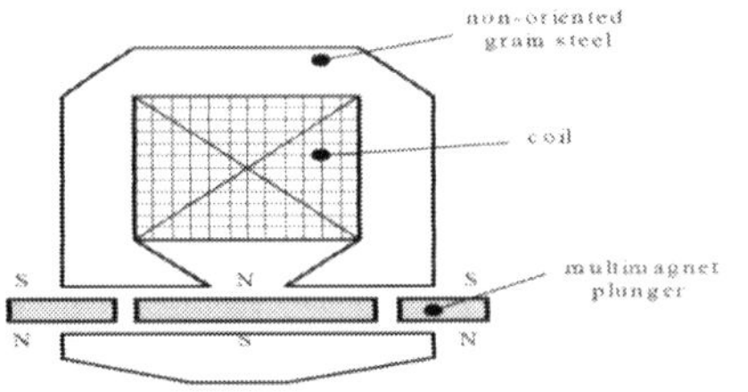

Fig. 5. A Cross Sectional View Of A MMLG [9], [10].

The MMLG presented is designed as cylindrical. The MMLG consists of a coil wound in a stator slot. The ferromagnetic stator has a small gap at the center. In that gap a ring shaped magnets are installed. These magnets are free to move horizontally with the external forces. The move in the air-gap backward and forward induces EMF due to changes of flux linkage with coils in the stator. However, the MMLG have some limitations. One of the crucial one is the structure of high strength magnets are fragile in structure. The structure strength of the magnet material may reduce because of the high-speed fatigue forced by high speed switching of translator. The magnetic properties of magnets decrease when exposed to high temperature. The removal of heat is also issue with this type of machines as the magnets are closed in the stator. The only way to remove heat is through radial radiation this will cause the loss of efficiency

3.2 Moving iron linear generator

This is a new type of linear generator. MILG is having design which is and most of the MMLG limitations are tried to remove. MILG can also have lot of variation in terms of design. The simplified MILG prototype is presented, a as shown in Fig.6.

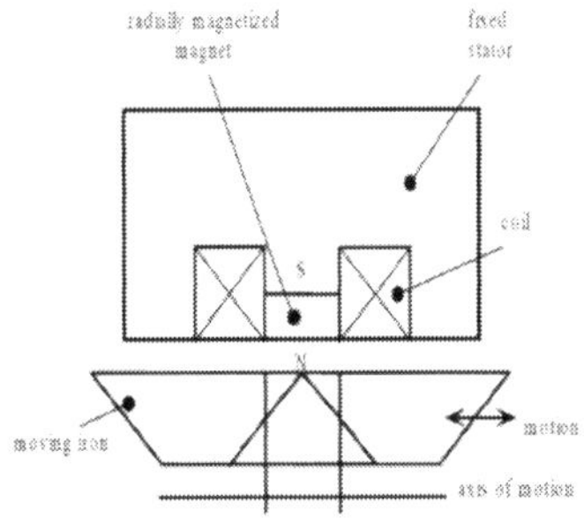

Fig .6. Cross Sectional View Of A MILG [9], [10].

This linear generator consists of iron plunger moving in the cylindrical stator center. The stator is housing the coils and magnets. The magnets are arranged in such fashion that the magnetic flux produced by them should link with stator coils. The iron moves with external forces and it changes the reluctance of the magnetic circuit. This motion causing the change in flux linkage in coils and inducing EMF in these coils.

MILG have some advantages like low cast and rugged design. There are some limitations in MILG for example, the ration of power to weight is less as compared to other machines. The weight of the translator and stator is very high due to which the high-speed oscillation have issues with structure stability.

3.3 Moving coil linear generator

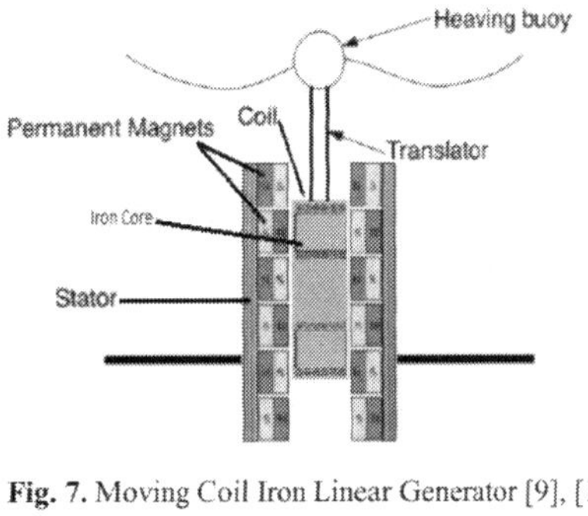

Fig. 7. Moving Coil Iron Linear Generator [9], [10].

Addressing these limitations in MMLG and MILG the type of linear generator is suggested that is, moving coil linear generator (MCLG). MCLG is presented to reduce most of mentioned problems. There are two types of design methodologies; Moving Coil Iron-cored Linear Generator (MCILG) and Moving Coil Coreless Linear Generator (MCCLG). The MCILG methodology is to design a translator made from iron. The translator have slots to accommodate the windings. The translator is coupled with oscillating force. The magnets are located on the stationary stator. When the translator coils moves in magnetic field of stator magnets the EMF is induced. The simplified MCILG is shown in Figure 7.

For the iron cored, EMF induced in the MCILG increases with increase in the coil thickness which will increase the thickness of stator. The EMF induced is increased also with number of turns in translator which results the increase in the weight of the translator. The flux linkage with coils decreases due to saturation caused by the stator core thinning. The cogging force for MCILG is very higher and it is one of the limitations of these machines.

The MCCLG is based on the concept of translator having no iron parts to decrease the cogging. The coreless translator winding is coupled with external translational forces. The winding will be magnetically coupled with stator magnets inducing EMF in the coreless coils. The MCCLG design is shown in Figure 8.

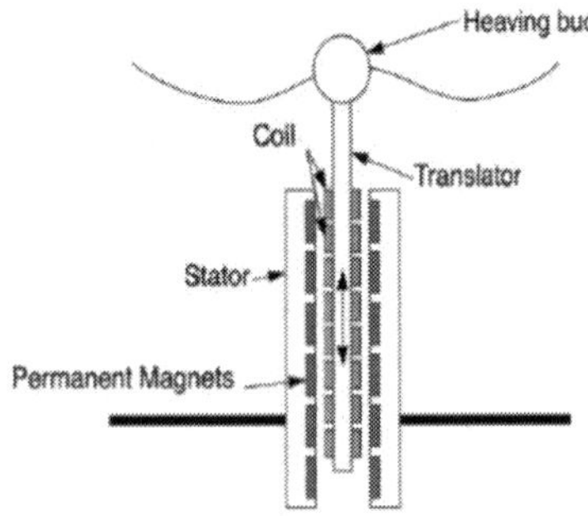

Fig. 8. Moving Coil Coreless Linear Generator (MCCLG) [9], [10].

The Moving Coil Coreless Linear Generator (MCCLG) is presented in the paper. Successively, MCCLG will be considered for use in the wave energy conversion system. The MCCLG is simulated and the operation of the generator is studied. The results from the simulation suggests that MCCLG is a suitable for the field renewable energy conversion system in Saudi Arabia. Much detail about MCCLG will be discussed along with simulations, advantages and drawbacks will be presented in the coming sections of this paper. Also, a detailed model of the MCCLG is presented and explained in Figs.15 and 16.

Referring to the previous discussion it is clear that MCLG are one of the better then other two types MMLG and MILG. MCLG coils are moving and magnet are placed on stationary stator hence magnets are easily cooled down. The magnets are also not subjected to the higher trust forces due reciprocating forces. Therefore, MCLG's are rugged and easy to design. The maintenance of magnets and other magnetic parts are easy and easy to install. These benefits MCLG are more desirable for industrial production. The light weight translator high frequency oscillation can be converted to energy. MCLG can compensate trust forces. The MCLG can achieve higher efficiency and specific power of nearly rotary generator. There are lot of advantages over other types of LG but MCLG have some limitations. The coils on translator requires the connection that can withstand translational motion which arise a problem of connector wear and tear problem in higher power applications.

4 Design Methodology

The design is based on data available from Yanbu off shore in which the average yearly wave's height is about 0.5 meters. Eight permanent magnet poles are placed on the stator made of plastic sheets. The coil is connected with a non-magnetic rod i.e. plastic. When the Buoy moves with the rod connected, the coil moves in the field of the magnets and thus produces an EMF with a value of 3 Vp-p and 6 Vp-p (simulated). The experiments, that were done, provide useful data for this design. This methodology will be applied using prototype design and tests to prove the theory of this type of machine. In addition, finite element simulations (FE) are implemented to obtain an FEM model of MCLG. Finite element software used is ANSOFT Maxwell, which is widely used in engineering designs of electromagnetic and electromechanical devices. Figure 9 shows experimental setup and arrangements of magnets.

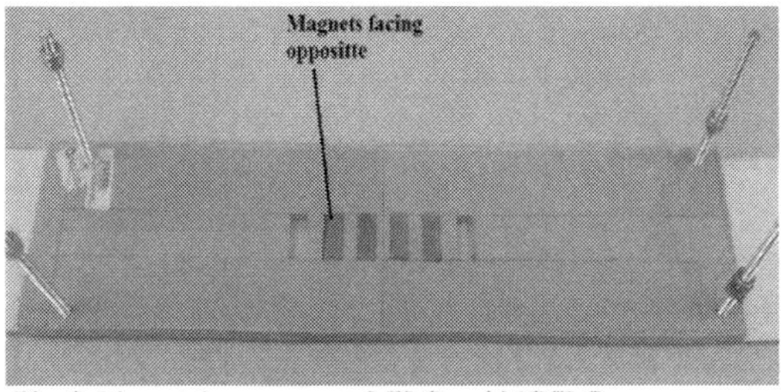

Fig. 9. Magnet Arrangements MILG and MCCLG.

5 Finite Element Method (FEM)

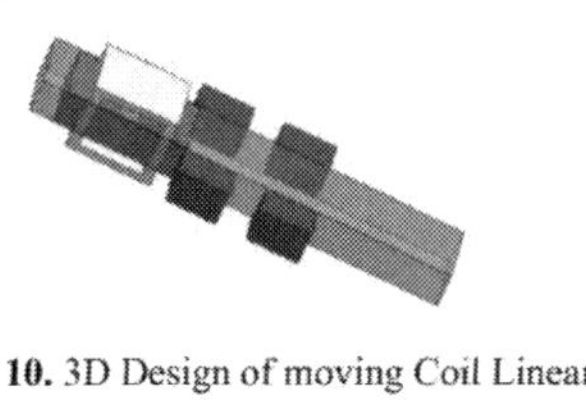

Fig. 10. 3D Design of moving Coil Linear Generator.

The analytical problem having complex solutions or need some approximation and to get very near solution the Finite Element Analysis (FEA) is used. This analysis is used to solve the problem of science and technology where complicated geometry is geometries are involved. FEM is used to solve the problem of engineering like load calculation and material properties calculations. Nowadays, Design geometry which is a commonly used method for multi physics problems is more complicated and the accurate solutions are desirable. In this paper The MCCLG and MCILG are setup in FEM. For MCILG the space between the coils is given properties as Iron and for

MCCLG the central area of coil got the properties of free space as shown in fig.10.

6 Prototype results

The In the start of process of prototype implementation, experimental results were low and not up to the expected level. A revision to the design calculation conducted with improvement on air gap of the machine and number of turns of the coil resulted in better outcomes as shown in Figure 11. Output voltage reached 4 Vp-p, which is still not up to the expected level. Some limitations such as; materials availability in local markets and limited local manufacturing facilities. Design is one of the limitations as it is approved in FEM analysis.

Fig .11. Prototype Result on Oscilloscope.

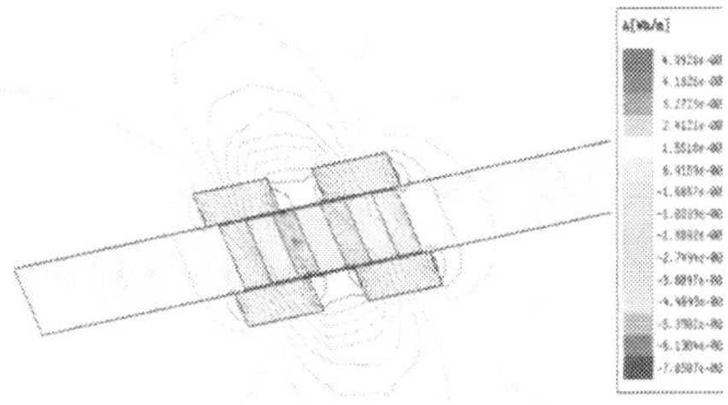

Fig. 12. MCILG Flux Lines.

Finite element method showed more promising results and provided wider vision to the project. Moreover, allowed the investigation of the deficiencies of the prototype model. The results of MCILG and MCCLG are given in figures list below.

From all aspects comparison conducted on Iron-core and Coreless Linear Generators. The comparison showed that both are usable for our needs with the following differences; Voltage and current are induced in both machines, but MCILG has higher values, Flux produced using MCCLG has lower values of flux line concentration and density than MCILG. For cogging force on the translator MCCLG for sure is in better condition than MCILG due to the elimination of ferromagnetic materials in the translator construction. However, both machines are applicable solutions for use. On the other hand, Moving Coil Coreless Linear Generator machine has been selected for its design simplicity and lighter weight.

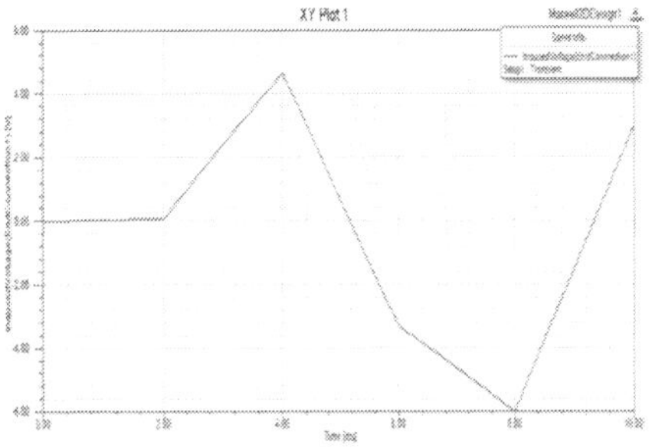

Fig. 13. MCILG Voltage Induced.

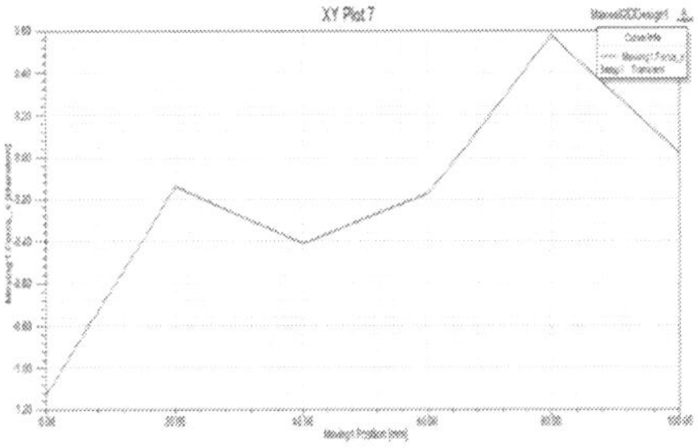

Fig. 14. MCILG Cogging Force.

7 Conclusion

This paper is focused on design and study the performance of a linear generator. The design analysis and simulation of Moving Coil Coreless Linear Generator MCCLG is presented in the paper. The study is focused to methodology to increase efficiency and power to weight ratio Machine air gap flux and cogging forces simulation is done by FEM analysis. The FEM software package is used to simulate MCCLG. The output voltage of generator is obtained by performing machine simulation. The output of generator is analyzed by FEM. The design parameters are calculated by the data locally and prototype is designed and tested. The fabrication of prototype by local market was a challenging.

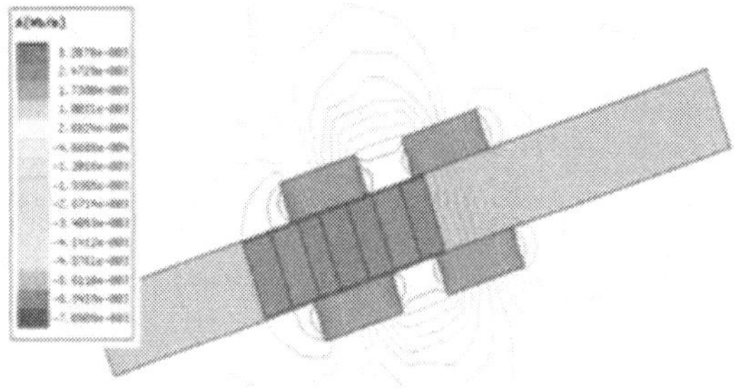

Fig. 15. MCCLG Flux Lines.

https://doi.org/10.1051/e3sconf/20186407005

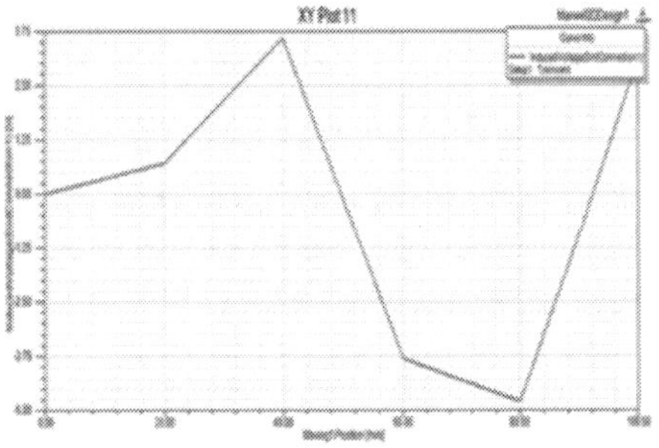

Fig. 16. MCCLG Voltage Induced.

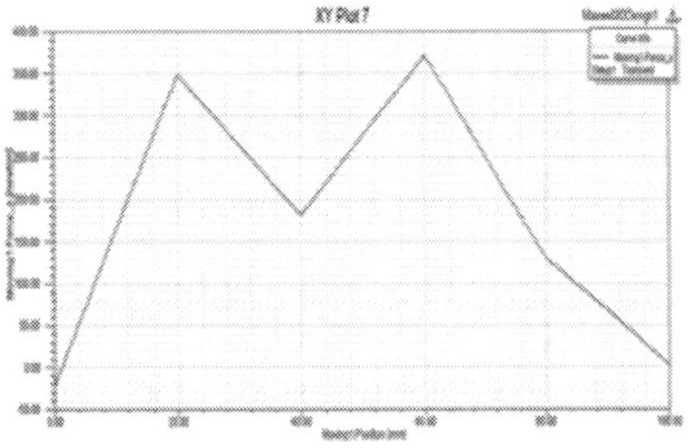

Fig. 17. MCILG Cogging Force.

References

1. Muhammad Rifa'I., Frasisco Danang.,Sarjiya. "Design and analysis of tri core permanent magnet linear generator for wave energy conversion in south coast of Java Island", 6th International Annual Engineering Seminar, InAES 2016, DOI: 10.1109/INAES.2016.7821944.
2. U.S. Department of the Interior "Technology White Paper on Wave Energy Potential on the U.S. Outer Continental Shelf," 2006. URL:
3. ww.ocsenergy.anl.gov/documents/docs/OCS_EIS_WhitePaper_Wave
4. Dulpichet Rerkpreedapong"Field Analysis and Design of a Moving Iron Linear Alternator for Use with Linear Engine"Master Thesis, College of Engineering and Minerals, West Virginia University 1999.
5. Loránd SZABÓ, Claudiu OPREA "FEM BASED ANALYSIS OF ADVANCED LINEAR GENERATORS FOR WAVE POWER PLANTS"MicroCAD 2007:International Scientific Conference, 22-23 March 2007
6. Kristof., Guillaume UGent, Bart Meersman et. al. , "A power take-off and control strategy in a test wave energy converter for a moderate wave climate "Proceedings of the International Conference on Renewable Energies and Power Quality (ICREPQ'16). 2016.
7. Siemens New Archives "Hat trick for SeaGen tidal current turbine" http://www.siemens.co.uk/en/news_press/index/news_archive/seagen-tidal-currant-turbine-milestone-hat-trick.htm 2017, http://hal.archives-ouvertes.fr
8. Thermal-Fluids Central "Ocean Thermal Energy Conversion" https://thermalfluidscentral.org/encyclopedia/index.php/Ocean_Thermal_Energy_Conversion access date 2018
9. Williams renewables "Biomass" http://www.williamsrenewables.co.uk/biomass/ (access date at 2018)
10. Thorpe, T. W. A brief review of wave energy, Technical report no. R120, Energy Technology Support Unit (ETSU), A report produced for the UK Department of Trade and Industry, 1999.
11. Fazal, Imran, Mohd Noh Karsiti, K.S. Rama Rao, and Saiful Azrin Zulkifli. "Modeling and Simulation of Moving Iron Linear Generator (MILG)", Applied Mechanics and Materials, 2011
12. Johannes Falnes. "A review of wave-energy extraction". Marine. Structure, Elsevier 2007, 185–201.
13. Pelc, R. and Fujita, R. M. "Renewable energy from the ocean." Marine. Policy, Elsevier 2002, 26(6), 471–479.
14. Arshad W. M., Bäckström T., Thelin P., and Sadarangani C., "Integrated free-piston generators: An Overview". IEEE NORPIE-2, Stockholm, 2002.

 https://doi.org/10.1051/e3sconf/20186407006

Power Distribution Design of MicroSat Power Control Unit Elegant Bread Board (EBB)

Che Cheng Huang, *Jia* Jing Yeh

Satellite Electrical Engineering Department, National Space Organization, Hsin Chu, Taiwan

Abstract. The Power Control Unit (PCU) in Micro-satellites program proposed to develop by NSPO, acts as the satellite power control and distribution center. One of the modules in the PCU, named the Power Distribution (PD), the main function provides the rated voltage and current of each power outlet for the satellite subsystem. When overload or short-circuit happens and persists for a time period, PD will limit the load current in a pre-designed value, and shut off the load to avoid damage on essential devices. In this paper, we will introduce the PCU EBB function and circuit design of PD.

1 Introduction

"Microsat" is usually applied to the name of an artificial satellite with a wet mass between 10 and 100 kg (22 and 220 lb). [1] However, this is not an official convention and sometimes those terms can refer to satellites larger than that, or smaller than that (e.g., 1–50 kg (2.2–110.2 lb)).

MicroSats can be built smaller than regulator satellites to reduce the large economic cost of launch vehicles and the costs associated with construction. Miniature satellites, especially in large numbers, may be more useful than fewer, larger ones for some purposes – for example, gathering of scientific data and radio relay. Technical challenges in the construction of MicroSats may include the lack of sufficient power storage or of room for a propulsion system.

Aimed at building a sustainable space industry, National Space Organization (NSPO) proposes to develop Micro-satellites program. The major goal for this program is to cooperate with domestic academia and industry developing the space components, subsystems, system, and forming space. This paper will focus on Power Control Unit architecture description for MicroSat satellite and the Power Distribution circuit design in the PCU Elegant Breadboard (EBB).

2 Power control unit

The Power Control Unit (PCU) in MicroSat is similar to a human heart. PCU is in charge of receiving solar power from solar arrays and regulating battery charging current according to On Board Computer (OBC) commands. It also controls and distributes power to various satellite load users upon the request of OBC command. A PCU block diagram is shown in Fig. 1.

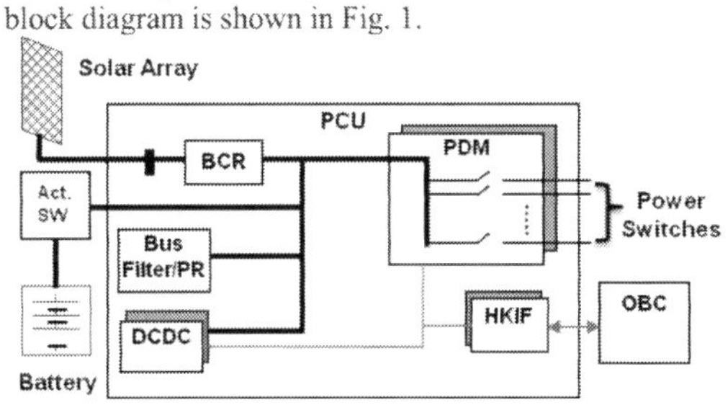

Fig. 1. Function Diagram of Power Control Unit

The PCU consists of 5 main functions, i.e. Power Distribution, Battery Charge Regulation, DC-DC Conversion, Power Relay, and Housekeeping Interface. Their main functions are as following:

• Convert input solar power to spacecraft required power by means of executing pulse width modulation switching commands received from OBC
• Provide Battery over voltage hardware protection
• Distribution of unregulated primary electrical power via over-current protected outlets to bus and instrument units as well as electrical heaters upon OBC command
• Distribution of thermostatic-control heater Activation of solar array Hold Down& Release Mechanism (HDRM) upon OBC command

• Activation of propulsion equipment upon OBC command
• Hardware protection of spacecraft main bus power under-voltage and preserve recovery capability
• Manage OBC main processor initialization and failure detection, recovery
• Provision of status monitoring and tele-command interfaces allowing the system and ground operate the power system, evaluate its performance and initiate appropriate countermeasures in case of abnormal conditions

3 Power distribution module

The Power Distribution (PD) in the PCU EBB provides the rated (MainBus) voltage and current of each power outlet for the satellite unit. MainBus voltage is the tie point for power from the solar arrays and the Battery [2]. Each power outlet for the satellite unit is controlled by power switches. The PDM contains 66x power switches of which 58 power outlets are available. For each power outlet, the power switches are implemented with power MOSFETs, and executed ON/OFF control by the FPGAs [3].

3.1 Architecture

According to sub-system requirements, The PD is separated from several outlet architectures. Some power outlets only have one switch; some outlets are configured in series for loads or with parallel outputs for special units. Architecture of Power Distribution module is shown in Figure 2.

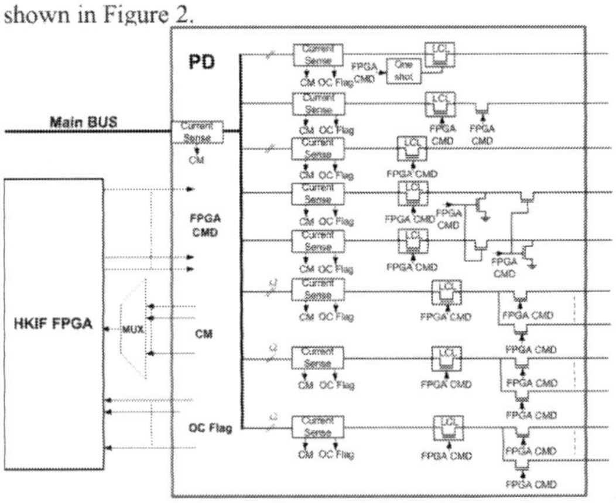

Fig. 2. Architecture of Power Distribution module

3.2 Circuit design

Whatever the sub-system voltage and load current requirements are, The PD outlets have the same functionality and similar circuit design, as shown in Figure3. Each power outlet both includes a power switch control circuit, current sensing circuit, the transistor feedback circuit, over current detector and current monitor. For convenient design, there are three level outlets in PDM.

Table 1. List of Outlet Types

#	Switch/Current Monitor	Current Monitor Range	Max Load		Latch Current Limitation	Trip-off current	Trip-off time
			Max Power	Current			
A	+	5.00A	50W	2.27A	6.82A	3.41A	5ms
B	+	5.00A	110W	5A	-	7.50A	5ms
C	+	1.00A	20W	0.91A	2.73A	1.36A	5ms

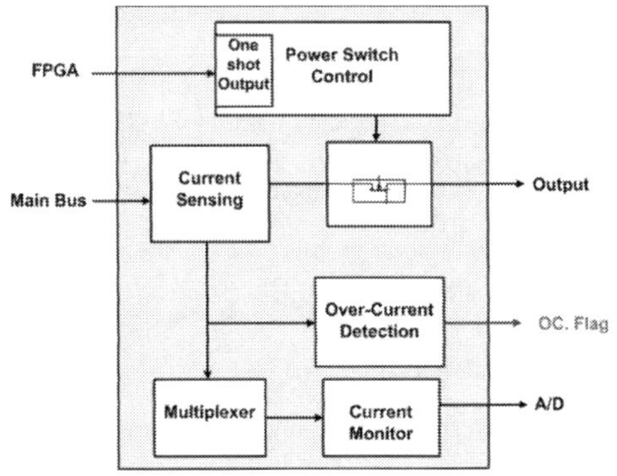

Fig. 3. Power Distribution function

3.3 Power switch control circuit

In PD module system, power switch control circuit is designed to on/off control the power output to the load. This circuit is controlled by FPGA to turn on/off the power MOSFET switch. In order to meet the requirements, the choosing of the power MOSFET should consider the characteristics of absolute maximum voltage and drain current rating, as well as on/off response time, and rise/falling time. Its design circuit is shown in Figure 4.

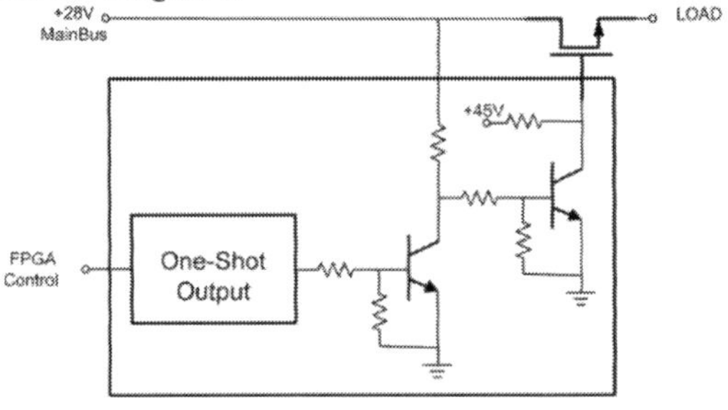

Fig. 4. Power switch control circuit

Without redundancy design in PCU EBB, when an over current occurred by OBC / S-band receiver, the protection process will power cycle OBC / S-band receiver by using a one-shot output which provide one second pulse to turn off and on again OBC / S-band receiver.

The one-shot output circuit is implemented with a mono-stable multi-vibrator featuring both positive and negative edge trigger inputs which can be used an inhibit input Complementary output pulse is provided.

There are three trigger inputs from the device. Inputs (A) are active-low trigger transition inputs and input (B) is an active-high transition Schmitt-trigger input. For OBC / S-band receiver protection, inputs (A) are fit in with our purpose to power cycle load.

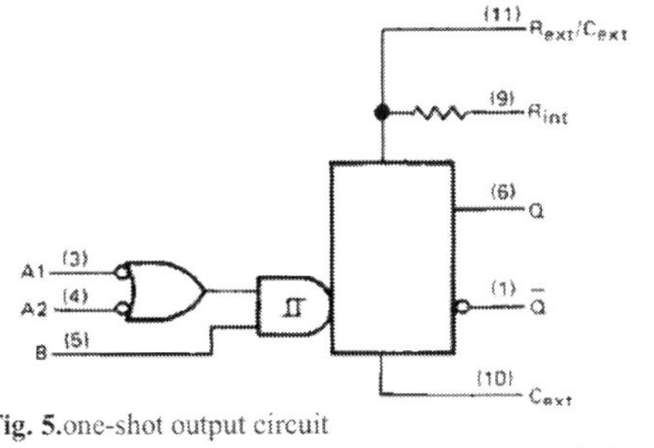

Fig. 5. one-shot output circuit

An internal timing resistor is provided for design convenience minimizing component count and lay-out problems. This device can be used with a single external capacitor

The basic output pulse width is determined by selection of an internal resistor R_{INT} or an external resistor (R_X) and capacitor (C_X). Once triggered the output pulse width is independent of further transitions of the inputs and is a function of the timing components. Pulse width can vary from a few Nano-seconds to 28 seconds by choosing appropriate R_X and C_X combinations.

Inputs			Outputs	
A1	A2	B	Q	$\bar{Q}$
L	X	H	L	H
X	L	H	L	H
X	X	L	L	H
H	H	X	L	H
H	↓	H	⊓	⊔
↓	H	H	⊓	⊔
↓	↓	H	⊓	⊔
L	X	↑	⊓	⊔
X	L	↑	⊓	⊔

H = High Logic Level
L = Low Logic Level
X = Can Be Either Low or High
↑ = Positive Going Transition
↓ = Negative Going Transition
⊓ = A Positive Pulse
⊔ = A Negative Pulse

Fig. 6. three trigger inputs

The pulse width is essentially determined by external timing components R_X and C_X. For C_X >1000 pF the output is defined as:

$$T_W = K\, R_X\, C_X \tag{1}$$

where

[R_X is in Kilo-ohm]

[C_X is in pico Farad]

[T_W is in nano second]

[K = 0.7]

Power cycle OBC / S-band receiver with one second pulse, RX is chose 300k ohm, CX is chose 4.7uF.

3.4 Current sensing circuit

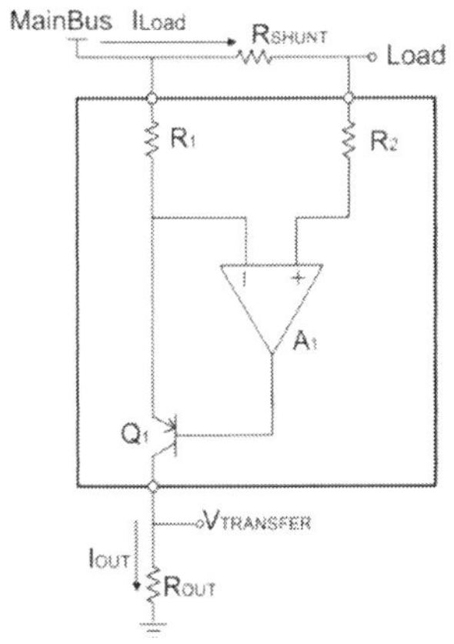

Fig. 7. High voltage current shunt monitor IC connection

Current sensing circuit is designed to catch load current value that provides for over current and current value evaluating. A high voltage current shunt monitor IC is chosen to achieve this purpose.

The current shunt monitor IC measures a small differential input voltage generated by a load current flowing through an external shunt resistor. The operational amplifier (A_1) is connected across the shunt resistor (R_{SHUNT}) with its inverting input connected to the MainBus side, and the non-inverting input connected to the load side of the satellite unit. Amplifier A_1 responds load current by causing Transistor Q_1 to conduct the necessary current through Resistor R_1 to equalize the potential at both the inverting and non-inverting inputs of Amplifier A_1. The current through the emitter of Transistor Q_1 (I_{OUT}) is proportional to the input voltage (V_{SENSE}), and, therefore, the load current (I_{LOAD}) through the shunt resistor (R_{SHUNT}). The output current (I_{OUT}) is converted to a voltage ($V_{TRANSFER}$) by using an external output resistor, the value of which is dependent on the input to output gain equation desired in sub-system requirements.

The transfer function of the current shunt monitor IC is

$$I_{OUT} = (gm \times V_{SENSE}) \tag{2}$$

$$V_{SENSE} = I_{LOAD} \times R_{SHUNT} \tag{3}$$

$$V_{TRANSFER} = I_{OUT} \times R_{OUT} \tag{4}$$

$$V_{TRANSFER} = (V_{SENSE} \times R_{OUT})/1000 \tag{5}$$

Where: gm = 1000 μA/V

3.5 Transistor feedback circuit

As overload or short-circuit takes place, the LCL[4] limits load current in a greatest acceptable value with trip-off time. LCL function is implemented with a simple Transistor feedback circuit. The BJT transistor is used to

https://doi.org/10.1051/e3sconf/20186407006

control the gate length of power MOSFET according to the $V_{TRANSFER}$ magnitude. As the load current increases rapidly, $V_{TRANSFER}$ is proportional immediately. Therefore, BJT transistor starts to decrease the gate voltage (length) of power MOSFET that reduces the load current, also means $V_{TRANSFER}$ decreases. Finally BJT transistor maintains a control balance to power MOSFET; the feedback circle forms a limitation to the load current in a latching value. In order to get different LCL types, resistor R_1 and R_2 value can be adjusted by requirement.

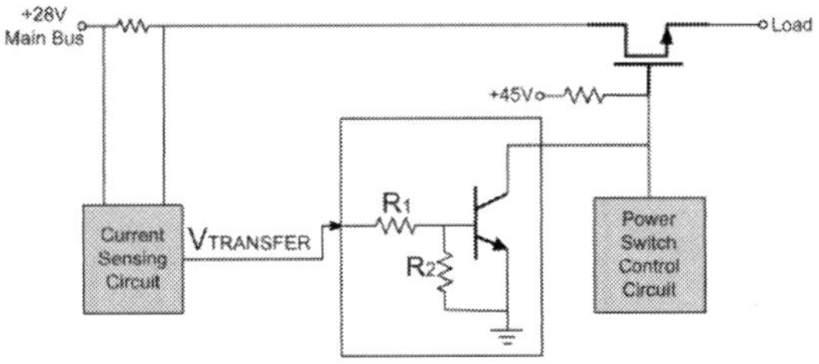

Fig. 8.BJT feedback circuit

3.6 Current monitor

Current monitor circuit can provide user to read load current value. The realizable method is using an OP amplifier that amplify $V_{TRANSFER}$ for corresponding current value. Each load has a max allowed current, current monitor output should within A/D converter max scale to avoid not appear desired current value.

In order to save PCB area, the same level outlets can share the current monitor by using an analog multiplexer. FPGA switches multiplexer to output every channel's current value to A/D converter.

3.7 Over current detector

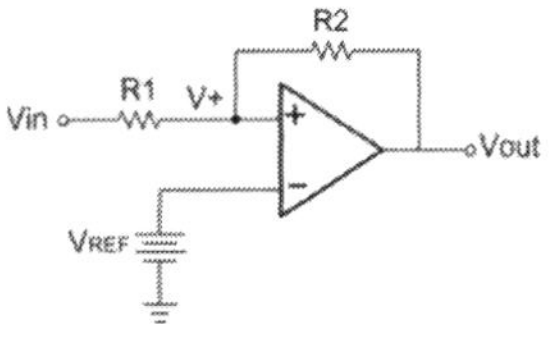

Fig. 9. Over current deector with Schmitt trigger

In Current Sensing Circuit, load current is transformed to $V_{TRANSFER}$ that can be compared in the Over Current detector. As $V_{TRANSFER}$ is higher than V_{REF} which means the load current is higher than the limit, Trip-off Load, the comparator [5] will output an over-current flag immediately to FPGA. The Schmitt trigger circuit also provides the Hysteresis [6] function to avoid error trigger by noise interference.

4 Measurement

A circuit evaluated board is designed to verify PD outlet function. This evaluated board includes power switch control circuit, current sensing circuit, transistor feedback circuit, over current detector and current monitor.

For functional testing, an electronic load is used to simulate a real satellite equipment load. Electronic load, Agilent N3300A (mainframe) and N3305A (500W E-Load module), can be set to sink arbitrary current from PD as desired.

According to each equipment power demands in satellite, PD outlet could be designed for different specifications, as showed in Table I. Chose Class C LCL be an example of testing. If overload or short-circuit happens, the LCL protection function will be trigged. For class C, the trip off time is 5ms ± 5% and LCL is set to 2.73 A± 20%.

Sent 3.3V TTL command to switch control circuit, power MOSFET can be switched and main bus voltage measured at outlet output. By using the electronic load to sink 3.0A from outlet, current sensing circuit transferred current to voltage, $V_{TRANSFER}$ that received by BJT feedback circuit. The LCL protected load current to keep current at 2.73A instantly as showed in Figure 8. At the same time, over current detector produced a flag which will be count by housekeeping interface (HKIF)'s FPGA in PCU EBB for Trip-off time countdown. After 5ms ± 5%, the FPGA can turns off the switch to terminate load current output.

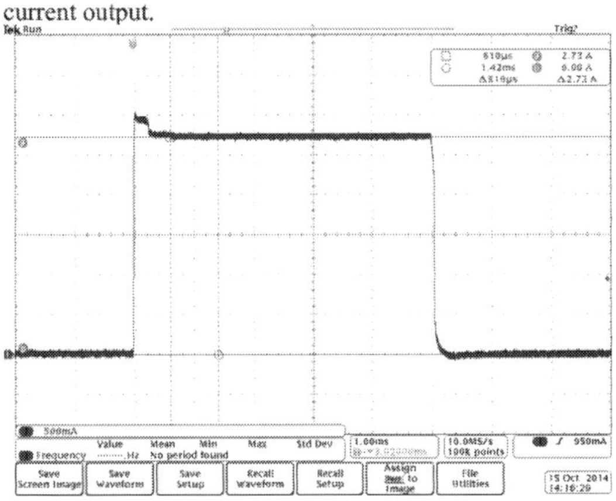

Fig. 10. Latching Current Limit measurement result

For current monitor test, using the electronic load to sink output and record the respond value of current monitor. These results of current monitor are linear, as table II shows. For PCU EBB, transmitting these analog data to a 12 bit A/D converter which can output digital hex data to HKIF's FPGA. According to Engineering Translation factor which wrote in Fly software, the current value can appear on display page.

Table 2. List of Outlet Types

Current	Measurement	Engineering Translation	Error
0A	0.05V	0.02A	NA
0.5A	1.103V	0.51A	2.0%
1.0A	2.176V	1.01A	1.0%
1.5A	3.257V	1.51A	0.60%
2.0A	4.340V	2.01A	0.50%
2.5A	5.467V	2.54A	1.60%
3.0A	6.600V	3.07A	2.30%

5 Conclusions

We present the PCU architecture description for Micro-Satellite and the PD circuit design in the PCU Elegant Breadboard (EBB) in this study. In order to verify the circuit design, an evaluated board was manufactured to provide measurement and fine-tuning for PD_EBB. According to evaluated board testing result, PD_EBB design has finished the circuit design, PCB fabrication is ongoing and functional checkout will execute sequentially.

PCU_EBB will be the first step for MicroSat mission, the follow-on PCU development will be engineer model (EM), Engineering Qualification Model (EQM) which will execute the Electromagnetic Interference (EMI) analysis, Thermal-Vacuum Testing, and vebration testing.

Through above Environmental testing, the PCU will continue to step in Fight Model (FM) which is real using in the MicroSat mission.

References

1 Networks [Online]. Available: https://en.wikipedia.org/wiki/Small_satellite#Microsatellites
2 Che-Cheng Huang, Jia-Jing Yeh, Zhe-Yang Huang, Chien-Kai Tseng "FORMOSAT-5 Satellite Power Protection Design," *Applied Mechanics and Materials,* Vol. 145, pp 536-541, Dec. 2011
3 Raymond A. Mack, Demystifying switching power supplies, Elsevier Inc. Austin, TX, USA,(2005), pp. 33
4 Earl Crandall, Power supply testing handbook: strategic approaches in test cost reduction, Kluwer Academic Pub, (1997), pp. 27
5 Herrick, Robert J., DC/AC circuits and electronics: principles & applications, Thomson Learning,(2002), pp. 378
6 U.A. Bakshi, A.P Godse, Analog Electronics, 1st ed., Technical Publications, (2009), pp. A7.

Analysis and Design of Filament Power Supply with Voltage-Fed Single-Switch ZVS Inverter

Yueh-Ru Yang

Ming Chi University of Technology, New Taipei City, Taiwan

Abstract. In this paper, an off-line voltage-fed single-switch zero-voltage-switching (ZVS) inverter is analyzed and designed for the filament power supply of cooker magnetrons. Due to the low resistance of cathode filament, an isolation transformer with large turn ratio is used. To obtain the ZVS operation, a parallel-resonant capacitor is connected to the primary of transformer. The capacitor, transformer and filament constitute a parallel resonant tank. Both the transformer leakage inductance and the transistor output capacitance are merged into the resonant tank. The resonant energy provides the ZVS operation of the inverter switch, and the secondary voltage drives the cathode filament. To verify the analysis, a prototype inverter is built and tested. Experimental results validate the circuit design.

1 Introduction

Microwave heating is an important contactless heating approach and the related techniques and applications in household, industrial, agricultural and medical fields are continuously developed [1-3]. The microwave generator of microwave ovens is a continuous-wave magnetron. Its basic components include a hollow metal cylinder, a cathode filament, two ring magnets, even-number anode resonant cavities and an output antenna. The cathode filament is located in the central of the hollow cylinder and is surrounded by the even-number anode cavities. The space between the anode and cathode is the working space of magnetron. In the space, the dc voltage between anode and cathode forms a horizontal electric field and the two ring magnets respectively located on the top and bottom of the hollow cylinder build a vertical magnetic field. As the cathode filament is heated to a critical temperature, the cathode filament emits thermal electrons to the working space. Then, the perpendicular electric and magnetic fields respectively exert forces to the space electrons. As the space electrons sweep across the opening of anode cavities, the polarities of adjacent anode vanes are alternately changed and hence the LC resonance of the anode cavities is caused. At proper conditions, the space electron clouds eventually form rotating spokes. As the spoke number is a half of the anode cavity number, the magnetron enters the pi mode operation. To easily enter the pi mode, the even and the odd anode vanes are respectively tied together. The anode output antenna intercepts the energy of anode resonant cavities and radiates microwaves. With the constant magnetic and electric fields, the anode power is determined by the average anode current.

The cathode filament is the heart of magnetrons. To prolong the usable life, the filament should be preheated to the thermionic emission temperature before applying negative high voltage to the cathode filament. In general, the pre-heating of cathode filament may take more than two seconds. Inappropriate start-up and filament temperature usually shorten the filament life and degrade the performance. In order to keep filament at proper temperature, the filament driving voltage should be decreased with the average anode current. Shown in Fig.1 is an example that the filament driving voltage decreases with the average anode current. Accordingly, as shown in Fig.2, besides the anode power supply (APS), a complete magnetron driver needs a separate filament power supply (FPS) to preheat and regulate the filament power. In this work, the voltage source V_d is rectified from the utility line. The anode of magnetron is grounded to the heat sink. The APS supplies a negative dc voltage to the cathode and the FPS heats the filament with a small ac voltage. To preheat the filament, switch S1 is turned on before switch S2.

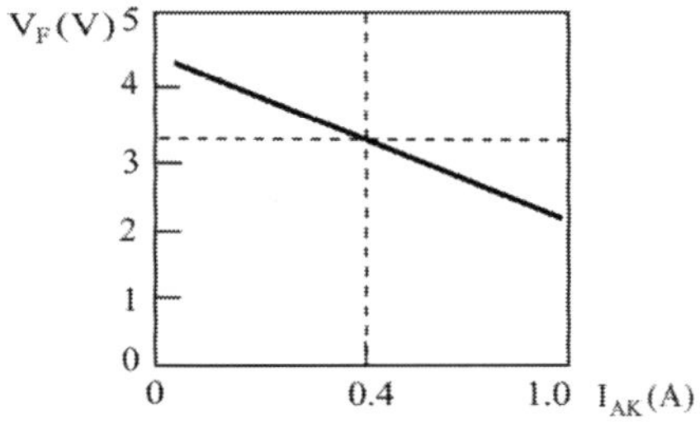

Fig. 1. The required filament voltage decreased with the average anode current.

Currently, there are many types of soft-switching circuits can be used to improve the circuit efficiency and reduce the volume and weight of power supplies. Due to simplicity and reliability, single-switch zero-voltage-switching (ZVS) circuits have been widely used in induction heating, cooking, LCD TV, EV chargers and other industrial applications [4-10].

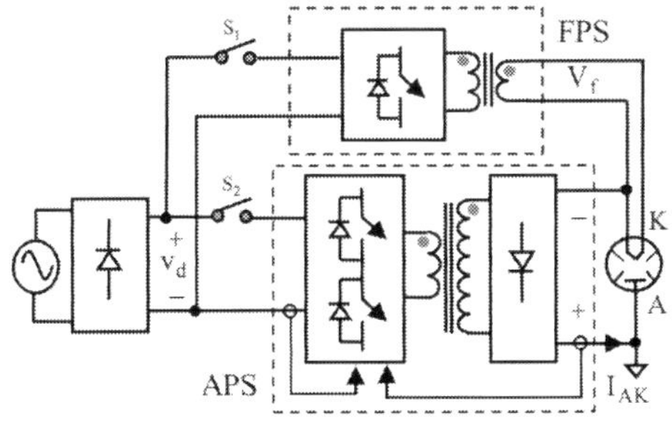

Fig. 2. The block diagram of an off-line magnetron driver.

2 Circuit operations

Fig.3 shows the circuit diagram of the FPS designed in this paper. It is a voltage-fed single-switch ZVS inverter. A step-down transformer is used to attain the required voltage conversion ratio and the secondary voltage drives the cathode filament.

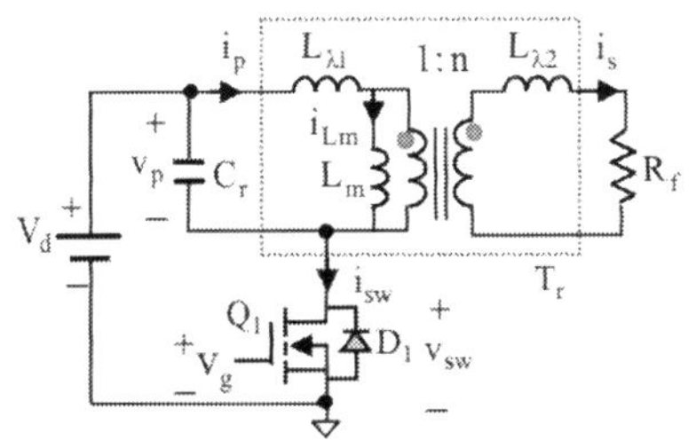

Fig. 3. The circuit diagram of single-switch ZVS inverter.

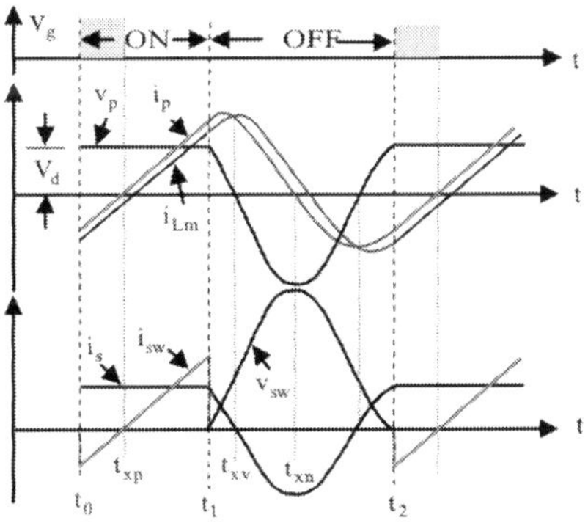

Fig. 4. The principal waveforms of single-switch ZVS inverter.

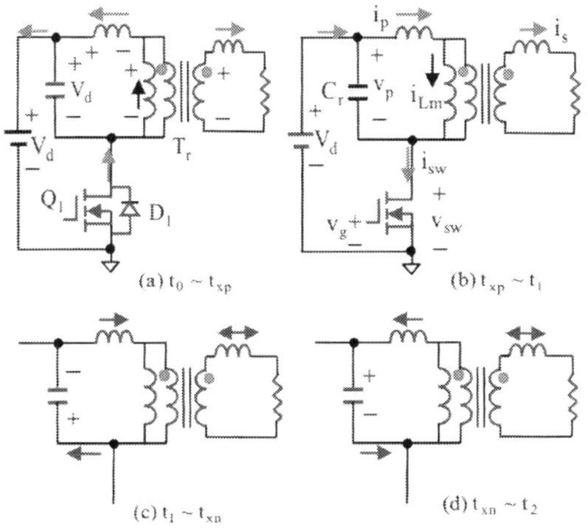

Fig. 5. The circuit operations of the single-switch ZVS inverter.

As switch Q_1 is turned off, transformer T_r and capacitor C_r comprise a parallel resonant tank. The primary leakage inductance and the transistor output capacitance are included in the resonant tank. By low-damped resonance, the resonant energy makes the resonant voltage reach the dc voltage V_d. Thus, switch Q_1 can be turned on at zero-voltage condition and hence switch loss is reduced. Fig.4 conceptually shows the current and voltage waveforms of the ZVS inverter. According to the primary current waveform, a switching cycle is divided into four stages. The detailed circuit operations are sketched in Fig.5.

At ON stage, $t_0 < t < t_1$, the primary voltage equals to V_d. The switch current i_{sw} is negative at t_0. Since D_1 is conducted, Q_1 is turned on at zero voltage condition. After t_{xp}, switch current rises from zero. As Q_1 is cut off at t_1, the OFF stage begins. At OFF stage, $t_1 < t < t_2$, T_r resonates with capacitor C_r, and the filament R_f consumes the resonant energy. At t_{xn}, the sinusoidal resonant current crosses zero and the resonant voltage reaches maximum. By small damping ratio, the primary resonant voltage rises to V_d at t_2. Then, D_1 naturally conducts and Q_1 can be turned on at zero voltage condition.

3 Circuit analyses

Fig.6 shows a simplified resonant circuit, where L_p is the primary inductance, $L_p = L_m + L_{\lambda 1}$, L_m is the magnetizing inductance, $L_m = kL_p$, k is the transformer coupling factor, $k = L_m/(L_m + L_{\lambda 1})$. C_{sw} is the switch output capacitance and C_p is the primary capacitance, $C_p = C_r + C_{sw}$. As the secondary leakage inductance is

 https://doi.org/10.1051/e3sconf/20186407007

ignored, the primary resistance can be approximated as $R_p \approx R_f / n^2$. For driving low resistance filament, the transformer turn ratio is much less than unity, $n = N_s / N_p << 1$. The initial inductor current and initial capacitor voltage are I_0 and V_0 respectively.

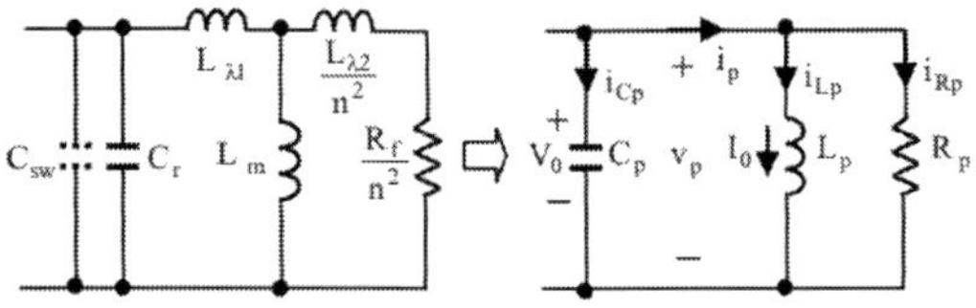

Fig. 6. The simplified equivalent circuit.

With coupling factor k and turn ratio n, the secondary voltage of ON stage can be approximately expressed as

$$v_s(t-t_0) \approx knv_p(t-t_0) = knV_d . \quad (1)$$

The transformer secondary current is

$$i_s(t-t_0) \approx \frac{knV_d}{R_f} + [i_s(t_0) - \frac{knV_d}{R_f}]\exp[-\frac{R_f}{L_{\lambda 2}}(t-t_0)] \quad (2)$$

At OFF stage, the characteristic equation of resonant tank is

$$s^2 + \frac{1}{R_pC_p}s + \frac{1}{L_pC_p} = s^2 + 2\alpha s + \omega_0^2 = 0 , \quad (3)$$

Here α is the damping factor, $\alpha = \frac{1}{2R_pC_p}$, ω_0 is the natural frequency, $\omega_0 = \frac{1}{\sqrt{L_pC_p}}$. The damping ratio is $\xi = \frac{\alpha}{\omega_0}$. With the loaded quality factor Q_L, (3) can be rewritten as

$$s^2 + 2\alpha s + \omega_0^2 = s^2 + \frac{\omega_0}{Q_L}s + \omega_0^2 = 0 . \quad (4)$$

$$Q_L = \frac{\omega_0}{2\alpha} = \frac{1}{2\xi} = \frac{R_p}{Z_0},\ Z_0 = \omega_0 L_p = \frac{1}{\omega_0 C_p} = \sqrt{\frac{L_p}{C_p}} .$$

For under-damped case, $\xi < 1$ or $Q_L > 0.5$, the characteristic equation has two conjugate complex roots,

$$s_{1,2} = -\alpha \pm j\omega_d ,\ \omega_d = \sqrt{\omega_0^2 - \alpha^2} . \quad (5)$$

The natural voltage response of the primary is

$$v_p(\tau) = e^{-\alpha\tau}[B_1 \cos(\omega_d \tau) + B_2 \sin(\omega_d \tau)] ,\ \tau = t - t_1 . (6)$$

With initial conditions,

$$v_p(0) = V_0 ,\ \frac{dv_p(0)}{dt} = \frac{V_0}{R_pC_p} + \frac{I_0}{C_p} ,$$

the constants B_1 and B_2 are determined.

$$B_1 = V_0 ,\ B_2 = V_0\frac{\alpha}{\omega_d} - (I_0 + \frac{V_0}{R})\frac{1}{\omega_d C_p} . \quad (7)$$

With the voltage response, the resistive current is

$$i_{Rp}(\tau) = v_p(\tau) / R_p . \quad (8)$$

The inductive current obtained by the KCL is

$$i_{Lp}(\tau) = -i_{Rp}(\tau) - i_{Cp}(\tau) ,\ i_{Cp}(\tau) = C_p\frac{dv_p(\tau)}{dt} . \quad (9)$$

For un-damped case, $\alpha = 0$, $B_1 = V_0$, $B_2 = I_0Z_0$, the primary voltage and current are respectively

$$v_p(\tau) = V_0 \cos(\omega_0\tau) - I_0Z_0 \sin(\omega_0\tau) . \quad (10)$$

$$i_p(\tau) = I_0 \cos(\omega_0\tau) + \frac{V_0}{Z_0}\sin(\omega_0\tau) \quad (11)$$

The primary voltage equation (10) can be written as

$$v_p(\tau) = -V_{pk}\sin(\omega_0\tau - \phi) \quad (12)$$

$$V_{pk} = V_0\sqrt{1+\gamma^2} ,\ \gamma = \frac{I_0Z_0}{V_0} ,\ \phi = \sin^{-1}(\frac{1}{\sqrt{1+\gamma^2}}) .$$

As shown in Fig.4, the parallel resonance begins at t_1 and ends at t_2. The resonant voltage crosses zero at t_{xv} and the phase angle between t_{xv} and t_1 is ϕ. The total angle of the OFF stage is $\pi + 2\phi$. As the angle of ON stage is expressed as $\delta\pi$, $0 < \delta < 1$. The relation between δ and γ can be found by the volt-second balance.

$$V_d\delta\pi + 2\int_0^{\phi} v_p(\theta-\phi)d\theta = \int_{\phi}^{\pi+\phi} v_p(\theta-\phi)d\theta - \pi V_d , \quad (13)$$

$\theta = \omega_0 t$. For undamped case, $\alpha = 0$,

$$\delta = \gamma\frac{2}{\pi} - 1 . \quad (14)$$

The voltage stress applied on the switch Q_1 is

$$V_{sw} = V_d + V_{pk} = V_d(1+\sqrt{1+\gamma^2}) \quad (15)$$

In case that $\delta = 0.3$, $\gamma = 2.04$, $V_{sw} = 3.27V_d$.

In order to reduce the switch voltage stress, the interval of the ON stage should be decreased and hence a smaller inductance and a larger capacitance are adopted at the desired natural angular frequency.

https://doi.org/10.1051/e3sconf/20186407007

To easily obtain the ZVS condition, the damping ratio of the RLC tank should be smaller than 0.1. Namely, the loaded quality factor should be greater than 5. In case of $\xi \le 0.1$, the primary minimum resistance is $R_p \ge 5Z_0$. With the minimum resistance, the condition of the characteristic impedance can be expressed as

$$\sqrt{\frac{L_p}{C_p}} \le \frac{R_{f,min}}{5n^2} \tag{16}$$

As the exponential term of (2) is ignored, the energy dissipated at the ON stage can be expressed as

$$W_{ON} = \int_{t_0}^{t_1} i_s^2(t-t_0)R_f dt, \quad i_s(t-t_0) \approx \frac{knV_d}{R_f} \tag{17}$$

At OFF stage, the dissipated energy can be expressed as

$$W_{OFF} = \int_{t_1}^{t_2} i_s^2(\tau)R_f dt, \quad i_s(\tau) \approx \frac{knv_p(\tau)}{R_f}, \quad \tau = t - t_1 \tag{18}$$

4 Simulation and experiment

An off-line single-switch ZVS inverter circuit is designed for a 33W/3V filament. The widely used 8-pin control chip L6561 is used to control the switch conduction time and detect the zero point of switch voltage. The switching frequency is about 20 kHz at rated power and input voltage. To easily obtain the ZVS condition, the loaded quality factor of the RLC parallel resonant tank is set larger than 5. The designed circuit parameters are listed as follows.
$P_{avg} = 33W$, $R_{f,min} = 0.16\Omega$, $R_f = 0.32\Omega$, $V_d = 300V$, $n = 0.02$, $k = 0.92$, $L_p = 480uH$, $C_p = 0.1uF$, $Z_0 = 69\Omega$, $f_{sw} \ge 20kHz$.

Fig.7 displays the natural responses of the primary voltage with initial voltage 300V and initial current 6A. The four traces respectively correspond to the 0.1, 0.16, 0.32 and 100 Ohms filaments. It illustrates that the RLC parallel circuit with above initial conditions can obtain the ZVS operation while the filament is greater than 0.1 Ohm, and the corresponding switch voltage stress at 100Ohms load is about 800V.

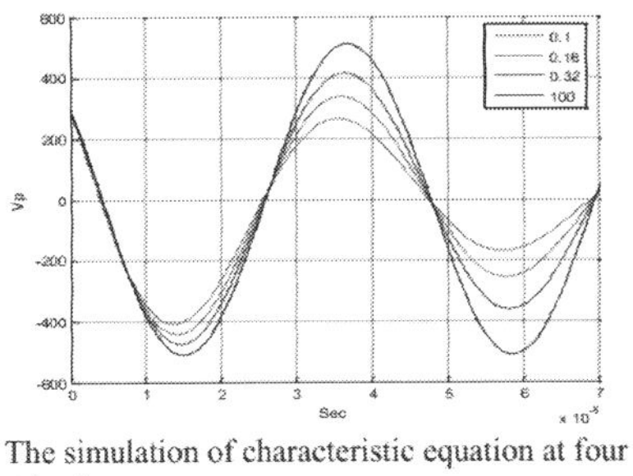

Fig. 7. The simulation of characteristic equation at four different loads.

Fig.8, Fig.9, and Fig.10 respectively show the waveforms simulated at $100\,\Omega$, $0.32\,\Omega$, and $0.16\,\Omega$ loads. It can be found that the reflected current increases with the decrease of the filament resistance. As shown in Fig.8, the transformer magnetizing current is equal to the primary current, and the negative switch current that flows through the anti-parallel diode is greater than that of Fig.9 and Fig.10. The switch voltage stress is about 800V. The simulated waveforms demonstrate the circuit analyses. The switch current is a part of the primary current, and the difference between the primary and the magnetizing currents is the reflected load current. The transistor switch obtains the ZVS operation and the sinusoidal resonant waveforms can be observed at the OFF stage. Fig.11 and Fig.12 display the practical waveforms of a scale-down inverter respectively measured at light-load and heavy-load conditions. The ac output voltage of the inverter is rectified and filtered to a dc signal and then fed back to the control chip L6561 to control the conduction time of switch. The output power of the ZVS inverter is dominated by the switch conduction time. And, the switch off time is determined by the resonant parameters. The measured waveforms coincide with the simulated waveforms. The waveforms shown in Fig.13 depict the relation between the filament current and the anode to cathode voltage. At start-up stage, the inverter heats the filament, and the preheating time lasts about 3 seconds. The anode to cathode voltage reaches about 7kV. In order to protect the magnetron from high voltage stress, the anode power supply should be turned on after the pre-heating process.

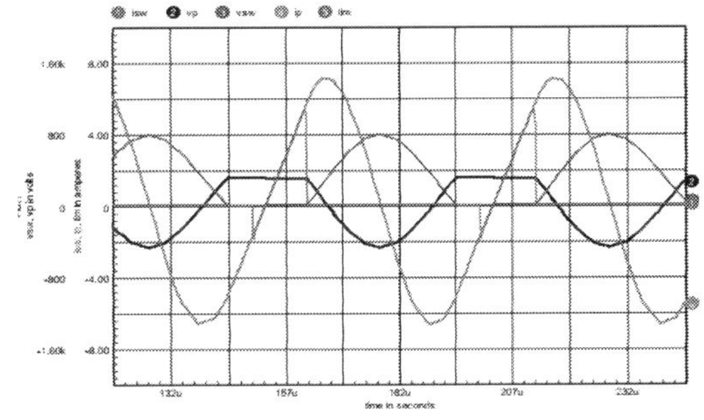

Fig. 8. The waveforms simulated with 100 Ohms load.

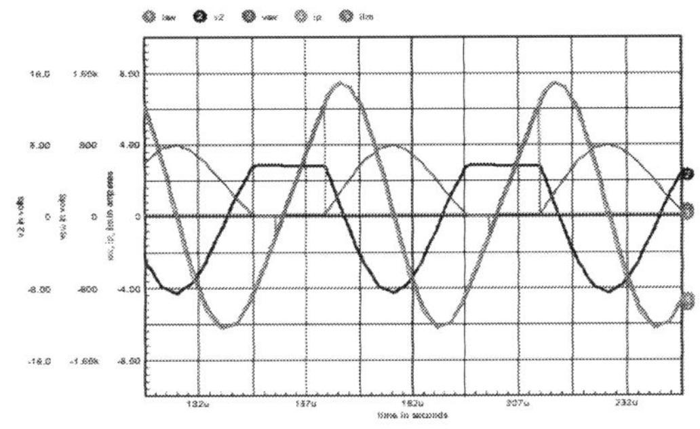

Fig. 9. The waveforms simulated with 0.32 Ohms load.

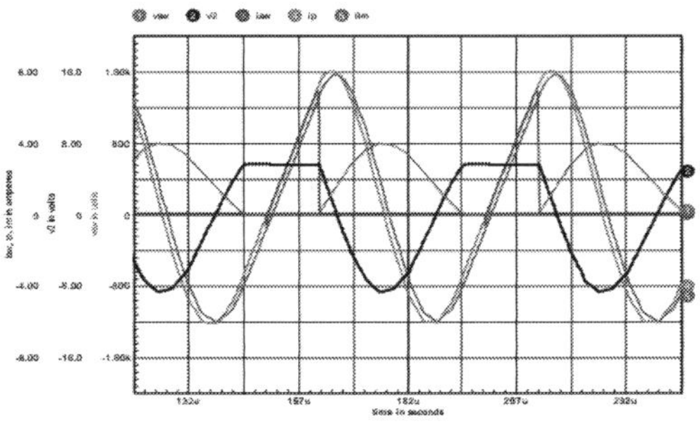

Fig. 10. The waveforms simulated with 0.16 Ohms load.

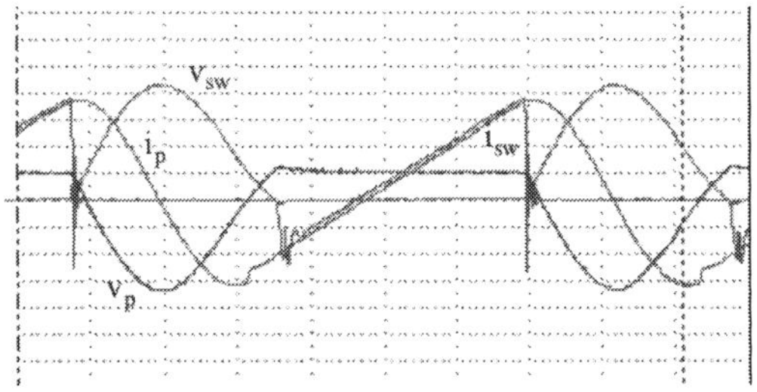

Fig. 11. The steady-state waveforms at light load conditions.

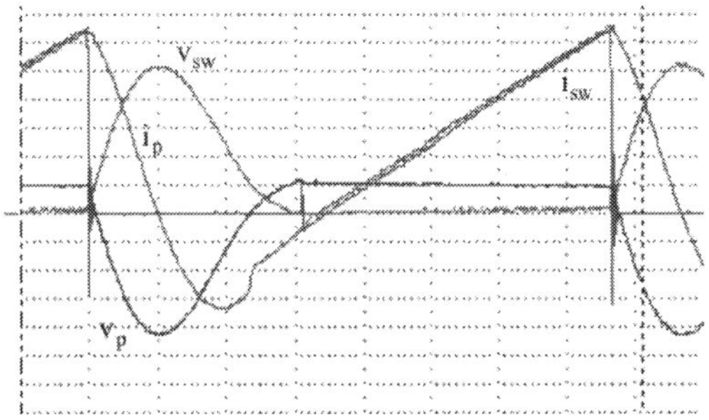

Fig. 12 The steady-state waveforms at heavy load conditions.

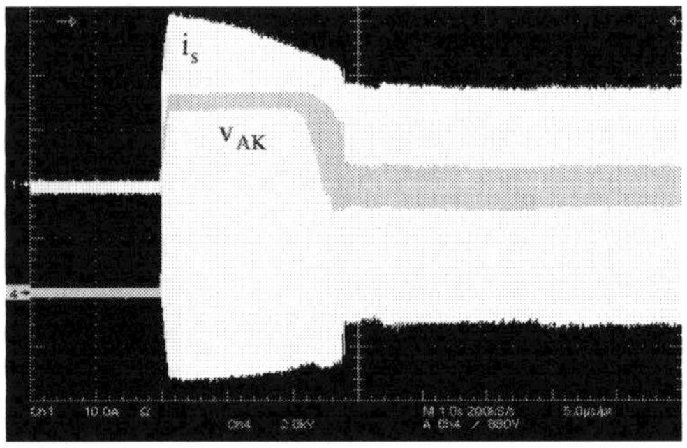

Fig. 13 The waveforms of filament current (10A/Div) and anode to cathode voltage (2kV/Div) at start-up stage, Time:1s/Div.

5 Conclusions

Before applying dc high voltage to a cooker magnetron, the cathode filament is preheated to reach the thermionic emission temperature. As the magnetron starts working, the filament power is regulated according to average anode current. This paper designs an off-line single-switch zero-voltage-switching inverter as the filament power supply. As the transistor switch of the inverter is turned on, the input voltage drives the cathode filament via a step-down isolation transformer. As the transistor switch is turned off, the transformer resonant voltage drives the filament. The output power is dominated by the switch conduction time. The resonant parameters determine the switch off time.

This work is supported by the MOST of Taiwan; the contract number is MOST 107-2637-E-131-003.

References

1. G.E. Georghiou, R. Meredith, A.C. Metaxas and D. Gurwicz, "Switch Mode Power Supply for Microwave Heating Based on the Boucherot Effect," Journal of Microwave Power and Electromagnetic Energy, vol.34, no.4, pp.206-215, (1999).
2. Luiz P. P Junior, Marcos E. T, Juliano S. B. and Leandro M., "Modeling and control of a high-frequency magnetron power supply for microwave heating applications," IEEE Brazilian Power Electronics Conference COBEP/SPEC, (2015), pp.1-6.
3. S-R Jang, H-J Ryoo, S-H Ahn, J-S Kim and G-H Rim, "Development and Optimization of High-Voltage Power Supply System for Industrial Magnetron," IEEE Trans. I.E., vol.59, no. 3, pp.1453-1461, March (2012).
4. S-H Ryu, J-H Ahn, K-S Cho and B-K Lee, "Single-Switch ZVZCS Quasi-Resonant CLL Isolated DC-DC converter for 32" LCD TV," J Electr Eng Technol., (2015); 10(4):1646-1654.
5. J-M Kwon, W-Y Choi, and B-H Kwon, "Single-switch Quasi-Resonant Converter," IEEE Trans., Industrial Electronics, vol.56, no. 4, April (2009).
6. M. Saoudi, D. Puyal, C. Bernal, D. Anton and A. Mediano, "Induction cooking systems with single switch inverter using new driving techniques," (2010) IEEE international Symposium on Industrial Electronics, ISIE, pp.878-883.
7. Shweta Hegde and Afshin Izadian "Control of single switch inverters," Power and Energy Conference at Illinois (PECI), (2014). pp.1-6.
8. C. Bernal, J. Avellaned and J. Milla, "Single switch quasi-resonant inverter on Silicon Carbide with regenerative snubber for induction heating appliances," (2014) 16th European Conference on Power Electronics and Applications, pp.1-5.
9. Yuya Kitano; Hideki Omori; Noriyuki Kimura; Toshimitsu morizane; Kunio Nakagawa; Mutsuo Nakaoka "A new wireless EV charger using single switch ZVS resonant inverter with optimized power transfer and low-cost PFC," (2015) International Conference on Electrical Drives and Power Electronics (EDPE), pp.515-521.
10. Lukasz Roslaniec, Alexander S. Jurkov, Anas Al Bastami, and David J. Perreault "Design of Single-Switch Inverters for Variable Resistance/Load Modulation Operation," IEEE Trans. Power Electronics, vol.30, no. 6, pp.3200-3214, (2015).

Numerical Studies on a NACA0018 Airfoil Blade HAWT with Trailing Edge Jet Flow

Uzu-Kuei Hsu[1], *Cheng-Hsien* Tai[2], *Chia-Wei* HSU[1] and *Jiun-Jih* Miau[3]

[1]Air Force Institute of Technology, No. 1, Jyulun Road, Gangshan Dist, Kaohsiung 82063, Taiwan.
[2]National Pingtung University of Science and Technology, No.1, Shuefu Road, Neipu, Pingtung 91201, Taiwan.
[3]National Cheng-Kung University, No.1, University Road, Tainan City 701, Taiwan

Abstract. This study analyzed an airfoil blade for a horizontal-axis wind turbine (HAWT) with a trailing-edge jet flow design. This design was realized by drilling a hole in the trailing edge of an NACA0018 blade of a conventional HAWT to serve as a pressure injection nozzle. Five inflow wind speeds and three trailing-edge jet flow conditions were examined in the test. The results revealed the efficiency differences between a HAWT with the new jet flow design and conventional HAWTs. The experimental methods employed involved a wind tunnel experiment and a computational fluid dynamics (CFD) simulation. The results revealed that when the inflow wind speed was low, the trailing-edge jet flow accelerated the initiation phase and increased the rotating speed of the HAWT; however, when the inflow wind speed was high, damping occurred and the rotating speed of the turbine blades decreased.

1 Introduction

Horizontal axis and vertical axis models constitute the two main types of wind turbine designs and can be distinguished by their appearances. Horizontal-axis wind turbines (HAWTs) can be divided into those with two blades, three blades, and over three blades. In real-world scenarios, turbines have either an upwind or downwind design. Currently, the most commonly employed design is the three-bladed HAWT. Designs of vertical-axis wind turbines (VAWTs) include Savonius rotor, Darrieus rotor, and H rotor models. Apart from the Savonius rotor, which rotates by drag force, the other rotors mentioned here operate by lift force [1], [2]. Numerous variations have been developed for HAWTs and VAWTs, and each variation exhibits strengths and weaknesses. For example, VAWTs require a higher startup wind speed and have low efficiency because their effective angle of attack varies considerably during each revolution. This impedes the design of an optimal blade shape [3]–[5]. However, HAWTs are not hampered by such limitations, and have become the main turbine design used worldwide. This is because following numerous improvements, they now approach the Betz limit. However, HAWT blades generate a considerable antitorque force during rotation, which leads to the requirement of additional reinforcements for the tower and base structure [6]. These requirements limit the flexibility regarding turbine location and usage. In addition, in case of strong winds, turbines usually have a cut-out speed design—typically at approximately 20–25 m/s—to prevent the breakage of blades due to excessive rotation speed [7].

In 1891, La Cour installed the first experimental wind turbine to study the application of wind power. Currently, both HAWT and VAWT designs are limited to harnessing the relationship between the atmospheric boundary layer and the ground level [8], which causes inconsistencies regarding wind power in wind fields and poor power generation efficiency. Moreover, due to structural issues, turbines require substantial space for construction, thereby hindering their application. According to [9], recent wind power devices are based on a tower or pillar structure and involve high production costs. These tower or pillar structures have a height limit of 150 m, and the height of such a structure cannot be adjusted according to the upper-air climate; therefore, both excessive and insufficient wind speeds compromise normal power generation mechanism.

The CP is close to 0.4 for a good-designed wind turbine. It is not only caused the structure failure of the wind turbine by the high wind speed, but also reduced the wind power transfer by the wind drag. The low efficiency bypass flow is induced by the reverse pressure because the wind flow is not easily through the wind power system. A kind of design, wind cup, was applied to deal with this problem, and it can increase the CP about 5~10%. However, the problems still exist in "reverse pressure" and "cut speed" in high wind speed. This study is to develop a device which produces high-pressure tailing jet. For the horizontal axis wind turbine, it can accelerate the rotation to increase power output in low speed, and produce a stable anti-torque force to wind turbine in high speed. It becomes the all day long operations, not only to overcome the problem of reversed pressure in high speed and cut off wind speed.

 https://doi.org/10.1051/e3sconf/20186407008

It greatly enhances the CP value, and structure of the wind turbine is more robust and easy to control. It provides a new thinking for wind turbine design.

2 Experimental apparatuses

This experiment was conducted in the wind tunnel laboratory of the Air Force Institute of Technology. This wind tunnel used has a length of 11.68 m, and is a low-speed, open-circuit, and suction-type wind tunnel. The main structure can be divided into a settling chamber, a contraction section, a test section, and a drive section. In the rear of entry segment placed with honeycomb nest and layers of sieve, its main function is to adjust the speed of the flow field, to reduce speed disturbance form the side, and to eliminate large-scale vortex and turbulent flow. Contraction section is used to accelerate fluid velocity and reduce speed disturbance from upstream. Experiment wind measured from the test section inlet cross-sectional center resulting uniform flow of low turbulence which is less than 1% and maximum test speed can be up to 30m/s. The test conditions are inflow speed and mass flow of the jet flow. There are nine kinds of inflow speeds (2~10m/s) and three kinds of trailing edge jet flow (0.5, 1.0, and 1.5 kgf/cm^2) in analysis. The wind tunnel setup and test model are shown in Fig. 1. The setup of the wind tunnel is as follows:

2.1 Settling Chamber

The settling chamber (130.8 cm long × 289 cm wide × 211 cm tall) at the entrance of the wind tunnel is a flow-straightening device containing a honeycomb and screens. The honeycomb acts as a filter that break up eddies from the downstream of the honeycomb into smaller eddies to reduce the static turbulence and cross turbulence of the airflow, which minimizes turbulent force in the flow field and increases the uniformity and stability of the airflow; in addition, this renders the separation of the boundary layer in the contraction section less likely to occur.

2.2 Contraction Section

This is the most crucial section for airflow straightening. Its main function is to further suppress the turbulence in the airflow that passes through the preceding flow straightener. This section is responsible for steady flow approximation in the test section; therefore, it requires special consideration.

2.3 Test Section

In this wind tunnel, this section is 215 cm long, 100 cm wide, and 65 cm tall. This section comprises fixed steel panels on the top and bottom and large transparent movable acrylic sheets on both sides, which enables users to directly observe wind tunnel flow.

2.4 Drive Section

This section comprises a contraction tube, fan, and diffuser. The contraction tube has a rectangular entrance that gradually shrinks into a round fan opening with a 1.3 m diameter. Because eddies generated by fan movement in a blowing-type wind tunnel may affect the uniformity and stability of airflow in the test section, the wind tunnel used in this study has a suction-type fan. The fan has a diameter of 1.3 m and contains ten fan blades. To prevent the vibration caused by motor and fan blade rotations affecting the experimental results, the drive section and the test section were connected by a piece of soft canvas; in addition, the drive section has a standalone foundation to prevent vibration transmission.

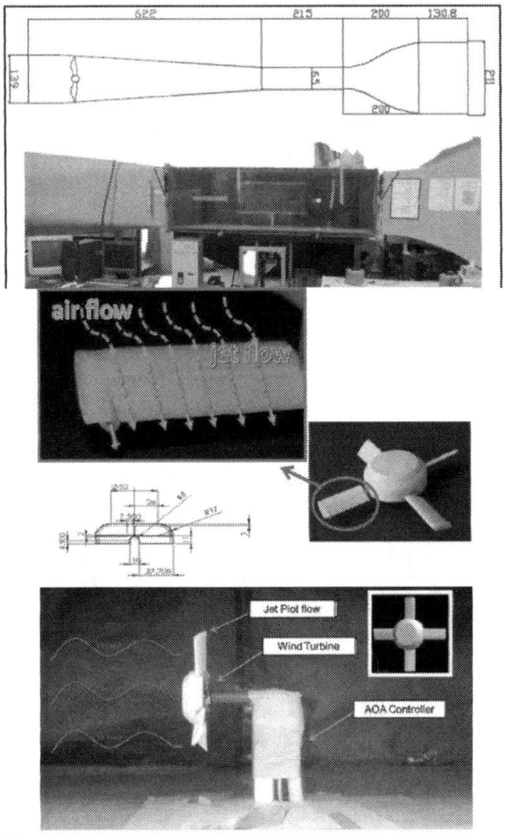

Fig. 1. Wind tunnel and test model.

3 Numerical method

3.1 Governing equations

This study computes the unsteady viscous flow-fields by solving the Reynolds-averaged Navier-Stokes equation (1) combined with the continuity equation (2), as follows.

$$\frac{\partial}{\partial t}(\rho u_i)+\frac{\partial}{\partial x_j}(\rho u_i u_j)=-\frac{\partial p}{\partial x_i}+\frac{\partial \tau_{ij}}{\partial x_j}+\rho g_i \tag{1}$$

$$\frac{\partial \rho}{\partial t}+\frac{\partial}{\partial x_i}(\rho u_i)=0 \tag{2}$$

The simulations are performed using the CFD scheme on the control-volume method. The aerodynamic characteristics for passing of the rotational wind turbine are associated with the dynamic unsteady flow field. This study will focus on the transient aerodynamic coefficient effects caused by the different rotating speeds and different inflow speeds. The flowfield is assumed to be turbulent and the conservative variables are solved sequentially. The terms representing convective and diffusive flux are evaluated using the second-order accuracy hybrid differential scheme. In addition, the coupling between the pressure and the velocity is achieved by means of the SIMPLE algorithm [10], while the time accuracy is improved by performing the FAS multi-grid method. The residual smoothing approach is also applied to accelerate the convergence of solutions within each physical time step, and the sliding mesh technique in dynamic mesh is used to analyze the different cases. Two hexahedron mesh blocks are established before starting the transient iterations. First of all, the sliding and non-sliding blocks are defined before setting up the interface zone that varies when the blocks move. During the iterations, the solver will interpolate the variables on the overlapping surface at both sides.

3.2 Turbulence Model

Because there are two air-conditioners operating inside the cabin, the flow field belongs to the turbulent status. In order to simulate the turbulence status inside a closed space, it will require a certain degree of preciseness for near-wall flow field simulation. So, in this research, we adopt RNG k-ε Turbulence Model to simulate the turbulent flow field induced by Reynold Stress terms in the momentum equation [11].

4 Physical model and boundary conditions

4.1 Physical model

The production of Grid System Calculating is the priority job in the application of hydrodynamics calculation. In this wind turbine case, ICEM/CFD software package is used to complete the physical model drawing and the fabrication of a 3-D structural hexahedron grid system. On the grid deployment, in order to effectively simulate the special phenomenon of the flow field, such as the encryption processing done in the area of wind blade and jet flow nozzle, one needs also to consider the matching between the grid spacing density and the aspect ratio tuning. Therefore, the grid system has been divided into 15 blocks with 2,218,759 grids. The calculation platform is a Lenovo P900 with a 2.4GHZ Xeon CPU E5-2630 (32 cores) and 64GB RAM.

4.2 Boundary conditions

Fig. 2 shows the computational domain. The boundary conditions are as following:

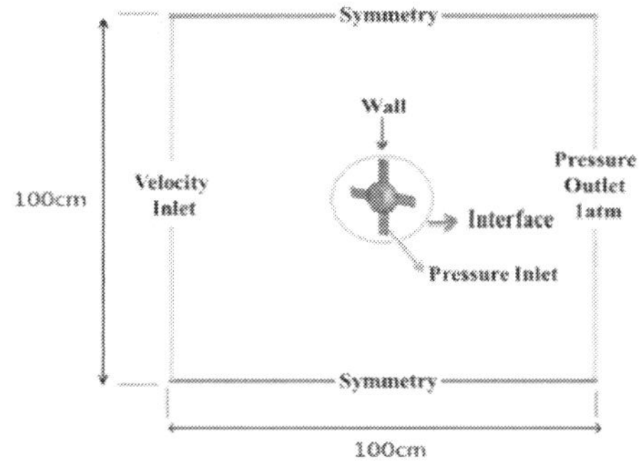

Fig. 2. Boundary conditions and computational domain in a NACA0018 Airfoil Blade HAWT (sliding grid in the interface, fixed grid out of the interface)

4.2.1 Velocity Inlet

The wind turbine usually reaches the rated capacity at a wind speed of between 10 to16 m/s, depending on the design of the individual turbine. The cut-out wind speed is when the turbine stops production and turns out of the main wind direction. Typically, the cutout wind speed is in the range of 20 to 25 m/s. A typical starting wind speed for HAWT is about 2~7 m/s. In this study, there are nine wind speeds are used: 2~10 m/s as shown in table 1.

4.2.2 Outflow

Assume a fully developed in the outlet, upper, and lower boundaries, because the pressure and velocity are unknown. In addition, consider the conservation, to make the mass flux out equal to the mass flux coming into the domain all the outlet velocity components ui,j are multiplied by the ratio M_{in}/M_{out} . Thus the outlet plane velocities with the continuity correction are given by

$$u_{i,j}=u_{i-1,j}\times\frac{M_{in}}{M_{out}} \tag{3}$$

4.2.3 Wall

In viscous fluid flow, the wall condition satisfies the no-slip condition.

4.2.4 Moving wall:

The rotating part is set according to the wind turbine working principle. The boundary of the simulated surface is the interface of the HAWT, and the boundary condition adjacent to the wind turbine is set as the sliding interface to generate a sliding grid system.

4.2.5. Mass flow of the trailing edge jet:

The jet flow on of the trailing edge on the wind blade has three cases to test: 0.5, 1.0, and 1.5 $^{kgf}/_{cm^2}$

5 Results and discussion

Fig. 3 and Table 1 show the experimental results for rotation speed distribution trends under different inflow wind speeds and trailing-edge jet flow settings. When the inflow wind speed was set at 2 m/s and trailing-edge jet flow was not applied, the blades were unable to overcome the static friction force; therefore, they did not rotate. However, with the introduction of trailing-edge jet flow, the blades started to rotate, and the rotation speed increased proportionally with the trailing-edge jet flow. Fig. 4 (a) and 4 (b) show the pressure distribution under an inflow wind speed of 2 m/s. When the trailing-edge jet flow was weak, higher pressure was noted at the front of the blades. By contrast, when the trailing-edge jet flow increased, the high-pressure zone decreased. The velocity vectors in Fig. 5 show that the trailing-edge jet flow enabled the rotor to form a high-velocity trailing-edge flow. This expanded the low-pressure zone at the back of the flow field and produced a suction affect at the front, which lowered the pressure in the high-pressure zone and increased the flow speed through the turbine, thereby increasing the rotation speed.

Table 1. Studies matrix and experimental results.

Inlet Velocity (m/s)	2	3	4	5	6	7	8	9	10	
$P_{jet\ flow}$ 0.0kgf/cm²	0	0	914.15	1089.5	1369.5	1584.63	1672.75	2289.75	2535.88	
$P_{jet\ flow}$ 0.5kgf/cm²	374.75	381.50	1047.75	1142.13	1435.38	1578.63	1648.28	1787.00	2518.88	
$P_{jet\ flow}$ 1.0kgf/cm²	485.38	468.63	1032.25	1111.75	1366.75	1525.13	1609.50	1676.38	2452.00	RPM
$P_{jet\ flow}$ 1.5kgf/cm²	630.00	727.88	1015.75	1078.25	1340.00	1485.13	1586.75	1635.38	2183.63	

When the flow rate at the inflow wind speed increased to 5 m/s, the trailing-edge jet flow did not have a marked effect on the rotation speed and the pressure distribution, as shown in Fig. 3 and Fig. 4 (c)–(e). The maximal rotation speed occurred when the trailing-edge jet flow was set at 0.5 kgf/cm^2. Increasing the trailing-edge jet flow beyond this point resulted in a lower rotation speed; similar results were observed in the experimental conditions with inflow wind speeds of 4, 5, and 6 m/s. Fig. 6 reveals that when the trailing-edge jet flow was set at 1.5 kgf/cm^2, a reverse eddy current formed at the back of the flow field and created a drag force in the upstream inflow. The aforementioned phenomenon can be further observed in Fig. 7; when the inflow wind speed was 10 m/s and the trailing-edge jet flow was set at 1.5 kgf/cm^2, the strong reverse eddy current collided with the inflow airflow, which decreased the rotation speed of the turbine blades. Moreover, as can be observed in Fig. 3 and Table 1, when the inflow wind speed was 10 m/s, the trailing-edge jet flow reduced the rotation speed, suggesting an inversely proportional relationship.

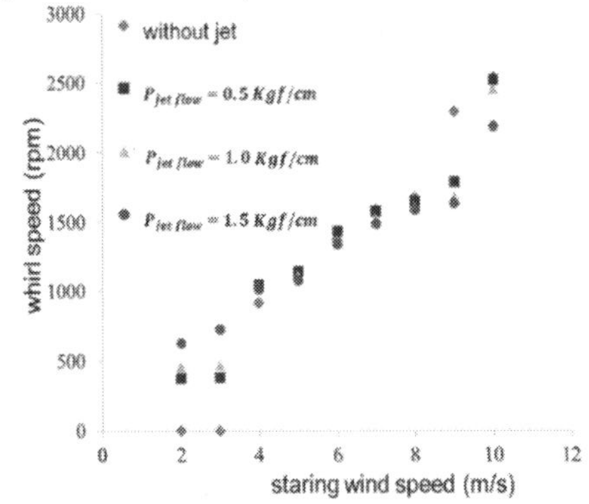

Fig. 3. Experimental results of the NACA0018 Airfoil Blade HAWT in different inflow and trailing edge jet flow.

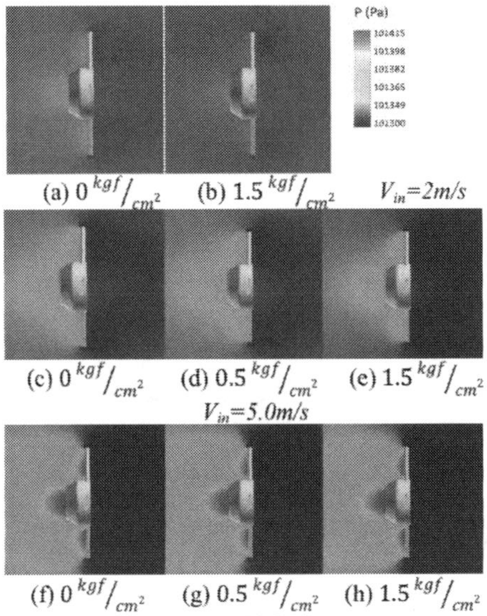

Fig. 4. Pressure distribution in Blade HAWT in different inflow and trailing edge jet flow

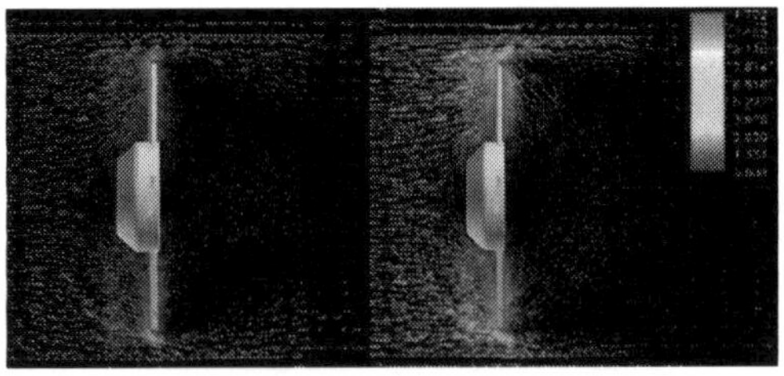

Fig. 5. Velocity vectors in Pressure distribution in V_{in}=2m/s ($L: V_{jet} = 0\ ^{kgf}/_{cm}$, $R: V_{jet} = 1.5\ ^{kgf}/_{cm}$)

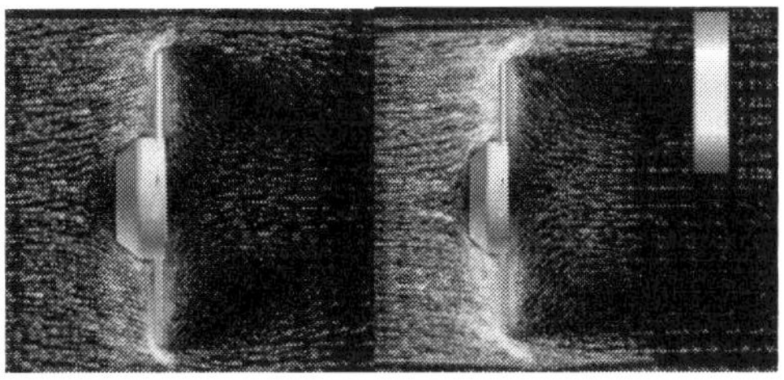

Fig. 6. Velocity vectors in Pressure distribution in V_{in}=5m/s ($L: V_{jet} = 0\, {}^{kgf}/_{cm}$, $R: V_{jet} = 1.5\, {}^{kgf}/_{cm}$)

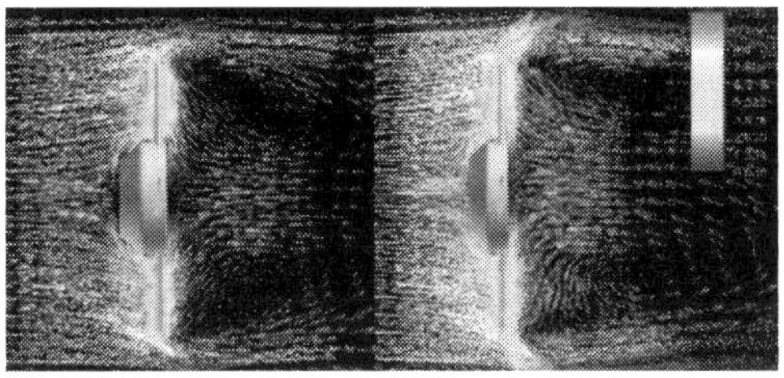

Fig. 7. Velocity vectors in Pressure distribution in V_{in}=10m/s ($L: V_{jet} = 0\, {}^{kgf}/_{cm}$, $R: V_{jet} = 1.5\, {}^{kgf}/_{cm}$)

The experimental and simulation results demonstrated that in general, turbine blades rotated due to inflow and formed a partial high-pressure zone at the front of the turbine. When trailing-edge jet flow was applied, the flow velocity at the back of the turbine increased thus reducing the regional pressure. Additionally, the relative pressure difference caused a suction effect that increased the rotation speed of the turbine. However, as the inflow wind speed increased, the airflow passing through the turbine blades also sped up, and the blades rotating at a high speed caused shielding, which impeded the trailing-edge jet flow from moving downstream. Subsequently, as the eddy system developed, a strong inverse eddy current was established that blocked the upstream airflow and reduced the rotation speed of the turbine; as a result, a natural damping effect was observed.

6 Conclusions

In this study, a HAWT was developed with a trailing-edge jet flow design. The experimental and simulation results revealed that with a low inflow wind speed, this design could assist in accelerating the turbine startup or and power output thereby overcoming the shortcomings of an HAWT. With a high inflow wind speed, this design caused a damping effect, which lowered the rotation speed and stabilized the turbine. These characteristics will allow this type of HAWT to operate efficiently under all weather conditions.

Acknowledgement

The author is grateful to the Ministry of Science and Technology of the Republic of China (Taiwan) for financial support under contract number MOST 107-3113-E-006 -011.

References

1. Savonius SJ, "The S-Rotor and Its Application," Mechanical Engineering, Vol.53, No.5, pp.333-338, 1931.
2. Tony Burton, David Sharpe, Nick Jenkins, Ervin Bossanyi: Wind Energy Handbook, 2005.
3. Hayashi T, Li Y, and HARA Y, "Wind Turnel Test on a Different Phase Three-Stage Savonius Rotor," JSME International Journal, Series B, Vol.48, No.1, pp.9-16, 2005
4. Fujisawa N and Shibuya S, "Observations of dynamic stall on Darrieus wind turbine blades," Journal of Wind Engineering and Industrial Aerodynamics, Vol.89, PP.201–214, 2001
5. Hsu UK, Tai CH, Lin JY, "Dynamic Simulation of a Vertical Axis Wind Turbine with NACA4412 Airfoil," Journal of Aeronautics/Astronautics and Aviation, Series B, Vol. 41, No. 1, pp. 11-16, 2009.
6. Cristina L. Archer ,Ken Caldeira, "Global Assessment of High-Altitude Wind Power ," Energies, 2(2), 307-319; doi:10.3390/en20200307, 2009.
7. Paraschivoiu, I., Wind Turbine Design, Polytechnic International Press, 2002.
8. Burton T, Sharpe D, Jenkins N, and Bossanyi E, "Wind Energy Hand Book," John Wiley and Sons, 2001.
9. Thresher R, Robinson M, and Veers P, "To capture the wind," IEEE Power Energy Mag., vol. 5, no. 6, pp. 34–46, 2007
10. Patankar, S. V. (1980). Numerical Heat Transfer and Fluid Flow. Taylor & Francis. ISBN 978-0-89116-522-4.
11. Yakhot, V. and Orszag, S. A., "Renormalization Group Analysis of Turbulence: I. Basic Theory," Journal of Scientific Computing, Vol. 1, pp. 1-51, 1986.

The Load Distribution of the Main Shaft Bearing Considering Combined Load and Misalignment in a Floating Direct-Drive Wind Turbine

Jingyang Zheng[1], *Jinchen* Ji[1], *Shan* Yin[2], and *Van Canh* Tong[3]

[1]School of MME, University of Technology Sydney, 2007 Broadway, Sydney, Australia
[2]State Key Laboratory of ADMVB, Hunan University, 410082 Changsha, Hunan, China
[3]Department of UPMS, Korea Institute of Machinery and Materials, 34103 Yuseong-gu, Daejeon, Republic of Korea

Abstract. The main shaft tapered double-inner ring bearing (TDIRB) of floating direct-drive wind turbine system (FDDWT) is one of the most critical components in FDDWT, and its failure accounts for a large proportion of wind turbine malfunctions and faults. Over the past decades, a significant number of methods have been proposed to calculate the contact load distribution along the roller length in TDIRB, since the contact load distribution of roller is the key factor of fatigue life of TDIRB. Most of methods, however, neglected the misalignment of inner ring with respect to outer ring and friction force. In this paper, with the help of comprehensive and accurate quasi-static mathematical method, the contact load distribution of internal loads in TDIRB are analysed by considering the effects of combined loads, angular misalignment and friction force at different wind speeds for FDDWT. The simulation results show that the amount of combined load has an apparent effect on the contact load distribution along the TDIRB raceways and flanges in both rows. Furthermore, the slight change of tilted misalignment has a great influence on the contact load distribution. In addition, the slight angular misalignment easily produces noncontact zone for the bearing raceway in both rows, which is significantly disadvantage for the external load uniform transmitting to each roller.

1 Introduction

Currently, wind power is one of the cheapest sources for large-scale renewable energy [1]. A historical record of 4,331 MW of new offshore wind power was installed across nine markets globally in 2017 [2]. The generator of FDDWT, unlike traditional geared wind turbine, operates at a varying frequency, directly proportional to the rotor speed. The advantage of this design is the complex high-speed geared drive train common in most conventional wind turbines [3] replaced by a direct-drive generator. Unfortunately, many generators of FDDWT are suffering premature TDIRB failure [4].

The fatigue life of bearing is closely associated with the contact load distribution along the raceways and flanges. Considering misalignment and heavy load conditions, Harris [5] carried out an early study on double-row CRB under radial and thrust load. Puneethkumar M V and Vennapusa V used the FEA and linear method to build up the tapered roller bearing model which were time-consuming and inaccurate [6, 7]. Van Canh [8] proposed a more accurately mathematical method to calculate the pressure distribution along roller length considering the centrifugal force and gyroscopic moment.

To enhance the modelling level and accuracy with respect to pressure distribution calculation, this paper proposes the model used combination of non-Hertzian elastic half-space theory and slicing technique considering the combined load, angular misalignment and friction force to solve the contact load distribution along the raceways and flanges under the heavy load and low-speed conditions.

2 Tapered double-inner ring bearing models

In this section, in order to obtain the contact load distribution and deformation along the roller length, a comprehensive quasi-static model subjected combined load, misalignment and friction is presented here. The typical cross-section construction of an TDIRB is shown in Fig. 1: α_i (semi-cone angle of inner raceway), α_m (semi contact angle), α_o (semi-cone angle of outer raceway), ε (the semi-cone angle of roller), l_{we} (effective length of roller), D_{max} (large end roller diameter), D_{min} (small end roller diameter), D_m (mean roller diameter), α_f (inner ring flange angle), r_i (semi-minimum inner raceway contact radius), r_m (pitch radius) and r_o (semi-maximum inner raceway contact radius). In order to simplify the analysis, the significant assumptions introduced as follows: the rolling elements are made from purely elastic materials; only elastic deformation associated with the local contact area of roller is considered; the outer ring is fixed in space and structural deformation of the bearing components is neglected; the influence of the cages and the roller skewing are also neglected; the components of TDIRB and lubricant keep the same temperature during operation.

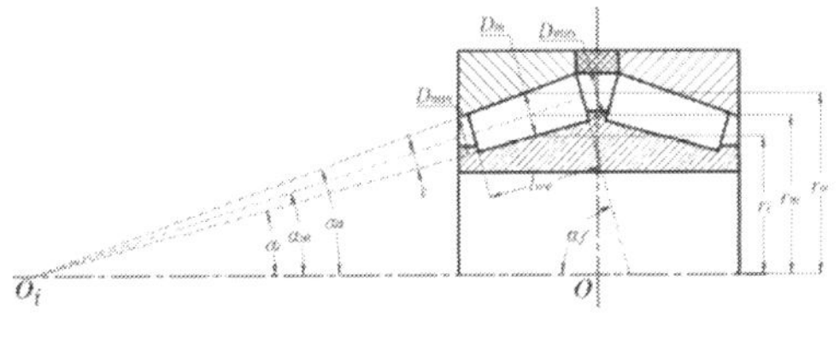

Fig. 1. Basic geometrical dimensions of a TDIRB

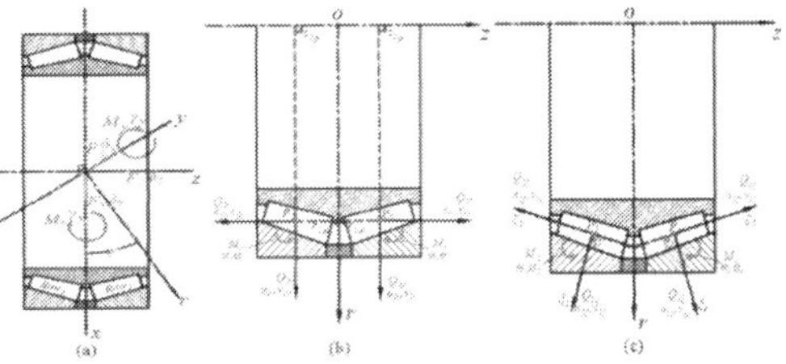

Fig. 2. TDIRB coordinate systems with loads, displacements and angular misalignment: (a) global coordinate system; (b) cylindrical coordinate system; (c) inclined coordinate system.

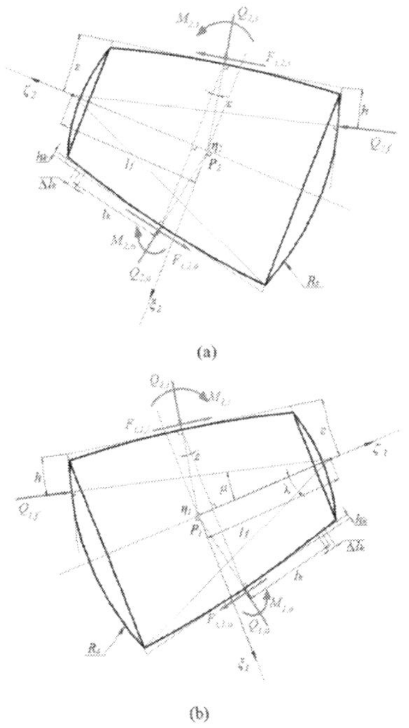

Fig. 3. Free body diagram of rollers: (a) a roller in row 2; (b) a roller in row 1.

2.1 Coordinate systems

The five degree-of-freedom (DOF) TDIRB model is established based on Van Canh's [9] model considering friction force. A right-handed global coordinate system *(x, y, z)* is introduced with origin located at the centre of TDIRB, where *x*-axis is the vertical axis with positive in a downward direction, *y*-axis is specified at a location angle ψ=900 from *x*-axis, and *z*-axis is the TDIRB rotational axis. This global system is fixed with respect to the bearing outer ring, as shown in Fig. 2 (a). Two local cylindrical coordinate systems (r_M, ψ, z_M) are also introduced, where the subscript M denotes the TDIRB row, therefore, M=1 refers to row 1 (right row) and M=2 refers to row 2 (left row). The reference points of inner ring cross-section are $P1(r_p, z_p)$ and $P2(r_p, -z_p)$, as shown in Fig. 2 (b). In addition, two inclined coordinate systems (ξ_M, ζ_M, η_M) are employed with the inclined angle α_m and $-\alpha_m$, as shown in Fig. 2 (c). Noted that η_1-axis is perpendicular to (ξ_1, ζ_1) plane in inward direction, while η_2-axis is normal to (ξ_2, ζ_2) plane in outward direction.

The TDIRB is subjected to an external force vector in the global coordinate system, as shown on Fig. 1,

$$\{F\}^T = \{F_x, F_y, F_z, M_x, M_y\} \qquad (1)$$

and the inner ring displacement vector is

$$\{\delta\}^T = \{\delta_x, \delta_y, \delta_z, \gamma_x, \gamma_y\} \qquad (2)$$

Meanwhile, the displacement vectors of inner ring cross-section in cylindrical coordinate system and inclined coordinate system are introduced by $\{u_M\}^T = \{u_{M,r}, u_{M,z}, \theta\}$ and $\{u_{M,\alpha_m}\}^T = \{u_{M,\xi}, u_{M,\zeta}, \theta\}$, respectively, where, θ is the angular displacement of inner raceway cross-section. From the Fig. 2, we can obtain

$$\{u_M\} = [R_M]\{\delta_M\} \qquad (3)$$

$$\{u_{M,\alpha_m}\} = [K_M]\{u_M\} \qquad (4)$$

Similarly, the displacement vectors of rollers in each local coordinate system are described by $\{v_M\}^T = \{v_{M,r}, v_{M,z}, \phi_M\}$ and $\{v_{M,\alpha_m}\}^T = \{v_{M,\xi}, v_{M,\zeta}, \phi_M\}$ respectively, where, ϕ_M are the angular displacement of rollers in both rows. From Fig. 2, we can calculate

$$\{v_{M,\alpha_m}\} = [K_M]\{v_M\} \qquad (5)$$

where, $[K_M]$ and $[R_M]$ are the transformation matrices referred to [8, 11].

2.2 Contact deformation

Fig. 3 shows the free body diagrams of rollers in row 1 and row 2. To commence the analysis, an alternative slicing technique [10] is employed, in which the contact length of roller is divided into ns slice. According to [9, 10], the contact deformations $\delta_{M,N,k}$ of the kth laminae in row 1 and row 2 can be calculated by

$$\delta_{M,N,k} = \delta_{M,N} + \beta_{M,N} \cdot l_k - h_k \qquad (6)$$

where, N is the bearing ring index, N=*i* for inner ring and N=*o* for outer ring. $\beta_{M,N}$ are the relative angular misalignment between rollers and inner ring, as well as rollers and outer ring respectively calculated by

$$\beta_{M,i} = \theta - \phi_M \qquad (7)$$

$$\beta_{M,o} = \phi_M \tag{8}$$

l_k is the axial position of lamina k, which ranges from $-l_{we}/2$ to $l_{we}/2$, as shown in Fig. 3. h_k is the crown drop at lamina k caused by the modified roller profile. According to [8, 11, 13], we employ the fully-logarithmic roller profile and flat raceway profiles. The mathematical function of roller profile adopted for TRB was given by in reference [12]. $\delta_{M,N}$ are the contact deformations between rollers and rings caused by their translation motion. $\delta_{M,f}$ are the contact deformation between big-end of rollers and inner ring flange. $\delta_{M,N}$ and $\delta_{M,f}$ are presented by, respectively [9, 10],

$$\delta_{M,i} = (u_{M,\xi} - v_{M,\xi}) \cdot \cos\varepsilon - (u_{M,\zeta} - v_{M,\zeta}) \cdot \sin\varepsilon \tag{9}$$

$$\delta_{M,o} = v_{M,\xi} \cdot \cos\varepsilon + v_{M,\zeta} \cdot \sin\varepsilon \tag{10}$$

$$\begin{aligned} \delta_{M,f} = (u_{M,\xi} - v_{M,\xi}) \cdot \sin\mu_0 + (u_{M,\zeta} - v_{M,\zeta}) \cdot \cos\mu_0 + \\ l_f \cdot \left[\cos\mu_0 - \cos(\mu_0 + \beta_{M,i})\right] \end{aligned} \tag{11}$$

where, μ_0 is the angle between roller centre line and roller-flange contact line, l_f is the distance between reference points $P1$ and $P2$ and the intersection of the flange force and the centre line of roller, as shown in Fig. 2 and Fig. 3.

2.3 Internal load and friction

As aforementioned, the alternative slicing technique is used for the non-Hertzian line contacts, and the contact forces $q_{M,N,k}$ at the kth lamina are calculated by the following equations

$$q_{M,N,k} = c \cdot \delta_{M,N,k}^{10/9} \cdot \Delta l_k \,; \qquad \delta_{M,N,k} > 0 \tag{12}$$

where, Δl_k is the width of a lamina. The contact stiffness c is a constant, which depends on the material properties and contact geometry at the roller and raceway [5]. Therefore, the total contact loads $Q_{M,N}$ and moments $M_{M,N}$ along the roller length can be obtained

$$Q_{M,N} = \sum_{k=1}^{n_s} q_{M,N,k} \tag{13}$$

$$M_{M,N} = \sum_{k=1}^{n_s} q_{M,N,k} \cdot l_k \tag{14}$$

In terms of the contact force between big roller-end and inner raceway flange, $Q_{M,f}$ are regarded as classical Hertzian force and conveniently calculated as follows

$$Q_{M,f} = c_f \cdot \delta_{M,f}^{3/2} \,; \qquad (\delta_{M,f} > 0) \tag{15}$$

where, the contact stiffness c_f is a constant, which depends on the material properties and radius of the big-end spherical shape of rollers [5].

Since the main shaft TDIRB in FDDWT running with low speed and heavy load condition, the effects of centrifugal and gyroscopic forces are considered insignificant, while the thrust frictional force in the roller-raceway interface is relatively important for the static equilibrium in (ξ_M, ζ_M) planes [11, 12], see Fig. 3. The friction force is simply described by

$$F_{t\,M,N} = Q_{M,N} \cdot \mu_{z,\max} \tag{16}$$

where, $\mu_{z,max}$ is the maximum of the traction coefficient in the axial direction.

3 Tapered double-inner ring bearing equilibirum

The quasi-static analysis model of TDIRB are described based on the equilibrium of all rollers and inner ring. In the following section, the modelling process and solution approach are presented for the aforementioned equilibrium equations.

3.1 Quasi-static equilibrium equations of rollers

Based on the free body diagrams and in [8], the rollers equilibrium equations can be calculated by

$$\{F_{M,r}\} = \begin{Bmatrix} F_{M,\xi} \\ F_{M,\zeta} \\ M_{M,\eta} \end{Bmatrix} = \begin{Bmatrix} Q_{M,i}\cos\varepsilon - Q_{M,o}\cos\varepsilon + Q_{M,f}\sin\mu_0 + (F_{t\,M,i} + F_{t\,M,o})\sin\varepsilon \\ -Q_{M,i}\sin\varepsilon - Q_{M,o}\sin\varepsilon + Q_{M,f}\cos\mu_0 + (F_{t\,M,i} - F_{t\,M,o})\cos\varepsilon \\ M_{M,i} - M_{M,o} + Q_{M,f} l_f \sin\mu_M + 0.5(F_{t\,M,i} + F_{t\,M,o}) D_m / \cos\varepsilon \end{Bmatrix} = \begin{Bmatrix} 0 \\ 0 \\ 0 \end{Bmatrix} \tag{17}$$

where, μ_M are the angles between inner ring flange contact line and roller centre line for both rows as the shift of the contact point on both inner raceway flange and roller big end under load. In equation (17), the nominal angle μ_0 is used for the force equations as the effect of its changes under load is very small. For moment equations, however, the actual value is considered since it is important for line contact rollers.

To solve the set of nonlinear equations (17), the iterative Newton-Raphson method is employed with Jacobian matrices $[J_{M,r}]$ and increments $\{\Delta v_{M,\alpha_m}\}$, as follow

$$\{F_{M,r}\} + \left[J_{M,r}\right]\{\Delta v_{M,\alpha_m}\} = \{0\} \tag{18}$$

3.2 Quasi-static equilibrium equations of inner ring

To the determine the equilibrium equations of rollers, the inner ring contact load vectors in inclined coordinate system (ξ_M, ζ_M, η_M) can be presented by

$$\{Q_{M,\alpha_m}\} = \begin{Bmatrix} Q_{M,\xi} \\ Q_{M,\zeta} \\ T_{M,\eta} \end{Bmatrix} = \begin{Bmatrix} -Q_{M,i}\cos\varepsilon - Q_{M,f}\sin\mu_0 - F_{t\,M,i}\sin\varepsilon \\ Q_{M,i}\sin\varepsilon - Q_{M,f}\cos\mu_0 - F_{t\,M,i}\cos\varepsilon \\ -M_{M,i} - Q_{M,f} l_f \sin\mu_M - 0.5 F_{t\,M,i} D_m / \cos\varepsilon \end{Bmatrix} \tag{19}$$

Subsequently, the correspondent contact load vectors in cylindrical coordinate systems can be found using transformation matrix $[K_M]$, by

$$\{Q_M\} = \left[K_M\right]^T \{Q_{M,\alpha_m}\} \tag{20}$$

Then, the contact load vectors can be transformed to the global coordinate system using the transformation matrix, resulting in equivalent load vectors as follows

$$\{f_M\} = [R_M]^T \{Q_M\} \tag{21}$$

Finally, the equilibrium equations of the inner ring in global coordinate system are presented with combining all the aforementioned equivalent loads and external loads as

$$\{F_g\} = \{F\} + \sum_{j=1}^{Z} \{f_M\}_j = \{0\} \tag{22}$$

where, Z is the number of rollers in each row. Similar to the rollers equations, the set of non-linear equations (22) are then solved by the numerical Newton-Raphson iterative scheme.

Table 1. TDIRB geometrical and material properties.

Z-Number of rollers in each row	40
α_i-Semi-cone angle of inner raceways (degree)	10.8
α_o-Semi-cone angle of outer raceways (degree)	12.5
ε-Semi-cone angle of roller (degree)	0.85
l_{we}-Effective contact length of tapered roller (mm)	158.75
D_{max}-Large end roller diameter (mm)	83.44
D_{min}-Small end roller diameter (mm)	78.49
d_{max}-Maximum inner raceway contact diameter (mm)	1,055.15
d_{min}-Minimum inner raceway contact diameter (mm)	994.95
E-Elastic modulus (GPa)	208
ν-Poisson ratio	0.3

3.3 Numerical analysis

The analysis of the axial force and misalignment influence of inner ring on the internal load between rollers and raceways is conducted on a TDIRB-LM287649D [15] used in a 5 MW FDDWT and manufactured by Timken. The bearing geometrical and material properties are presented in Table 1. The rated generator speed of the 5 MW FDDWT is 12.1 *rpm*. The external load, F_x, applied on generator shaft along x-axis onto the inner ring is 1090 *KN*. A zero axial preload is employed for the purpose of simulation. In realistic operation conditions, certain preload magnitude can be applied on TDIRB.

First, as the short-term wind speed increased by 1 m/s per 10 minutes from 4 *m/s* to 25 *m/s*, hence, the axial force is assumed to increase linearly from 350 *KN* to 650 *KN* and the misalignment angle, γ_y, is 0 *mrad*. The results of the quasi-static model consist of series of simulations carried out by Matlab, as shown in Fig. 4. The comparisons in Fig. 4 show that the contact loads between roller and inner ring are similar to those between rollers and outer ring in both rows. Then, the contact loads of row 1 are obviously bigger than those in row 2. A similar situation occurs for the roller and flange contact load distribution. Furthermore, the maximum contact loads between rollers and raceways as well as roller end and flange occur at the roller location angle ψ=0° (bottom of the TDIRB), while the minimum contact loads occur at the roller location angle ψ=180° (top of the TDIRB). Obviously, the number of roller that comes into contact with raceways gradually increases around the top of row 1, while it reduces around the same location angle of row 2. The roller to inner and outer raceways contact loads at the azimuth angle ψ=0° in row 1 experience obvious increasing from 89.98 *KN* to 108.2 *KN*, whereas the contact loads in the same location in row 2 show an apparent decrease form 34.84 *KN* to 9.148 *KN*. Similarly, the roller end-flange contact loads increase from 2.769 *KN* to 3.331 *KN* for the rollers in bottom of row 1 and reduce from 1.072 *KN* to 0.2816 *KN* for the rollers in the same position of row 2. In comparison to the results presented in Fig. 1 and Fig. 2 of [16] and Fig. 11 of [13], the aforementioned results are in good agreement, and the presented numerical quasi-static model follows the same trends as in [13] and [16]. In summary, this model shows a good agreement with the reference model.

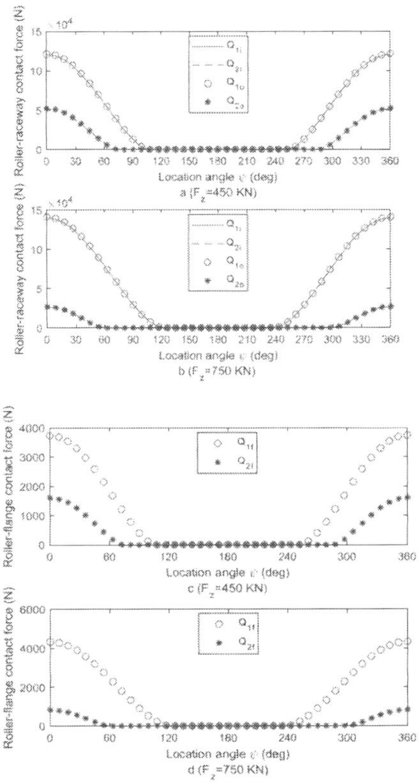

Fig. 4. Contact forces distribution under different axial forces.

Second, to investigate the influence of misalignment on the contact load distribution, we specify the misalignment angle ranging from 1 *mrad* to 4 *mrad* with the constant radial and axial forces, F_x= 1090 *KN* and F_z=550 *KN* respectively, and the results are shown in Fig. 5. A slight increase of the angular misalignment produces a significant change on the contact load distribution of rollers. It is found from Fig. 5 (a) to Fig. 5 (d) that at the bottom of bearing, the roller-raceway contact loads increase from 46.21 *KN* to 140 *KN* for row 1 and from 233.71 *KN* to 281.6 *KN* for row 2, respectively, with the misalignment angle growing from 1 *mrad* to 4 *mrad*. Moreover, the contact loads in row 2 are much larger than those in row 1. Similarly, at the location of ψ=180°, the contact loads show an obvious increase from 11.85 *KN* and 0 *KN* to 249.8 *KN* and 40.49 *KN*. Meanwhile, there occurs noncontact zone in inner raceway and the corresponding outer raceway, which decrease by more than 50% with the increase of angular misalignment. At the location around ψ=90° and ψ=270°, the contact loads show the same decline trend from 26.57 *KN* to 15.93 *KN* in row 1, whereas the noncontact exists in row 2. Compared with the contact loads along the raceways, the roller-flange contact loads experience the same trend as shown in Fig. 5(c) and 5(d), and the maximum loads come up to 7.688 *KN* at the top of bearing in row 1 and 8.667 *KN* at the bottom of bearing in row 2.

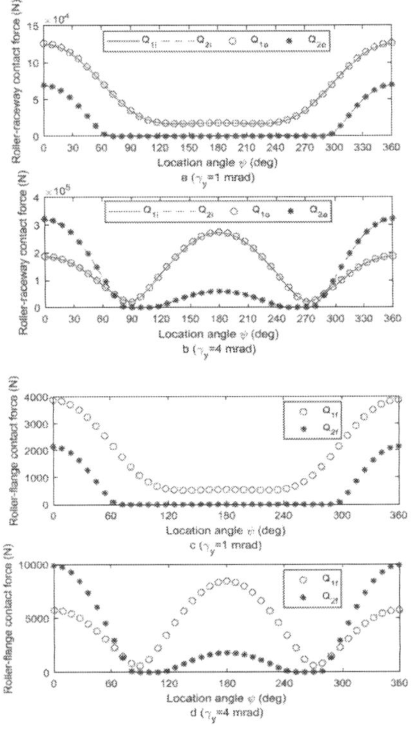

Fig. 5. Contact forces distribution under different angular misalignments.

4 Conclusions

This paper puts forward a comprehensive quasi-static model that utilizes the precise alternative slicing technique and non-Hertzian contact estimation to accurately simulate the contact load distribution along the raceway and flange, especially for heavy load and low speed operating conditions. The detailed contact load results are obtained via simulation, which take into account the influence of combined loading and tilted misalignment between the raceways. The simulation results show that the amount of combined load has an apparent effect on the contact load distribution along the TDIRB raceways and flanges in both rows.

Furthermore, the slight change in tilted misalignment has a great influence on the contact load distribution. The variation of contact loads distribution in row 1 and 2 show a contrary tendency. With the increasing of misalignment angle, the contact loads of rollers around the bottom of TDIRB in row 1 and the top of TDIRB in row 2 increase significantly. In addition, a slight angular misalignment produces noncontact zone for the bearing raceway in both rows, which is significantly disadvantage for the external load uniform transmitting to each roller. These aforementioned results also potentially presented the importance of combined load and misalignment influences analysis for the pressure distribution of rollers and fatigue life of TDIRBs in the following work.

Acknowledgement

The author thanks the University of Technology Sydney and Chinese Scholarship Council for funding the research work.

References

1. CEC. Wind energy. Clean Energy Council; 2018.
2. GWEC Global Wind 2017 Report - A Snapshot of Top Wind Markets in 2017: Offshore Wind. 2017
3. Zhao M, Ji JC. Nonlinear torsional vibrations of a wind turbine gearbox. Applied Mathematical Modelling. 2015; 39:4928-50.
4. Matthew W, Jon T, Won S, Peter T. Improving wind turbine drivetrain bearing reliability through pre-misalignment. Wind Energy. 2014; 17:1217-30.
5. Roller Cpot. Analysis of contact pressure distribution of the straight and crowning profiles of tapered roller bearing. 2014.
6. Sivareddy VV. Modelling and analysis of deformation on a flexure bearing in linear compressor. 2014.
7. Harris TA, Kotzalas MN. Advanced concepts of bearing technology: rolling bearing analysis: CRC Press; 2006.
8. Tong V-C, Hong S-W. Characteristics of tapered roller bearing subjected to combined radial and moment loads. International Journal of Precision Engineering and Manufacturing-Green Technology. 2014; 1:323-8.
9. Tong V-C, Hong SW. Effects of Roller Profile on the Stiffness of Tapered Roller Bearings. Journal of Automation and Control Engineering. 2015; 3:151-6.
10. Andréason S. Load distribution in a taper roller bearing arrangement considering misalignment. Tribology. 1973; 6:84-92.
11. Warda B, Chudzik A. Effect of ring misalignment on the fatigue life of the radial cylindrical roller bearing. International Journal of Mechanical Sciences. 2016;111-112:1-11.
12. ISO/TS-16281. Rolling bearings-Methods for calculating the modified reference rating life for universally loaded bearings. 2008.

13. Nelias D, Bercea I, Paleu V. Prediction of Roller Skewing in Tapered Roller Bearings. Tribology Transactions. 2008; 51:128-39.
14. Oh KP. Analysis of a Needle Bearing. Journal of Tribology. 1984; 106:78-87.
15. Timken Tapered Roller Bearing Catalog. 2018.
16. Nélias D, Bercea I, Mitu N. Analysis of Double-Row Tapered Roller Bearings, Part II – Results: Prediction of Fatigue Life and Heat Dissipation. Tribology Transactions. 2003; 46:240-7.

Impact of *df/dt* Inertia Control on Doubly Fed Induction Generator Type Wind Turbine

Hua Huang[1], *Xin* Tao[2], *Xu* Tian[2], *and Fei* Peng[2]

[1]Huazhong University of Science and Technology, 430074, Wuhan, China
[2]State Grid Qinghai Electric Power Company, Xining 810000, Qinghai Province, China

Abstract. the supplemental *df/dt* control is the common method for wind power generation to support system inertial response. This paper investigates the impact of the *df/dt* inertia controller on doubly fed induction generator (DFIG) type wind turbine. The controller is usually a high pass filter. Two parameters which are the gain and the time constant are considered. By constructing the linearized model of DFIG, the impacts of the two parameters of the controller are discussed. It shows that with different phase locked loop (PLL) parameters and current control parameters, different time constants and gains are required to ensure the DFIG stability. The time domain simulation validates the result.

1 Introduction

As more and more wind power generation injecting to power system, the lack of inertia becomes significant. The solution is to make the wind power generation support inertial response. The common method is to detect the change rate of grid frequency (*df/dt*) to make wind power generators generate extra active power [1-7]. This control imitates the rotor dynamic process of synchronous generator (SG). The gain between *df/dt* and the extra active power is considered as the inertia constant from wind power generation [5]. Since it is difficult to implement the derivative control, a high pass filter is often used. Due to the simple physical significance and control logic, it is expected to widely use in more and more wind power generators.

Obviously, the dynamic properties of wind power generator in electromechanical time scale are changed due to the supplemental inertia control. Researches naturally focus on the response of wind power generator and its impact on power system stability in electromechanical time scale. For example, in [3], the impact of different parameters of the *df/dt* controller was studied, and it concluded that the high pass filter time constant will reduce the effect of the inertial response. In [8-9], the impact of doubly fed induction generator (DFIG) inertia control on SG's small signal stability was studied.

The researches are based on the condition that the wind power generations are stable when applied the supplemental control. However, in some situations, improper setting *df/dt* controller parameters will lead to instability. Few researches have focused on the influence of the *df/dt* control on wind power generator itself. In [10], it mentioned that the *df/dt* control will deteriorate the DFIG shafting dynamic. However, the detailed analysis is not given.

In this paper, we focus on the impact of the supplemental *df/dt* control on DFIG's electromagnetic dynamics. By analyzing the linearized model of DFIG in electromagnetic time scale, it is found the *df/dt* control will influence the stability of the DFIG's rotor side current control. The smaller time constant and larger gain of the *df/dt* controller tends to lead to instability. The analysis reveals that different phase locked loop (PLL) parameters and current control parameters will require different time constant and gain to ensure the DFIG stable.

The rest of the paper is recognized as follows: section II describes the control and establishes the linearized electromagnetic model. Section III analyzes the stability of *df/dt* on DFIG current control with consideration of different impact factors. Section IV gives the simulation results. The conclusion is in section V.

2 Control strategy and modelling of DFIG

2.1 Control Strategy of DFIG

The control strategy of a DFIG is shown in Fig. 1 [11]. The maximum power point tracking (MPPT) algorithm generates the desired rotor speed, and the rotor speed controller generates electromagnetic torque reference. The output of *df/dt* controller, as shown in the red block in Fig. 1, is added to the torque reference to compose the final electromagnetic torque reference. As the stator voltage is approximately equal to 1 in per unit, the value of electromagnetic torque is approximately equal to the active power component of rotor side current. For simplicity, the d-axis rotor current reference, which represents the active power component, is equal to the final electromagnetic torque reference. The frequency detection is come from the phase locked loop (PLL). The grid side control is in normal control mode and for simplicity is not plotted in Fig. 1.

The rotor side current control adopts the common vector control strategy. It is also shown in Fig. 1.

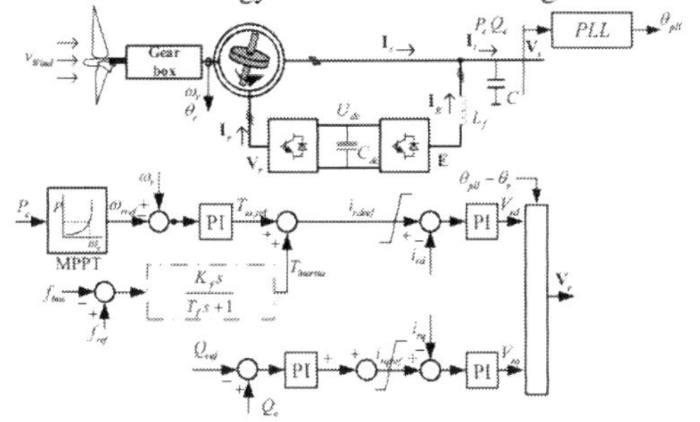

Fig. 1. DFIG rotor side control with df/dt inertia control.

2.2 Model of DFIG Rotor Side Current Control with Inertia Control

The analyzed system is shown in Fig. 2 that a single equivalent DFIG connects to an infinite bus through a long transmission line. As the outer speed control loop and reactive power control loop are relatively slow, when considering the rotor side current control, its influence can be neglected. Hence, the d-axis current reference is only changed by the *df/dt* control, while the q-axis current reference is unchanged.

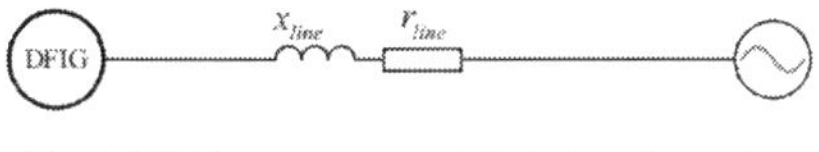

Fig. 2. Single DFIG connects to a infinite bus through transmission lines.

2.2.1 Electric Circuit Modeling

Define the stator current injecting to grid is the stator current positive direction, and the rotor current injecting to rotor is the rotor current positive direction. Hence, the stator voltage vector $\mathbf{V}_s$ and rotor voltage vector $\mathbf{V}_r$ can be expressed as:

$$\begin{aligned} &\mathbf{V}_s + R_s\mathbf{I}_s + j\omega_1 L_s\mathbf{I}_s + dL_s\mathbf{I}_s/dt = dL_m\mathbf{I}_r/dt + j\omega_1 L_m\mathbf{I}_r \\ &\mathbf{V}_r - R_r\mathbf{I}_r + j\omega_{slip}L_m\mathbf{I}_s + dL_m\mathbf{I}_s/dt = dL_r\mathbf{I}_r/dt + j\omega_{slip}L_r\mathbf{I}_r \end{aligned} \quad (1)$$

Where $\mathbf{I}_s$ and $\mathbf{I}_r$ are stator current and rotor current vector. L_s is the stator inductor, L_r is the rotor inductor, L_m is the mutual inductor, R_s is the stator resistor, and R_r is the rotor resistor. For simplicity, the grid side converter current injection is ignored as the mainly characteristics are reflected in rotor side converter. However, the capacitor is considered as it significantly affects the stability. According to the grid circuit, the stator voltage can be also expressed as:

$$\mathbf{V}_s = \mathbf{V}_b + R_l\mathbf{I}_{s1} + j\omega_1 L_l\mathbf{I}_{s1} + dL_l\mathbf{I}_{s1}/dt \quad (2)$$

$$j\omega_1 C\left(\mathbf{V}_s - \mathbf{I}_{s2}R_c\right) + dC\left(\mathbf{V}_s - \mathbf{I}_{s2}R_c\right)/dt = \mathbf{I}_{s1} \quad (3)$$

$$\mathbf{I}_s = \mathbf{I}_{s1} + \mathbf{I}_{s1} \quad (4)$$

Where $\mathbf{V}_b$ is the voltage of the infinite bus. $\mathbf{I}_{s1}$ and $\mathbf{I}_{s2}$ are the current in line and in capacitor respectively. C is the capacitor and R_c is the resistor on it. The equations are all in a public constant speed rotating reference. Let the *d*-axis of the public constant speed rotating reference overlaps with the stator voltage vector in steady state. Eliminate the stator current and ignore the rotor speed variation, in small signal equation, by replacing the differential operation by Laplace operator s, the rotor current and stator voltage can be expressed as in Eq. (5), the expression can be seen in Appendix.

$$\begin{aligned} \Delta\mathbf{V}_s &= \mathbf{G}_{VV}(s)\Delta\mathbf{V}_r \\ \Delta\mathbf{I}_r &= \mathbf{G}_{VI}(s)\Delta\mathbf{V}_r \end{aligned} \quad (5)$$

2.2.2Control System Modeling

In the control system, the variables are in PLL reference frame. In steady state, the d-axis of the public constant speed rotating reference overlaps with the d-axis of PLL. However, in dynamic process, they are inconsistent. The relation can be seen in Fig. 3.

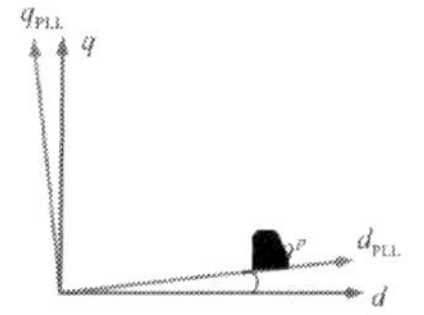

Fig. 3. The axis relation.

Hence, according to the phase relation shown in Fig. 3, we can get:

$$\begin{aligned} i_{rd}^P &= i_{rd}\cos\Delta\theta^P + i_{rq}\sin\Delta\theta^P \\ i_{rq}^P &= i_{rq}\cos\Delta\theta^P - i_{rd}\sin\Delta\theta^P \end{aligned} \quad (6)$$

Where the superscript *P* means the variables are in PLL reference frame. θ^P is the output phase angle of PLL. Then the linear small disturbance equation of rotor current can be expressed as:

$$\begin{aligned} \Delta i_{rd}^P &= \Delta i_{rd} + i_{rq0}\Delta\theta^P \\ \Delta i_{rq}^P &= \Delta i_{rq} - i_{rd0}\Delta\theta^P \end{aligned} \quad (7)$$

The subscript 0 means initial value. With the same process, we can get the linear small disturbance equation of the rotor voltage in public constant speed rotating reference expressed as:

$$\begin{aligned} \Delta V_{rd} &= \Delta V_{rd}^P - V_{rq0}\Delta\theta^P \\ \Delta V_{rq} &= \Delta V_{rq}^P + V_{rd0}\Delta\theta^P \end{aligned} \quad (8)$$

$\Delta\theta^P$ is come from PLL. The equivalent construction of PLL is shown in Fig. 4, where the input is the stator voltage phase θ_{grid}, and the output is θ^P.

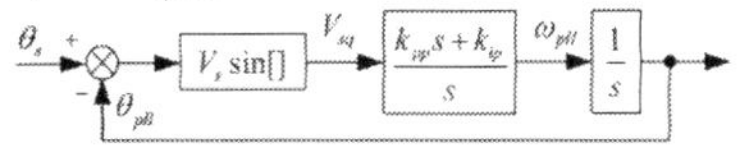

Fig. 4. Diagram of PLL.

Considering that the stator voltage magnitude in per unit is approximately equal to 1, and in steady state, θ_{grid}, is equal to θ^P, from Fig. 4, we can get the linear small disturbance equation of $\Delta\theta^P$ is expressed as:

$$\Delta\theta^P \approx \frac{k_{pp}s + k_{ip}}{s^2 + k_{pp}s + k_{ip}} \Delta\theta_{grid} = G_{pll}(s)\Delta\theta_{grid} \quad (9)$$

In the public constant speed rotating reference, the stator voltage phase angle variation can be expressed as:

$$\Delta\theta_{grid} \approx \Delta V_{sq} / V_{s0} = \Delta V_{sq} / V_{sd0} \quad (10)$$

A reality should also be paid attention is the control delay effect. It is caused by the digital control. To simplify the analysis, the sampling time delay and output time delay are unified modeled that is replaced by a unit time delay. To linearize the discrete unit time delay, the Pade equivalent method is adopted. Synthetically considering the simplicity and accuracy, the second order Pade equivalent is adopted as shown in (11).

$$\frac{1}{z} \approx G_d(s) = \frac{1 - \frac{T_d}{2}s + \frac{T_d^2}{12}s^2}{1 + \frac{T_d}{2}s + \frac{T_d^2}{12}s^2} \quad (11)$$

Where T_d=1.5T_s, and T_s is the sampling time. Hence, the rotor side current control linear small disturbance block diagram can be plotted in Fig. 5.

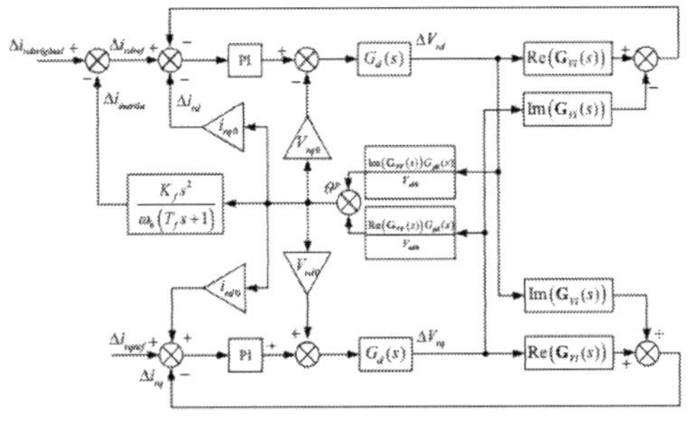

Fig. 5. Small signal analysis diagram of DFIG rotor current control loop.

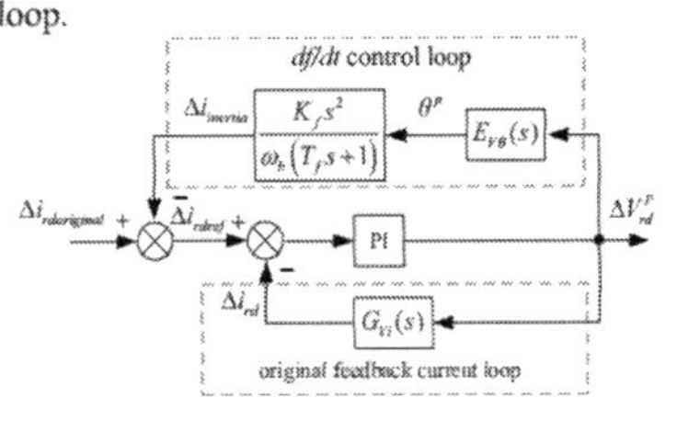

Fig. 6. Small signal diagram of DFIG *d*-axis rotor current control.

3 Stability Analysis

The *df/dt* control is added at d-axis rotor current control loop, so the study object should focus on the *d*-axis current control loop. In order to understand the impact of the *df/dt* control, according to Fig. 5, we separate the analysis loop into two parts as shown in Fig. 6. The first part named as original feedback current loop represents the normal current control process. The other part named *df/dt* control loop represents the supplemental impact introduced by the *df/dt* control. The following study is based on Fig. 6.

3.1 Impact of *df/dt* parameters

In Fig. 6, the *df/dt* control loop can be seen as an external part of the feedback rotor current. The bode diagram of this external loop and the original feedback current loop is plotted in Fig. 7. We can see that, the *df/dt* control's property is similar with the original feedback current loop, especially in phase properties. The main difference is the magnitude. Hence, after adding *df/dt* control, it equivalently increases the rotor current control loop's bandwidth.

In Fig. 7(a), the gain of the *df/dt* controller K_f is set to 10, and the time constant of the *df/dt* controller T_f is set to 0.1, 0.5, 1 separately. The dotted line represents the original current feedback loop. It shows that with the time constant of the *df/dt* controller decreasing, the magnitude of the *df/dt* control loop increases. Hence, according to the classical linear control theory, the bode diagram of the total open loop of the current control shows that with the time constant of the *df/dt* controller decreasing, the system tends to be unstable.

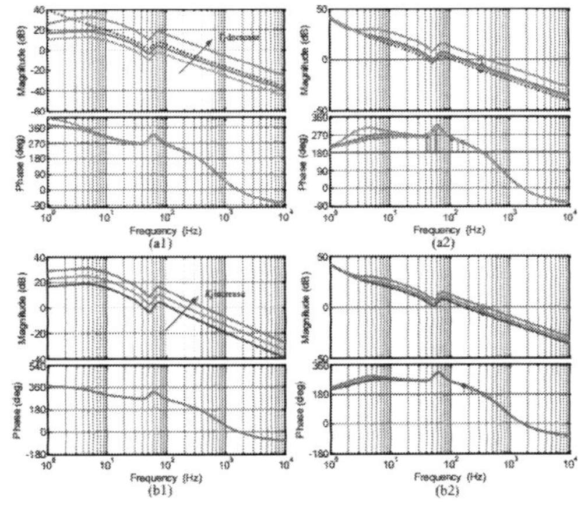

Fig. 7. Bode diagram of df/dt control loop and original feedback current loop.

In Fig. 7(b), the time constant of the *df/dt* controller T_f is set to 0.5, and the gain of the *df/dt* controller is K_f set to 10, 20, 40 separately.it shows that with the gain of the *df/dt* controller increasing, the *magnitude* of the *df/dt* control loop increases. Hence, the bode diagram of the total open loop of the current control shows that with the gain of the *df/dt* controller decreasing, the system tends to be unstable.

3.2 Impact of PLL parameters

The frequency detection is from PLL. So the transfer function of PLL will also influence the dynamic of *df/dt* control loop. Meanwhile, the original feedback current loop is also influenced by PLL.

The bode diagrams with different PLL parameters are plotted in Fig. 8. The dotted lines represent the original feedback current loop. In Fig. 8(a), it can be seen that the with different PLL parameters, the original feedback current loop almost remains unchanged. However, the magnitude of the *df/dt* control loop will be larger when the PLL parameters are larger. Consequently, in Fig. 8(b), the phase margin of the open loop of the current control become smaller, with PLL parameters become larger, the system tends to be unstable.

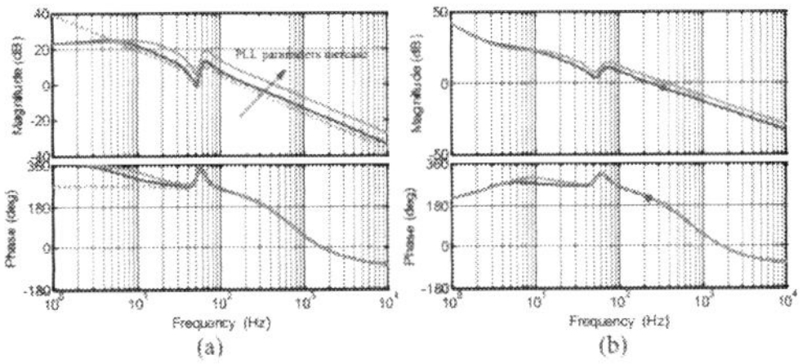

Fig. 8. Bode diagram with different PLL parameters.

3.3 Impact of Current controller parameters

The open loop gain of the rotor current control is also depended on the current controller. Meanwhile, the current control parameters will also influence the *df/dt* control loop and the original feedback current loop.

Fig. 9 shows the bode diagrams with different current control parameters. The dotted lines represent the original feedback current loop. In Fig. 9(a), it can be seen that with different current control parameters, the original feedback current loop and the *df/dt* control loop are almost unchanged. Hence, it the larger current control parameters will directly increase the gain of the current control's open loop. The stability tends to decrease.

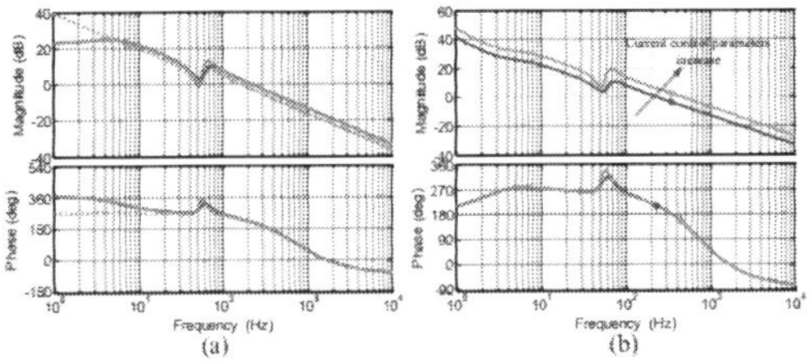

Fig. 9. Bode diagram with different current control parameters.

4 Time Domain Simulation Results

Table I. DFIG parameters for basic case

Items	value
Grid impedance and resistor	6.6858e-04H, 0.042Ω
df/dt controller	K_f=20, T_f=0.5
k_p and k_i of PLL	60, 1400
k_p and k_i of current controller	0.6, 8

The time domain simulation is carried out in MATLAB/SIMULINK. The simulation step size is 5e-6s to guarantee the discrete simulation will not affect the stability phenomena. The switching frequency of the DFIG control system is 2000Hz.

A basic case is used to compare the different impact factors. In this basic case, the core parameters are shown in Table I. Other parameters are shown in Appendix.

Different factors are compared in time simulations. The *d*-axis rotor current in control system and stator current are plotted. At t=5s, the *df/dt* control is added.

In Fig. 10, the result of different *df/dt* control parameters are shown. Fig. 10(a) is the basic case. In Fig. 10(b), the K_f is 40 and T_f is 0.5. In Fig. 10(b), the K_f is 20 and T_f is 0.1. In the basic case, it is stable. When increase the gain K_f, the system become unstable, the oscillation in i_{rd} gradually divergence. When reduce the time constant T_f, it can also lead to the system unstable. The results fit the analyzed above. The stator current shows the stable waveform and the unstable waveform. When it becomes unstable, the waveform become obviously distortion.

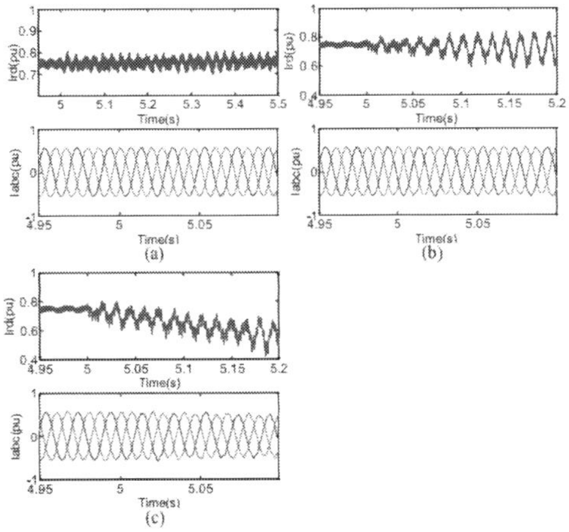

Fig.10. DFIG d-axis rotor current in control system and stator current with different df/dt controller parameters.

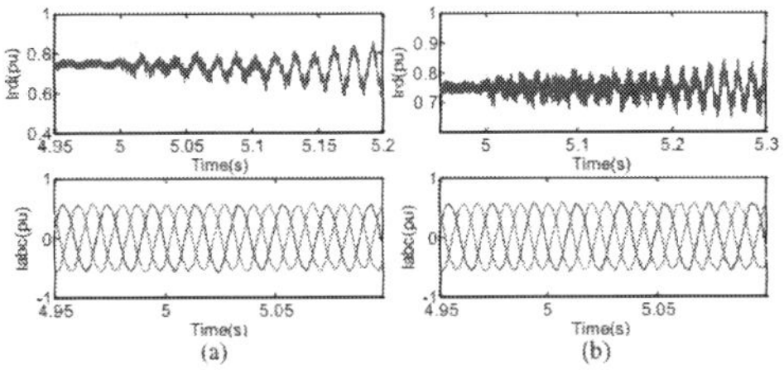

Fig. 11. DFIG d-axis rotor current in control system and stator current (a) different PLL parameters; (b) different current control parameters.

In Fig. 11(a), the different PLL parameters are compared. In this case, the k_p and k_i of PLL are tuned to 120 and 6400. Compared to the basic case, the larger PLL parameters tends to lead to unstable. The phenomenon is similar that the oscillation gradually divergence.

In Fig. 11(b), the different current control parameters are compared. In this case, the k_p *and* k_i of current controller are tuned to 1.2 and 16. As analyzed above, with the gain of the current controller increase, the stability becomes worsen.

The four comparisons very well fits the analyzed conclusions before.

5 Conclusions

The classical *df/dt* control in some circumstances will leads to undesired instability issues in DFIG current control. In this paper, the reason why it will bring instability is analyzed, and the impact factors including grid strength, PLL parameters and current controller parameters are considered. The results show that the gain and time constant of *df/dt* controller cannot be arbitrary set. When the PLL parameters are large, and current controller parameters are large, the gain of *df/dt* should be smaller and the time constant of *df/dt* should be larger to guarantee the stability. The time domain simulation validates the results. The results of the paper are desired to offer a limitation for setting the *df/dt* controllers.

Acknowledgments:

The work presented in this paper was fully supported by the grant from Basic Research Project of Qinghai Province of China (2016-ZJ-767), and State Grid Corporation Science and Technology Project (52283015000M).

APPENDIX

Table II. DFIG parameters

Symbol	Description	value
R_s	Stator resistance	0.023 p.u.
R_r	Rotor resistance	0.016 p.u.
L_s	Stator self-inductance	3.08 p.u.
L_r	Rotor self-inductance	3.06 p.u.
L_m	Mutual inductance	2.9 p.u.
P_N	Rated power	1.0MW
U_N	Rated stator voltage	690 V
U_m	Rated open loop rotor voltage	1950V
f_N	Rated frequency	50 Hz

References

1. J. Ekanayake and N. Jenkins, "Comparison of the response of doubly fed and fixed-speed induction generator wind turbines to changes in network frequency," IEEE Trans. Energy Conversion, vol. **19**, no. 4, pp. 800-802, Dec. 2004.
2. J. Morren, S. W. H. de Haan, W. L. Kling, and J. A. Ferreira, "Wind turbines emulating inertia and supporting primary frequency control, " IEEE Trans. Power Systems, vol. **21**,no. 1, pp. 433-434, Feb. 2006.
3. M. Kayikçi and J. V. Milanovic', "Dynamic contribution of DFIG-based wind plants to system frequency disturbances," IEEE Trans. Power Systems, vol. **24**, no. 2, pp.859-867, May. 2009.
4. G. Ramtharan, J. B. Ekanayake and N. Jenkins, "Frequency support from doubly fed induction generator wind turbines" IET Renewable Power Generation, vol. **1**, no. 1, pp. 3–9, Mar. 2007.
5. J. Mauricio, A. Marano, A. Expósito, and J. Ramos. "Frequency regulation contribution through variable speed wind energy conversion systems", IEEE Trans. Power Systems., vol. **24**, no. 1, pp. 173-180, Feb. 2009
6. N. W. Miller, K. Clark, M. E. Cardinal, R. W. Delmerico "GE wind plant dynamic performance for grid and wind events" AEE TECH WIND GRID, Apr. 2009.
7. N.W. Miller, K. Clark, M. Shao, "Frequency responsive wind plant Controls: Impacts on Grid Performance" IEEE Power and Energy Society General Meeting, Jul. 2011.
8. X. Xi, H. Geng, and G. Yang, "Small signal stability of weak power system integrated with inertia tuned large scale wind farm," IEEE PES Innovative Smart Grid Technologies (ISGT), Kuala Lumpur, pp.514-518, 2014.
9. J. Ma, Y. Qiu, Y. Li, W. Zhang, Z. Song and J. S. Thorp, "Research on the impact of DFIG virtual inertia control on power system small-signal stability considering the phase-locked loop," IEEE Trans. Power Syst., to be published.
10. G.Tsourakis, B. M. Nomikos, and C. D. Vournas, "Contribution of doubly fed wind generators to oscillation damping", IEEE Trans. Energy Convers., vol. **24**, no. 3, pp. 783-791, Sept. 2009.
11. K. Clark, N. W. Miller, J. J. Sanchez-Gasca. "Modeling of GE wind turbine generators for grid studies", version 4.5, (2010).

Particle Swarm Optimized–Support Vector Regression Hybrid Model for Daily Horizon Electricity Demand Forecasting Using Climate Dataset

Mohanad S AL-Musaylh[1,2], *Ravinesh* C Deo[1], and *Yan* Li[1]

[1]School of Agricultural, Computational and Environmental Sciences, University of Southern Queensland, QLD 4350, Australia
[2]Management Technical College, Southern Technical University, Basrah, Iraq

Abstract. This paper has adopted six daily climate variables for the eleven major locations, and heavily populated areas in Queensland, Australia obtained from Scientific Information for Land Owners (SILO) to forecast the daily electricity demand (*G*) obtained from the Australian Energy Market Operator (AEMO). Optimal data-driven technique based on a support vector regression (SVR) model was applied in this study for the *G* forecasting, where the model's parameters were selected using a particle swarm optimization (PSO) algorithm. The performance of PSO–SVR was compared with multivariate adaptive regression spline (MARS) and the traditional model of SVR. The results showed that the PSO–SVR model outperformed MARS and SVR.

1 Introduction

Electricity demand (*G*) forecasting is a purely fundamental yet a challenging optimisation task for improving business efficiency of the electricity industry. A relationship between the *G* data and temperature is clearly evident in winter and summer [1]. Hence, it would be significant to develop a forecasting model employing both the *G* and related climate input datasets.

In recent years, support vector regression (SVR), PSO algorithm, and multivariate adaptive regression splines (MARS) have been widely adopted in energy demand forecasting [1]. Those methods have been used to forecast *G* in [1, 2], however, the influences of the climate datasets are not incorporated yet.

The main contribution of this research paper is to improve the *G* forecasting accuracy by involving climate datasets and integrating the merits of the PSO algorithm with the SVR model. To evaluate the PSO-SVR model, the traditional methods of the SVR and MARS algorithms are also developed.

2 Theoretical Background

2.1 Support vector regression

A nonlinear regression problem can be solved by a SVR model, which is a machine learning method and pioneered by [3], below:

$$y = f(X) = \omega \cdot \emptyset(X) + b \tag{1}$$

where $X = \{x_i\}_{i=1}^{i=n} \in \mathcal{R}^N, y = \{y_i\}_{i=1}^{i=N} \in \mathcal{R}$ are the predictors and target variables, respectively. b is a constant, ω is the weighted vector, and $\emptyset(X)$ represents the mapping function employed in the feature space. A minimisation technique is used to estimate the coefficients ω and b as follows [3]:

$$\text{Minimize } \frac{1}{2} \| w \|^2 + C\frac{1}{N}\sum_{i=1}^{N}(\xi_i + \xi_i^*) \tag{2}$$

$$\text{Subject to} \begin{cases} |y_i - (\langle w, x_i\rangle + b) \geq \varepsilon + \xi_i \\ \langle w, x_i\rangle + b - y_i \leq \varepsilon + \xi_i^* \\ \xi_i, \xi_i^* \geq 0 \end{cases} \tag{3}$$

where the smoothness of the function is determined by $\frac{1}{2}\| w \|^2$, C and ε are the model's parameters and the nonnegative slack variables (ξ and ξ^*) demonstrate the distance between actual and equivalent boundary values of a function approximation. A nonlinear regression function can be expressed by Eq. 4 after applying Lagrangian multipliers and optimising conditions [3]:

$$f(X) = \sum_{i=1}^{i=N}(\alpha_i - \alpha_i^*)K(x_i, x_j) + b \tag{4}$$

where x_i and $x_j \in X$, and the term $K(x_i, x_j)$ denotes the kernel function. α_i and α_i^* are Lagrangian multipliers [3].

In this study, the radial basis function (RBF) was used in the processing of the SVR model as follows [4]:

$$K(x_i, x_j) = exp\left(\frac{-\|x_i - x_j\|^2}{2\sigma^2}\right) \quad (5)$$

where the kernel width and inputs are represented by σ and x_i, x_j, respectively. The critical task for developing the SVR model with a good accuracy is to determine the three parameters which are kernel width (σ), the loss function (ε) and regulation (C) during the training period [5]. This is achieved through a hybrid method called particle swarm optimization (PSO) in section 2.3 below.

2.2 Multivariate adaptive regression splines

The relationship between X and y is demonstrated by the MARS model as follows [6]:

$$y = \hat{f}(X) = a_0 + \sum_{m=1}^{M} a_m BF_m(X) \quad (6)$$

where X and y are offered in Eq. 1, a_0 is a constant, $\{a_m\}_1^M$ are the model coefficients, M is the number of basis functions in MARS, and $BF_m(X)$ is a spline function defined as $C(X|s, t_1, t, t_2)$, where $t_1 < t < t_2$, and $s = \pm 1$.

The number of the basis functions for developing MARS model is determined through the Generalized Cross-Validation criterion (GCV) based on the mean square error (MSE) [6]:

$$GCV = MSE/[1 - \frac{\tilde{C}(M)}{N}]^2 \quad (7)$$

where $MSE = \frac{1}{N}\sum_{i=1}^{N}[y_i - \hat{f}(X_i)]^2$. $\tilde{C}(M) = C(M) + v \cdot M$, where v is a penalty factor with a characteristic value of $v = 3$. $C(M)$ is the number of parameters being fitted. In the training dataset, the lowest value of the GCV refers to the optimal MARS model.

2.3 Particle swarm optimization (PSO)

To select the best parameters of the SVR model, which are the regulation function (C), kernel width (σ) and loss function (ε), the PSO algorithm, first introduced by Kennedy and Eberhart [7, 8], is employed in this paper using the mean square error (MSE) as the fitness function as follows [9-11]:

$$V_{i,j}(k+1) = \omega * V_{i,j}(k) + c_1 * \text{rand}() * \left(P_{\text{best}_{i,j}}(k) - X_{i,j}(k)\right) + c_2 * \text{rand}() * \left(g_{\text{best}_j}(k) - X_{i,j}(k)\right) \quad (8)$$

$$X_{i,j}(k+1) = X_{i,j}(k) + V_{i,j}(k+1) \quad (9)$$

where $X_i = (X_{i1}, X_{i2}, \dots, X_{iD})^T$ is the ith particle from the initial population of the size of $i = 1,2,\dots,N$ and a dimension $of\ j = 1,2,\dots,D$. $V_i = (V_{i1}, V_{i2}, \dots, V_{iD})^T$ is the velocity of each particle X_i in the population. According to [9, 11], rand() represents a random number between zero and one while the individual and global extreme values are represented by $P_{\text{best}_{i,j}}$ and g_{best_j}, respectively. The two values of c_1 and c_2 are usually within [2, 2.05], whereas ω can be defined as follows [9, 10]:

$$\omega = \omega_{min} + \frac{(T_{max} - T) * (\omega_{max} - \omega_{min})}{T_{max}} \quad (10)$$

where ω_{min} and ω_{max} usually equal to 0.4 and 0.9; T and T_{max} are the current and maximum iteration numbers, respectively [9].

3 Materials and Methods

3.1 Electricity demand data (G)

In this study, the G data were recorded half-hourly (48 times per day) in Megawatts (MW) for the state of Queensland, and these data were acquired from the Australian Energy Market Operator (AEMO) [12] for the period of 01-01-2015 to 31-12-2016 (dd-mm-yyyy). The 30-minute data periods were converted to daily terms by obtaining total values for each day. Figure 1 showed the plots of the actual G data series.

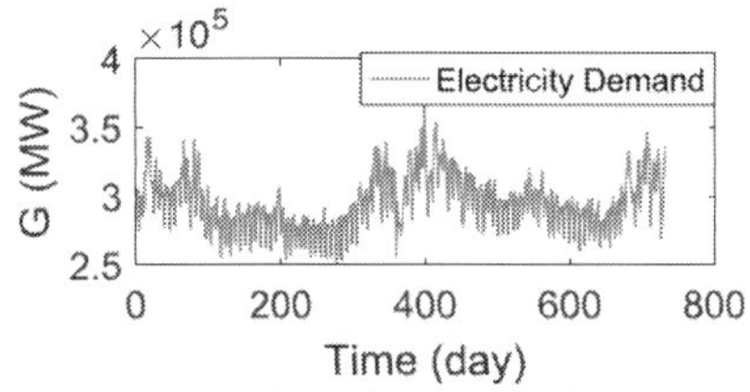

Fig. 1. Time-series of electricity demand (G MW) data.

3.2 Climate dataset

Historical climate datasets for the same period of the G data were obtained from Scientific Information for Land Owners (SILO) [13]. The data were collected for the main eleven stations, which contain the majority of the population of Queensland, that were shown in Fig. 2 and Table 1.

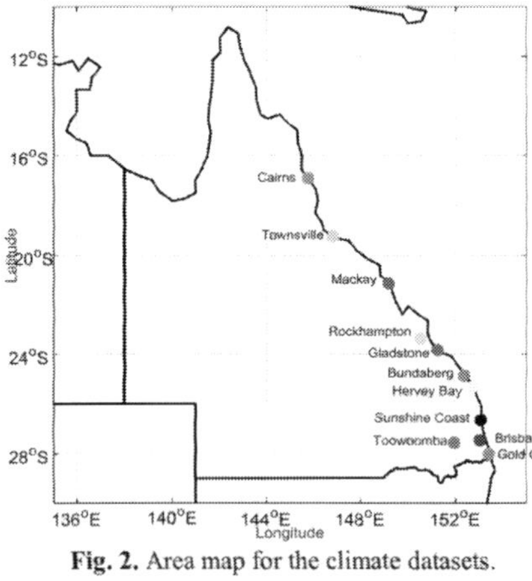

Fig. 2. Area map for the climate datasets.

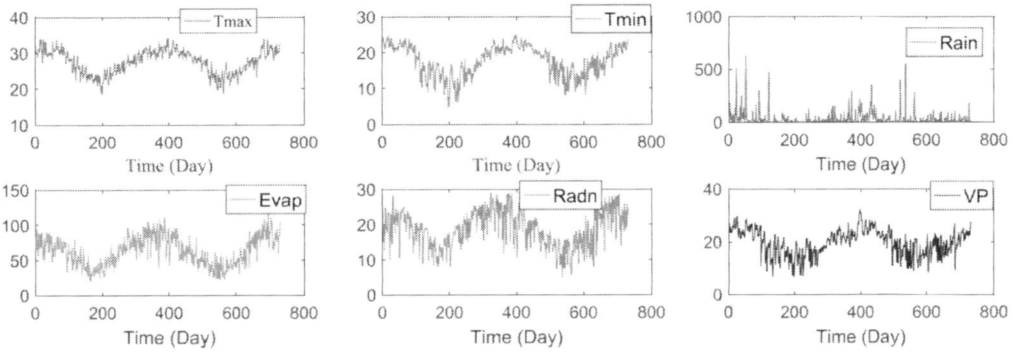

Fig. 3. Time-series of climate datasets used in this study.

Table 1. Population for each climate station

No.	Station	Population number
1	Cairns	240,190
2	Townsville	229,031
3	Mackay	43,364
4	Rockhampton	79,726
5	Gladstone	61,640
6	Bundaberg	92,897
7	Hervey Bay	53,035
8	Sunshine Coast	346,522
9	Brisbane	2,270,800
10	Gold Coast	555,721
11	Toowoomba	149,512
	Total	4,122,438

The population numbers were obtained from Australian Bureau of Statistics [14] where the total number of population resulted from the eleven stations in Table 1 is very close to the population of whole Queensland (4,883,739).

The input data were comprised of the time-series of maximum and minimum air temperature (T_{max} and T_{min}), rainfall (Rain) evaporation (Evap), solar radiation (Radn) and vapour pressure (VP). The datasets for whole Queensland were obtained by getting the average of T_{max}, T_{min}, Radn and VP and the total values of Rain and Evap of the eleven stations datasets. Those were used as the inputs of the models. Figure 3 showed the plots of those actual time series.

3.3 Forecast model development and validation

The climate variables in section 3.2 above were used to forecast the *G* data by developing three models: PSO-SVR, SVR and MARS. As there is no a single method for splitting data into training, validation and testing [5], the data were divided into subsets of 70% for training, 15% for validation and 15% for testing.

MATLAB-based Libsvm toolbox (version 3.22), developed by Chang and Lin [15], was used to build the SVR model. To develop a hybrid SVR model, the PSO algorithm (section 2.3) was used to select the optimal parameters based on the smallest value of *MSE*. To evaluate the accuracy of the SVR model, the software packages version 1.13.0 was employed [16] for the MARS model.

The models were validated in Table 2 using the root-mean square error (*RMSE*, Eq. 11). The PSO–SVR model yielded the lowest *RMSE*, which indicated the best accuracy compared to the other models.

Table 2. Root-mean square error (*RMSE*; MW) in the validation dataset

Model	*RMSE* (MW)
PSO–SVR	**12730.52**
SVR	12930.53
MARS	16701.61

3.4 Model performance evaluation

This study adopted a wide range of statistical error criteria in the testing period based on statistical indicators. Those are mean absolute error *(MAE)*, root-mean square error (*RMSE*), relative error (%) based on *MAE* and *RMSE* values (*MAPE* and *RRMSE*), Willmott's Index *(WI)*, the Nash–Sutcliffe coefficient (E_{NS}), and the Legates and McCabe Index (E_{LM}) [17-26], represented below:

$$RMSE = \sqrt{\frac{1}{n}\sum_{i=1}^{i=n}\left(G_i^{for} - G_i^{obs}\right)^2} \quad (11)$$

$$MAE = \frac{1}{n}\sum_{i=1}^{i=n}\left|G_i^{for} - G_i^{obs}\right| \quad (12)$$

$$RRMSE = 100 \times \frac{\sqrt{\frac{1}{n}\sum_{i=1}^{i=n}\left(G_i^{for} - G_i^{obs}\right)^2}}{\overline{G^{obs}}} \quad (13)$$

$$MAPE = 100 \times \frac{1}{n}\sum_{i=1}^{i=n}\left|\frac{G_i^{for} - G_i^{obs}}{G_i^{obs}}\right| \quad (14)$$

$$WI = 1 - \left[\frac{\sum_{i=1}^{i=n}\left(G_i^{for} - G_i^{obs}\right)^2}{\sum_{i=1}^{i=n}\left(\left|G_i^{for} - \overline{G^{obs}}\right| + \left|G_i^{obs} - \overline{G^{obs}}\right|\right)^2}\right], \quad 0 \leq WI \leq 1 \quad (15)$$

$$E_{NS} = 1 - \left[\frac{\sum_{i=1}^{i=n}\left(G_i^{for} - G_i^{obs}\right)^2}{\sum_{i=1}^{i=n}\left(G_i^{obs} - \overline{G^{obs}}\right)^2}\right], -\infty \leq E_{NS} \leq 1 \quad (16)$$

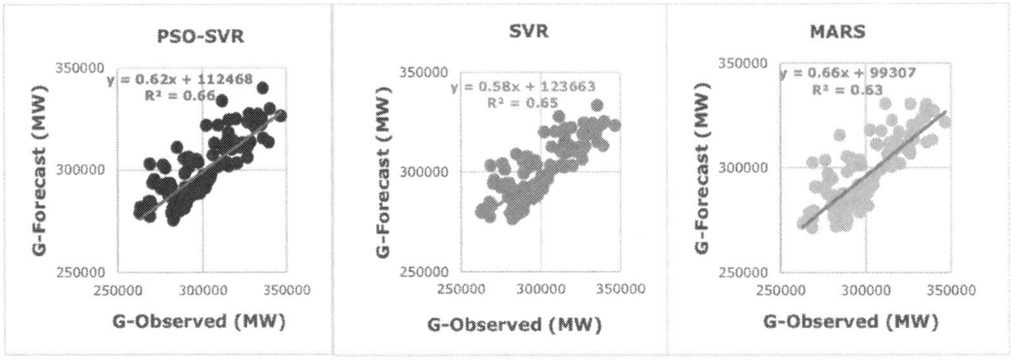

Fig. 4. Scatterplot of the *G*-forecasted vs. *G*-observed of electricity demand data in the testing period using the three models. The equation of linear regression line and the coefficient of determination are incorporated.

$$E_{LM} = 1 - \left[\frac{\sum_{i=1}^{i=n}\left|G_i^{obs} - G_i^{for}\right|}{\sum_{i=1}^{i=n}\left|G_i^{obs} - \overline{G^{obs}}\right|}\right], -\infty \leq E_{LM} \leq 1 \quad (17)$$

where G_i^{for} and G_i^{obs} are the ith forecasted and observed values of G in the testing period, respectively; n is the total number of G_i^{for} or G_i^{obs} values, $\overline{G^{for}}$ and $\overline{G^{obs}}$ are the means of forecasted and observed values, respectively.

4 Results and Discussions

The performance of the PSO–SVR model for the daily forecast *horizon* was compared with traditional SVR and MARS models in *the* testing period. The results of the comparison indicated that the PSO–SVR yielded better performances (lowest *RMSE*, and *MAE*, as well as the largest *WI*, E_{NS}, and E_{LM}) than SVR and MARS models. Those values were summarized in Table 3.

Table 3. The performance evaluation of the models in the test period

Model	*WI*	E_{NS}	*RMSE* (MW)	*MAE* (MW)	*MAPE* (%)	*RRMSE* (%)	E_{LM}
PSO–SVR	**0.88**	**0.66**	**11716.97**	**9668.15**	**3.27**	**3.92**	**0.41**
SVR	0.87	0.64	11909.00	9895.09	3.34	3.99	0.40
MARS	0.87	0.60	12612.22	11003.20	3.69	4.22	0.33

The scatterplots of G_i^{for} vs. G_i^{obs} and the model forecasted errors, $|FE| = |G_{FOR,i} - G_{OBS,i}|$ in the testing period for the three models were shown in Figs. 4 and 5, respectively. The lowest forecasted errors ($|FE|$) were shown by the PSO-SVR model in this study (Fig. 5). On the other hand, the highest correlation of determination (R^2) was achieved by the PSO-SVR model (Fig. 4). Overall, a significantly greater accuracy was attained by the PSO–SVR model than the other models.

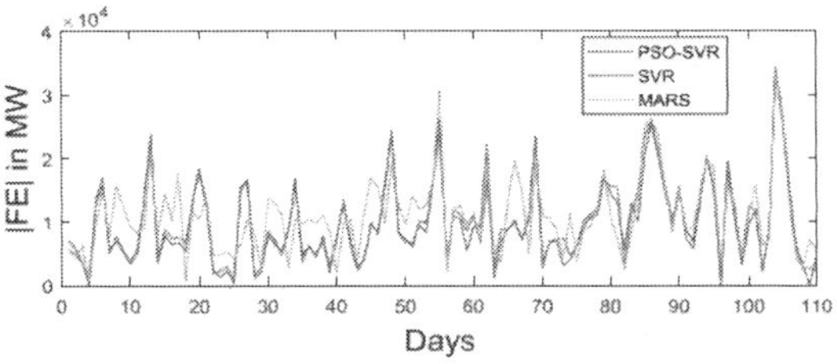

Fig. 5. Model forecasted errors, $|FE| = |G_{FOR,i} - G_{OBS,i}|$ in the testing period using the three models.

5 Concluding Remark

In this paper, a hybrid PSO–SVR model was proposed for daily *G* forecasting horizon in Queensland, Australia, where the model used data from Australian Energy Market Operator (AEMO) and Scientific Information for Land Owners (SILO). The MARS and the traditional method of SVR were also used in this research study to evaluate the main model. The results showed that the PSO–SVR outperformed the MARS and SVR models. As a result, the data-driven tool constructed by the PSO–SVR model is a powerful forecasting framework which can support the National Electricity Market (*e.g.*, AEMO). Although the PSO–SVR model performed well in this paper, some challenges in model development section could be appeared. As the PSO algorithm needs a longer time to produce the SVR parameters, alternative methods, such as multi-swarm PSO and sine cosine algorithm may need to be used. In addition, the model could be improved using ensemble-based uncertainty testing by a bootstrapping technique. Those should be addressed in future studies.

Acknowledgments

Firstly, we would like to offer our thanks to the Australian Energy Market Operator (AEMO), Australian Bureau of Statistics, and Scientific Information for Land Owners (SILO) which provided all the required data. We are very grateful to the Ministry of Higher Education and Scientific Research in the Government of Iraq for funding the first author's PhD project.

References

1. M. S. Al-Musaylh, R. C. Deo, J. F. Adamowski, and Y. Li, "Short-term electricity demand forecasting with MARS, SVR and ARIMA models using aggregated demand data in Queensland, Australia," *Advanced Engineering Informatics*, vol. 35, pp. 1-16, 2018.
2. M. S. Al-Musaylh, R. C. Deo, Y. Li, and J. F. Adamowski, "Two-phase particle swarm optimized-support vector regression hybrid model integrated with improved empirical mode decomposition with adaptive noise for multiple-horizon electricity demand forecasting," *Applied Energy*, vol. 217, pp. 422-439, 5/1/ 2018.
3. V. N. Vapnik, *Statistical learning theory* (no. Book, Whole). New York: Wiley, 1998.
4. J. A. K. Suykens, J. De Brabanter, L. Lukas, and J. Vandewalle, "Weighted least squares support vector machines: robustness and sparse approximation," *Neurocomputing*, vol. 48, no. 1–4, pp. 85-105, 10// 2002.
5. R. C. Deo, X. Wen, and F. Qi, "A wavelet-coupled support vector machine model for forecasting global incident solar radiation using limited meteorological dataset," *Applied Energy*, vol. 168, pp. 568-593, 2016.
6. J. H. Friedman, "Multivariate adaptive regression splines," *The annals of statistics*, pp. 1-67, 1991.
7. J. Kenny, "Particle swarm optimization," in *Proc. 1995 IEEE Int. Conf. Neural Networks*, 1995, pp. 1942-1948.
8. R. Eberhart and J. Kennedy, "A new optimizer using particle swarm theory," in *Micro Machine and Human Science, 1995. MHS'95., Proceedings of the Sixth International Symposium on*, 1995, pp. 39-43: IEEE.
9. X. Li and C. Li, "Improved CEEMDAN and PSO-SVR Modeling for Near-Infrared Noninvasive Glucose Detection," *Computational and mathematical methods in medicine*, vol. 2016, 2016.
10. M. N. Alam, "Particle Swarm Optimization: Algorithm and its Codes in MATLAB," ed: ResearchGate, 2016.
11. M. Hasanipanah, A. Shahnazar, H. B. Amnieh, and D. J. Armaghani, "Prediction of air-overpressure caused by mine blasting using a new hybrid PSO–SVR model," *Engineering with Computers*, vol. 33, no. 1, pp. 23-31, 2017.
12. AEMO. (2016). *Aggregated Price and Demand Data - Historical*. Available: http://www.aemo.com.au/
13. S. J. Jeffrey, J. O. Carter, K. B. Moodie, and A. R. Beswick, "Using spatial interpolation to construct a comprehensive archive of Australian climate data," *Environmental Modelling & Software*, vol. 16, no. 4, pp. 309-330, 2001.
14. A. B. o. Statistics. (2018). *Census*. Available: http://www.abs.gov.au/
15. C.-C. Chang and C.-J. Lin, "LIBSVM: a library for support vector machines," *ACM Transactions on Intelligent Systems and Technology (TIST)*, vol. 2, no. 3, p. 27, 2011.
16. G. Jekabsons, "Adaptive Regression Splines toolbox for Matlab/Octave," *Version*, vol. 1, p. 72, 2013.
17. K. Mohammadi, S. Shamshirband, M. H. Anisi, K. A. Alam, and D. Petković, "Support vector regression based prediction of global solar radiation on a horizontal surface," *Energy Conversion and Management*, vol. 91, pp. 433-441, 2// 2015.
18. C. J. Willmott, S. M. Robeson, and K. Matsuura, "A refined index of model performance," *International Journal of Climatology*, vol. 32, no. 13, pp. 2088-2094, 2012.
19. C. J. Willmott, "On the evaluation of model performance in physical geography," in *Spatial statistics and models*: Springer, 1984, pp. 443-460.
20. C. J. Willmott, "Some comments on the evaluation of model performance," *Bulletin of the American Meteorological Society*, vol. 63, no. 11, pp. 1309-1313, 1982.
21. C. J. Willmott, "On the validation of models," *Physical geography*, vol. 2, no. 2, pp. 184-194, 1981.
22. C. W. Dawson, R. J. Abrahart, and L. M. See, "HydroTest: a web-based toolbox of evaluation metrics for the standardised assessment of hydrological forecasts," *Environmental Modelling & Software*, vol. 22, no. 7, pp. 1034-1052, 2007.
23. D. R. Legates and G. J. McCabe, "Evaluating the use of "goodness-of-fit" measures in hydrologic and hydroclimatic model validation," *Water resources research*, vol. 35, no. 1, pp. 233-241, 1999.
24. P. Krause, D. Boyle, and F. Bäse, "Comparison of different efficiency criteria for hydrological model assessment," *Advances in Geosciences*, vol. 5, pp. 89-97, 2005.
25. T. Chai and R. R. Draxler, "Root mean square error (RMSE) or mean absolute error (MAE)?–Arguments against avoiding RMSE in the literature," *Geoscientific Model Development*, vol. 7, no. 3, pp. 1247-1250, 2014.
26. J. E. Nash and J. V. Sutcliffe, "River flow forecasting through conceptual models part I—A discussion of principles," *Journal of hydrology*, vol. 10, no. 3, pp. 282-290, 1970.

Life cycle energy and carbon footprint analysis of large MW scale grid connected wind power systems in India

Jani Das , and *Rangan* Banerjee

Department of Energy Science and Engineering, Indian Institute of Technology Bombay, Mumbai, India

Abstract. The centralized grid is the main electrical supply to the country. About 80% of the installed generation capacity is fossil fuel fired power plants, though there is increase in share of renewable generation over the years. India has significant wind potential, and a major share in the renewable generation of the country. The energy and carbon footprint sustainability of wind power systems in Indian conditions is the focus of this research. This paper presents a study of large grid connected wind storage systems with reference to the cradle to grave primary energy requirements and carbon emissions in an Indian scenario.

1 Introduction

Renewable power is clean sources of energy. When energy is converted from different sources, whether depleting or non-depleting, to usable form, there are environmental impacts. Methods for analyzing sustainability in terms of energy and environmental factors and subsequent comparison between different options, can direct future energy plans and policies.

Wind energy capacity over the world is roughly doubling every three years, with a global installed capacity of 539 GW. India ranks fourth among the world nations with a capacity of 33GW as on December 2017[1].Wind power constitute 55% of the total renewable power in the country [2]. But the energy and environmental impact of wind turbine systems has to be throughout its life time has to be accessed in Indian situations. Life cycle study analyses material and energy flows in the life time of a system, to ascertain the energy and environmental sustainability.

Several life cycle assessment studies of wind turbine systems have been done and published in literature. Some studies have been done by wind turbine producing companies [3-5] and in scientific articles [6-9].These studies mainly focus on energy pay back and environmental impact of large wind power systems at different locations on the globe.

This paper presents a research related to two large scale grid connected wind farms located at two different locations in India, with different turbine manufacturers. Life cycle energy and carbon footprint analysis has been done on the two systems. Energy and environmental sustainability of the systems have been compared by energy payback time, net energy ratio and emission factor.

2 Life Cycle Analysis Methodology

The life cycle assessment (LCA) used in the paper is consistent with the ISO standards [10-12]. A wind farm and its subsequent auxiliaries are analysed during all phases of its life cycle from cradle to grave, material production, manufacturing of its main components, transportation to the wind farm site, the installation, the startup ,maintenance and final dismantling and stripping down to waste materials which are either recycled, incinerated or landfilled. The end of life scenario of the system is considered as a separate stage. Material production refers to extraction of raw materials from earth or recyclates and refining them to usable materials which were used to manufacture the grid connected wind farm. Manufacturing represents all the processes required to convert the raw materials into wind farm components and assembling it into the final product. The cradle to grave primary energy requirements E in MWh_{th} is given by

$$E_{ctg} = E_{mp} + E_{mnf} + E_{Inst} + E_{oper} + E_{tran} + E_{eol} \quad (1)$$

and carbon emission CE in $kgCO_2$ for wind farm is given by

$$CE_{ctg} = CE_{mp} + CE_{mnf} + CE_{Inst} + CE_{oper} + CE_{tran} + CE_{eol} \quad (2)$$

ctg stands for cradle to grave; *mnf* for manufacturing; *inst* for onsite installation; *oper* for operation and maintenance; *tran* for transportation and *eol* for end of life.

2.1 System boundary

Within the limits of the boundary, are the components of the wind turbine, the transportation of the turbine components and the transmission cable to the farm site, the installation of the farm, the operation and maintenance and dismantling after the life time of the

https://doi.org/10.1051/e3sconf/20186408002

system. Outside the limits of the boundary, are the transformer substation and the national electricity network.

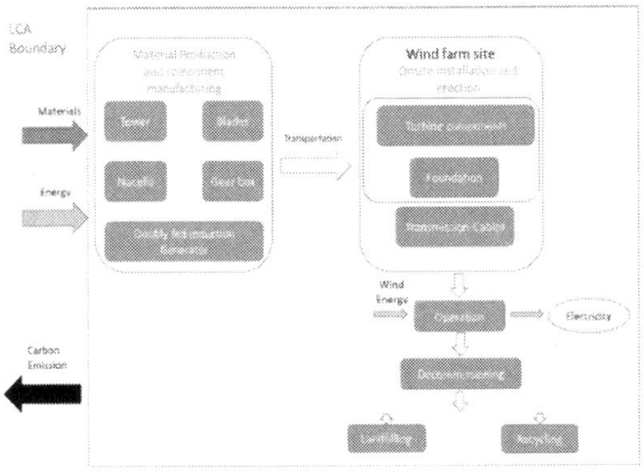

Fig. 1. LCA boundary of the system.

3 Wind Generation System Under Study

The wind farm considered for study consists of five onshore 2 MW wind turbine systems of two different manufacturers (Vestas and Seimens), transmission system connecting the farm to the grid and the substation which connects the system to the electricity network. The specifications of the turbine systems are given in Table 1.

Table 1. Specifications of the wind turbines.

Turbine Model	Vestas V90	Seimens Gamesa G80
Rated Power(kW)	2000	2000
Cut in wind speed(m/s)	4	3.5
Rated wind speed(m/s)	13	12
Cut out wind speed(m/s)	25	25
Generator type	4-pole (50 Hz)/6-pole (60 Hz) doubly fed generator, slip rings	

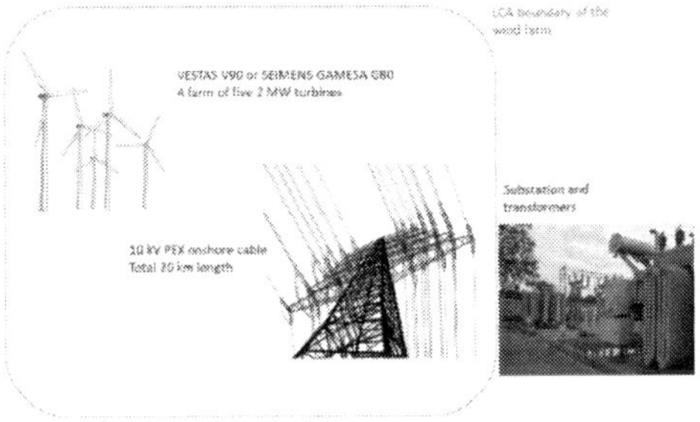

Fig. 2. Sketch of the wind farm.

The sketch of structure of wind farms considered is illustrated in Fig. 2. The wind farms are assumed to be located at Tuticorin, Tamil Nadu and Kutch, Gujarat.

3.1 The turbines

The turbines considered for study are Vestas V90 and Seimens Gamesa G80. The various components of the wind turbine are manufactured at different parts of the world and transported to India for the wind farm installation. The manufacturing points of the two systems are illustrated in Fig. 3. The map helps to identify the magnitude of transportation energy on the overall energy impact of the system.

The first step in life cycle analysis is the preparation of the life cycle inventory. The wind turbine manufacturing materials are quantified as part of the preparation of the life cycle inventory. The materials which are used for manufacture can be directly from the ore or can be from recycled materials. The term "virgin" corresponds to materials which are taken directly from earth. The Virgin/Recycled ration gives the percentage of both types of materials in the overall structure. Manufacturing includes the material and energy input for assembling the different components into the final product or a part of it. The various components are transported to the site and the energy is quantified as transportation energy. Onsite installation also requires a significant amount of energy investment. The decommissioning phase of the wind energy system should also be considered with implications on the system and environment.

3.1.1 Vestas V90 Turbine

Vestas Wind systems is of Danish origin and have manufacturing points at various places in India. The turbine under study is a 2MW system with a doubly fed induction generator.

3.1.2 Seimens GAMESA G80 Turbine

Seimens Gamesa Renewable Energy is a Spanish wind turbine manufacturing company mainly located in Spain. A 2MW turbine with similar characteristics is used for comparison.

4 Energy and Carbon Footprint Analysis of the Wind Generation System

The components of life cycle energy are quantified for both the turbines situated at the locations considered for study. Two different scenarios are analyzed for the decommissioning phase of the system;

a) Scenario1-100% landfilling of the system
b) Scenario2-Partial recycling and landfilling of the system

The energy sustainability is measured in terms of Energy Pay Back Time (EPBT) given by

$$EPBT = \frac{E_{ind}}{E_{out}} \qquad EPBT > 0$$

E_{ind} is the total of indirect energy given to the system (MWh_{th}) and E_{out} is the annual electricity generated expressed in thermal units.

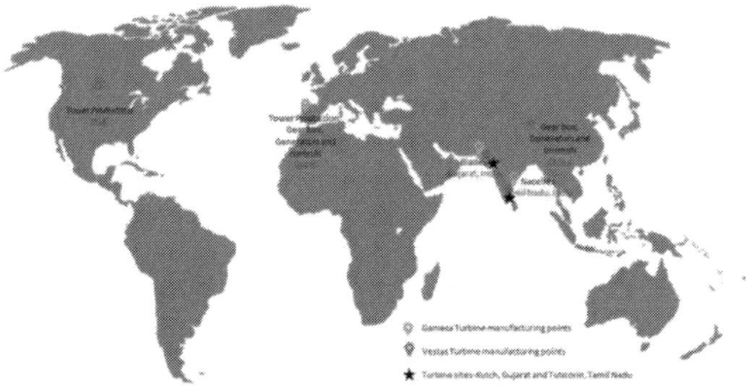

Fig. 3. Manufacturing points of the two turbine systems.

The transportation energy constants calculated for road transport is 341.5 kJ/ton-km and for sea transport is 120.3 kJ/ton-km for India. The corresponding carbon emission factors are 2.124 x10-4 and 8.713E x 10-6 $kgCO_2$/kg-km respectively. The conversion factor for electricity to Primary Fossil energy is taken as 0.38, owing to the electricity generation mix of India dominated by thermal power, primarily coal [13]. The embodied carbon for electricity generation in India is taken as 0.94 kgCO2 per unit [14]. The conversion factor for heat to Primary Fossil Energy is 1.

The carbon footprint of the wind farm is quantified as the emission factor (EF) expressed as carbon emission per unit generation of electricity scaled to a year .The wind farm life time is 20 years. Primary energy requirements and the emission factor of the two decommissioning scenarios for both locations are dealt with separately in the following sections.

4.1. 100% landfilling of the system

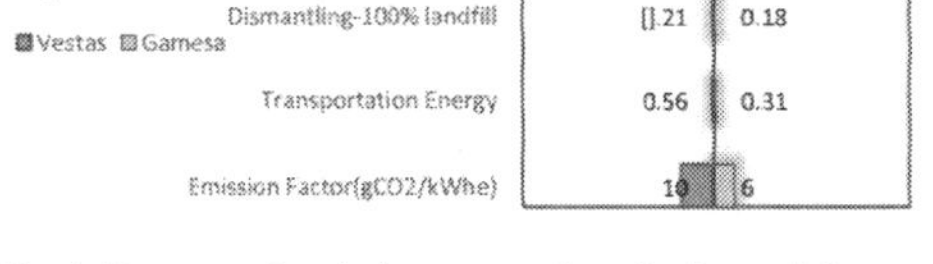

Fig. 4. Energy and emission comparison for Scenario1 for Tuticorin.

The entire wind farm is landfilled including the foundation and the turbine components. None of the components are recycled. The values for Tuticorin are shown in Fig. 4. It is assumed that 1.1% of the total primary energy is used for landfilling of the system and the landfilling site is 5km away from the farm site.

4.2. Partial Recycling and landfilling of the system

The alternate dismantling scenario considered for comparison is partial recycling of materials used for the production of the turbine and the transmission system. Tables 2 and 3 give the details of the material type and the disposal method considered for analysis for the two turbines. The metals are recycled, while the concrete used for manufacturing the foundation of the turbines are landfilled completed. Plastics, fiber glass and resin are incinerated. The primary energy requirements and the emission factor comparison is done.

Table 2. Disposal Scenario of V90.

Material type	Disposal method
Steel	Recycled (90%)
Copper	Recycled (90%)
Iron	Recycled (90%)
Aluminium	Recycled (90%)
Plastic	Incinerated (100%)
Concrete	Landfill (100%)

Table 3. Disposal Scenario of G80.

Material type	Disposal method
Steel	Recycled (90%)
Copper	Recycled (90%)
Iron	Recycled (90%)
Resin	Incinerated (100%)
Fibre glass	Incinerated (100%)
Plastic	Incinerated (100%)
Concrete	Landfill (100%)

The primary energy requirements and the emission factor for the second scenario for Kutch is shown in Fig. 5.

 https://doi.org/10.1051/e3sconf/20186408002

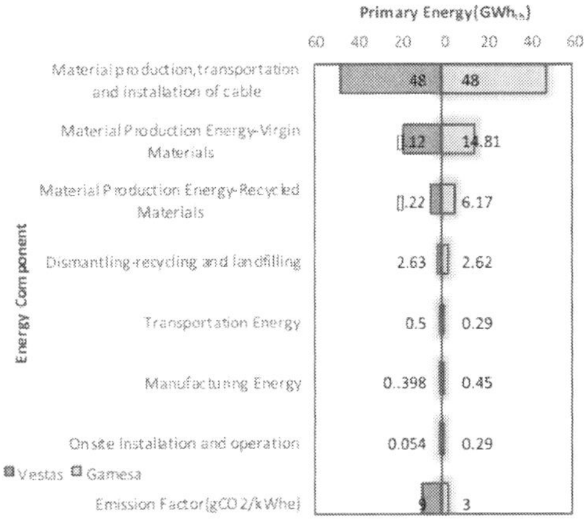

Fig. 5. Energy and emission comparison for Scenario2 for Kutch.

5 Results and Discussion

The EPBT and EF have been calculated for the four operating conditions of the wind farm-100% landfill and partial recycling and landfilling scenarios for the two locations, Tuticorin and Kutch. The sensitivity of EPBT to the various factors are discussed in this section.

5.1. Impact of Plant Load Factor

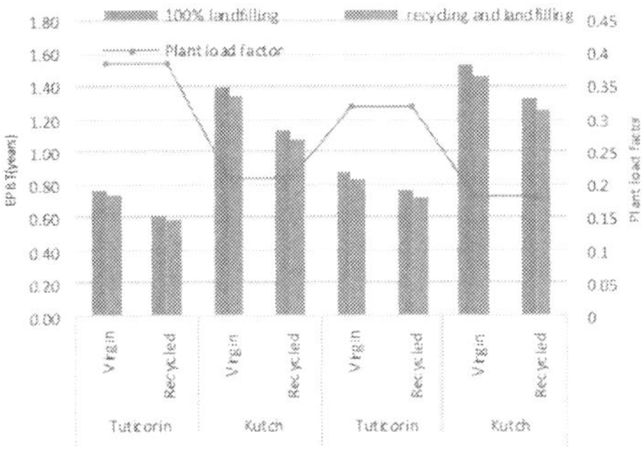

Fig. 6. Variation of EPBT and PLF.

Plant load factor is the ratio of available wind power potential to the rated power of the plant. It depends on the geographic location and the plant power characteristics Variation of EPBT and PLF for the two locations for the two disposal scenarios is shown graphically in Fig .6. It can be observed that, EPBT is less for a system with high PLF. The emission factor is found to be high for Kutch owing to the less PLF. The variation is shown in Fig. 7.

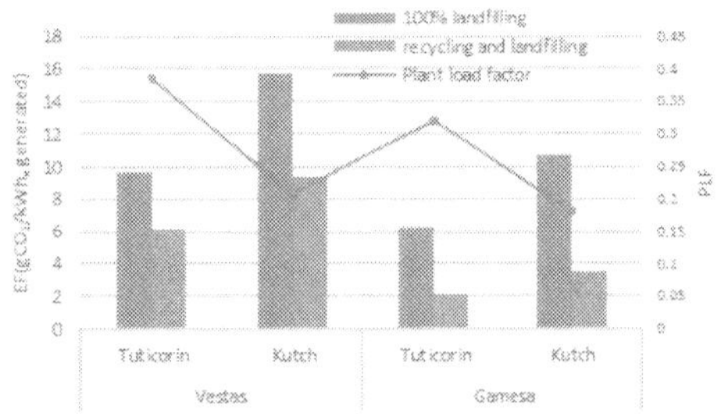

Fig. 7. Variation of EF and PLF.

5.2. Impact of Recycling of materials for component manufacture

Metals, which constitute more than 70% of the wind turbine and cable, can be recycled for secondary usage. It can be seen that there is a reduction of 19% in the total primary energy requirement for the Vestas turbine when calculated for the two locations for the 100% landfill. Condition. Similar analysis shows a reduction of 13% for the Gamesa turbine. The values are 20% and 14% for the second disposal scenario.

5.3. Impact of Transportation Energy

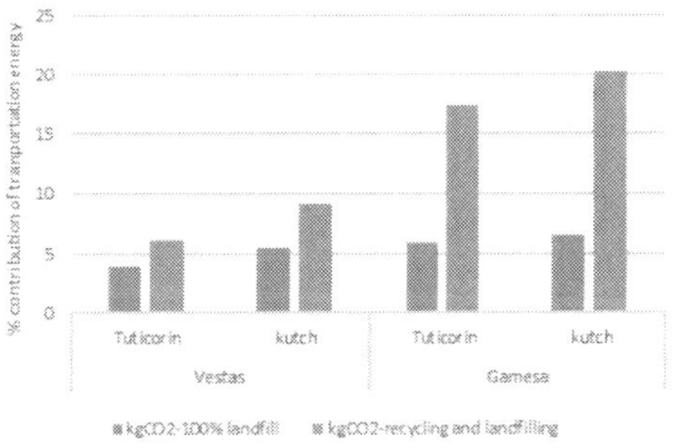

Fig. 8. Sensitivity to transportation energy.

Components of the wind farm are transported from different parts of the globe to the sites. It can be seen that they contribute less than 1% to the total primary energy requirement of the wind farm system. The Vestas turbine has more transportation component (0.8%), than the Gamesa turbine (0.4%) for both the disposal scenarios. The percentage contribution of transportation energy to the total carbon footprint of the system and the sensitivity to turbine type, location and disposal scenario is shown in Fig. 8. The total carbon footprint is more for the 100% landfilling option, thereby reducing the percentage contribution of transportation energy. Moreover, the carbon footprint is found to be high for Gamesa Turbine in comparison with the Vestas turbine, as the overall footprint is less. The overall results are tabulated in Table.4.This analysis helps in selection of turbine make and wind farm location, in terms of energy and environmental sustainability.

Table 4. Consolidated results for the two disposal scenarios for the two turbines.

Location	Model	V/R	Disposal Scenario-1		Disposal Scenario-2		PLF
			EPBT	EF	EPBT	EF	
Tuticorin	Vestas	V	0.8	9.7	0.7	6.2	0.4
		R	0.6		0.6		
	Gamesa	V	0.9	6.2	0.9	2.1	0.3
		R	0.8		0.7		
Kutch	Vestas	V	1.4	16	1.3	9.4	0.2
		R	1.1		1.1		
	Gamesa	V	1.6	11	1.5	3.5	0.2

6 Conclusion

This work proposes a comparison in terms of energy and carbon emission for large scale wind farms located in different locations in India. The energy generated is calculated using the wind speed data for the location and the wind turbine power curve. The impact of plant load factor, recycled materials and transportation energy on the sustainability of wind farms is analysed. Plant load factor has a negative relation with the energy pay back time and emission factor. The pay back time for the systems is in the range of 0.5-1.5 years. The cable transmission length has a major share of energy investment to the whole system. These are definitely site specific. Usage of recycled materials for manufacturing of components will decrease the material production energy of the system. Two disposal scenarios have been considered, out of which the partial recycling and landfilling option seems to be energy and environmental friendly. Transportation Energy doesn't play a major role in this study, as the other energy components are comparatively more. This factor is also turbine manufacture and location specific. This study can be a reference for energy and carbon footprint analysis of large grid connected wind systems in Indian conditions. The methodology can be made use for similar studies of different turbines at different locations of the country to examine their viability.

References

1. World wind energy association, 2018 report. Accessed: www.wwindea.org on 30-03-2018.
2. "Wind Power Development in India-An Overview",August-October 2017, Akshay Urja. Accessed: http://mnre.gov.in/file- manager/akshay-urja/july-october- 2017/EN/Images/20-25.pdf on 30-03-2018.
3. Elsam Engineering A/S, "Life Cycle Assessment of Offshore and Onshore Sited Wind Farms", Report by Vestas Wind Systems A/S of the Danish Elsam Engineering, 2004.
4. "Life cycle assessment of offshore and onshore sited wind turbines" Vestas Wind Systems A/S, Alsvij 21, 8900 Randus, Denmark (www.vestas.com),2005.
5. S.R. Carrascal, "Life Cycle Assessment of 1 Kwh energy generated by Gamesa G114-2.0MW On-shore wind farm "Technology Department, Gamesa Corporation,2014.
6. K.R. Haapaala and P. Prempreeda "Comparative life cycle assessment of 2.0 MW wind turbines," International Journal of Sustainable Manufacturing, vol. 3(2), Jan. 2014, pp.170-185.
7. J. Yang and B. Chen "Life Cycle Assessment of Wind Power Generation System", In Handbook of Clean Energy Systems, J. Yan (Ed.), July 2015.
8. L. Xu, M. Pang, L. Zhang, W.R. Poganietz, S.D Marathe , "Life cycle assessment of onshore wind power systems in China" Resources, Conservation and Recycling, vol.132, May 2018, pp.361-368.
9. N. B Munir, Z Huque and R.R Kommalapati, "Impact of Different Parameters on Life Cycle Analysis, Embodied Energy and Environmental Emissions for Wind Turbine System". Journal of Environmental Protection, vol.7,2016, pp.1005-1015.
10. ISO (1997) ISO International Standard, ISO/FDIS 14040, Environmental Management — Life Cycle Assessment — Principles and Framework.
11. ISO (1998) ISO International Standard, ISO 14041, Environmental Management — Life Cycle Assessment — Goal and Scope Definition and Inventory Analysis.
12. ISO (2000) ISO International Standard, ISO 14042, Environmental Management — Life Cycle Assessment — Life Cycle Impact Assessment.
13. CEA Annual Reports-Executive Summary-Feb 2018, http://www.cea.nic.in/reports/monthly/executivesummary/2018/exe_summary-02.pdf; Accessed on: Feb 2018.
14. CO_2 Baseline and Database for the Indian Power Sector Central Electricity Authority. Accessed: http://www.cea.nic.in/reports/others/thermal/tpece/cdm_CO2/user_guide_ver5.pdf;Accessed on: June 2017.

The Innovative Process in the Interruption of Wind Power in Portugal

António Brito , and *Luís* Neves

Energy Purchase Department, EDP Serviço Universal SA, Portuguese Last Resort Supplier, Coimbra, Portugal

Abstract. The production of energy under a special regime of renewable origin has had a sustainable evolution in Portugal. Since the 1990s, the percentage of renewables has been growing steadily, with special emphasis being given to wind, photovoltaic, mini-hydro, biogas and high-efficiency cogeneration [1]. More recently there has been a strong push in promoting small scale production and self-consumption [2]. There are several periods of time when the country's electricity load is 100% supplied by renewable energy. There are also periods when surplus renewable energy is exported to Spain. However, there are some periods when production exceeds consumption and it is then necessary to reduce wind power production. In Portugal there is an innovative process underway to carry out this reduction, which was systematized by the General Department of Energy and Geology in Order No. 8810/2015, of August 10. Thus, in the case of wind power plants that receive power reduction orders, the remuneration equivalent to that which is lost is paid by the other producers, through the Last Resort Supplier. Those power plants that have not been interrupted pay a percentage of their power produce to the Last Resort Supplier, to compensate those that have been interrupted. The total of the payments made to the producers whose power plants were interrupted must equal the receipts from those plants that continued to produce energy. This new concept is exemplified in this paper by the wind power cut that occurred on March 12 and 13, 2017 in Portugal. An explanation of what occurred on that day is presented, to understand why this cut was made, having been reached the limits of energy exports to Spain. To implement this new interruptible compensation model it was necessary to develop a mathematical algorithm and include it in the computer application named GPCE - Producers' Management and Energy Purchase, that belongs to the Last Resort Supplier. When the interruption occurred in March 2017, the computer system worked correctly. The sum of the payments made equaled the sum of the receipts. It should be noted that there was a high degree of discipline shown by the producers.

1 Introduction - Evolution of wind energy in Portugal

Production under a special regime in Portugal began with the publication of Decree-Law nº 189/88, of May 27. With the publication of Decree-Law nº 339-C / 2001, of December 29, there was an improvement in the remuneration for wind power generation. This fact has led to a significant increase in the number of installed wind power plants in Portugal. Thus, it gave an initial boost to the new technology that was emerging. There have been progressive reductions in the cost of new power plants through the scale effect.

As wind technology has been disseminated, energy acquisition prices have been reduced, namely through Decree Law 33-A / 2005, of February 16, and 225/2007, of May 31. The graph below shows the significant increase in the production of wind energy in GWh, whose yearly production is represented in green.

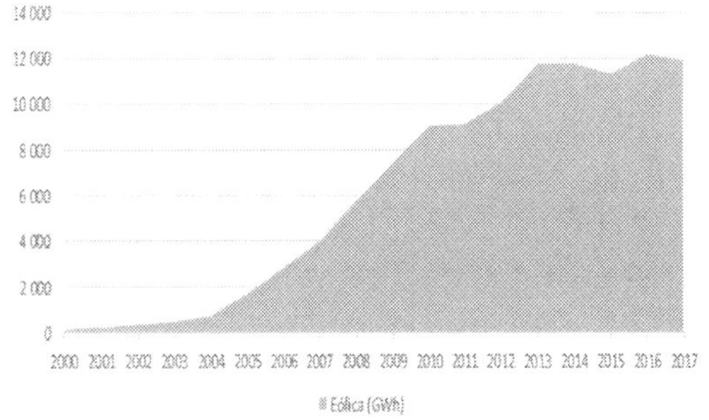

Graph 1. Evolution of wind power energy generation in Continental Portugal, in GWh, during the period from 2000 till 2017.

At the end of 2017 the total power installed in wind power plants was 5 578 MVA.

The next graph represents the evolution of installed power since the year 2000, in apparent power (MVA). At present, new wind power plants with a total power of 232.4 MVA are under construction.

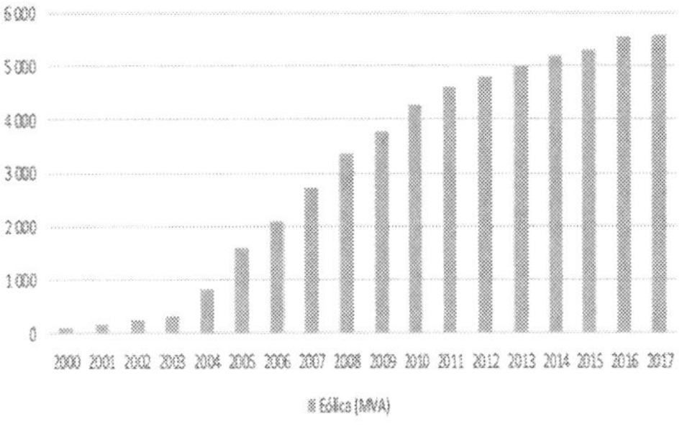

Graph 2. Evolution of wind installed power in Continental Portugal, in MVA, during the period from 2000 till 2017.

Later, in 2010, through Decree-Law no. 51/2010, of May 20, over-capacity of wind power plants was allowed. This was done through the reduction of global remuneration. The placement of equipment to support tensioning was encouraged. Thus, all wind turbines of a wind power plant that were connected to the transmission or to the distribution network were required to have installed equipment installed to support voltage tanks and provide reactive power during troughs.

The producers' representatives carried out various studies demonstrating that the probability that all wind turbines were producing energy at maximum power was very low. Thus, not overcoming the value of the bonding power, the Decree-Law 94/2014, of June 24, allowed them to produce this additional energy at a low price of 60 €/MWh.

Considering the great expansion of wind power, it became necessary to limit the reactive energy emitted to be able to control the voltage levels in the transmission and distribution networks.

More recently, under certain circumstances, there has been a need to interrupt the supply of electricity. The problem has arisen because it was needed to reduce the power or to disconnect certain wind power plants, while others are producing. This would cause complaints that led to the question: Why was my wind power plant cut off? To be solves this problem, the new concept of interrupting arose, in which no one is impaired.

2 What happened on march 12 and 13, 2017

From 3:00 p.m. to 6:30 p.m. on March 12 (Sunday), the Transportation Network Dispatcher had to reduce wind production in Portugal for the first time. The magnitude of the reduction was about 90 MW (less than 3% of the wind production that occurred at that time) [3]. At 9:25 p.m. it was again necessary to reduce the wind generation that was being injected into the grid. This time the magnitude of the reduction was around 80 MW. As the mentioned reduction was not enough, at 22h 02, an additional 80 MW was reduced, leading to a total reduction of 160 MW of wind inject power. At 4:24 am on March 13, the reduction instructions were canceled [3].

The reductions in power mentioned above preserved the balance between production and consumption, since it was not possible to export energy to Spain. In the chart following it can be regarded that the instructions issued by the Transmission System Operator were accomplished by the wind farms and their dispatch centers [3].

Thus, at the beginning of the first limitation period, the power concerning only wind production was 4 510 MW and the consumption was 4 828 MW. Considering the procedures mentioned above, it was possible to reduce the power by 90 MW. At the end of the first limitation period the power being produced had dropped, so that production and consumption could be balanced. At the start of the second limitation period, power increased again. It was then necessary to reduce power by 80 MW. Unfortunately, this reduction was not enough and it was necessary to reduce again by another 80 MW.

Graph 3. [3] – Power produced by wind farms, in MW, on March 12 (9 am, 12 am., 6 pm., 9 pm.) and on March 13 (0 am, 3 am, 6 am, 9 am)

Immediately before the decision to reduce power the transportation network was in an extreme situation, always needing to safeguard the security of supply and the balance between production and consumption.

The power cuts were decided as a last resort to maintain the balance of the system, considering that [3]:

- The consumption was reduced as it is typical of a Sunday with moderate temperature.
- The market program had put all thermal power plants out of service.
- The import programs resulting from the wholesale market had already been canceled.
- Hydro generation had already been reduced to the minimum.
- Hydro pumping had already been mobilized and was partially depleted.

The reduction instructions were limited, in both duration and quantity, to the strict minimum needed to keep the system balanced. The fact that the special regime production had exceeded consumption over a long period of time can be seen in the graph below. The black line represents consumption. As it can be seen, the production of renewable energy exceeded the country's consumption needs and part of that energy was exported to Spain.

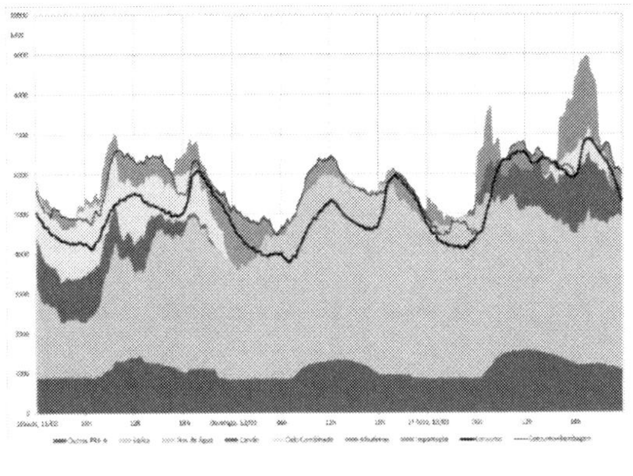

Graph 4. [3] - Diagram of power production, in MW, from March 11 to March 13, 2017.

3 The concept of interruption

Electricity production centers that use renewable energy sources and high-efficiency cogeneration benefit from priority access to networks, according to European Union directives. The Portuguese legislation also provides that Operators of the Public Utility Electricity Network must take appropriate operational measures to prevent or minimize the establishment of limitations in the transportation and distribution of electricity from these power generation centers. The legislation also contemplates the possibility of imposing limitations on the transportation and distribution of electricity from power plants under a special regime, due to reasons related to the reliability and security of supply of the networks. The interruption of production under a special regime may occur in which the Portuguese electricity system is exploited, as it is preview in the Regulations of the Transmission and Distribution Networks. These regulations provide for the possibility of congestion due to the unavailability of equipment or the inability to dispose of the injection of additional energy and over-production from wind power plants. Therefore, it has been necessary to lay down arrangements for interrupting production under a special regime that profits from a system of support for remuneration - known as feed-in tariffs.

Thus, the agents involved in this cut off process are the following:

a) Last Resort Supplier

b) Global System Technical Manager

c) Transport System Operator

d) Distribution System Operator

e) Producers under a special regime who benefit from a support to remuneration (feed-in tariffs), excluding small distributed production, normally connected to the low voltage network.

In exceptional situations of exploitation of the Portuguese electrical system, particularly when congestions occur or when safety in the production-consumption balance are in danger and the continuity of the electric power supply is in question, power reduction orders may be issued. In these cases, the production of special production power plants (linked to the transmission or distribution network) must be monitored so that these installations do not exceed a given power value.

The issuing of power reduction orders is the exclusive task for the Global System Technical Manager. The Distribution System Operator may request that the manager issue a power reduction order, when the operating situations in the distribution network require it. Power reduction orders are mandatory for producers under a special regime, specifying the beginning and end of the interruption period and the maximum power value that can be produced by the power plant. The power reduction order is issued and sent as far in advance as possible, with at least a minimum of 5 minutes before the beginning of the reduction period. The power plants that are subject to power reduction orders that are issued by the Global System Technical Manager or, when applicable, by the Distribution System Operator, have a maximum period of 15 minutes to adjust their production to the value requested in that order.

Producers who have been subject to a power reduction order have the right to receive remuneration equivalent to the production that would have result if no reduction order had been issued. This remuneration is paid by those producers who have not received power reduction orders.

All producers not subject to power reduction orders are obliged to pay a contribution to the producers whose production have been interrupted.

Thus, the Global System Technical Manager, or the Distribution System Operator must inform the Last Resort Supplier, whenever they issue a production reduction order, identifying for each interruption:

a) Producers who have been targeted for the reduction of power, as well as the start and end date and the time of the reduction;

b) Producers that were producing when the reduction order was issued and the electric energy they produced during the reduction order.

Thus, the Global System Technical Manager and the Distribution System Operator agreed the format of the computer file that is to be sent to the Last Resort Supllier containing the information necessary to perform the calculations of the remuneration to be paid and to receive from the producers. A protocol was signed between the Global System Technical Operator and the Last Resort Supplier relating to the operationalization of the information exchange, so that calculations of the remuneration to be paid and to be received from producers can be made in the software application GPCE - Producers' Management and Energy Purchase.

In the following figure the format of part of the mentioned file, which is emitted in the format csv - comma separated values, can be seen.

https://doi.org/10.1051/e3sconf/20186408003

The index key of the file´s registers into the databases is achieved through a delivery point code. This codification complies with European Union standards and is unique for each European Union installation.

Inicio;2017;03;12;15;00;
Fim;2017;03;12;15;30;
PRE_Limitacao;PT0001000000000168NC;84832;
PRE_Limitacao;PT0001906000100036VH;10629;
Racio;2017;03;12;15;15;1.000000;
Racio;2017;03;12;15;30;1.000000;
Dados;PT0001000000000168NC;2017;03;12;15;15;0.000;0.000;0.000;N;
Dados;PT0001000000000168NC;2017;03;12;15;30;0.000;0.000;0.000;N;
Dados;PT0001906000100036VH;2017;03;12;15;15;0.000;0.000;0.000;N;

Fig. 1. Example of a file format used in interruption calculations

The above file must be sent to the Last Resort Supplier within the first 10 working days of the following month (m + 1) for the reduction of production. This information should be used for the billing of the energy purchased from producers in month m+2 or m+3.

The Last Resort Supplier will have to proceed as follows: for educated producers, where there are limitations imposed, the estimated production not carried out by each insulation is multiplied by the average energy sales price of the installation verified in month m, for each identified producer. The amount obtained shall be added to the billing of each producer in month m+2 or m+3.

The Last Resort Supplier will have to proceed as follows:

a) for instructed producers, where there were limitations imposed, the estimated production not carried out by each installation is multiplied by the average energy sales price for that power plant verified in month m. The amount obtained shall be added to the billing of each producer in month m+2 or m+3;

b) The aggregate value of the amounts to be paid to the identified producers shall be allocated to producers who have not suffered a power reduction and that produced energy during the period in which limitation was imposed. The amount obtained is deducted from the billing of each producer in month m+2 or m+3.

4 Computer development

To carry out the mentioned calculations it was necessary to make a new module in the GPCE - Producers' Management and Energy Purchase computer application. At present, the number of producers under a special regime in Portugal is 1025, with about 75% of them adhering to electronic self-billing. Electronic billing has simplified the process of reconciliation with producers, where some receive money and others pay. The computer system must ensure that the total paid is equal to the total received.

The following figure shows the screen shot of the interruption module, which is divided into four entries: *a)* file loading, *b)* retribution calculations, *c)* control reports and *d)* circular sending. The process is completely automated, as it would be difficult to perform such calculations in a short period of time.

INTERRUPTION
ENTRANCE SCREEN

- File loading
- Retribution Calculations
- Control Reports
- Circular Sending
- Come back

Fig. 2. Input interrupt module screen of the GPCE Application

c) File Loading - Files from the Global System Technical Manager or from the Distribution System Operator are sent to the Last Resort Supplier in the format shown in Figure 1. These files are inputed into the computer system using the screen shown in figure 3 below. Using this screen, the various versions that have been sent are controlled, as well as the corrections that have been made.

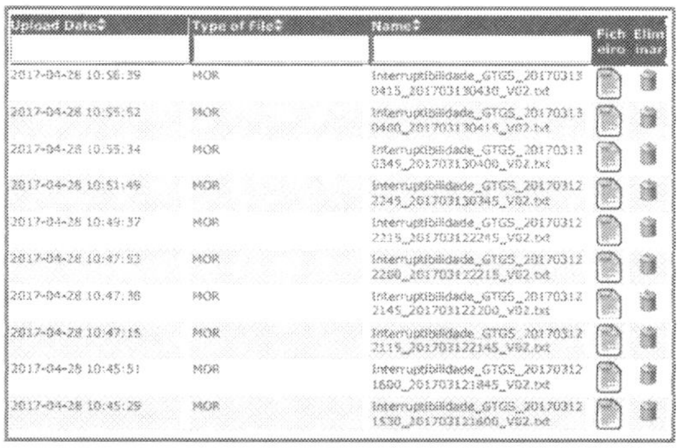

Fig. 3. Example of a file loading screen.

d) Retribution calculations - Calculations are performed and controlled through the screen shown in figure 4 (on the next page). The estimated undone production, in kWh, corresponds to the estimate of production that would have rsulted if a producer had not received reduction orders. This estimate, for each power plant, corresponds to the average value of power that is reduced every 15 minutes, and is verified in the period immediately preceding the first period in which the power reduction order was issued and the 15 minutes power comfirmed in the period immediately after that in which the limitation ends. If there are very long periods of limitation, this way of estimating production then does not reflect reality. The Global System Technical Manager will then calculate the energy estimation using the best information available, in order to approximate this estimate to reality.

e) Reporting - It is necessary to obtain several reports to control the calculations made. These calculations and the amounts obtained are communicated to the General Department of Energy and Geology for validation. After explaining the values obtained and getting authorization, the calculations are closed (see Figure 5 below).

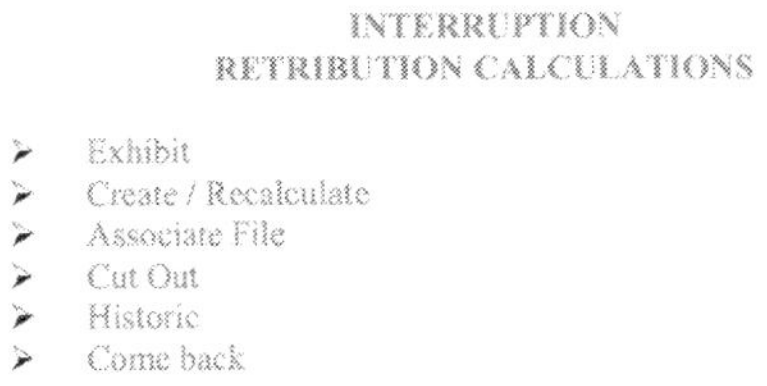

Fig. 4. Retribution calculations screen.

Fig. 5. Reporting and controlling of calculations screen.

d) Circular sending - After the calculations have been aproved, the Last Resort Supplier sends circulars to the producers involved, indicating the amounts to be added to or to be deducted from the billing of the month m+2 or m+3.

Fortunately, 75% of producers have adhered to electronic self-billing, which significantly simplifies the billing process. For those who still send paper invoices it was necessary to establish a control of the amounts paid and received. Circulars are sent automatically using email (see Figure 6 below).

Fig. 6. Circular sending screen.

5 Conclusions

The interruption process described in this paper is already being used and it is an innovation, that allows power that is injected into the network to be reduced, when necessary. The producers affected do not suffer a loss of revenue. The amounts paid by the remaining producers is relatively small, that does not cause any disturbance in their remuneration. The interruption described is very satisfactory: The Global System Technical Manager achieves his goal of reducing power. The producers whose power plants suffered a reducing of power receive the compensation and the other producers participate in that compensation. There had been no complaints till now. It can be concluded that the implementation of this new process has been a success. We strongly recommend the use of this new concept that has been described in this paper.

Acknowledgment

The authors would like to express their gratitude to the colleague Pedro Lourenço who, as business analyst, developed the software system and tested it. The system worked as expected when the interruption occurred on March 12 and 13, 2017.

References

1. A. Brito, L. Neves (2013), *The feed-in tariffs structure of renewable and cogeneration energy in Portugal*, 13º Congresso Luso Espanhol de Engenharia Eletrotécnica, Valência, 2013
2. P. Lourenço., C. Almeida (2018), *Production under feed-in tariffs in Portugal*, International Conference on Renewable Energies and Power Quality (ICREPQ'17), that was held in the University of Malaga on 4 till 6 April, 2018.
3. *Incident Report. Wind Power Cut, March 12 and 13* (2017). REN – Redes Energéticas Nacionais, Portugal.

Characterization and Classification of Daily Electricity Consumption Profiles: Shape Factors and k-Means Clustering Technique

Milton Mora-Alvarez[1], *Pedro* Contreras-Ortiz[1], *Xavier* Serrano-Guerrero[1], and *Guillermo* Escrivá-Escriva[2]

[1]Grupo de Investigación en Energías, Universidad Politécnica Salesiana, Cuenca, Ecuador
[2]Institute for Energy Engineering, Universitat Politècnica de València, Valencia, Spain

Abstract. This paper exposes a method to classify the electric consumption profiles of different types of consumers, based on patterns given. The direct characteristics method is used in this paper, this method is also known as shape factors deduction (SFs) to easily define consumption profiles by using the load patterns resulting from measurements in the time domain, considering weekdays and time ranges. After the characterization of load profiles, k-means clustering technique is applied to SFs. The SFs are segmented in such a way that, in each group similar SFs are gathered. The characterization and classification of electric profiles has important applications, such as the application of specific tariffs according the consumer type, determination of optimal location of generation resources in electrical distribution systems, detection of anomalies in transmission and distribution of electricity or classify geographical areas according to electricity consumption and perform an optimum balance of feeders in electrical substations.

1 Introduction

Due to the liberalization of the electricity market, electricity trading companies have more freedom to set tariffs, by following technical and economic legal regulations. The electricity companies are developed in a profit-oriented context and they are interested in developing market strategies for the tariffs formulation. Detailed knowledge of customer consumption is essential to design specific tariff options, in which tariffs are aligned to the effective electricity use [1], [2].

The generation, transmission, distribution and commercialization companies currently have an extensive measurements database [3]. The analysis of these data allows consumer characterization, however, this task becomes complicated when the appropriate tools are not available [4].

The characterization of consumers is used for the integral planning of the system, considering alternative electricity demand management (EDM). In EDM, the effectiveness of each strategy must be evaluated by characterizing the load profile and identifying the characteristics of the demand in each type of consumer [1]. If the electric companies know in detail the demand behavior, they can improve their tariffs offer [5].

The clustering of consumers based on their activity information or commercial codes seems inefficient, since the load patterns show important differences. The classification based on the consumption pattern similarity provides more effective results [6]. The procedure to classify consumers has different stages, i) define the information that can be collected in the field, ii) choose the characteristics that are used to execute the clustering techniques, iii) use clustering techniques, iv) evaluate the effectiveness of clustering by calculating validity indicators and v) establish the final load profiles that represent a limited number of final consumers types. In order to establish a consumer clustering based on similarity aspects, there are many techniques [6].

The importance of the characterization of consumers is highlighted in [7]. In this review, an extensive review of concepts and methods for the classification in competitive electricity markets is made. In addition, a general description of classification techniques, including classical approaches, is presented. Finally, this paper is focused on developed methods that describe the electrical behavior of customers based on real cases.

In [1], an integral approach is proposed to investigate the market's strategies development based on consumers' clustering into well-defined and non-overlapping classes.

Aiming to make the classification, two sets of indexes are considered, the first set is the priority indexes, based on contractual and historical data, which are stored in the public services database, and the second set is the measured indexes, obtained through field measurements.

For the management of the historical data of the different electrical consumption patterns (ECPs). We choose unsupervised procedures or also called clusters to find characteristics to be used in the categorization. This allows obtaining an idea of the nature or structure of the data [8].

All the clustering techniques are based on mathematical criteria that assess the quality of clustering models that comprise prototypes and data partitioning. The mathematical criteria serve as cost functions that must be minimized to obtain optimal cluster solutions [9]. Such cost functions define an optimization criterion of the clustering techniques.

These techniques can be used to simplify calculation, accelerate convergence and obtain an acceptable computational performance. All these steps give way to an elementary and approximate method that is described in this paper.

This paper shows a method to characterize and clustering the ECPs through the use of SFs proposed by the authors and the application of *k*-means clustering technique. The proposed methodology stands out for its simplicity and efficiency.

These SFs allow the reduction of sensitivity to outliers when they are grouped. Additionally, based on unsupervised learning and in an optimize cost function the number of clusters is found, which allows obtaining patterns to understand, summarize, and categorize the electricity consumption of consumers.

This work is segmented into four sections. In section II the applied methodology is presented, in section III obtained results are analyzed. Finally, in section IV, conclusions are presented.

2 Methodology

2.1 Information gathering

The methodology requires electricity consumption data in 15-minute intervals from electricity meters.

2.2 Data characterization

2.2.1 ECPs recognition

Table 1. Time ranges.

Time Period	Description	Number of measurements taken by the electricity meter every 15 minutes
07h00 to 18h00	Ecuadorian electric commercialization time ranges for commercial and industrial consumers [11].	44
18h00 to 22h00		16
22h00 to 07h00		36
06h00 to 12h00	Time ranges are determined by the residential, commercial and industrial load curves in Ecuador.	24
12h00 to 15h00		12
15h00 to 18h00		12

Each ECP is represented by the use of *N* characteristics according to the measure data and the desired detail of the load pattern representation. In this case, the number of characteristics to represent each daily ECP is 96, since each data corresponds to the measurements average taken during every 15 minutes of an entire day.

It is necessary to define time ranges and weekdays according to the electric tariff and working patterns as a previous step to classify the ECPs [10]. Table 1 shows the time ranges and the number of measurements in each time period for commercial and industrial consumers according to Ecuadorian regulations.

2.2.2 Shape factors definition for ECPs

The SFs are normalized active power values; these are obtained in different time periods depending of time ranges previously defined in Table 1. Each factor value depends on the behavior of each consumer [12].

To obtain the SFs of ECPs, the maximum, minimum and average power values are used ($P^{MAX,\Delta_t}, P^{MIN,\Delta_t}, P^{AV,\Delta_t}$) for each defined time range [13]. A total of 23 SFs are proposed, which characterize each ECP. Table 2 summarizes the SFs proposed by the authors. The SFs are considered to determine the characteristics of ECPs in each defined time range and according to the weekday it is analyzed.

Table 2. Shape factors.

SFs	Equation	Time Period
Load Factor (f_1)	$f_1 = \frac{P_{average}}{P_{maximum}}$	All day
Non-uniformity coefficient (f_2)	$f_2 = \frac{P_{minimum}}{P_{average}}$	All day
Average Impact	$f_x = \frac{P_{average_{time\ range}}}{P_{average}}$	For every time range in Table 1
Maximum Impact	$f_y = \frac{P_{maximum_{time\ range}}}{P_{average}}$	
Minimum Impact	$f_z = \frac{P_{minimum_{time\ range}}}{P_{average}}$	
All day's maximum impact (f_{21})	$f_{21} = P_{maximum}$	All day
All day's average impact (f_{22})	$f_{22} = P_{average}$	
All day's minimum impact (f_{23})	$f_{23} = P_{minimum}$	

Table 3. SFs' association.

Associated Factors	Dimensions	Time Period
f1-f2	2D	All day
f3-f4-f5	3D	06h00 to 12h00
f6-f7-f8	3D	12h00 to 15h00
f9-f10-f11	3D	15h00 to 18h00
f12-f13-f14	3D	07h00 to 18h00
f15-f16-17	3D	18h00 to 22h00
f18-f19-f20	3D	22h00 to 07h00
f21-f22-f23	3D	All day

There is one limitation when the direct form characteristics method is applied; this limitation takes place when visualizing the relationship between SFs in only one graphic. Therefore, there is a need to associate them in a way that they can be clearly observed as they are related (Table 3).

2.3 Clustering

The k-means clustering technique is a partitioning procedure based on a number of iterations [14]. The SFs are clustering in k mutual exclusive clusters and they return the cluster index to which each observation has been assigned. This method is applied on real observations and it creates a unique level of clusters. The k-means clustering technique is used since it represents a robust technique when large amounts of data are analyzed [15].

Each observation has in-space location, this allows the method clustering them in a way that, inside of each cluster the observations are as close as possible to each other and as far as possible from the other observations of the other clusters [16].

The k-means clustering technique calculates the centroids of each cluster, initially in an arbitrary way from the data in analysis [17], [18]. These centroids are used to classify such observations into clusters according to a selected metric. The clusters' centers are recalculated until the minimum cost function is found from the applied metric inside the method. The k-means technique uses the Euclidean distance (Ed) as the metric for data clustering [16]. The Euclidean distance between the SFs that are in the n-dimensional space is calculated. In a general form, the Euclidean distance between the SFs of $ECP_1(f_1, f_2, \dots, f_{23})$ and $ECP_2(f_1, f_2, \dots, f_{23})$, is define as:

$$E_d(\mathrm{ECP}_1, \mathrm{ECP}_2) = \sqrt{\sum_{i=1}^{23} [f_{i(ECP_1)} - f_{i(ECP_2)}]^2} \qquad (1)$$

To advance with the clustering process, it is necessary to find the maximum number of clusters for analyzed data. Then an adequate number of clusters is defined based on the use of silhouettes coefficients (SCs) of each ECP [19].

2.3.1 Maximum number of clusters

To determine the maximum number of clusters in which the SFs can be segmented, the methodology uses the Sturges' rule [20]:

$$k = 1 + \log_2(N) \qquad (2)$$

where:

- $\boldsymbol{N}$ is the number of samples (number of daily ECPs).
- $\boldsymbol{k}$ is the maximum number of clusters in which SFs can be segmented.

1) Output arguments

When the k-means clustering technique is applied the output arguments obtained are [12]:

- Indexes of clusters for each ECP.
- Location of the centroids according to the number of clusters specified.

2) Silhouette coefficients

The SCs are used to know the optimal number of clusters for the data. Then the SC is calculated for each ECP based on SFs. It is a measure that combines the agglomeration and separation on how close an ECP is to other ECPs in the same cluster. The SC for each ECP can vary from -1 to +1. Based on this criterion, an ECP is correctly grouped when the SC is close to 1 [14].

If majority of ECPs have a high silhouette value, the clustering is correct. The SC of each ECP calculation is based on the equation (3) [21], [22]:

$$S_i = \frac{bi - ai}{max(ai, bi)} \qquad (3)$$

where:

- $\boldsymbol{a_i}$ represents the average distance value determined from the i-sample to all data in the same cluster.
- $\boldsymbol{b_i}$ represents the minimum value of the average distances determined from the i-sample to all data in other clusters.
- $\boldsymbol{S_i}$ represents the SC for the i-sample.

From the equation (3), the average value of SCs is obtained according to the number of clusters with which the segmentation procedure was performed. The highest average value indicates the appropriate number of clusters to be applied [23] as shown in Figure 1.

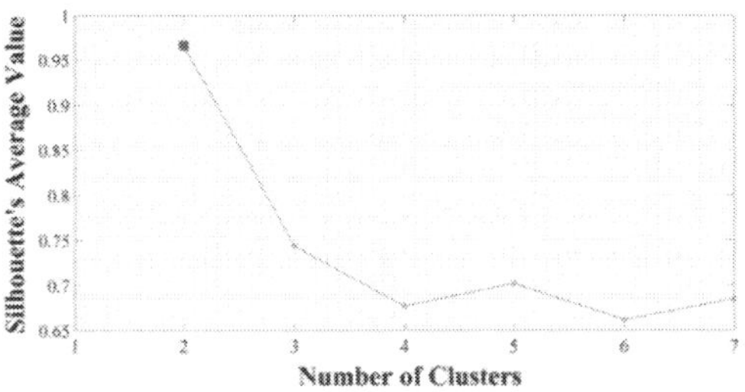

Fig. 1. Average silhouette values according to the clusters' number.

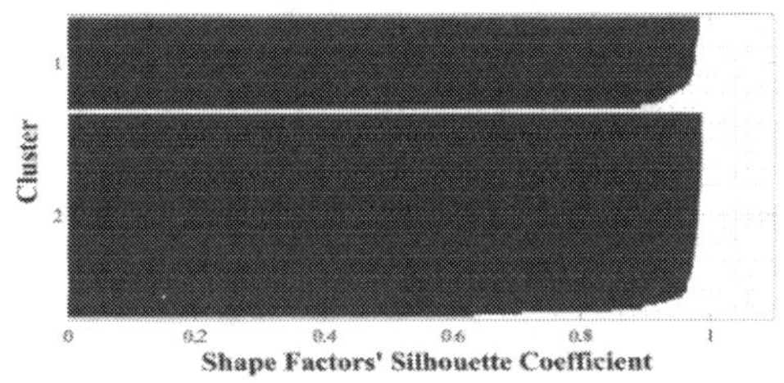

Fig. 2. SCs for each observation with 2 clusters.

3 Results analysis

The results analysis is made based on the measurements obtained from the main connection of the *Universidad Politécnica Salesiana (UPS), Sede Cuenca, Ecuador*.

3.1 Characterization of ECPs

In Figure 3, Mondays ECPs are show; while Figures 4 and 5 show the association of the SFs characterize every single ECPs in 2D and 3D correspondingly, as show in Table 3, for the reason that the 23 SFs cannot be displayed in a single figure.

In this way, Figure 4 and Figure 5 show the association of SFs in two elementary clusters. This is corroborated by determining the highest average value of the SCs.

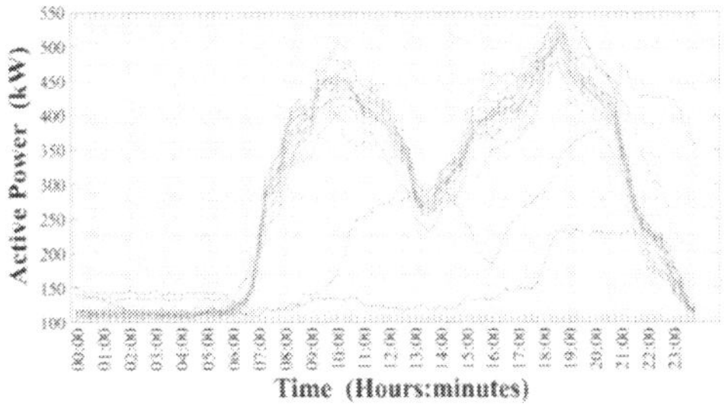

Fig. 3. Electric load profiles (Mondays).

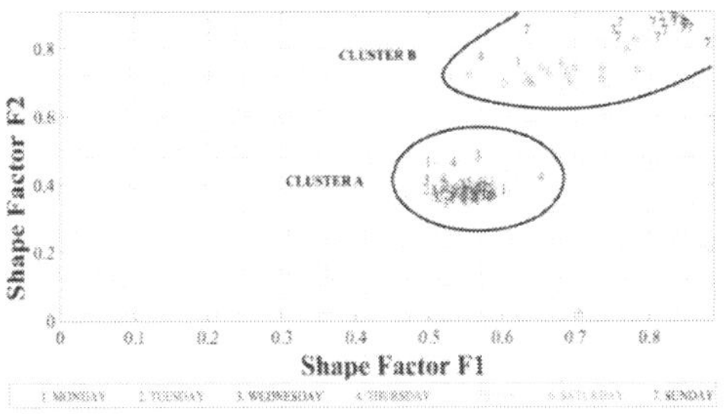

Fig. 4. SFs association (F1-F2).

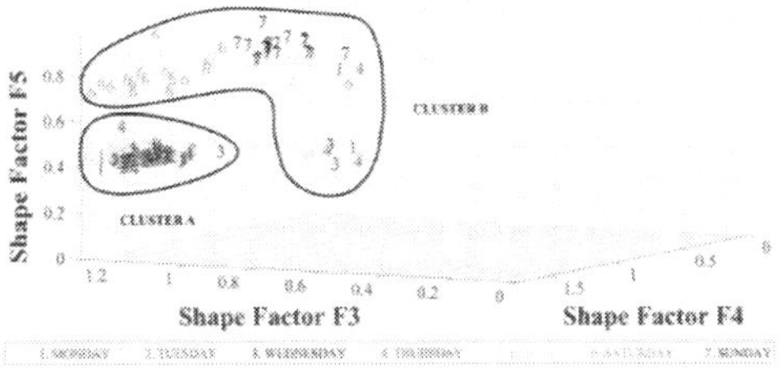

Fig. 5. SFs association (F3-F4-F5).

3.2 The k-means clustering technique application

The *k*-means application is carried out with 126 ECPs. According to the available data and the Sturges Rule, it is determined that the maximum number of possible clusters is 7.

As follows, the average values of the SCs for each possible cluster are obtained. This allows the determination of the appropriate number of clusters for the analyzed ECPs. Table 4 shows the average values of the SCs according to the number of clusters. The highest average value of the SCs shows the most appropriate number of clusters. Therefore, the 126 ECPs analyzed are segmented into 2 clusters.

Table 4. Average values of the SCs according to the number of clusters.

Clusters	Average value of the SCs
1	-
2	0.9663
3	0.7435
4	0.6768
5	0.6789
6	0.6618
7	0.6847

In this case, cluster A contains 87 profiles, while cluster B contains 39 profiles. These profiles represent workdays and nonworking days correspondingly.

3.3 ECPs classification

In order to carry out a correct ECPs classification, outliers are identified inside each cluster [24].

After applying unsupervised support vector machines and obtaining the best separation hyperplane between all the data inside each cluster [18, p. 53], 11 atypical profiles are detected in cluster A (Figure 6) and 10 in cluster B (Figure 7).

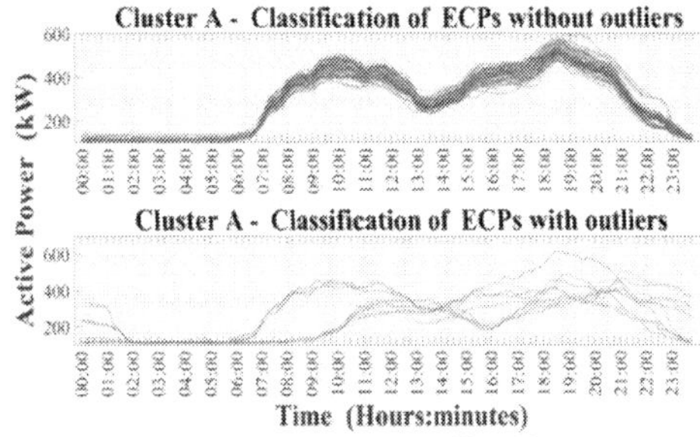

Fig. 6. Cluster A's ECPs, 76 typical profiles (above), 11 atypical profiles (below)

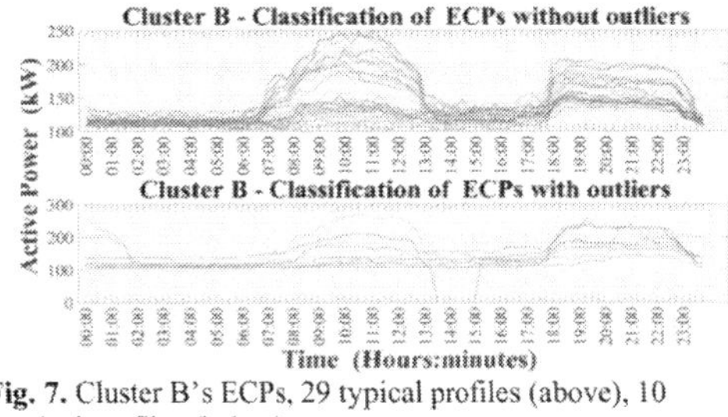

Fig. 7. Cluster B's ECPs, 29 typical profiles (above), 10 atypical profiles (below).

3.4 Obtaining patterns

The set of typical ECPs define the electrical consumption pattern. Figure 8 shows the pattern of workdays (cluster A), while Figure 9 shows the pattern of nonworking days (cluster B).

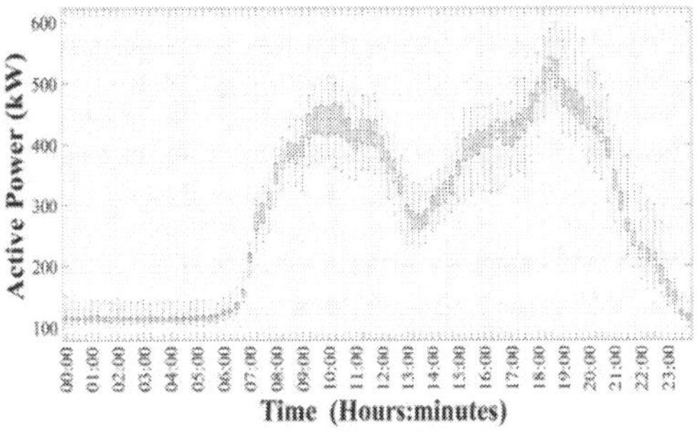

Fig. 8. Cluster A ECPs' pattern box-plot.

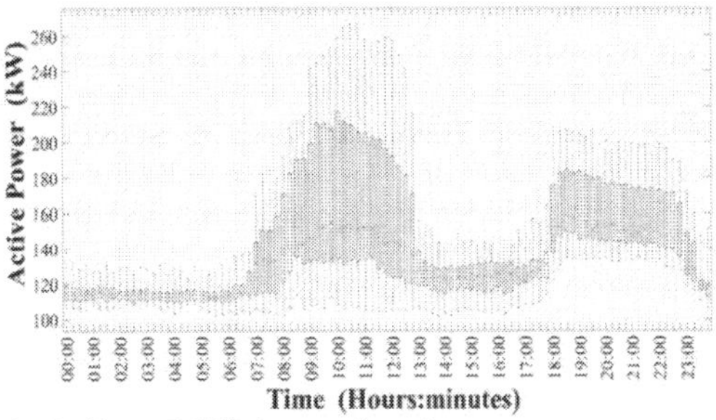

Fig. 9. Cluster B ECPs' pattern box-plot.

The electric consumption characterization allows the ECPs recognition at a measurement point. From each ECP, their respective SFs are defined, which are the basis for the clustering process. Next, atypical profiles are identified and separated. The typical profiles of each cluster represent the usual demand of the installation, that is, a consumption pattern. Consumption patterns are useful in several areas, such as forecasting and predicting demand [25], energy management, efficiency policies implementation and energy intelligence, improving the supply of tariffs, etc.

4 Conclusions

In this paper, a characterization method was applied for clustering of ECPs at UPS Cuenca, Ecuador by using SFs and the application of the *k*-means technique.

The SFs definition allows the characterization of the ECPs. This requires criteria and intuition. The *k*-means technique allows clustering the SFs in such way that the ECPs are clustered according to their similarity.

The appropriate number of clusters in which the ECPs are segmented is defined through the SCs criterion, which allows to simplify the process and improve the computational performance.

The anomalies detection process allows to obtain SCs values closer to 1 in each cluster, which makes it possible to define clearly the electric consumption characteristics. This way the understanding of the structure that they possess improves considerably.

Traditionally, most electric companies used to classify their consumers using minimum electrical parameters and some commercial regulations. This way, current electricity markets need to classify their consumers by indicators capable of characterizing their true electrical behavior.

The method allows obtaining the average daily energy consumption pattern of each cluster. This pattern is obtained from the typical profiles of each cluster and allows the determination of the kind of consumer to which it belongs; as well as defining tariffs that are clear, transparent and easy to understand, but flexible enough to follow the variations in the consumer's load pattern induced by specific rates.

If the consumption patterns are known by the consumer, they could be interested in managing the energy they use, by improving their energy performance, achieving economic benefits and decreasing their environmental footprint.

One of the limitations of the k-means method is the sensitivity to outliers.

In addition, a problem presented by the method applied in this paper is summarized in the precision with which the data can be clustered. The method works efficiently with large databases but when there are abrupt changes in the ECPs, the method can become imprecise, since SFs will be affected.

References

1. G. Chicco, R. Napoli, P. Postolache, M. Scutariu, and C. Toader, "Electric energy customer characterisation for developing dedicated market strategies," *2001 IEEE Porto Power Tech Proc.*, vol. 1, pp. 371–377, 2001.
2. G. Chicco, "Customer Behaviour and Data Analytics," *2016 Int. Conf. Expo. Electr. Power Eng.*, no. Epe, pp. 771–779, 2016.
3. J. Molina and J. García, "Técnicas de Análisis de Datos," p. 266, 2006.
4. U. Fayyad, G. Piatetsky-Shapiro, and P. Smyth, "From Data Mining to Knowledge Discovery in Databases," *AI Mag.*, pp. 37–54, 1996.
5. G. Chicco, R. Napoli, P. Postolache, M. Scutariu, and C. Toader, "Customer Characterization Options for Improving the Tariff Offer," *IEEE Power Eng. Rev.*, vol. 22, no. 11, p. 60, 2002.
6. G. Chicco, "Overview and performance assessment of the clustering methods for electrical load pattern grouping," *Energy*, vol. 42, no. 1, pp. 68–80, 2012.
7. G. Chicco, R. Napoli, and F. Piglione, "A Review of Concepts and Techniques for Emergent Customer Categorisation," pp. 51–58, 2002.
8. R. O. Duda, P. E. Hart, and D. G. Stork, "Pattern Classification," *New York: John Wiley, Section.* p. 654, 2000.

9. J. Valente de Oliveira and W. Pedrycz, *Advances in Fuzzy Clustering and its Applications Advances in Fuzzy Clustering and its Applications*. 2007.
10. X. Serrano-Guerrero, G. Escrivá-Escrivá, and C. Roldán-Blay, "Statistical Methodology to Assess Changes in the Electrical Consumption Profile of Buildings," *Energy Build.*, vol. 164, pp. 99–108, 2018.
11. Consejo Nacional de Electricidad, "Estudio y gestión de la demanda eléctrica," *Plan Maest. Electrif. 2013-2022*, vol. 2, pp. 41–42, 2013.
12. G. Chicco, R. Napoli, and F. Piglione, "Application of clustering techniques to load pattern-based electricity customer classification," no. June, pp. 6–9, 2005.
13. M. Piao, J. B. Lee, H. S. Shon, E. J. Cha, K. Ah Kim, and K. H. Ryu, "Identification of temporal interval relation of frequent patterns during incremental phase," *Legacy*, pp. 497–502, 2011.
14. G. K. Gupta, *Introduction To Data Mining With Case Studies*. PHI Learning Pvt. Ltd., 2014.
15. G. Chicco, R. Napoli, and F. Piglione, "Application of Clustering Algorithms and Self Organising Maps to Classify Electricity Customers," *2003 IEEE Bol. PowerTech - Conf. Proc.*, vol. 1, pp. 373–379, 2003.
16. MathWorks, "k-Means Clustering," 2017. [Online]. Available: https://www.mathworks.com/help/stats/k-means-clustering.html.
17. J. Han, M. Kamber, and J. Pei, *Data Mining: Concepts and Techniques*. 2012.
18. S. M. H. Jansen, "Customer Segmentation and Customer Profiling for a Mobile Telecommunications Company Based on Usage Behavior," *Proc. - 3rd Int. Conf. Data Min. Intell. Inf. Technol. Appl. ICMIA 2011*, no. July, pp. 308–313, 2007.
19. P.-N. Tan, M. Steinbach, A. Karpatne, and V. Kumar, *Introduction to Data Mining*. 2012.
20. J. J. González, "El histograma con la TI-92: optimización de clases," *Revista de Didáctica de las Matemáticas*, vol. 61, pp. 67–72, 2005.
21. P. J. Rousseeuw, "Silhouettes: A graphical aid to the interpretation and validation of cluster analysis," *J. Comput. Appl. Math.*, vol. 20, no. C, pp. 53–65, 1987.
22. F. Scarlatache, G. Grigora, G. Chicco, and G. Câr, "Using k-Means Clustering Method in Determination of the Optimal Placement of Distributed Generation Sources in Electrical Distribution Systems," *IEEE Optim. Electr. Electron. Equip. (OPTIM), 2012 13th Int. Conf.*, pp. 953–958, 2012.
23. MathWorks, "silhouette," 2017. [Online]. Available: https://www.mathworks.com/help/stats/silhouette.html?s_tid=doc_ta.
24. D. M. J. Tax, "One-class classification," 2001.
25. X. Serrano-Guerrero, R. Prieto-Galarza, E. Huilcatanda, J. Cabrera-Zeas, and G. Escrivá-Escrivá, "Election of Variables and Short-term Forecasting of Electricity Demand Based on Backpropagation Artificial Neural Networks," *IEEE Int. Autumn Meet. Power, Electron. Comput.*, 2017.

E3S Web of Conferences **64**, 08005 (2018) https://doi.org/10.1051/e3sconf/20186408005
ICPRE 2018

Overcurrent Protections in MV Grid with Local Energy Sources

Józef Lorenc , *Bartosz* Olejnik, and *Aleksandra* Schött

Institute of Electrical Power Engineering, Poznań University of Technology, Poznań, Poland

Abstract. The article presents typical methods of connecting various types of local power sources to a medium voltage network. Describes the advantages and disadvantages of individual solutions. The authors focus particularly on the issues of power system protection, with particular emphasis on overcurrent protection (ANSI 50 and 67). The key part of the article is the presentation of the selection methodology, along with examples, of setting this type of protection for a frequently encountered case - connecting local energy sources with busbars with a line with receipts.

1 Introduction

The article applies only to networks with local power sources connected to medium voltage lines derived from HV/MS stations, whose main task is to supply consumers, and the value of power absorbed from the 110 kV network is usually much higher than the power generated by local sources (the local power plant designation LPP will be used later in the text). The properties of such systems do not depend on the type of primary energy used (water, wind, biomass), but on the type of generators and network parameters.

The principles given in this article should be treated as a proposal, and the substantive value should be assessed independently. It seems that only exploitation will show the superiority of some theories after the failure of the network with LPP and their correct or incorrect liquidation. An additional difficulty is often poor recognition of the automation installed in the LPP itself due to the lack of data or deliberate operation of the manufacturer.

Relay settings in networks with LPP are a big problem for most engineers dealing with power system protection. In this publication, the author's methodology for selecting the overcurrent protection settings in Poland will be presented. It is noted that due to different network configurations, the dependencies given here may not be appropriate for cases in other countries.

Due to the author's approach to the topic, the number of publications cited is small.

2 Types of local energy sources and their connection methods

From the point of view of the type of generators, the following division can be made [1,2]:

- synchronous generators working synchronously with the network, i.e. connected directly to the network (without rectifying-wave circuits) - the most dangerous, because they cause an increase in short-circuiting powers and the danger of maintaining the so-called load island,
- asynchronous generators - less dangerous, but still generate short-circuit current - usually in a relatively short time, because they basically require cooperation with a source of passive power,
- synchronous generators working asynchronously with the network, i.e. indirectly connected (with rectifying-wave systems) - the most safe, but rather found in large wind farms, the article does not apply,
- DC power sources with inverters, i.e. photovoltaic sources, just as safe as listed in point c and this article is not considered.

Regarding the sources given in point b, it should be noted that various behaviors are possible. In experimental studies, in which Poznan University of Technology participated, it was shown that in such systems short-circuit current can be generated even for a few seconds.

In this work, only connections according to the diagrams given in Figures 1 and 2 will be considered. In Figure 1 LPP are connected to the line with receivers, i.e. MV/LV stations. The connection method can be very different, in Fig.1a there is one power plant with one common transformer and several generators (this is often the case for hydroelectric plants), and in Fig.1b several power plants connected in different places of the line (this may be for small wind farms)). White rectangles mean "protection point" hereinafter referred to as PP, which consists at least of transformers (CT, without/and/or VT), protection and circuit breaker.

In Figure 2, LPPs are connected to the so-called subscriber line, it has no load. There may be systems that differ from the number of connected sources or other details, but these differences will not affect the overall distribution.

Equipment with power protection can also be different. For example - for the system in Fig. 1a just before the PP-E protection point from the line side, a remotely controlled recloser may be installed [3], which is also equipped with power protections. In the same example, there is no need to be a PP-T security point.

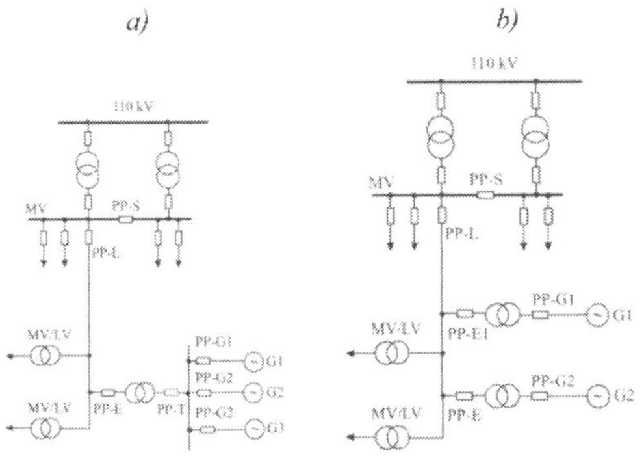

Fig. 1. Connection of LPP to the line with receipts.
Explanations for Figures 1 and 2: PP-L - the protection point of the line in the transformer-switch station, PP-S - trunk connector, PP-E - local power plant, PP-G - generator.

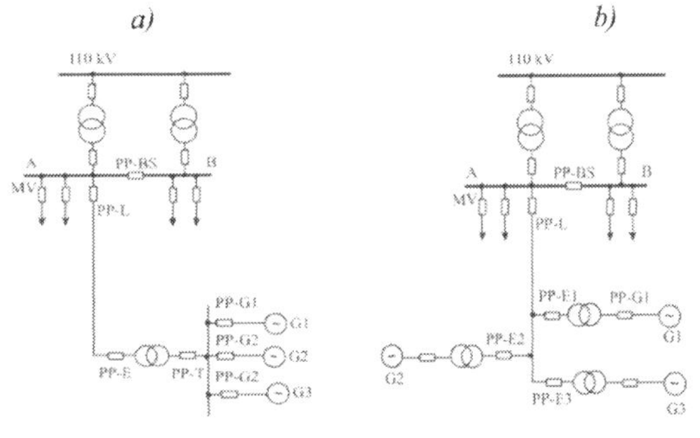

Fig. 2. Connection to the subscriber line.

3 General requirements

Power system protection systems in LPP networks put the basic requirements, i.e. selectivity and discrimination, speed and reliability of operation, sensitivity, economy, i.e. the same as typical systems for classic networks. In addition, they also put up some specific requirements:

- enabling the execution of a correct autoreclosure (ARC) cycle in nearby overhead lines,
- enabling the execution of the automatic transfer switching (ATS) cycle between the 110 kV/SN station sections,
- correct operation of busbar protection (BP) and local circuit breaker (LCBR) reserve in HV/MV station,
- protection against the effects of island work (especially in terms of lowering or increasing voltage or the quality of electricity).

During the analysis, the assumption was made that the fields in the HV/MV substation are equipped with typical security and automation: ARC, ATS, automatic load shelding, LCBR and overcurrent logic busbar protection. Short-circuit and short-circuit overcurrents are used against the effects of phase-to-phase short-circuits, and they are based on zero current and voltage components and their associated quantities.

Depending on the technical details, including the type of protections used, but also their settings, there may be the following risks from local power plants [4]:

- creation of a load island with improper electricity parameters; the increase of electric voltage is particularly dangerous,
- the possibility of unnecessary line disconnections with local energy sources during short circuits in other parts of the network,
- obstructing ARC automation by supplying voltage and burning of an electric arc,
- obstructing the implementation of the ATS automation cycle, if the MV switchgear section is switched,
- unnecessary blocking of busbar protection during short-circuits on busbars,
- difficulties in implementing the LCBR.

In this work, the principles of selection of overcurrent protections and their settings for networks with lines to which both receipts and LPP are connected are presented.

4 Phase-to-phase short-circuits in the MV network

4.1 Overcurrent protections

The risk assessment and some suggestions regarding the possibility of avoiding them during phase-to-phase short-circuits will be carried out on the basis of the diagram shown in Figure 3. The short circuits for which the analysis will be carried out are marked with symbols from K1 to K4. In each of these short-circuits short-circuit current is the result of superposition of two components: flowing from the power system side (I_k^S) and from the LPP side (I_k^E).

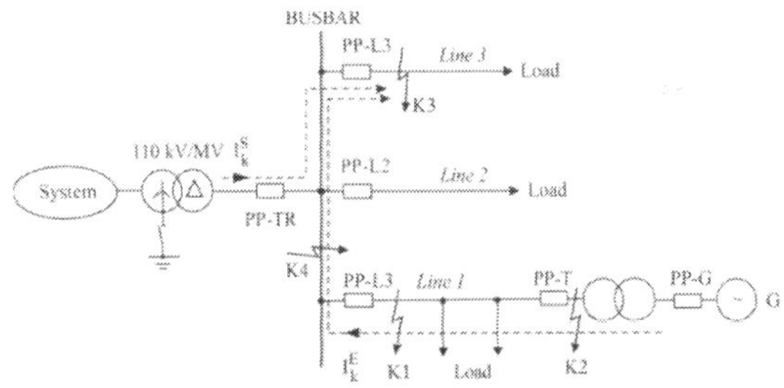

Fig. 3. Operation of overcurrent protection against the effects of phase-to-phase faults.

It is proposed to use directional blocks and exemplary time settings according to the proposal shown in Fig. 4, where each short circuit in the line and LPP will be switched off selectively. At points PP-L3, PP-T and PP-G, two delayed-current protection are proposed: one with a longer time without directional blocking (ANSI 50) and the other with a shorter time, with a directional block (ANSI 67) [5,6]. The problem may be to provide directional blocks for overcurrent protection. IEDs are generally equipped with this function, but not always in every place of the network there are voltage transformers necessary for its proper operation

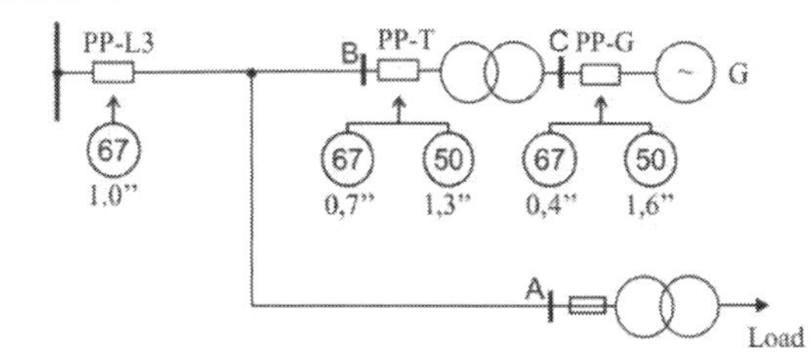

Fig. 4. Proposal of selecting delayed overcurrent protection (50/67) for the line with generation and consumers.

At point PP-L3, the current setting value should be selected according to the following:

$$I_{set} \geq \frac{k_b k_r I_{max}}{k_p \vartheta_i} \tag{1}$$

where:

- I_{nast} - current setting of the protection,
- k_b - a safety factor that can be taken equal to 1.1-1.2,
- k_r - coefficient of self-starting of the motors supplied from the protected section of the network, depending on the content of the motor load in the overall load, the range of its value is 1-4,
- I_{max} - the maximum load current of the protected line segment, depending on the LPP power, it may be the current resulting from its rated power or the current resulting from the load parameters, for example inactive LPP,
- k_p – resetting ratio,
- ϑ_i –current transformer ratio.

Sensitivity should be checked according to (2), where:

- I_{kmin} - the minimum current flowing through the protection point during a short circuit at the end of the protected section (it is always a two-phase short-circuit current for the network system when it is the smallest - eg in summer),
- k_c - sensitivity factor which must be taken equal 1.5 for the basic protection, and 1.2 for the reserve protection.

$$I_{set} < \frac{I_{kmin}}{k_c \vartheta_i} \tag{2}$$

Sensitivity should be checked for the following cases:

- in places A and B with the required kc = 1,5,
- in place C at kc = 1,2 (may be a problem with this requirement).

For PP-T and PP-G security you should use (1) and check the sensitivity in (2). For PP-T it is necessary to check the sensitivity for point A taking into account the current flowing only from the G side and for shorting on the busbars 110 kV/MV at k_c = 1.5 and for point C taking into account the current flowing from the power system side also at k_c = 1.5. For PZ-T, check the sensitivity for short-circuit in G considering the short-circuit current flowing from the system, and in points C and B from the side G.

If near PP-T, the so-called recloser, it will not be possible to get selectivity, because there will be no time grading [5]. You can increase the time delay in PPL-3 to, for example, 1.3 or 1.5 s and get this option. It seems that both the recloser and the PP-T security can be set the same way and do not require selectivity between these points, because the physical distance between them will be very small. The task of the recloser is usually to obtain the ability to remotely disable LPP by the disposition without interfering with devices not owned by the network operator [7].

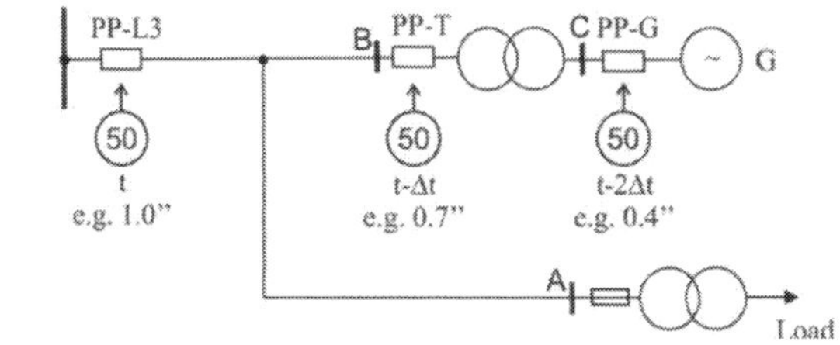

Fig. 5. Simplified variant of the selection of delayed overcurrent protections (50) and their time settings for the line with loads.

If it is not possible to use directional interlocks, it is suggested to select the protections as in Fig.5. At the same time, more general expressions were given to the selection of the time delay. Grading started from the setting in the linear field of the switchboard, because as already mentioned, increasing this setting is extremely difficult and inadvisable. The given values are just an example. In this situation, any phase-to-phase fault in the cooperating MV network may result in the LPP being turned off. The setting values of the currents should be selected in accordance with (1) and (2). This is not a solution consistent with the classic principles of protection automation, because it has no selectivity, but it is simple and the short-circuit will be turned off. It should be anticipated that in the case of a phase-to-phase short-circuit in another line, the LPP will first be switched off, and only then the damaged line will be disconnected.

4.2 Short-circiut protections (50)

Short-circuit protection is required in PPL-3 for the line and at least in PP-T for the power plant transformer.

The short-circuit protection in PPL-3 should be set so that its range ends before the PP-T point. This requirement can be achieved by choosing the setting according to the formula

$$I_{set} \geq \frac{k_b I_{kmax}}{\vartheta_i} \tag{3}$$

where:

- k_b - safety factor, selected from 1.2 to 1.6,
- I_{kmax} - maximum short-circuit current flowing through the PPL-3 from the power system side during short-circuit near PP-T, it will be a three-phase short-circuit current at the highest short-circuit power on the 110 kV/MV station busbars.

The time delay of this protection can be in the range of 0.1-0.3 s, it is difficult to indicate a clear selection rule. To obtain the maximum value of this time, the heat resistance of overhead lines is sometimes calculated.

It is also necessary to check if in case of shorting on busbars current flowing to them from the G side will not trigger the short-circuit protection at point PP-L3. It is not about the missing operation during shorting on busbars, because it will lead to switching off the LPP, but during short circuits at the beginning of the other lines led out from the 110 kV/SN station. Therefore, the dependence (3) should be checked when inserting as I_{kmax} the value of the maximum short-circuit current flowing from the LPP side when shorted on busbars. If the requirement is not met, which is rare, use the directional interlock (ANSI 67) for the short-circuit protection at PPL-3. Failure to meet the requirements will only occur if the current flowing from the LPP side is greater by at least 20% higher than the current flowing from the power system side when shorted on busbars. The occurrence of this condition is possible with a small electrical distance LPP from busbars and high power (on the order of several MW).

The short-circuit protection of the transformer at the PP-T point, which is its protection during internal faults, should be set according to the higher value obtained from two dependencies:

$$I_{set} \geq \frac{k'_b I_{nT}}{k_p \vartheta_i} \quad (4)$$

$$I_{set} \geq \frac{k''_b I_{kmax}}{k_p \vartheta_i} \quad (5)$$

in which:

- I_{nT} - transformer rated current,
- I_{kmax} - maximum short-circuit current on rails behind the transformer,
- k_b' - safety factor including the impact of the magnetizing current at switching the transformer to idling (= 4-8),
- k_b'' - safety factor for resetting the current setting of this protection from the delayed protections located behind the transformer, i.e. practically in PP-G (= 1.2-1.6).

Dependence (5) is to protect the transformer from unnecessary disconnections from the impact of the magnetizing current. It is assumed that it is energized only from the network side and the generator (s) are synchronized to it. The value of the coefficient is dependent on the time delay, which should be within 0.1 - 0.3 s, the longer the time, the coefficient may be smaller.

If the system does not have a directional block for delayed overcurrent protections, the principle of selecting short-circuit protection does not change.

5 Summary

The article presents typical LPP connection systems with the medium voltage network used in Poland, also giving the advantages and disadvantages of specific solutions. It seems that in other countries, the problem of connecting LPP, especially with relatively low power (up to several MW), will be based on similar assumptions.

Due to limitations in the publication, the focus was only on the problem of setting the overcurrent and short-circuit overcurrent protection (ANSI 50), also with the directional components (ANSI 67).

The connection of LPP to the MV line and switchgear carries many threats to the correct operation of power system protection in the cooperating network. The article presents a few of these threats, of which it is necessary to be aware.

When selecting protection and their settings, special attention should be paid to the protection of recipients. It seems that the worst disruption would be an increase in voltage that can cause damage - especially electronic equipment.

In relation to delayed overcurrent protection in MV networks, the principle should be formulated that in the case of using digital constructions (terminals or field controllers) it is recommended to graduate temporarily every 0.3 s, and not as for older electromechanical or static analog solutions - 0.5 s. This principle is not related to the existence of the LPP network, but has a high degree of generality. Its introduction will shorten the duration of phase-to-phase faults, and sometimes improve the selectivity of security in the network.

The conclusion from the main thread of the article is the thesis that the connection of a local power plant to the line with the recipients worsens the conditions of their supply, especially by increasing the probability of interruptions in energy supply.

References

1. J. Martínez, P. C. Kjær, P. Rodriguez and R. Teodorescu, "Short circuit signatures from different wind turbine generator types," 2011 IEEE/PES Power Systems Conference and Exposition, Phoenix, AZ, 2011, pp. 1-7.
2. E. Muljadi, N. Samaan, V. Gevorgian, Jun Li and S. Pasupulati, "Short circuit current contribution for different wind turbine generator types," IEEE PES General Meeting, Minneapolis, MN, 2010, pp. 1-8.
3. J. Lorenc, J. Andruszkiewicz, B. Olejnik, B. Staszak and P. Balcerek, "Earth fault detection and isolation system for MV network," 2015 Modern Electric Power Systems (MEPS), Wroclaw, 2015, pp. 1-7.
4. T. Gjengedal, "Integration of wind power and the impact on power system operation," Large Engineering Systems Conference on Power Engineering, 2003, 2003, pp. 76-83.
5. IEEE Standard Electrical Power System Device Function Numbers, Acronyms, and Contact Designations," in IEEE Std C37.2-2008 (Revision of

https://doi.org/10.1051/e3sconf/20186408005

IEEE Std C37.2-1996) , vol., no., pp.1-48, Oct. 3 2008.

6. S. A. Gopalan, V. Sreeram, H. H. C. Iu, "A review of coordination strategies and protection schemes for microgrids", *Renewable and Sustainable Energy Reviews*, vol. 32, pp. 222-228, 2014.
7. P. I. Santos e Abreu and A. G. Martins, "Assessment of the behavior of protection systems in radial networks with distributed generation," 2016 51st International Universities Power Engineering Conference (UPEC), Coimbra, 2016, pp. 1-6.

Outlier Detection Methods for Uncovering of Critical Events in Historical Phasor Measurement Records

André Kummerow , *Steffen* Nicolai, and *Peter* Bretschneider

Department of Energy, Advanced System Technology (AST) Branch of Fraunhofer IOSB, Illmenau, Germany

Abstract. The scope of this survey is the uncovering of potential critical events from mixed PMU data sets. An unsupervised procedure is introduced with the use of different outlier detection methods. For that, different techniques for signal analysis are used to generate features in time and frequency domain as well as linear and non-linear dimension reduction techniques. That approach enables the exploration of critical grid dynamics in power systems without prior knowledge about existing failure patterns. Furthermore new failure patterns can be extracted for the creation of training data sets used for online detection algorithms.

1 Background and main objectives

Phasor measurement units (PMU) are widely used in largescale power grids to enhance static SCADA information with high-resolution (up to 60 f.p.s.) phasor and frequency data. PMUs are interconnected into wide area monitoring systems (WAMS) and increase situational awareness of modern control centres by tracking dynamic events (e.g. abrupt changes, critical oscillations) and perform real-time analytics – see Fig. 1. The detection and classification of grid disturbances (e.g. generator loss, load loss, line outages, system oscillations) from real field measurements is currently a main application field of PMUs in transmission and distribution power systems. [1–5]

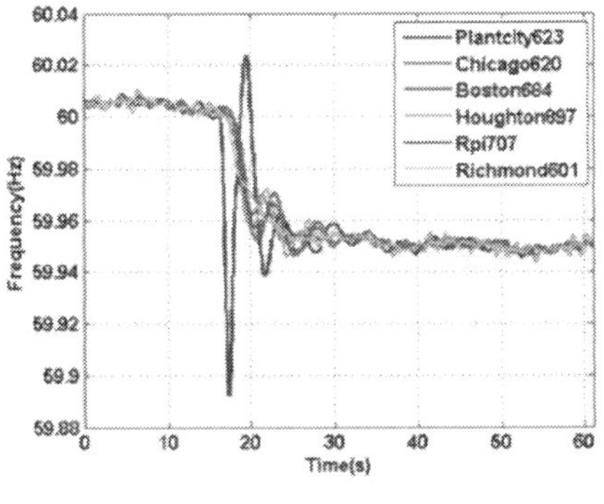

Fig. 1. Example of a generator loss recorder by PMUs from [4].

In literature [6–9] several concepts have been proposed for the identification of grid disturbances from PMU data. Several methods from multivariate statistics (e.g. Principle Components Analysis or K-Means clustering) are combined with techniques from frequency transform (e.g. Wavelet Decomposition, Stockwell Transform) and modal analysis (e.g. Prony Analysis, Matrix Pencil Method, Empirical Mode Decomposition). Those studies heavily rely on sufficient disturbance samples for different failure types. Usually, the training samples are generated from dynamic simulations or from the analysis of disturbance record files.

This survey presents a new method for the uncovering of arbitrary grid disturbances from historical PMU data sets when no prior knowledge about grid dynamics or critical events is available. In contrast to previous works this approach is developed for the analysis of mixed and unbalanced PMU data sets, where only a small number of samples is assumed to be critical. The automated analysis procedure deals with data sets from single PMU sensors and uncovers potential failure patterns (e.g. voltage sags, oscillations, frequency drops) from real field measurements. Thereby grid dynamics and disturbances can be explored as well as potential relationships between critical events.

The general concept for the use of outlier detection methods to extract disturbances from mixed PMU data sets is shown in part 2. Additionally, some theoretical background is provided for the most important learning algorithms within this survey. Part 3 presents and discusses the main results from the extraction of time and frequency features as well as the different outlier detection methods.

2 General concept and outlier detection methods

The uncovering of critical disturbances from mixed and imbalanced PMU data sets can be addressed to the field of score-based, unsupervised outlier detection.

For this task several algorithms exist to identify data which differs from expectations or majority behaviour. Successful applications include e.g. fraud or thief detection. The available techniques can be distinguished roughly into probabilistic, distance-based, reconstruction-based, domain-based and information-theoretic approaches [10, 11]. The general concept for PMU outlier detection is shown in Fig. 2.

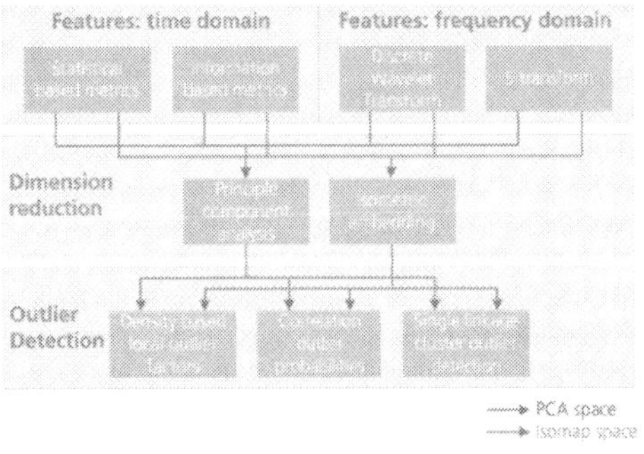

Fig. 2. Three-phase concept for PMU outlier detection.

Table 1. Comparison of outlier detection methods.

Method	Main principle	Metric used as outlier score
LOF	local densities of data points and its neighbourhood	local outlier factor
COP	deviation within local correlation using robust PCA	correlation outlier score
SiLiOd	hierarchical clustering using shortest distances	path lengths to final cluser

As a preprocessing step the historical data set is split into PMU samples of equal length. For each sample features are extracted in time and frequency domain using statistical and information based metrics as well as coefficients from Discrete Wavelet Transformation (DWT) and Stockwell transformation or S transform [12–14]. The features have to be designed carefully to capture the signal behaviour in presence of critical events. In a second step the feature space is reduced with linear and nonlinear dimension reduction techniques. This increases the performance of the outlier detection algorithms and redundant features can be eliminated. To preserve global data structures Principle Component Analysis (PCA) and Isometric Embedding (Isomap) are used. In the last step the outlier detection algorithms compute an outlier score for each PMU sample. Within this survey three methods are used: Local Outler Factors (LOF), Correlation Outlier Probabilities (COP) and Single Linkage Outlier Detection (SiLiOd). All of those methods generate different metrics as outlier scores. A short summary is given in Table I. As an unsupervised approach the number of outliers to be extracted has to be defined in advance. This can be seen as the amount of expected contamination of the PMU dataset and has to be selected by trial and error.

The first method is a well-established technique for outlier detection and calculates an outlier factor for each data point [15, 16]. The local densities of the data points are compared with the densities of its neighbours.

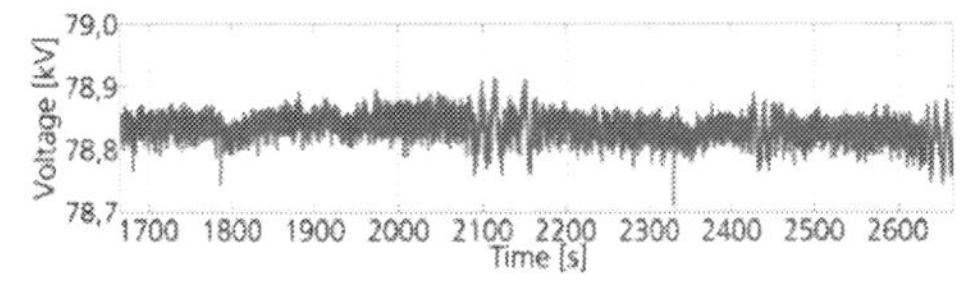

Fig. 3. Voltage magnitude raw signal.

Using the LOF method, outliers are treated as data points with high differences between the density of that point and its surroundings, so that LOF >>1. In contrast to the previous approach, COP (correlation outlier probabilities) is a subspace-based outlier detection technique using PCA as local correlation model [17]. Outliers are defined as data points that significantly differ from the local correlation within a defined neighbourhood. In addition to the two above mentioned techniques an own implementation of a cluster-based outlier detection algorithms is applied using hierarchical agglomerative clustering (HAC). At each iteration step the two clusters are merged which are closest to each other. This procedure is repeated until one final cluster is left. In single linkage clustering the similarity between two clusters c_i and c_j is defined as the distance D between the two closest cluster members x_i and x_j – see (1):

$$D(c_i, c_j) = \min(|\underline{x}_i - \underline{x}_j|) \tag{1}$$

The generated cluster tree can be seen as a bipartite graph G with vertices V and edges E. From that, the corresponding adjacency matrix A is computed according to (2):

$$\boldsymbol{A} = [a_{i,j}]\ with\ a_{i,j} = \begin{cases} 1\ when\ (i,j) \in \mathcal{E} \\ 0\ otherwise \end{cases} \tag{2}$$

That adjacency matrix can be used to detect outliers by calculating the shortest paths or minimum number of edges between the data points and the final cluster group. Those data points with short path lengths are treated as outliers. This method is called Single-Linkage Based Outlier Detection (SiLiOd). Further literature for agglomerative clustering methods and single linkage clustering can be found in [18].

3 Analysis results and discussion

3.1 Experimental setup

This study uses a record from real measured PMU data which includes voltage magnitudes from different PMU sensors [19]. Fig. 3 shows an excerpt from the measurements of one PMU sensor. The reporting rate is 30 frames per second. In a pre-processing step the PMU raw data is filtered for high-frequency noise and normalized. Table II gives an overview of the main parameters and specifications for the used methods. The general calculations and discrete Wavelet decompositions are done in [20]. For the S Transform an open-source code is used [13]. The Isometric embedding (Isomap) is taken from an open-source Java based machine learning library [21].

TABLE I. CHOOSEN PARAMETERS FOR OUTLIER DETECTION

Parameter	Value
Number of outliers	15
S Transform max. frequency	1 Hz
DWT decomposition level	12
DWT wavelet function	db5
Isomap number of neigbours	50
SiLiOd distance function	Mahalanobis
LOF / COP distance function	Euclidean
LOF / COP number of neighbours	50

The Local Outlier Factor (LOF) and Correlation Outlier Probabilities (COP) method is taken from the Java based ELKI (Environment for Developing KDD-Applications Supported by Index-Structures) data mining framework [22].

3.2 Feature extraction in time and frequency domain

According to Fig. 2 in the first stage features are extracted from the PMU samples in time and frequency domain. Within this study the total feature space consists of 16 different features F1 to F16. Typical disturbance patterns (e.g. oscillations, line or generator trips, voltage sags) are characterized by high signal variations and abrupt signal changes.

Those signal traits can be captured by statistical moments like variance (F1), skewness (F3) and kurtosis (F2), model-based metrics like linear slope (F4) or information based metrics like sampled entropy (F5). Additional information are gained by time-frequency representation of the PMU samples. The features are derived from the energy values of the absolute Stockwell Transform (S-Transform) coefficients and the energy values of the detailed coefficients from multi-level Wavelet decomposition. Fig. 4 shows an example for the S-Transform of a PMU sample. The S-Transform derived features include mean energy (F6), maximum energy along time axis (F7) and frequency axis (F8) as well as energy variance along time axis (F9) and frequency axis (F10).

The DWT derived features include mean energy (F11), energy variance (F12) and maximum energy (F13) over all decomposition levels. For a better understanding Fig. 5 and Fig. 6 show some exemplary time and frequency domain features.

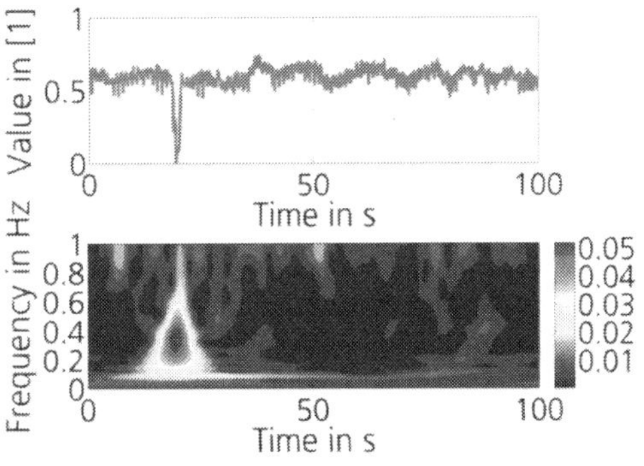

Fig. 4. S-Transform absolute coefficiencts for a PMU sample

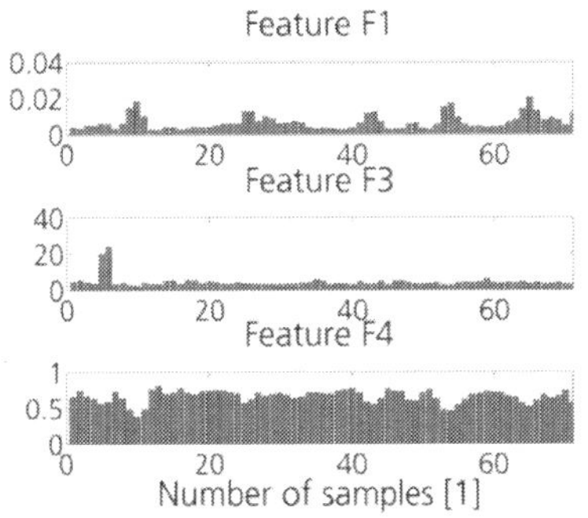

Fig. 5. Results for time domain features F1, F3 and F4

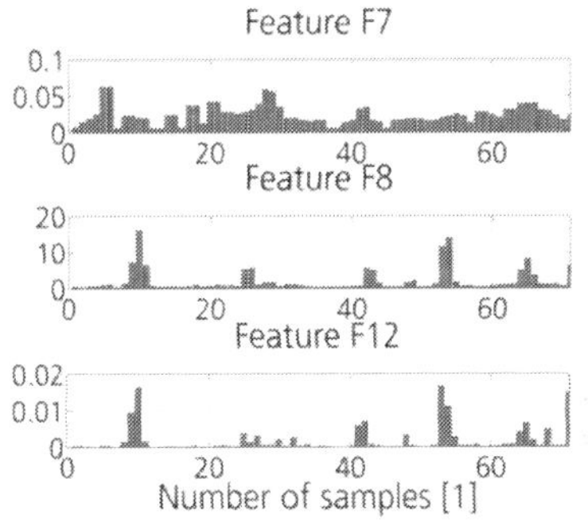

Fig. 6. Results for frequency domain features F7, F8 and F12.

3.3 Outlier scores and disturbance patterns

The following investigations refer to the extracted features in time and frequency domain for one PMU sensor. The results are mapped into a two-dimensional space using PCA and Isomap – see again Fig. 2. Afterwards the three outlier detection methods LOF, COP and SiLiOd calculate an outlier score for each PMU sample. Given that, for each feature space there are 6 different outlier detectors:

- Detector 1,2: PCA / Isomap + LOF
- Detector 3,4: PCA / Isomap + COP
- Detector 5,6: PCA / Isomap + SiLiOd

As noted in Table II the number of outliers to be extracted is set to 15 given a total number of 71 PMU samples. In case of LOF and COP the number of nearest neighbours substantially determines the results. Fig. 7 show the outlier scores from detector 1 for different numbers of nearest neighbours using the time-domain features.

The reduced two-dimensional space is spanned by the first and second principle component (PC) of the PCA. The detected outliers are highlighted in red. High outlier scores with LOF >> 1 indicate a high distance between the data point and its surroundings. Far-away outliers are detected successfully by LOF. When increasing the number of nearest neighbours the outlier scores are changing especially for data points with a low distance to its surrounding. Within this survey the number of nearest neighbours is set to 50 which corresponds to more than half of the total PMU samples.

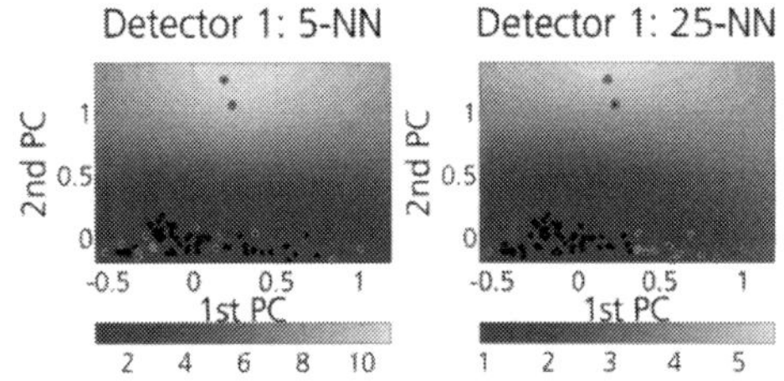

Fig. 7. Outlier scores for detector 1 (time domain features).

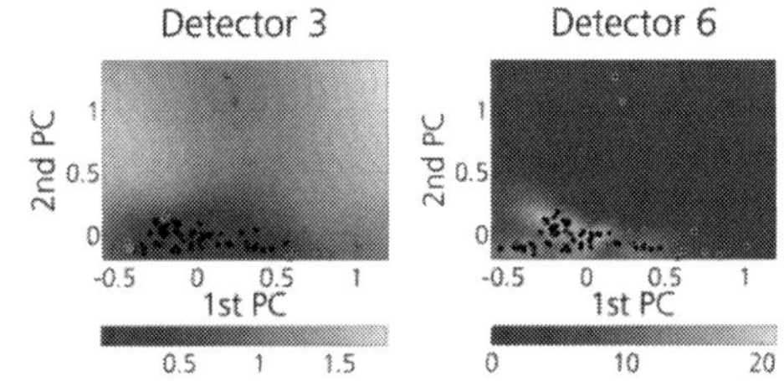

Fig. 8. Outlier scores for detector 3 and 6 (time domain features).

Fig. 8 presents the COP and SiLiOd outlier scores for the same data set based on the PCA mapping (detector 3 and 6). The COP outlier scores show similar results compared to the LOF outlier scores using a number of nearest neighbours of 50. This parameter is fixed for all subsequent investigations. In case of SiLiOd low outlier scores or small path lengths correspond to a high outlier degree. The metric is inversely proportional to the metrics from LOF or COP. Within this survey the Mahalanobis distance is used to create the cluster tree – see again Table II. In Fig. 9 the aggregated results from the 6 outlier detectors are presented within ranking matrices. A low rank indicates a high outlier degree and vice versa. The ranking matrices show the top 15 outliers over all PMU samples. For far-away outliers with ranks from 1 to 4 low variations between the results can be observed which indicates high certainty. This accounts for the sample numbers 6, 7 and 65, 66. In contrast to that, near-by outliers with high ranks are widely spread among the PMU samples indicating low certainty of the results. Some high-rank examples can be detected for the sample number 1 and 60. Different results are obtained using features from frequency domain analysis – see Fig. 10. In this case high outlier ranks can be observed inter alia for the sample numbers 28 and 29.

In a last step the results of the different outlier detectors can be aggregated via rank transformation. With that the PMU samples with the highest outlier ranks are selected among all detectors. The results are presented in Fig. 11 for the time excerpt from 200 to 1800 s for different outlier numbers. The whole voltage signal (blue) and the identified outlier samples (red) are shown. As it can be seen in both cases a voltage sag between 200 and 300 s and a voltage oscillation between 450 and 580 s can be detected successfully. Additionally some other outliers are identified in the lower figure between 1200 and 1500 s including other voltage oscillations. As it can be seen, when choosing a low number of outliers not all disturbances or unusual pattern might be captured. When choosing a high number of outliers more disturbances can be identified but also a certain amount of non-relevant patterns. For a better illustration Fig. 12 compares some outlier (left picture) and non-outlier (right picture) PMU patterns. Sample 6 corresponds to a voltage sag and sample 10 corresponds to a voltage oscillation. Both samples have a high outlier rank whereas samples 20 and 60 have the lowest outlier ranks – see again in Fig. 9 and Fig. 10.

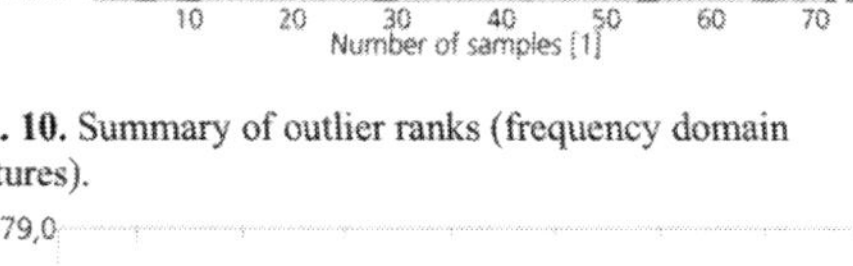

Fig. 9. Summary of outlier ranks (time domain features).

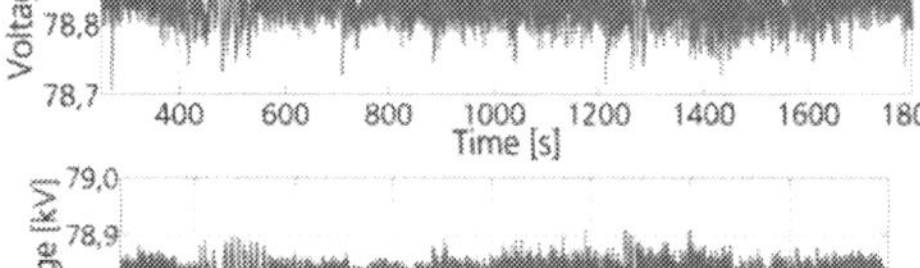

Fig. 10. Summary of outlier ranks (frequency domain features).

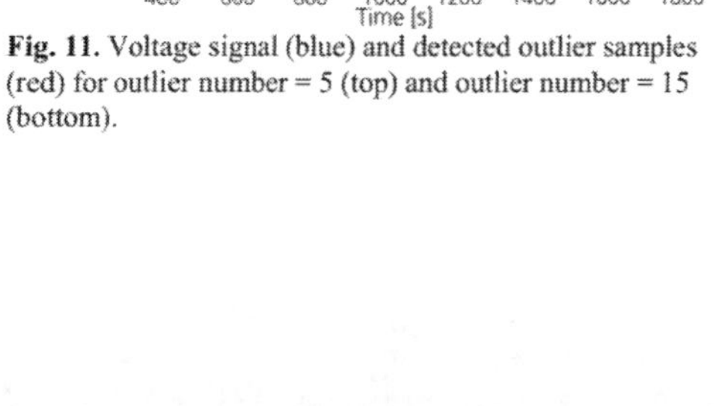

Fig. 11. Voltage signal (blue) and detected outlier samples (red) for outlier number = 5 (top) and outlier number = 15 (bottom).

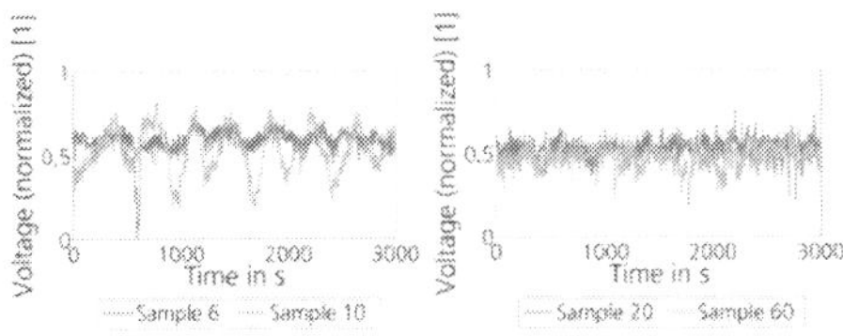

Fig. 12. Outlier samples (left) and non-outlier samples (right).

4 Summary and outlook

This study introduces a new method for the uncovering of critical disturbances from historical phasor measurements. Within a three-phase concept outlier scores are computed for each PMU sample to detect critical disturbances. The relevant features are generated in time and frequency domain including e.g. statistical moments and multi-level Wavelet decomposition. Different outlier detection methods are combined using voltage magnitudes from a real phasor measurements record. The algorithms include established techniques like Local Outlier Factors (LOF) and Correlation Outlier Probabilities (COP) as well as an own implementation of a single linkage based outlier detection method (SiLiOd). The results of 6 different outlier detectors are compared which each other. It is shown that typical disturbance patterns like voltage sags and voltage oscillations can be extracted with the use of outlier detection methods. With this concept mixed PMU data sets can be explored and analyzed by uncovering unusual or potential critical events. This can support further decision making and serve as training data for online detection and classification algorithms.

For further investigations additional test data is required to validate the results and to extract other disturbance types like line or generator trips as well as analyzing other measurements like frequencies or voltage angles. Also, the extracted potential disturbances have to be clustered or grouped in a post-processing step to better select relevant and non-relevant signal patterns. Apart from that new techniques for feature extraction and outlier detection (e.g. Isolation Forests or Deep Autoencoder) have to be tested and compared to the existing approach. Also an procedure has to be developed for the automated extraction of critical disturbances or events from mixed and imbalanced PMU data sets. With such an approach large PMU data sets can be explored and filtered for subsequent event analysis or classification tasks.

References

1. J. F. Hauer, D. J. Trudnowski, and J. G. DeSteese, "A Perspective on WAMS Analysis Tools for Tracking of Oscillatory Dynamics," in *2007 IEEE Power Engineering Society General Meeting*, Tampa, FL, USA, pp. 1–10.
2. V. Terzija *et al.*, "Wide-Area Monitoring, Protection, and Control of Future Electric Power Networks," *Proc. IEEE*, vol. 99, no. 1, pp. 80–93, 2011.
3. L. F. d. Santos, G. Antonova, M. Larsson, and S. Fujii, "The Use of Synchrophasors for Wide Area Monitoring of Electrical Power Grids," 2013.
4. J. Guo *et al.*, "Design and Implementation of a Real-Time Off-Grid Operation Detection Tool from a Wide-Area Measurements Perspective," *IEEE Trans. Smart Grid*, vol. 6, no. 4, pp. 2080–2087, 2015.
5. S. Balischewski *et al.*, "A new development platform for the next generation of power system control center functionalities for hybrid AC-HVDC transmission systems," in *IEEE PES General Meeting Chicago July 2017*.
6. S. Brahma *et al.*, "Real-Time Identification of Dynamic Events in Power Systems Using PMU Data, and Potential Applications—Models, Promises, and Challenges," *IEEE Trans. Power Delivery*, vol. 32, no. 1, pp. 294–301, 2017.
7. Q. H. Alsafasfeh, "PATTERN RECOGNITION FOR FAULT DETECTION, CLASSIFICATION, AND LOCALIZATION IN ELECTRICAL POWER SYSTEMS," Dissertation, Department of Electrical and Computer Engine, Western Michigan Univ, Kalamazoo, Michigan, 2010.
8. M. Biswal and P. K. Dash, "Detection and characterization of multiple power quality disturbances with a fast S-transform and decision tree based classifier," *Digital Signal Processing*, vol. 23, no. 4, pp. 1071–1083, 2013.
9. Y. Ge *et al.*, "Power System Real-Time Event Detection and Associated Data Archival Reduction Based on Synchrophasors," *IEEE Trans. Smart Grid*, vol. 6, no. 4, pp. 2088–2097, 2015.
10. A. Chiang, E. David, Y.-J. Lee, G. Leshem, and Y.-R. Yeh, "A study on anomaly detection ensembles," *Journal of Applied Logic*, vol. 21, pp. 1–13, 2017.
11. V. Chandola, A. Banerjee, and V. Kumar, "Anomaly detection," *ACM Comput. Surv.*, vol. 41, no. 3, pp. 1–58, 2009.
12. C. S. Burrus, R. A. Gopinath, and H. Guo, *Introduction to wavelets and wavelet transforms: A primer*. Upper Saddle River, NJ: Prentice Hall, 1998.
13. R. G. Stockwell, L. Mansinha, and R. P. Lowe, "Localization of the complex spectrum: The S transform," *IEEE Trans. Signal Process.*, vol. 44, no. 4, pp. 998–1001, 1996.
14. J. S. Richman and J. R. Moorman, "Physiological time-series analysis using approximate entropy and sample entropy," (eng), *American journal of physiology. Heart and circulatory physiology*, vol. 278, no. 6, H2039-49, 2000.

 https://doi.org/10.1051/e3sconf/20186408006

15. J. H. Janssens, "Outlier Selection and One-Class Classification," Dissertation, Tilburg University, Tilburg, 2013.
16. M. M. Breunig, H.-P. Kriegel, R. T. Ng, and J. Sander, "LOF: Identifying Density-Based Local Outliers," in *Proceedings of the 2000 ACM SIGMOD international conference on Management of data*, Dallas, Texas, United States, 2000, pp. 93–104.
17. H.-P. Kriegel, P. Kroger, E. Schubert, and A. Zimek, "Outlier Detection in Arbitrarily Oriented Subspaces," in *IEEE 12th International Conference on Data Mining (ICDM), 2012: 10 - 13 Dec. 2012, Brussels, Belgium ; proceedings*, Brussels, Belgium, 2012, pp. 379–388.
18. E. Alpaydin, *Introduction to machine learning*, 2nd ed. Cambridge, Mass: MIT Press, 2010.
19. A. Allen, M. Singh, and E. Muljadi: NREL and S. Santoso: University of Texas at Austin, "PMU Data Event Detection: A User Guide for Power Engineers," Technical Report, NREL National Renewable Energy Laboratory, The University of Texas at Austin NREL/TP-5D00-61664, 2014.
20. I. The MathWorks, Ed., "MATLAB 7.5 and Signal Processing Toolbox 6.1, Statistics Toolbox 6.1, Wavelet Toolbox 4.1," Natick, Massachusetts, United States, 2007.
21. E. Raff, "JSAT: Java Statistical Analysis Tool, a Library for Machine Learning," *Journal of Machine Learning Research*, vol. 18, no. 23, pp. 1–5, http://jmlr.org/papers/v18/16-131.html, 2017.
22. E. Achtert, S. Goldhofer, H.-P. Kriegel, E. Schubert, and A. Zimek, "Evaluation of Clusterings - Metrics and Visual Support Washington, DC, USA (Arlington, Virginia), 1-5 April, 2012," pp. 1285–1288.

Industrial Energetic Districts: Impact Analysis on the Global Energy Efficiency and Business Competitiveness

Lesme Corredor[1], José Avendaño[2], Robert Bello[1], Álvaro Redondo[1], José Calle[1], and Jesús Viloria[1]

[1]Universidad del Norte, Barranquilla, Colombia
[2]Gracetales S.A., Barranquilla, Colombia

Abstract. Process industries located in emerging economies have relative low levels of production to similar ones located in developed countries, this fact influences the implementation feasibility of cogeneration and/or tri-generation systems that allow a substantial increase in the plant global energy efficiency. In this paper, an energy and economic analysis of several alternatives of cogeneration was done for a company located in Barranquilla (Colombia, South America) that produces vegetable oils and derivatives and its energy matrix is approximately 90% thermal and 10% electric. In this investigation two type of analysis were done, both supported by process simulation software, these are: 1) Taking the plant as the control volume and evaluating the entire electrical demand supply with natural gas engine and turbine – generator, plus exhaust gases heat recovery for refrigeration and/or preheating of thermal oil or water in boilers. 2) As an energy-industrial district, where the company takes advantage of the residual heat of a gas turbine and sells the excess of electrical power to nearby plants, a concept introduced by the authors as Sustainable Energetic Industrial District in Emerging Economies (SEIDEE). The input variable considered for this analysis was electric demand which restricts the technology implementation. It was found that the investment return period is notably lengthy when the thermal machine supplies the electric power demanded by the industrial plant. This period is considerably reduced when the SEIDEE concept is implemented, this period reduction is between 57% and 65%.

1 Introduction

Industrial plants that have an energy matrix composed by electricity and thermal sources, are very attractive to analyze an eventual cogeneration and tri-generation systems implementation, this is because of the fossil fuels' high heating value. To industrial sector, these alternatives evaluation are very interesting, this will decrease electric and thermal consumption during the year. In this paper, authors evaluate gas turbine and internal combustion engine implementation, considering economic and technical aspects required by the industrial plant.

These plants have a yearly maximum electric consumption; moreover, this consumption is variable during the year, the reason why equipment and machines inside the plant must be able to work in partial and nominal conditions. Electricity produced by gas turbine and internal combustion engine is evaluated in the following scenarios:

1. Thermal machines supply electric energy, lower than maximum electric consumption. In this scenario, the plant must buy electricity, which does not develop plant self-sufficiency.

2. Thermal machines supply electric energy equal to maximum electric consumption. This scenario develops plant self-sufficiency.

3. Thermal machines supply electric energy, higher than maximum electric consumption. This scenario does not only mean plant self-sufficiency, but also will make interesting electricity sale to nearby plants, introducing the proposed concept by authors: Sustainable Energetic Industrial District in Emerging Economies (SEIDEE).

Second and third scenarios are very attractive to analyze, that is why they will be studied in detail. The first scenario is not interesting, it is because the industrial plant will depend on electric grid and electricity costs will not decrease considerably. The third scenario is the most attractive one, it provides the highest economic benefit, but is the most complex because of the technical requirements.

As well as electric consumption, industrial plants have a maximum thermal consumption which tends to be partialized in natural gas, coal or diesel; in this case the thermal consumption is totally produced by natural gas. Natural gas taxes are a very important topic in the region that is why a reduction in its consumptions will be interesting for any industry. These conditions make cogeneration and tri-generation systems implementation a seemly alternative.

Authors considered the following aspects to implement these systems:

- To identify which equipment demands the highest thermal consumption. The objective is to implement

 https://doi.org/10.1051/e3sconf/20186408007

the proposed technologies (internal combustion engine or gas turbine) to reduce this thermal consumption.

- To analyze industrial plant electric demand fluctuation. This aspect will identify which technology satisfies technical requirements to be integrated to the industrial plant.

This article discusses cogeneration (taking advantage of energy produced by the technologies proposed to reduce natural gas consumption) and/or tri-generation (cold production using an absorption chiller) systems implementation. These alternatives will be evaluated for the second and third scenario. Economic and technical aspects will be considered and analyzed to implement the best alternative.

2 District environment and load definition

The performance of the industrial district in the analysis is affected by different factors that define its energy efficiency and competitiveness. The district is located in the northern coast of Colombia, specifically in the city of Barranquilla. The average temperature of this region is 33°C and the relative humidity of air 82% [1]. Adding to these, the power demand of the plant is notably variable, something that depends on the season and the requirements of clients. For this reason, it is possible to observe that the maximum power consumed in a month is 910 kWhe, while the average demand of the plant is 785 kWhe. This demand is not so significant in comparison to the consumption of thermic energy that this district uses in its productive process. The following figures show the hourly behavior of electric energy demand in twelve months in a specific year.

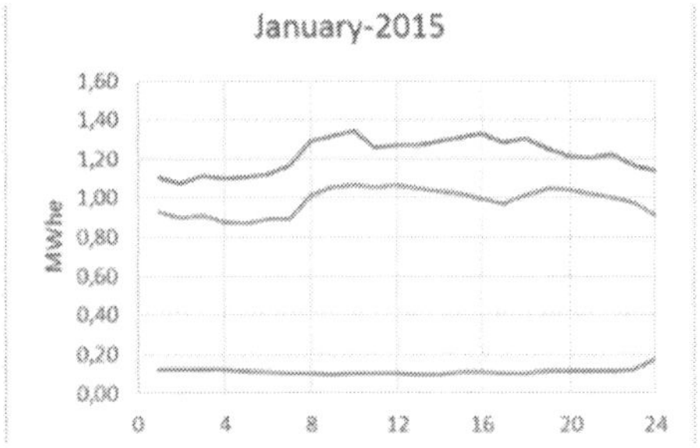

Fig. 1. Hourly electric consumption profiles for January.

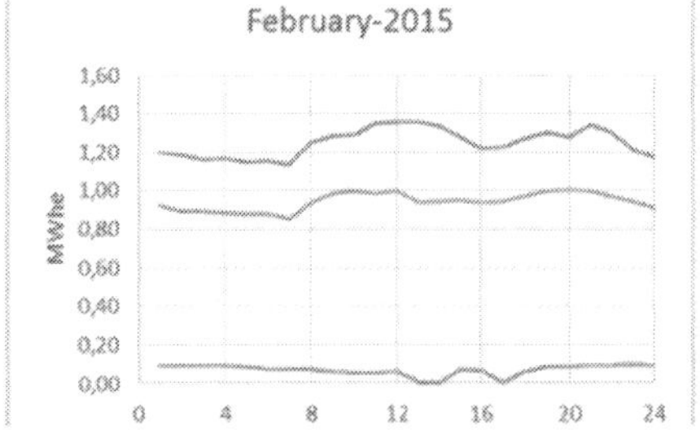

Fig. 2. Hourly electric consumption profiles for February.

Fig. 3. Hourly electric consumption profiles for March.

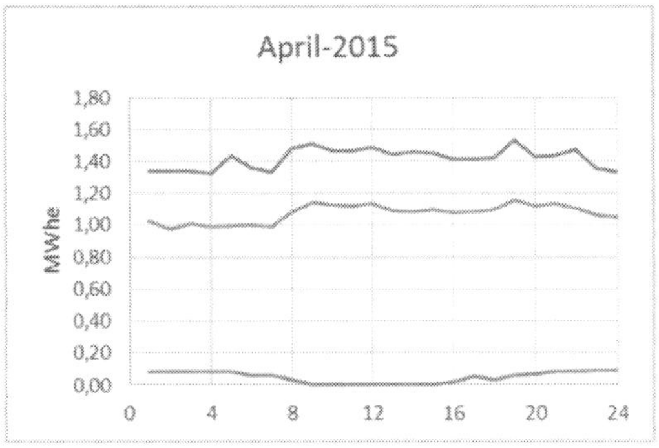

Fig. 4. Hourly electric consumption profiles for April.

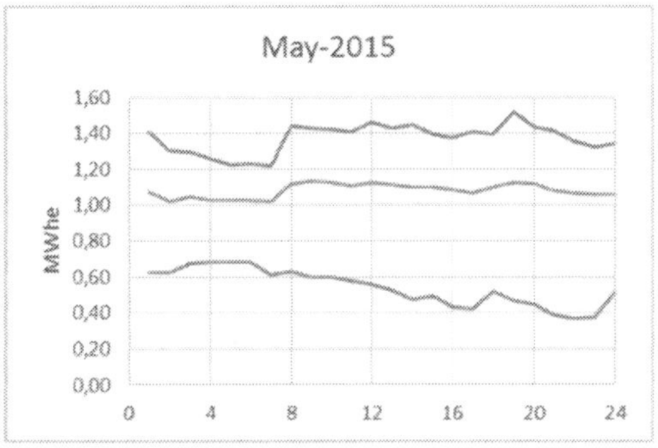

Fig. 5. Hourly electric consumption profiles for May.

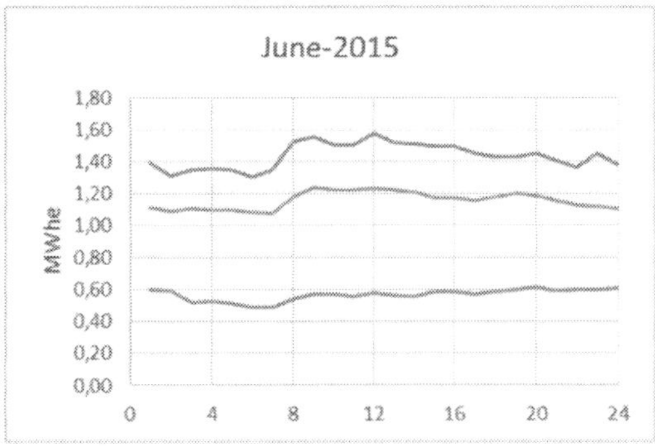

Fig. 6. Hourly electric consumption profiles for June.

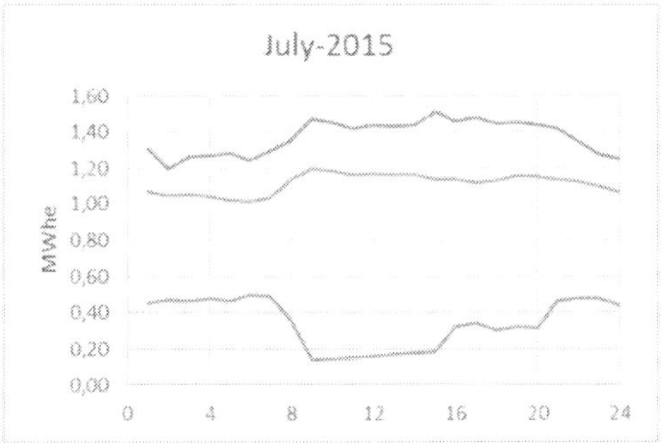

Fig. 7. Hourly electric consumption profiles for July.

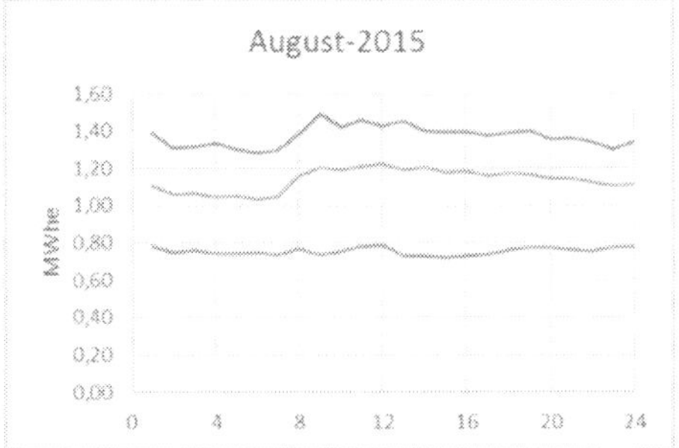

Fig. 8. Hourly electric consumption profiles for August.

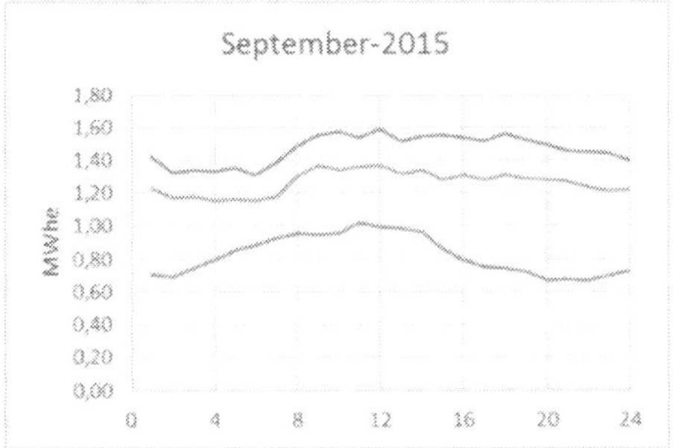

Fig. 9. Hourly electric consumption profiles for September.

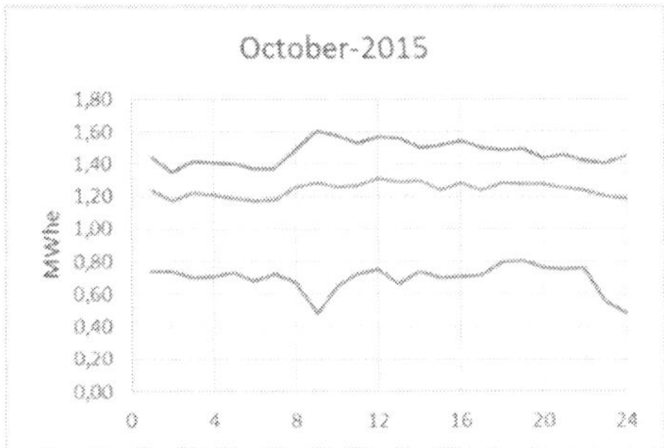

Fig. 10. Hourly electric consumption profiles for October.

For the following group of figures from Figure 1 to Figure 12, the colors presented in the graphs correspond to:

- Orange stripe: hourly minimum consumption registered in the month.
- Blue stripe: hourly maximum consumption registered in the month.
- Gray stripe: hourly consumption for a characteristic day in the month.

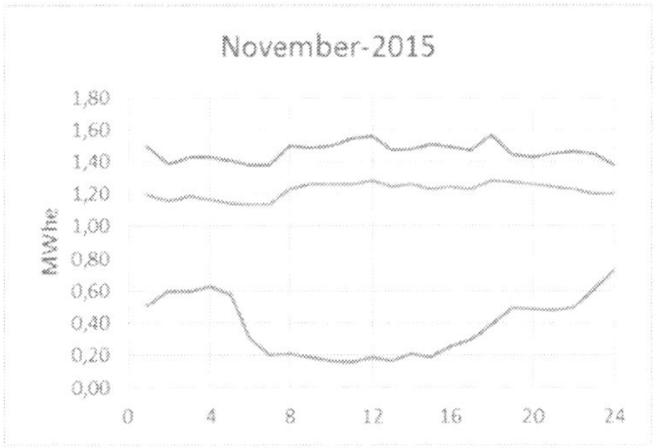

Fig. 11. Hourly electric consumption profiles for November.

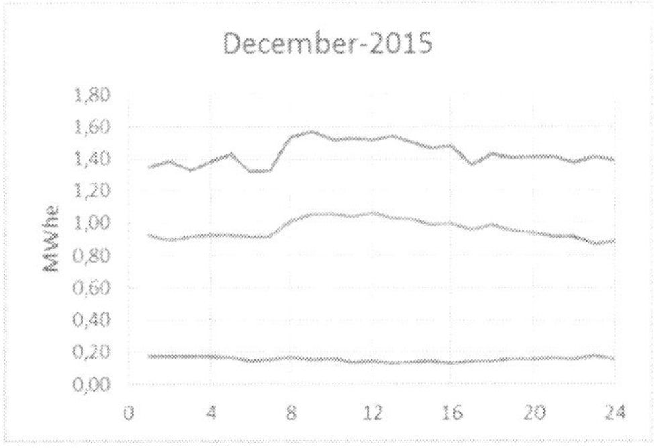

Fig. 12. Hourly electric consumption profiles for December.

The thermic energy demand of this district reaches a 90% in relation to the total energy consumed by the plant. The highest demands present when clients make high requirements reaching a monthly consumption of 17,054 kWht and an average consumption for this same period is 14,850 kWht.

3 Problem statement

Searching for improved energy efficiency and to raise the competitiveness of the plant, the analysis of the implementation of power generation systems turns into an interesting alternative. The high demand of thermic energy that the district shows, leads to considerate it as a "thermal host". The area where the company is located turns into a potential client for generated power since the neighbors are plants with similar consumptions to this district. Therefore, the implementation of cogeneration and tri-generation technologies appear to be viable alternatives.

The implementation of these technologies is tied to variables such as the behavior of energy requirements, the place where they are to be installed and prices of

https://doi.org/10.1051/e3sconf/20186408007

natural gas. To implement these systems, it is necessary to carry out corresponding studies and select the most favorable configuration taking into consideration the interests of the project.

4 Historical background

In Colombia, the tendency to start implementing cogeneration technology or at least to study its viability has spread in several companies. This is why a large number of companies have been found that have dared to investigate and investigate how this technology works and how it would impact its costs. The most representative is the experience lived on behalf of the pasta company Doria. In this company a cogeneration system was implemented in which an internal combustion engine based on natural gas is used as a thermal machine. In this company as in many others there is the problem in production times as a consequence of the instability of the public service network. Due to this the installation of the cogeneration system with a Cummins gas engine with a capacity of 1,750 kW was evaluated. With which it is able to supply the thermal demand of the plant and additionally generate electricity.

Thanks to this system implemented by the aforementioned company, savings of 60 % have been achieved in the purchase of electricity to the public grid and, most importantly, the electricity generated is highly reliable. And in terms of fossil fuels, the company has improved its fuel consumption achieving savings of 70 % which also translates into $50,000 a month in energy costs. The most interesting of this is that the motor-generator system works 24 hours increases the productive times of the company [1].

Other companies have opted to start similar to the one that was presented by Pastas Doria. Companies like Empacor, Corona, Monómeros have dared to bet on this technology not only for the benefits it brings but also because it becomes much more profitable from the economic point of view in addition to contributing to the reduction of CO2 emissions towards the atmosphere. Finally, no evidence has been found that within all of the aforementioned companies an energy district is developed. In this article, the authors propose the development of an industrial energy district, which increases the efficiency of the cogeneration process and additionally increases the reliability of the electrical energy generated by a thermal machine [7].

5 Simulation specifications

To simulate, the conversion from ISO conditions was made to analyze the expected performance of the machines in the correspondent room temperature (35°C). Calculations of important engineering variables and the study of the systems in general were all made using the datasheets from the equipment in use. To calculate, it was assumed that electric demand values are equal to industrial plant instantaneous power; it is because there is no power measurement. In order to select the most suitable thermal machines (in terms of capacity), electric demand profile was directly taken from company's consolidate.

Flows, temperatures, pressures and other thermodynamic variables used to simulate were provided by the company. This information will not be mentioned in this paper because of confidentiality terms. All data used to simulate thermal machines proposed was extracted directly from manufacturers' datasheet.

To evaluate new technology incorporation to the process, authors used Aspen HYSYS®. The software models industrial processes and was used to simulate operational equipment conditions. The equipment which demands the highest thermal consumption in the industrial plant is a 75,000 lb/h Nebraska® boiler. Its consumption represents an 85% of the plant thermal demand. The following figure shows the boiler simulation.

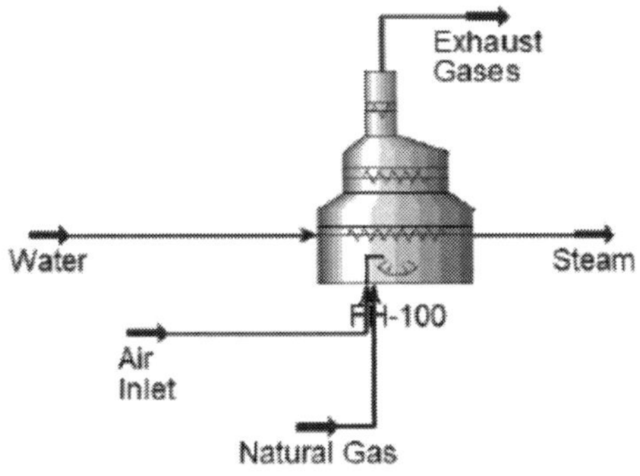

Fig. 13. Boiler simulation.

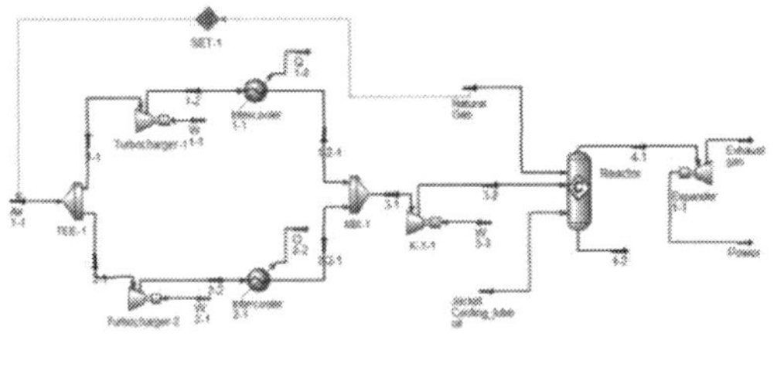

Fig. 14. Internal combustion engine simulation.

A natural gas internal combustion engine implementation is an alternative proposed. The internal combustion engine was analyzed in three cases:

1. Internal Combustion Engine Cummins® 1,000 N6C [3] to implement a cogeneration system (Scenario 1).
2. Internal Combustion Engine Deutz® 1,600 TCG 2,020 [4] to implement a cogeneration system (Scenario 2).
3. Internal Combustion Engine Deutz® 1,600 TCG 2,020 [4] to implement a tri-generation system (Scenario 2).

To simulate internal combustion engines, authors used [5]. The following figures show Cummins® engine simulation and engine efficiency vs natural gas mass flow.

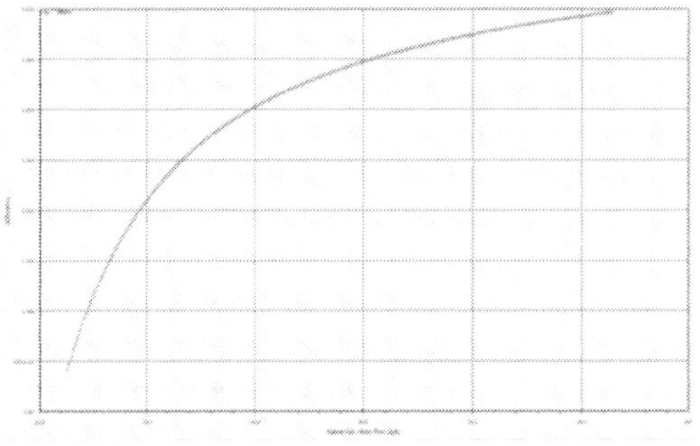

Fig. 15. Internal combustion engine efficiency vs natural gas mass flow (kg/h).

Natural gas turbine is the other alternative proposed for the analysis. The following, are the three cases evaluated to this thermal machine.

1. Solar Turbines® Saturn 20 [11] to implement a tri-generation system (Scenario 2).
2. Solar Turbines® Centaur 50 [10] to implement a tri-generation system (Scenario 3).
3. Solar Turbines® Taurus 70 [12] to implement a tri-generation system (Scenario 3).

To the simulation, the article [9] was used as a guide. The following figure shows the simulation.

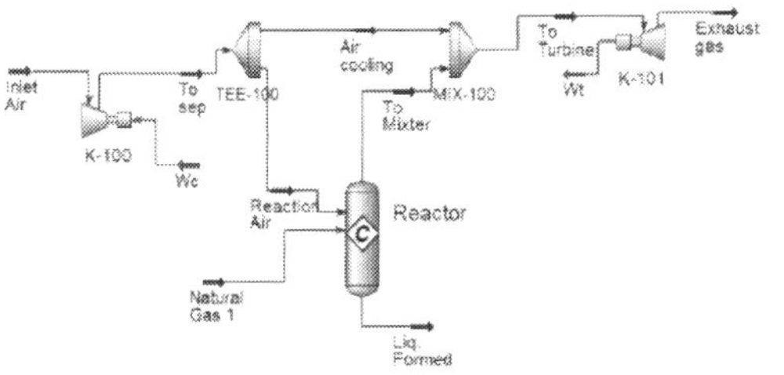

Fig. 16. Gas turbine simulation.

An absorption chiller implementation, which will supply the cold demand required by the plant was considered in this paper. Therefore, two Thermax® absorption chillers were considered, the first one has a capacity of 296TR and the other 501TR [13].

6 Configuration

To obtain energetic indicators of the process and to analyze the new technology performance, it is necessary to develop a configuration that includes a relationship between all the equipment mentioned before. This relationship could be by heat recovery systems or by power supply. Because of the viability of the project, it is important to define it by the prior items. Real equipment simulation is obligatory and some relevant information of them is presented in Table 1 and Table 2.

Table 1. Simulated and real gas turbines comparison.

	SOLAR GAS TURBINES®					
	Output Power (kW)			Exhaust Temperature (K)		
	Datasheet	Simulated	Deviation	Datasheet	Simulated	Deviation
Saturn® 20	1100	1160	5,455%	778,1	834,5	7,248%
Centaur® 50	4000	4028	0,700%	783,1	788,6	0,702%
Taurus™ 70	6600	6624	0,364%	778,1	746	-4,125%

Table 2. Simulated and real internal combustion engines comparison.

	INTERNAL COMBUSTION ENGINE					
	Output Power (kW)			Exhaust Temperature (K)		
	Datasheet	Simulated	Deviation	Datasheet	Simulated	Deviation
Cummins® 1000	1031	979,6	-4,985%	687	697,7	1,557%
Deutz® 1600	1555	1402	-9,839%	713	776,3	8,878%

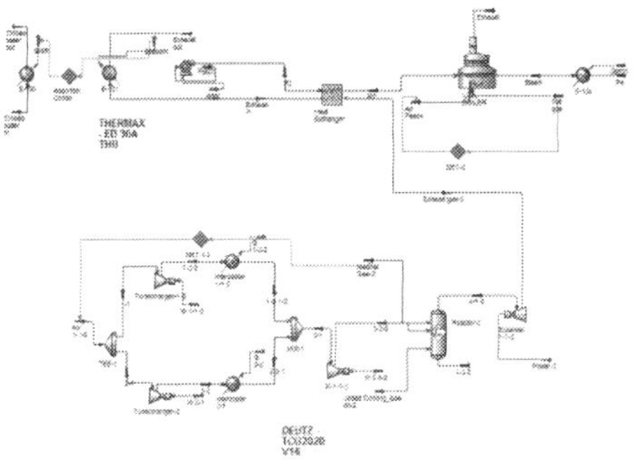

Fig. 17. Industrial plant with a tri-generation system composed by a 1.6 MW internal combustion engine and a 296TR absorption chiller.

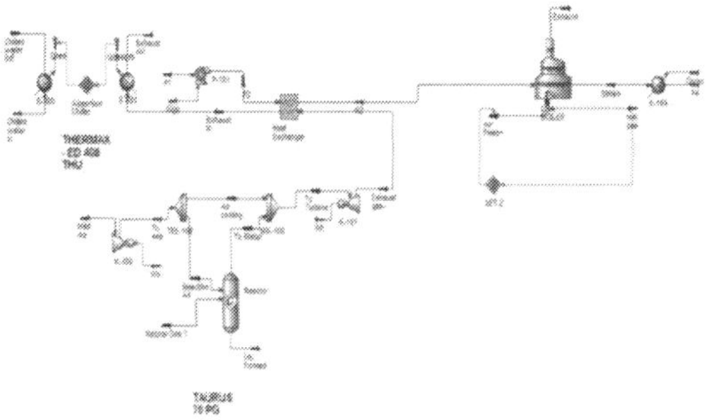

Fig. 18. Industrial plant with a tri-generation system composed by an 8 MW gas turbine and a 501TR absorption chiller.

Taking the equipment information as the base, a configuration of them has been proposed taking a positive performance as the main consideration for the implementation. The composition of each configuration is presented in the following figures.

The following figures present the configuration for the third case using an internal combustion engine (Deutz® TCG2020 V16 Internal Combustion Engine and 296 TR Thermax® Absorption Chiller).

In addition, the next figure shows the configuration with the implementation of the third case of gas turbine (Solar 6 Turbines® Taurus 70 and 501TR Thermax® Absorption Chiller).

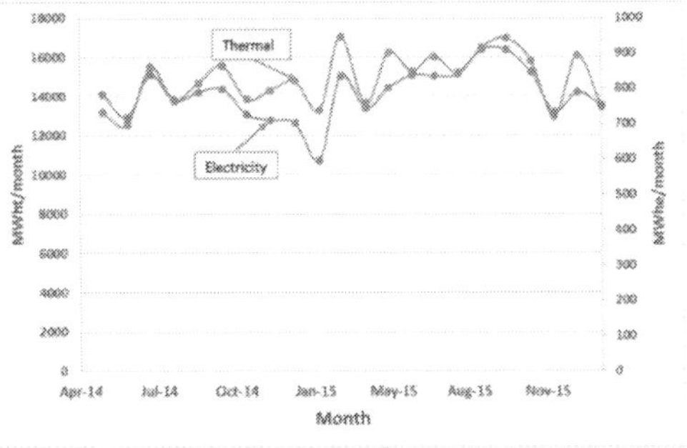

Fig. 19. Energetic demand profile.

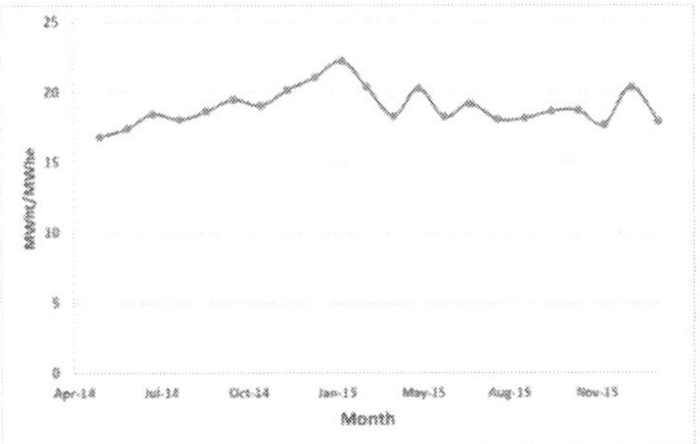

Fig. 20. Energetic indicator: Thermal and electric consumption ratio.

7 Statistic analysis

The charge simultaneity and the statistic superposition between the thermal and electrical demand are not considered in this paper. The reason is that the data registers available have been presented for different periods of time where the thermal demand (Nm3 of natural gas consumed) is registered month by month and the electrical consumption is presented hour by hour. In addition, the company shared data of a short period of time (May 2014 to February 2015) and this leads to the no application of statistical analysis by stochastic methods that generates a great reliability on the energetic demand of the firm. Sau Man Lai and Chi Wai Hui propose this method [8]. For this analysis, the electric consume behavior play the main roll. Because of the interests for this work, the electric demand is presented showing the maximums and minimums points of the electric and thermal consumption for each hour of the months that are under analysis (2015). It includes the atypical data produced by the energy interruption by the local supplier company. Besides, this paper analyses occurrence deleting the aberrant data. For this, representative data is selected for each hour and then the consume profile is developed to represent each month with a confidence of 60%. The electrical data is presented as energy consumption in kWhe. For practical purposes, constant energy consumptions are assumed for each hour. Energy consumption profiles are presented in the following pictures.

8 Cost analysis

For the cost analysis of the plant, the 3 different scenarios were studied taking into consideration the possibilities of implementing the 2 different technologies: gas turbine and internal combustion engine Actual prices for electricity and gas were obtained directly from the company to calculate potential savings. It is important to mention that available heat recovered was assumed to be used at maximum, later studies about companies' privacy would later need to be studied.

To calculate the costs of the different projects approximations were made using values from CHP's characterization of turbines and engine technologies [2]. The costs calculated on the project in study were later converted into present value assuming an annual inflation of 3% in USD till year 2016. The following table shows the costs approximated by the American entity and used in the current document:

Table 3. Combustion engine project approximate costs.

Combustion Engine Project Approximate Costs			
Capacity [kW]	500	1000	3000
Project Cost [$USD/kW]	1940	1640	1130
Maintenance Costs [$USD/kWh]	0.016	0.013	0.01

Table 4. Gas turbine project approximate costs.

Gas Turbine Project Approximate Costs			
Capacity [kW]	1000	5000	10000
Project Cost [$USD/kW]	3324	1314	1298
Maintenance Costs [$USD/kWh]	0.0111	0.0074	0.007

To summarize the whole economic analysis, the results will show: A) The approximate investment calculated for the project in study. These costs are a safe gross approximation to evaluate the viability of the project. B) The potential annual savings calculated taking into consideration heat recovery, refrigeration, power sale and others depending on the case. C) The capital investment return in years. D) The Internal Return Rate calculated on a life of 10 years. The detailed results of the project could not be showed because of confidentiality matters with the company.

As mentioned before the first scenario was analyzing the possibility of generating electricity below the companies' consumption needs. A Cummins® 1,000 kW engine [3] was used to study this scenario, where potential heat recovery was analyzed (cogeneration). The company would still need to buy electricity from the market to supply their needs when consumption is high, so the statistical data was used to approximate these costs (frequency of occurrence). The following table shows the results for the first scenario:

Table 5. Cost results scenario 1.

Co-generation Cummins® 1000 kW	
Gas Consumption [m3/h]	248.5
Electrical Output [kW]	948.22
Electrical Efficiency	41.55 %
Recovered Heat [kW]	802.5
Monthly Electrical Power Supply [MWh]	2250.6
Approximate Project Cost [k$USD]	3529
Potential Annual Savings [k$USD]	387.88
Investment Return [years]	9.1
Internal Rate of Return	4.63 %

Table 6. Cost results scenario 2.

	Co-generation Deutz® 1600	**Tri-generation Deutz® 1600**	**Tri-generation Saturn® 20**
Gas Consumption [m3/h]	306.1	286.1	472
Electrical Efficiency	38.81 %	38.81 %	23.53 %
Recovered Heat [kW]	962	469.3	677
Refrigeration Max [kW]	N/A	1,041	1,762
Approximate Project Cost [k$USD]	4,631	5,085	7,061
Potential Savings [k$USD/y]	547.7	308.3	258
Investment Return [years]	8.46	16.49	27.39
Internal Rate of Return [10 y]	5.93 %	-10.18 %	-19.28 %

The numbers do not seem to be too attractive for this alternative, mainly because fuel consumptions end making the operation expensive which shows that low power cogeneration is not viable with the high gas prices present in the region.

In the second scenario, the possibility of generating the maximum electric needs of the company was analyzed. In this scenario 3 cases were studied, the possibility cogeneration and tri-generation with Deutz® 1,600 kW [4] engine and tri-generation with Solar Turbines® Saturn 20 [11]. A Thermax® 300 refrigeration tons Absorption Chiller was used for the engine's case while a 500 RT was use for the gas turbine (this one would satisfy the whole refrigeration needs) [13]. Refrigeration data

Having to work at partial load, the engine's cases were analyzed with data from the original datasheet and frequencies of operation obtained from statistical data. The frequencies of operation were used to approximate engineering variables such as fuel consumption, and therefore to calculate corresponding costs.

Results for Cogeneration with Deutz® engine are far more satisfactory than tri-generation ones. The reasons for this are the low refrigeration needs of the company and the high costs of turbines compared to engines. The results of the scenario for the turbine let us see that in general the low scale of the plant does not help on the economical aspect of the project. Turbines tend to be inefficient when designed for low power values, which is of course the problem in the studied case. This problem results in high gas consumptions for low power availability which leads to a non-viable investment. Still for all cases of scenario 2, numbers do not sum up (Cogeneration returns at least positive income values). Adding to these the electrical demand of the company is not constant at all, which does not benefit the Turbines operation since it is technically vulnerable to abrupt power changes.

Finally, in the third scenario, the implementation of a Sustainable Energetic Industrial District in Emerging Economies (SEIDEE) is introduced. A SEIDEE is defined by authors as a small power generator inside a city that could supply power to nearby companies. This power generator company will be a "thermal host" that can take advantage of thermic energy to generate electricity in excess to sell at discount prices to nearby users.

For this final scenario, only gas turbines are used because of their good efficiency at higher loads. Specifically, cases with Solar Turbines® Centaur 50 [10] and Taurus™ 70 [12] were analyzed. For this scenario, the possibility of selling power to nearby companies was studied taking into consideration a probable sale with a 15% discount on commercial value (from 320 to 270$COP/kWh or 0.1 to 0.0844$USD/kWh). Comparisons between a regular sale (Approx. 0.047$USD/kWh) and a sale in what a SEIDEE would allow are showed in tables below for each case. Values of consumption, efficiency, recovered heat and others do not change for the SEIDEE proposal, just the price of a kWh.

Table 7. Cost results scenario 3.

Regular Generation	**Tri-generation Centaur® 50**	**Tri-generation Taurus™ 70**
Gas Consumption [Nm3/h]	1,368	2,112
Electrical Output [kW]	3,800	6,600
Electrical Efficiency	28.57 %	32 %

Recovered Heat [kW]	4,091	5,942
Refrigeration Max[kW]	1,762	1,762
Revenue for Power Sale [k$USD/y]	882	1,789
Approximate Project Cost [k$USD]	13,106	19,628
Potential Savings [k$USD/y]	497.56	738
Investment Return [years]	26.34	26.58
Internal Rate of Return	-20.73 %	-21.69 %

Table 8. Cost results scenario 3 SEIDEE.

SEIDEE	Tri-generation Centaur® 50	Tri-generation Taurus™ 70
Revenue for Power Sale [k$USD/y]	1,587.8	3,220
Potential Savings [k$USD/y]	1,203.2	2,170
Investment Return [years]	11.23	9.05
Internal Rate of Return	-0.98 %	3.57 %

Economic results end up being not that promising as expected mainly because as mentioned before the low scale of power generation does not allow the turbines to perform at high efficiencies. Nevertheless, a big improvement is seen when selecting the Taurus™ 70 and selling electricity close to commercial value. Therefore, the important variables are the price of sold kWh and the cost of natural gas consumption.

High prices of natural gas and low prices for the electric selling market do not allow numbers to show the real potential of tri-generation systems. These problems are characteristic from the region, and this is the reason the authors propose the creation of SEIDEE's in the city to encourage the implementation of cogeneration and tri-generation systems that will increase energy efficiency in the region and off course reduce contaminating emissions. Different governmental policies such as lower gas consumption tariffs, higher power generation incentives, tax reductions, and the creation of an open electrical market would all increase chances of a satisfactory implementation of these systems in the Caribbean coast of Colombia.

9 Conclusions

The first scenario proposed by authors is not interesting to industrial plants. Internal combustion engine supplies an electric demand fraction and the investment return period is lengthy. Moreover, the plant will still depend on electric grid. In the second scenario, in which the plant is self-sufficient, the investment return period is also lengthy, this is not an acceptable condition for companies in the region because it represents a high investment. These periods are restricted by the plant low-production levels, which impact directly the electric consumption; if the plant had a higher electric consumption, the investment return period would be shorter. Furthermore, industrial plant electric consumption is very variable during its operation, restricting proposed technologies. When the gas turbine is evaluated, these permanent variations in electric demand will represent demanding operational conditions that will shorten its lifetime; this will not allow its implementation. Referring to tri-generation, industrial plant cold production is medium-high, but this production is not constant, this condition will reduce absorption chiller performance, reason for which authors propose a cold-water storage tank, that ensures a constant cold production.

The third scenario is the most important, some interesting results were found. The investment return period is notably reduced when the concept introduced by authors (SEIDEE) is implemented, reducing this period between 57 % and 65 %. For this scenario, gas turbine will be the most suitable technology when SEIDEE is implemented. To get a shorter investment return period, the gas turbine must generate a greater electric power than the gas turbine analyzed (Solar Turbines® Taurus 70). Another reason that makes gas turbine more reliable than internal combustion engine, is that gas turbine will supply all the energy required by the energetic district. Finally, internal combustion engine is the most suitable thermal machine for this project, it is more inexpensive than gas turbine which represents the shortest investment return period and it can operate reliably when the industrial plant electric consumption fluctuates.

Acknowledgment

To all the work team it was a real pleasure that the industrial plant opened their doors, to meet and to practice all the knowledges referred to their processes and equipment.

Engineers Carlos Silva, Adalberto Salazar, Fabrizio Linás and Rafael Percy, for their help in the execution of activities that end up in the development of this paper.

References

1. Características Climatológicas de Ciudades Principales y Municipios Turísticos, "Documento IDEAM", 2016.
2. Combined Heat and Power Partnership, "Catalogue of CHP Technologies", 2008.
3. Cummins Power Generation, "C1000 Generator Set Datasheet", 2015.
4. Deutz, "TCG 2020 The Gas Engine", 2015.
5. Ekwonu, M. C., Perry, S., & Oyedoh, E. A. (2013). Modelling and Simulation of Gas Engines Using Aspen HYSYS. Journal of Engineering Science and Technology, 6(3), 1-4.

6. Gasnaturalfenosa.com.co. (2016). Cogeneración. [online] Available at: http://www.gasnaturalfenosa.com.co/co/grandes+clientes/soluciones+energeticas/1297102582386/cogeneracion.html [Accessed 18 Nov. 2016].
7. La Republica. (2014). Con cogeneración, las empresas reducen 50% el gasto de energía. [online] Available at: http://www.larepublica.co/con-cogeneraci%C3%B3n-las-empresas-reduce-50-el-gasto-de-energ%C3%ADa_153511 [Accessed 18 Nov. 2016]
8. Lai, S. M., & Hui, C. W. (2009). Feasibility and flexibility for a trigeneration system. Energy, 34(10), 1693-1704.
9. Saddiq, H. A., Perry, S., Ndagana, S. F., & Mohammed, A. MODELLING OF GAS TURBINE AND GAS TURBINE EXHAUST AND ITS UTILISATION AS COMBINED CYCLE IN UTILITY SYSTEM.
10. Solar Turbines, "Centaur 50
11. Solar Turbines, "Saturn 20 Gas Generator set", 2013.
12. Solar Turbines, "Taurus 70 Gas Generator set", 2013.
13. Thermax, "Multi-energy Vapour Absorption Machines" Catalogue, 2016.

Power Scheduling for Renewable Energy Connected to the grid

Ismail El kafazi [1], *Rachid* Bannari [2], *Abdellah* Lassioui [3] *and My Othman* Aboutafail [4]

[1,2]Laboratory Systems Engineering, Ensa, Ibn Tofail University Kenitra, Morocco;
[3]ESIT Team, LGS Laboratory, Ibn Tofail University Kenitra, Morocco;
[4]Laboratory of Electrical, Engineering Computing Sciences EECOMAS, Ibn Tofail University Kenitra, Morocco.

Abstract. In this work, a scheduling strategy considering an Energy Storage System (ESS) is proposed to manage the distributed energy resources (DERs) optimally in a grid-connected hybrid PV-wind- microgrid. In order to increase the use of the renewable energy sources and ensure a full charge in the battery for the next day. The optimization problem of this study is addressed through a linear mathematical model, to minimize the cost of energy bought from the utility grid and maximize the income for selling electricity generated by hybrid PV-wind. The optimization model is tested by using a simulation of the model. The achieved results confirm the effectiveness of the proposed scheduling strategy.

1 Introduction

Renewable energy such as Wind Turbine (WT) and Solar energy are the current trend in microgrids (MG). Moreover, Energy Storage Systems (ESSs) supply energy support and provides store energy during times of high generation. An MG has two modes of operation; an islanded mode or connected to the main grid [1]. The ESS coordinates with the RESs and loads in order to avoid excessive charge or discharge which may affect the performance and life-span of the ESS. Furthermore, an Energy Management System (EMS) with the operational conditions of the MG, can define proper operating conditions for the DG based on renewable energy sources (RESs) and loads, looking for an optimal dispatch and commitment between distributed units which ensures reliable operation of the MG regarding specific objectives and generation capacity [2], [3].
Energy management system (EMS) in microgrids is formulated as a nonlinear optimization problem. Diverse methods have been proposed to solve it in the literature, including particle swarm optimization [4], neural networks [5], mixed integer programming [6], and sequential quadratic programming [7].
This paper proposes an EMS to minimize the cost of buying energy from the main grid and maximize the revenue due to the renewables (photovoltaic (PV) energy generation and wind turbine (WT)) while preserving the lifetime of the ESS based on batteries by avoiding overcharge, excessive discharge, and discharge cycles. Additionally, assuring periods of full charge of the ESS, within cycles of discharge based on a 24-h ahead optimal scheduling.
In this proposal, the problem is modeled as a mixed integer linear programming (MILP) for scheduling the power of the distributed energy resources. The proposed optimization model is implemented under commercial optimization software such as GAMS.

2 Operation of the Microgrid

The microgrid is composed by 10 PV generators with a capacity of 2.4 KW and 1 wind turbine with a capacity of 1 KW as shown in Fig.2. The EMS schedules the generators in interactive and non-interactive way since their power generation may be curtailed from the EMS [8].

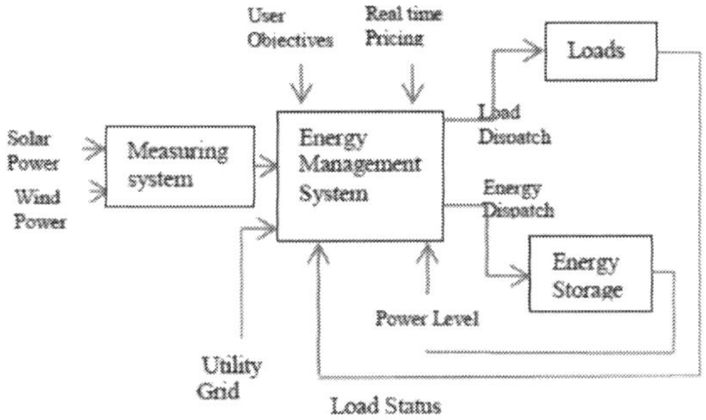

Fig.1. Scheme of the microgrid.

Furthermore, to improve the performance of the ESS a two-level steps which avoid battery overcharge is recommended by the manufacturers.

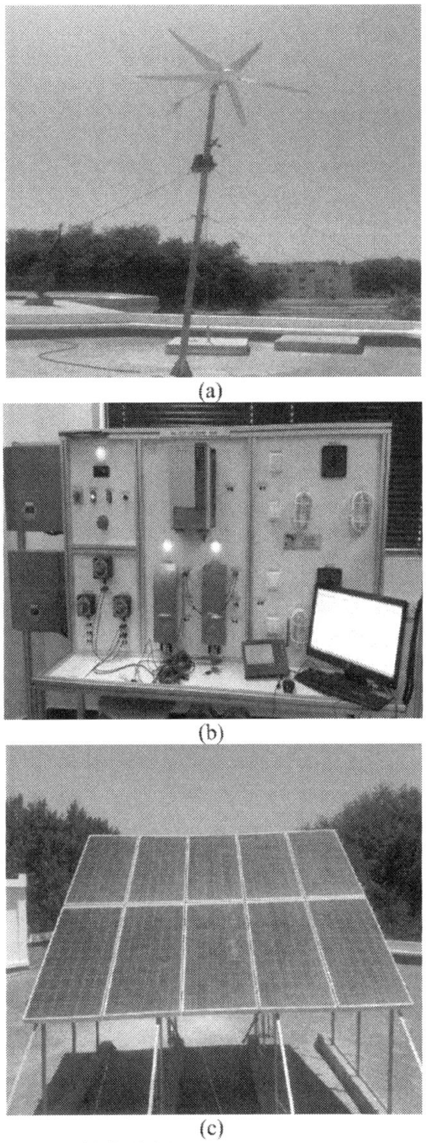

Fig. 2. Microgrid site in Morocco.

Another thing to avoid excessive discharge of the battery, the state of charge (SoC) of the battery is restricted to 20% for ensuring a larger lifetime of the battery array [9, 10]. In this reason, to implement this objective the EMS should ensure proper scheduling of the other distributed energy resources including the main grid to get an optimal commitment between battery charge and the energy bought from the network. Additionally, the EMS minimizes the operation of the system to reduce the number of discharge cycles. Apart from that, EMS schedules the operation of the distributed energy resources complementing the microgrid [2]. The proposed EMS has been composed as a modular architecture in which each module performs different functions independently and exchanges information, as can be seen in Fig. 1.

3 Optimization Model

The optimization model has been developed as Mixed Integer Linear Programming (MILP) for minimizing the cost of buying energy from the utility grid, maximizing energy efficiency by using ESS and maximizes the revenue obtained by selling power generating by renewable energy.

The parameters and variables used in this model are described in table II. All the generators are assumed to be connected to the grid supplying the total system demand. Therefore the network, losses and load ramp constraints are not taken into account. The model proposed in this paper simultaneously solves the optimization problems of minimizing the cost of buying energy and economic scheduling. Renewables such as PVs and WTs, their output power is dependent on the availability of the primary sources (i.e., solar irradiance or wind). Therefore, the forecast is required to consider them in the energy management optimization as mentioned in Fig.1. Forecasting methods for PV and WT can be used [11- 13]. In our proposal model, we assume that the power of a renewable energy beyond the production scheduling is given and there is no operation cost for renewables.

3.1 Collecting Data

The first step in this study is collecting data. Data was collected using Vantage Pro 2 Weather Station [14]. The input data assessed for one day (24h) to be used in order to test the system presented in Figure 3.

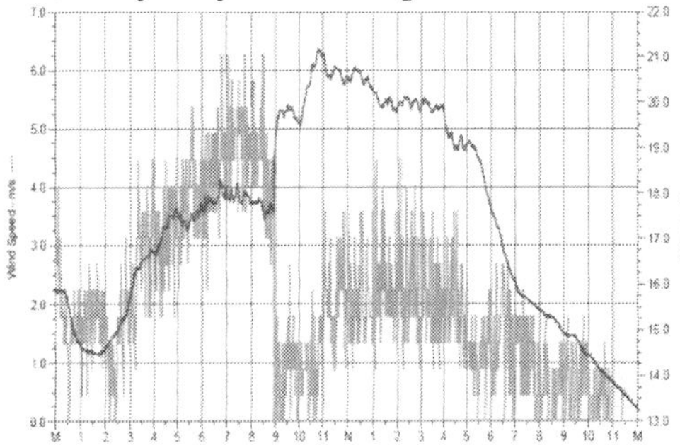

Fig. 3. Hourly wind speed (m/s) and temperature (C) for one day.

The proposed scheduling strategy enables to determine the quantity of energy required and the amount of energy which need to produce by the available powers. The load demand is an input; we analyze the availability of renewable energy. If this energy is equal or superior to demand we can just use it only whether to transport, sell or storage. On the opposite, if the energy is inferior to the load demand, we need to buy the energy from the utility grid.

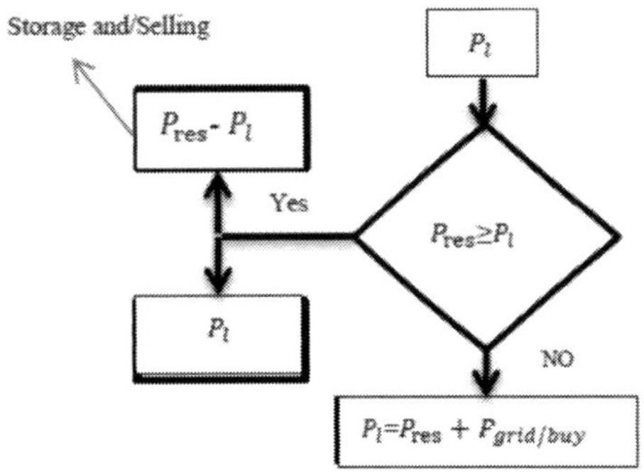

Fig.4. Proposed scheduling strategy.

TableI. Distribution energy per period in the day (24h).

Period (t)	00 :00-08 :00	08 :00-18 :00	20 :00-00:00
Energy product	P_{res} Sell energy to grid	$P_{grid} + P_{res}$ Buy energy from grid	P_{res} Sell energy to grid

Table I provides deciding the primary source of energy used for all the periods in the day. The scheduling from a power supply to the other one is performed at the time of the necessary. Additionally, the table facilitates calculating the production cost every time t.

3.2 Modeling

The production operational planning consists in taking decisions to optimally schedule internal production by renewable energy, to cover the local demand, and minimize the cost of imported electricity from the utility grid in the next hours or day. The index t is the elementary unit of time, t = 1, 2, 3. . . . T. Moreover, the indexes i is used to represent renewable generators. The objective function has been described as follow:

$$\text{OF} = \sum_{t=1}^{T}(P_{sell}^{grid}(t) * C_{sell}^{grid}(t) - P_{buy}^{grid}(t) * C_{buy}^{grid}(t)) \quad (1)$$

This objective function (1) aims to minimize the cost of buying energy from the utility grid and maximizing the revenue obtained by generating energy from the RESs. The two terms satisfies the requirements of the optimization problem for the time horizon. Additionally, are related to the energy absorbed from the grid and injected to the grid respectively.

Constraint

Energy balance (2) is the first condition that the model should include between generation, storage systems and demand in the microgrid.

$$\sum_{t=1}^{T}(P_{sell}^{grid}(t) - P_{buy}^{grid}(t)) + \sum_{i=1}^{2}\lambda_{grid}(t) * P_{res,i}(t) + P_{bat}(t) = P_l(t) \quad (2)$$

The first two terms are related to the energy absorbed from the grid and injected into the grid respectively. In turns, the third expression is the energy provided by the renewables. The variable $\lambda_{grid}(t)$ enables to manage the energy by sending ON/OFF controls. The last term corresponds to the energy of the battery, to complete the balance, the energy required by the load has to equal to the addition of the previous expressions. Moreover, the boundaries related to the output of the renewable energy are defined as follow:

$$0 \leq P_{res,i}(t) \leq P_{res,i}^{max}(t) \Longleftrightarrow \quad P_{res} = P_{pv} + P_w \quad (3)$$

The constraints of the utility grid are also described as follow:

$$0 \leq P_{sell}^{grid} \leq \sum_{i=1}^{2}\lambda_{grid}(t) * P_{res,i}^{max}(t) \quad (4)$$

$$0 \leq P_{buy}^{grid} \leq (1 - \lambda_{grid}(t)) * P_{max}^{grid}(t) \quad (5)$$

In addition, the constraints related to the ESS are defined as follow:

$$(1- B_t) + B_t = 1 \quad \forall t \in T \quad (6)$$

$$0 \leq P_{ch,t}^{bat} \leq B_t * P_{ch.max,t}^{bat} \quad B_t \in \{0,1\} \forall t \in T \quad (7)$$

$$0 \leq P_{dis,t}^{bat} \leq (1- B_t) * P_{dis.max,t}^{bat} \quad B_t \in \{0,1\} \forall t \in T \quad (8)$$

$$ES_t^{bat} = ES_{t-1}^{bat} + P_{ch,t}^{bat} * \eta^{bat,ch} - P_{dis,t}^{bat}/\eta^{bat,dis} \quad (9)$$

$$ES_{min}^{bat} \leq ES_t^{bat} \leq ES_{max}^{bat} \quad (10)$$

$$ES_{t=0}^{bat} = ES_{t=24}^{bat} \quad (11)$$

4 Results

The scheduling process is executed by applying real data of wind speed and solar irradiance of a summer day. The input data of the obtained RES energy are presented in Fig. 3.

The optimization problem is solved by using the software GAMS as algebraic model language and the solver CPLEX is used in order to compute the power references obtained for the proposed model. The system is implemented to minimize the cost of buying energy from the utility grid and to generate the charging and discharging schedule of the battery.

4.1 Scheduling Results

The optimization is solved in order to prove the effectiveness of the objective function. The parameters used in the model are presented in Table II as well as the generation profiles of a summer day with renewable energy generation (Wind and PV) (Fig. 5 and Fig. 6).

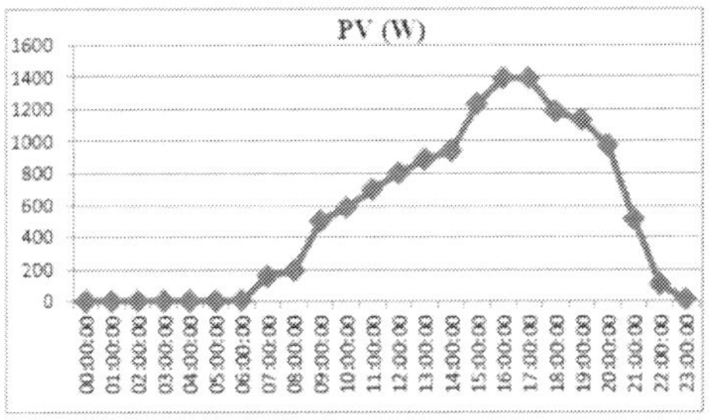

Fig. 5. Energy production from solar energy for one day.

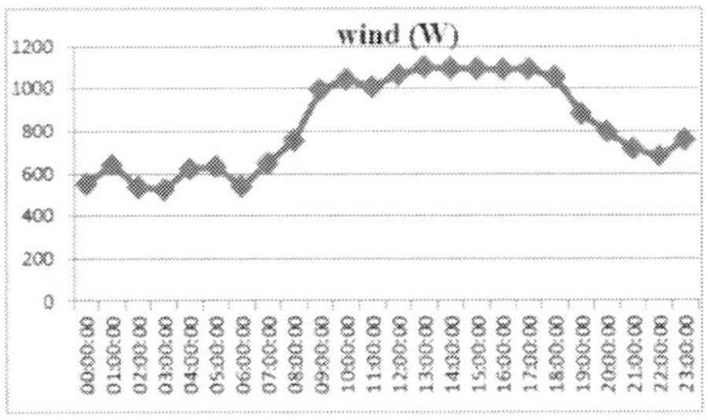

Fig. 6. Energy production from wind energy for one day.

As can be seen, there are a surplus of energy during the day and just part of the energy is employed while some reduction is used. Furthermore, the power profile exchanged with the grid is presented. In this case, the profile is positive when microgrid injects power into the grid (sells) and negative when absorbs power (buys).

On the opposite, the profiles of selling and buying power to the utility are shown in Fig. 7. It is possible to see in Fig. 7 that the surplus of RES energy is sold to the utility where the local generation is enough to supply the load and thus, it is not needed to buy energy from the grid. In addition, from 08h the part of the energy generated by wind energy is used in the local consumption and, at the same time, it is required to buy energy from the utility because the energy provided by the PV is not enough to supply the demand.

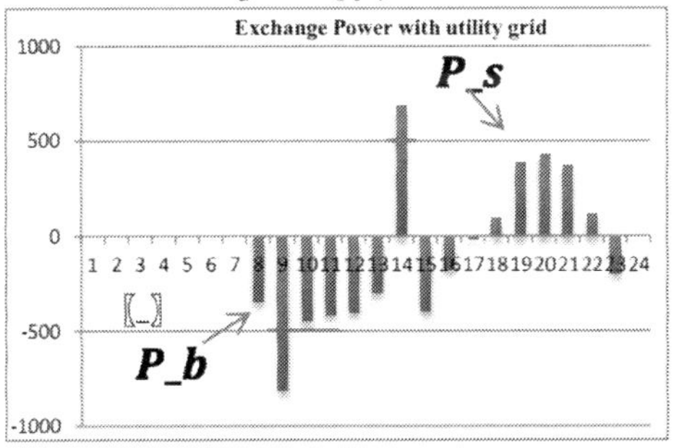

Fig. 7. Exchange power with utility grid.

Concerning the battery, the scheduled power profile and the charging/discharging of the battery during the day are presented in Figs. 8 and 9. As can be seen, the battery is discharged when the power stored from the RES is low while it is charged when there is high power in the RES.

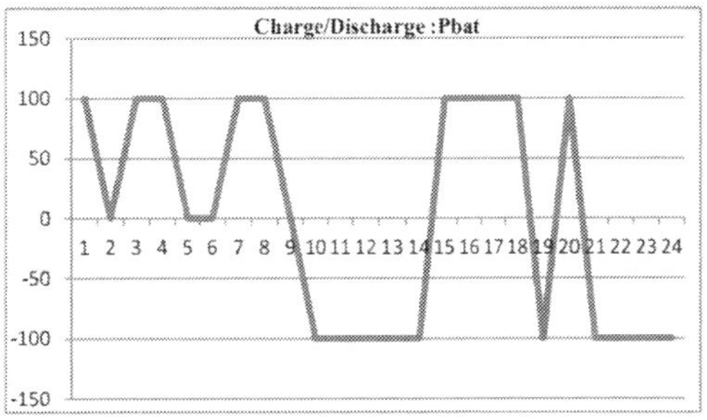

Fig.8. Charging/Discharging of battery during the day (24h).

The charging/discharging of the battery for the working day is presented in Fig8. The energy stored in the battery varies at different hours of the day.

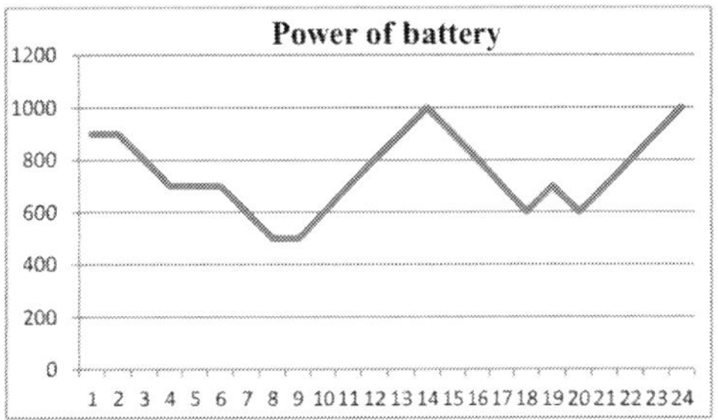

Fig.9. Power of battery during the day (24h).

As can be seen, in Fig.9, the scheduling keeps charging the battery to use the energy at the end of the day when the WT and PV do not generate energy.

Table 2. Parameter of the Model

Name	Description
T	Time scheduling
i	Index of units of Res
C_{sell}^{grid}	Cost of selling energy
C_{buy}^{grid}	Cost of buying energy
P_{res}	Power from renewables
P_{grid}	Power from the utility
P_{buy}^{grid}	Power bought from the utility
P_{sell}^{grid}	Power sold to the utility
P^{bat}	Power of the battery
ES_{max}^{bat}	Maximum battery energy level
ES_{min}^{bat}	Minimum battery energy level
$\eta^{bat,ch}$	Battery charging efficiency
$\eta^{bat,disch}$	Battery discharging efficiency
$\lambda_{grid}(t)$, B_t	Binary variable

Table 3. Parameter of the Battery

Battery unit	$Energy_{min}$	$Energy_{max}$	P_{max} charge	P_{max} $Dischar$
1	100	1000	100	100

5 Conclusion

A mixed integer linear programming model has been used for a grid-connected hybrid PV-Wind- ESS microgrid system in order to minimize cost of buying energy from the grid and maximize profits for selling energy generated by RESs. The optimization strategy has been used to set optimal power references for the distributed resources of the microgrid. The simulation results confirm that the proposed strategy of EMS is effective in grid-connected hybrid PV-Wind- ESS. Future work includes implementing the proposed EMS in a real system, power losses and analyzing its performance and their influence.

References

1. Q. Jiang, M. Xue, and G. Geng, "Energy management of microgrid in grid-connected and stand-alone modes," IEEE Transactions on Power Systems, vol. 28, no. 3, pp. 3380–3389, Aug 2013.
2. M. Iqbal, M. Azam, M. Naeem, A. Khwaja, and A. Anpalagan, "Optimization classification, algorithms and tools for renewable energy: A review," Renewable and Sustainable Energy Reviews, vol. 39, no. 0, pp. 640 – 654, 2014.
3. J. de Matos, F. S.F.e Silva, and L. de S Ribeiro, "Power control in ac isolated microgrids with renewable energy sources and energy storage systems," IEEE Transactions
4. S. Pourmousavi, M. Nehrir, C. Colson, and C. Wang, "Real-time energymanagement of a stand-alone hybrid wind-microturbine energy system using particle swarm optimization," IEEE Trans. Sustain. Energy, vol. 1, no. 3, pp. 193–201, Oct. 2010.
5. P. Siano, C. Cecati, H. Yu, and J. Kolbusz, "Real time operation of smart grids via FCN networks and optimal power flow," IEEE Trans. Ind. Informat., vol. 8, no. 4, pp. 944–952, Nov. 2012.
6. S. Choi, S. Park, D.-J. Kang, S.-J. Han, and H.-M. Kim, "A microgrid energy management system for inducing optimal demand response," in Proc. IEEE Int. Conf. Smart Grid Commun. (SmartGridComm), Brussels, Belgium, Oct. 2011, pp. 19–24.
7. C. Cecati, C. Citro, and P. Siano, "Combined operations of renewable energy systems and responsive demand in a smart grid," IEEE Trans. Sustain. Energy, vol. 2, no. 4, pp. 468–476, Oct. 2011.
8. F. Katiraei, R. Iravani, N. Hatziargyriou, and A. Dimeas, "Microgrids management," Power and Energy Magazine, IEEE, vol. 6, no. 3, pp.54–65, May 2008.
9. S. C. C. 21, "Guide for optimizing the performance and life of leadacid batteries in remote hybrid power systems," IEEE Std 1561-2007, pp. C1–25, 2008.
10. F. Marra and G. Yang, "Decentralized energy storage in residential feeders with photovoltaics," in Energy Storage for Smart Grids, P. D.Lu, Ed. Boston: Academic Press, 2015, pp. 277 – 294.
11. R. Huang, T. Huang, R. Gadh, and N. Li, "Solar generation prediction using the ARMA model in a laboratory-level micro-grid," in Proc. IEEE Int. Conf. Smart Grid Commun. (SmartGridComm), Tainan, Taiwan, Nov. 2012, pp. 528–533.
12. I. El Kafazi, R. Bannari, A. Abouabdellah, My O. ABOUTAFAIL and Josep M. Guerrero. Modeling and Forecasting Energy Demand.5th International Renewable and Sustainable Energy Conference (IRSEC'17December 04-07, 2017). (ACCEPTED).
13. G. Sideratos and N. Hatziargyriou, "An advanced statistical method for wind power forecasting," IEEE Trans. Power Syst., vol. 22, no. 1, pp. 258–265, Feb. 2007.
14. http://www.ges-lyon.fr

AUTHOR INDEX

AUTHOR INDEX